ENVIRONMENTAL GEOTECHNICS AND PROBLEMATIC SOILS AND ROCKS

PROCEEDINGS OF THE SYMPOSIUM ON ENVIRONMENTAL GEOTECHNICS AND PROBLEMATIC SOILS AND ROCKS / BANGKOK / DECEMBER 1985

Environmental Geotechnics and Problematic Soils and Rocks

Edited by
A.S.BALASUBRAMANIAM, S.CHANDRA, D.T.BERGADO
& PRINYA NUTALAYA
Asian Institute of Technology, Bangkok

A.A.BALKEMA / ROTTERDAM / BROOKFIELD / 1988

*The texts of the various papers in this volume were set individually
by typists under the supervision of each of the authors concerned.*

Published by
A.A.Balkema, P.O.Box 1675, 3000 BR Rotterdam, Netherlands
A.A.Balkema Publishers, Old Post Road, Brookfield, VT 05036, USA
ISBN: 978-90-5410-903-7

Environmental Geotechnics and Problematic Soils and Rocks, Balasubramaniam et al. (eds)
© *1987 Balkema, Rotterdam. ISBN 90 6191 785 9*

Contents

Section B: Environmental geotechnical aspects of major infra-structure projects

Section C: Environmental geotechnical aspects of natural hazards

Section D: Wastes and their use

Section E: Problematic soils and rocks

Environmental Geotechnics and Problematic Soils and Rocks, Balasubramaniam et al. (eds)
© *1987 Balkema, Rotterdam. ISBN 90 6191 785 9*

Preface

This volume contains forty-eight papers and is a part of a continuing series produced at the Asian Institute of Technology on the development of Southeast Asia. The papers contained in this volume are arranged in five sections: Geotechnical control and environmental protection, Environmental geotechnical aspects of major infra-structure projects, Environmental geotechnical aspects of natural hazards, Waste and their use and Problematic soils and rocks.

Section A deals with geotechnical control and environmental protection aspects of environmental geotechnics. There are nineteen papers in this section covering various aspects like evolution of slope due to mining, stabilization of soils by different additives, effect of earthquakes and ground freezing and the effect of environment on geotechnical properties of soils. Ten papers in Section B are related to major infra-structure projects. These papers are on tunnels, underground openings, sewer storage facilities, river bank structures and retaining walls. One paper is on geotechnical modelling and another one is more on general foundation system for soft soils.

Section C contains six papers on the environmental geotechnical aspects of natural hazards. The papers are related to the effect of hill roads on water shed functions in mountains, mudflow problems, landslides and detection of prospective sites for landslides and mechanics of deposition of sand. The papers on waste and their use are put together in section D, and deal with models for waste disposal systems, sanitary landfill treatment, colliery waste disposal and utilization of slag as a construction material.

The last section of the book contains eight papers related to the problematic soils and rocks, including dispersive and expansive soils. Various aspects of the strength of these are also discussed. The last paper is on the detection of instability in rocks.

In presenting these papers, the organisers of the related Symposium and the editors of the present volume are very pleased to record the help and support they have received in carrying out this work from several sponsoring organizations and individuals. Sincere gratitude is expressed to the General Committee members of the Southeast Asian Geotechnical Society, Dr. Za-Chieh Moh, Prof. Chin Fung Kee, Dr. Tan Swan Beng, Dr. E.W.Brand, Dr. Ting Wen Hui, the late Dr. Chai Muktabhant, Prof. S.L.Lee, Dr. Ooi Teik Aun, Mr. Yu Cheng, Mr. D.Greenway and Mr. J.R.R.Santos. Also, several institutions of engineers and geotechnical societies within and outside this region have helped in ensuring the continuation of this series of geotechnical events.

At the Asian Institute of Technology, sincere thanks are expressed among others to the President of AIT, Prof. A.M.North, Dr. K.Miyamoto, Dr. I.Towhata, Mr. Sataporn Kuvijitjaru, Mrs. Vatinee Chern, Mrs. Uraivan Singchinsuk, and Miss Nualchan Sangthongsttit, without whose enthusiastic support it would not have been possible to bring out this volume. The meticulous and untiring work of Mr. Vasantha Wijeyakulasuriya and Mr. Sanjay Govil in checking and proofreading of the text, figures and tables has made it possible to have this volume in the present form.

Finally, this volume is dedicated to Mr. Lek Kanchanaphol for his services and contributions to the geotechnical activities in Thailand.

A.S.Balasubramaniam
S.Chandra
D.T.Bergado
Prinya Nutalaya

Environmental Geotechnics and Problematic Soils and Rocks, Balasubramaniam et al. (eds)
© *1987 Balkema, Rotterdam. ISBN 90 6191 785 9*

Lek Kanchanaphol, 1925-1985

On 23rd June 1985 Khun Lek Kanchanaphol, the former Deputy General Manager for Hydro-electric Construction, Electricity Generating Authority of Thailand (EGAT) died: only a few months after his retirement from EGAT and after a period of ill-health of about nine months. His passing represents a tragic personal loss, not only to his family but also to the profession of large dam building and hydro-power development. Khun Lek was a strong supporter of geotechnical engineering activities in Thailand, particularly through the Southeast Asian Geotechnical Society and the Asian Institute of Technology.

Khun Lek, a man who was ever present on the scene, had his high school education at Wat Trimitr Witiyalai, Bangkok. A self-made man, Khun Lek had his engineering training in the Irrigation Technical School of the US Bureau of Reclamation and at the Water Resources Engineering Center, Roorkee, India. After a successful period of nearly 20 years of service to the Royal Irrigation Department, Khun Lek moved to EGAT when that state enterprise was formed in 1964. He was appointed Director of the Electricity Production Division in EGAT as early as 1966 and then became the Director of Hydro-electric Construction in 1969. The period 1969 to 1982 became vital formative years for hydro-power development in Thailand, and Khun Lek was a key figure in this advancement. The first major dam to be constructed in Thailand was Bhumipol Dam, finished in 1964. Subsequently, sixteen large dams of various types and heights have been built and put into operation in the years from 1964 to 1985. These large power projects were needed to cope with an increase in the consumption of electricity in the Kingdom of the order of 20% per year. In 1982 Khun Lek was appointed Deputy General Manager for Hydro-electric Construction in EGAT as a mark of his very talented and successful contributions to the profession of hydro-power development. In recognition of the continuous and growing contributions of Khun Lek to the development of the Kingdom of Thailand, His Majesty the King saw fit to confer upon Khun Lek the decorations of The Most Noble Order of the Crown of Thailand (from the Fourth through the Second Class), The Most Exhaulted Order of the White Elephant (from the Fifth through the Second Class), the Rattanaporn Medal, the Commemorative Model for H.M.Rama IX, and a Commemorative brooch. These honours were conferred on various occasions during the period 1950 to 1974.

Khun Lek has participated and presented technical papers in many of the Large Dam Conferences (ICOLD), particularly those held in Australia, Mexico and Sweden. Also, he made technical visits and study tours to a large number of countries. Khun Lek, always believed in international cooperation, and thus was a very strong supporter of the activities of the Southeast Asian Geotechnical Society in Thailand. He was instrumental in arranging the very successful international symposium held in 1980 at the Asian Institute of Technology on 'Problems and Practice of Dam Engineering', a Symposium which was attended by many of the Presidents and past-

Presidents of the International Commission on Large Dams (ICOLD), the International Society for Soil Mechanics and Foundation Engineering (ISSMFE), and the International Society for Rock Mechanics (ISRM). So, the Southeast Asian Geotechnical Society, the Asian Institute of Technology, and the Electricity Generating Authority of Thailand have lost a true friend and a professional who contributed greatly to the profession and the engineering community in Thailand and elsewhere.

A.S.Balasubramaniam

Section A
Geotechnical control and environmental protection

Environmental Geotechnics and Problematic Soils and Rocks, Balasubramaniam et al. (eds)
© 1987 Balkema, Rotterdam. ISBN 90 6191 785 9

Slope evolution on coal-mine disturbed land

Martin J.Haigh
Oxford Polytechnic, UK

SYNOPSIS. Discovering the parameters of a coal-spoil derived landform which will not suffer accelerated erosion remains a fundamental problem in reclamation research. This study examines patterns of erosion and slope evolution on coal-mine disturbed lands in Wales and North America. The evolution of raw technogenetic slopes is effected by the mid-slope-ward migration of two zones of accelerated erosion, one originating on and extending the upper convexity, the other located at the upper limits of the lower concavity of the slope. On the mature profile, maximum erosion occurs at the midslope. The speed of erosion is affected by the presence of vegetation and by the slope aspect but only the activities of the basal control seriously affect the pattern of erosion. Large-scale basal deposition may suppress the development of the lower-slope erosion peak. The pattern of erosion on the slope profile may be predictable as exponential functions of two morphological parameters; distance from the crest, and average slope angle.

1 INTRODUCTION

Erosion control is one of the most serious problems in land reclamation. Soil losses from surfaced-mined coal land can remain several orders of magnitude higher than those from undisturbed areas (Law 1984:57, Mann 1979). Paone et al. (1979) place the design of surface topography first in their list of the problems of reclamation, ahead even of revegetation (Down and Stocks, 1977). There is a need for regrading designs which will help reduce erosion (Barr and Rockaway, 1980, White and Plass 1974).

Unfortunately, little progress has been made in the design of land reclamation slopes. Shaefer et al. (1979) represent the leading edge of this research. Their strategy is founded on the principle that the reclaimed landforms should recapitulate the morphologies of the pre-mining landscape. Toy (1984:144) confirms that since it is currently impossible to reconstruct an entirely balanced condition between surface form and geomorphological process, surface features "should resemble those of the premining period". Grandt (1985:168) recalls U.S. Federal Bill H.R. 10079, introduced by Rep. E. Dirkson of Illinois

in 1940 which required surface mine operators "to make the contour of the land approximately the same as before the mining was begun".

One may debate the aesthetic justications for such a policy (Commission on Energy and Environment, 1981; Mallary and Carlozzi, 1976; Chironis, 1978), but it is wrong to assume that premining landforms are appropriate to land reclamation. Compared to the natural landscape, newly exposed spoils tend to be more fragmented, more compacted, more erodible, and less well clothed with vegetation (Doubleday and Jones, 1975; Smith and Sobek, 1978; Taylor and Spears, 1970). They are more prone to physical and chemical changes, and they have a radically different hydrology (Evangelou and Karathanasis, 1985; Touysinhthiphonexay and Gardner, 1984; Pole et al. 1980). In sum, these new lands require new landforms designed for the erosional properties of the mine spoils themselves.

The problem addressed, then, is the control of erosion on coal-mine spoils, and the ambition is to establish criteria for the design of reclaimed landscapes which will resist accelerated erosion. In Britain in 1971, 48 thousand hectares of coal-mined land required treatment, and despite the reclamation of more than 3 thousand hectares each year, that figure may be doubled today (Clouston, 1974). In the United States, coal mining activity affected 800 thousand hectares between 1930 and 1977. Some 150 thousand hectares still require treatment (U.S. Bureau of Mines 1979). The problem is far from being insignificant.

2 LOCATION OF STUDY SITES

Two of the study sites are located in Wales and one in northern Illinois, U.S.A.

2.1 Milfraen and Waunafon, S. Wales (U.K.), 51°13'N3°7'W, Rainfall 1255 mm/a, Frost 100 cycles/a.

The project is based within the post-industrial landscape of northern Gwent in South Wales. For Britain, this is a harsh environment. Blaenafon lies on a bleak, windswept, almost wholly tree-less moorland some 432 m O.D. Rainfall is heavy and persistent. Much of the area is water logged peat bog.
  Mining began at Blaenafon around 1584. Coal mining commenced in 1782, but this increased rapidly after 1789 when the first of Blaenafon's three ironworks commenced production. In 1906 a local miner, L. Browning, was able to list thirty collieries in a town whose fortunes were founded in coal and steel. The area began to decline in the early twentieth century. By the late 1930s, the last of the iron and steel enterprises failed. The Second World War gave a short reprieve, and ushered on a new era of surface mining; but subsequently industrial decline continued. By the close of the 1960s, only one active mine remained amidst an area of dereliction, which covered several square kilometres, and where scarcely any vestige of the original landscape remained. Attention came to be directed towards landscape rehabilitation. The 1970s saw a last flurry of activity, a last drift mine was sunk in 1973, but in 1974 deep mining ceased. An open-pit mine functioned through the mid-1970s, but by 1980 all mining had finished except for the reworking of old spoil tips in association with reclamation (Thomas 1981).
  A feature of land reclamation at Blaenafon has been its lack of success. Three attempts have been made to restore former open-pit mines. All reclamations are marred by severe erosion. Even the

most recent includes gully channels, some of which incise to depths of 3 m from time to time. The reclamations of deepmined wastes have hardly been more successful. On land reclaimed during the later 1970s, vegetation remains thin and patchy and there is obtrusive gully erosion.

2.2 Utica, N. Illinois (U.S.A.), 40°19'N88°59'W, Rainfall 981 mm/a, Frost 76 cycles/a.

The second study site lies in the heart of continental North America. The abandoned strip-mine ridges at Utica are part of the 400 h of coal-mine disturbed land in La Salle County, Illinois. Coal mining was active in this area in 1818, though surface 'strip-mining' for coal only began a hundred years later. This, in turn, was long finished by 1962 when the State of Illinois introduced its first surface-mine reclamation legislation. The mine-dumps of La Salle County are unreclaimed and abandoned ("orphan") sites. There has been little effort directed to reclamation (Bowden, 1974).

3 METHODOLOGY

The technique employed for this study was simple and direct. Erosion pins, 600 mm x 5 mm mild steel rods, were driven into the spoil slopes at carefully selected locations. Initially, the rods were allowed an exposure of about 20 mm. Data collection began some months after emplacement when signs of the original disturbance had vanished. Changes in the exposure of the pins were measured regularly by means of a depth gauge. Results are interpreted as changes in the elevation of the ground surface and have proved to be accurate to circa. 1 mm (Haigh, 1978).
  The data from individual erosion pin stations can be quite variable. This variation is due to measurement errors, to the migrations of pulses of sediment across the slope, to fluctuations in soil moisture, to frost heaving, rill incision, and the occasional tramplings of grazing animals. The influence of such local disturbances has been overcome by averaging the data from several erosion pin stations. The data cited by this study represent cumulative changes in the average exposure of three to five erosion pins. Data collection has proceeded on a semi-annual basis since 1972 in South Wales and 1978 in Illinois.

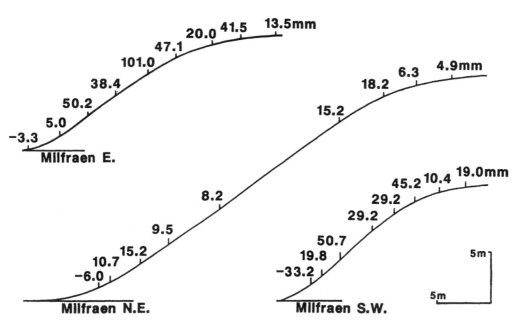

Fig. 1. Ground surface retreat on plateau-type colliery
spoil mounds at Milfraen, Wales. Vertical lines indicate
erosion pin sites, and adjacent numbers denote total retreat
(mm) for the period March 1973-March 1979. Milfraen 'S.W.'
and 'N.E.' profiles are vegetated. Milfraen 'E.' is not.

Initial experiment designs attempted to
encompass the impact of three variables:
basal control, vegetation cover, and slope
aspect. It was discovered that all three
factors seriously affect the rate of slope
retreat but, eventually, that only basal
control has any major impact on the
pattern of slope evolution.

4 EXPERIMENTS AT MILFRAEN, WALES

These deep-mine spoil-tips stand on open
moorland above Blaenafon in Wales (Haigh,
1978). Three slope profiles have been
instrumented with erosion pins. The
profiles are identified by their
orientation. The (warm) south west-facing
profile; Milfraen 'S.W.' and the (cold)
north east-facing profile, Milfraen 'N.E',
both support a thin grass cover. Milfraen
'E' has a neutral aspect and no
vegetation. The slopes are all
free-standing, there are no stream
channels at the slope foot, and there is
nothing to constrain the accumulation of
sediment. Fig. 1 summarises the results
from these three experiments on true-scale
representations of the slope profiles.
The ticks on the profile indicate the

location of rows of erosion pins, and the
adjacent numbers are cumulative ground
loss for the six years of observation.
These data update those published in Haigh
(1979a).
  The Milfraen profiles, share age and
genesis, and all are composed of the same
fine textured (D10 0.3-0.4 mm D50 1.0-2.0
mm) carbonaceous coal shale. However,
they are evolving at very different rates.
Milfraen 'N.E.', the cold aspect,
vegetated profile, is undergoing the least
ground loss. The average annual retreat
for the part of the slope above the point
where the visible deposition starts (King,
1977) is 1.9 mm. The pattern of retreat
is bimodal. There is a peak high on the
upper convexity and a smaller peak on the
uppermost part of the lower concavity.
Milfraen 'S.W.', the warm aspect,
vegetated profile is much shorter but it
is evolving more rapidly. The average
annual retreat rate for the slope above
the point of visible deposition is 5.1 mm.
Here slope processes are operating at 2.7
times the rate of those on Milfraen 'N.E.'
and should have accomplished
proportionally more morphological
alteration. In morphogenetic terms,
Milfraen 'S.W.' is older than Milfraen

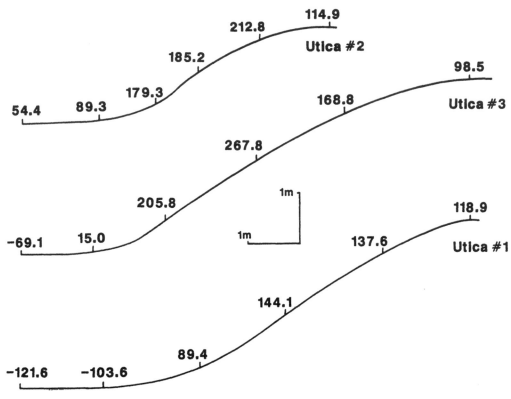

114.9

212.8

185.2

Utica #2

179.3

98.5

54.4   89.3

168.8

Utica #3

267.8

1m

205.8

1m

118.9

-69.1   15.0

137.6

Utica #1

144.1

89.4

-121.6   -103.6

Fig. 2. Ground surface retreat on strip-mine dumps at
Utica, Illinois. Vertical lines indicate erosion pin sites
and adjacent numbers denote total retreat (mm) for the
period May 1978-May 1983. Profile #1 faces south.
Profiles #2 and #3 face east.

'N.E.' Once again, however, the pattern
of ground retreat is bimodal. Maximum
retreat is recorded on the lower part of
the rectilinear mainslope and on the upper
convexity.

Milfraen 'E.' has an intermediate slope
length and a neutral slope aspect, but it
is unvegetated. Its annual rate of
retreat exceeds that for the average of
the two vegetated slopes by a factor of
1.9. Milfraen 'E.' has a rate of ground
loss of 6.6 mm p.a., which is 30% greater
than that of Milfraen S.W.; so Milfraen
'E.' is, morphogenetically, the oldest of
these three slopes.

The pattern of retreat on Milfraen 'E.'
is quite complex but it is dominated by a
massive score on the upper part of the
mid-slope. Two further peak scores, of
less than half this value are discovered
high on the lower concavity and at the
upper convexity.

5 EXPERIMENTS AT UTICA, ILLINOIS

The mine dumps at Utica, Illinois, are of
similar age to those at Milfraen. The
slopes have a similar form but they are
smaller and composed of a much more clayey
spoil (D50 ca.0.001 mm). They also exist
in a climatic context which is far more
harsh and erosive than that of South
Wales.

However, in morphogenetic terms, the most
important difference between these slopes
and those at Milfraen is in their basal
control. The slopes at Milfraen are
free-standing on open moorland, but those
at Utica belong to parallel, closely
packed, strip-mine ridges. Slope foot
deposition occurs in a very constricted
area between spoil ridges, and its impact
upon the slope foot is proportionately
increased.

Three profiles have been instrumented.

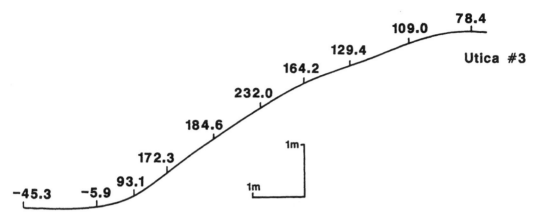

Fig. 3. Ground surface retreat at Utica #3. Vertical lines indicate erosion pin sites and adjacent numbers denote total retreat (mm) for the period May 1979-May 1983. Utica #3 faces east.

Utica #1 faces south. Utica #2 and Utica #3 face east and have a colder micro-climate. All three slopes are unvegetated. Utica #1 and Utica #3 share a similar slope length. Utica #2 is shorter. The results from the three experiments are portrayed against true-scale representations of the slope profile, as shown in Fig. 2. The numbers show cumulated annual ground loss in millimetres for the period May 1978 to May 1983, and update those published by Haigh and Wallace (1982).

The results from Utica indicate two patterns of erosion. That of Utica #3 closely resembles Milfraen 'E.' The profile includes the largest single ground loss record, which is at the midslope. Utica #1 and Utica #2 peak erosion scores are (more or less) shared between the midslope and upper convexity. Nowhere is there any suggestion that retreat is bimodal.

A pioneering study by Croxton (1928) showed that young strip-mine dumps in Illinois suffer maximum erosion at the slope crest. Comparative studies of strip-mine dumps of different ages in eastern Oklahoma (Goodman and Haigh, 1981) demonstrated that as strip-mine dumps age, the zone of peak erosion migrates from the slope crest towards the midslope (Fig. 3). In combination, the results from Utica and Milfraen tend to confirm this tendency, since, those from Milfraen 'N.E.' and 'S.W.' show erosion peaks at the upper convexity, and those from Milfraen 'E.' and Utica demonstrate peak erosion scores,which settle down around

midslope. However, the strip-mine results show little evidence for the existence of a lower slope zone of peak erosion. There are three possible explanations. The first is that it exists, but that instrumentation is too coarse to detect its presence. Work is currently in progress at Oklahoma and Utica to check this hypothesis. In May 1979, additional rows of erosion pins were established on Utica #3. The results overall (Fig. 3), do not indicate the presence of a lower slope zone of peak erosion (cf. Haigh, 1984). So, at present, it seems more reasonable to suggest that this second zone is absent. Possibly, the development of the lower slope zone of peak erosion is suppressed by the accelerated burial of the slope foot (Goodman and Haigh, 1981; McKenzie and Studlick, 1978). However, it is also possible that the relatively rapid retreat of these profiles has allowed the two zones of accelerated erosion to merge at the midslope (cf. Milfraen 'E.').

6 EXPERIMENTS AT WAUNAFON, WALES

A third set of experiments is located in Wales. Cefn Garn-erw is an artificial valley created as part of the reclamation of the Pwll Du "Opencast" (i.e. surface) coal mine at Waunafon, near Blaenafon, North Gwent, and around 1.5 km from Milfraen. The reclamation was completed in the 1950s, but only part of the area has revegetated, and much is suffering severe rill and gully erosion.

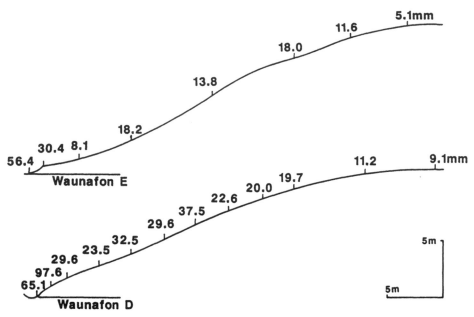

Fig. 4. Ground surface retreat on regraded surface-mine
dumps at Waunafon, Wales. Vertical lines indicate
erosion pin sites and adjacent numbers denote total retreat
(mm) for the period March 1973 - March 1979. Waunafon
'E.' is vegetated, Waunafon 'D.' is not, both profiles face west.

The slopes examined at Waunafon are more gentle than those at Milfraen, and while the spoil is rather more coarse it is also more clayey (D10: 0.3 mm D50: 3 mm), and infiltration rates are substantially lower. More important, however, is the fact that the valley at Cefn Garn-erw includes a small ephemeral stream channel which is undergoing active vertical incision as well as lateral migration to undercut certain valley-side zones.

As at Milfraen, some of the Waunafon instrumentation is established on slopes which are unvegetated, whilst some is on slopes which support a thin vegetation of tuft grass. Fig. 4 portrays the results from two representative profiles, one of which, Waunafon 'E.', is vegetated and the other of which, Waunafon 'D.' is not. The two grids share a neutral aspect, they are on the east side of the valley and face west. As before, Fig. 6, portrays true-scale slope profiles and the locations of rows of erosion pins. The numbers indicate ground retreat in millimetres cumulated for the six-year period March 1973-March 1979, and they update those published by Haigh (1979b).

The rates of ground loss on the slopes at Waunafon (above the zone affected by channel incision) are less than those on the steeper slopes at Milfraen. Average ground loss on unvegetated Milfraen 'E' was 6.6 mm p.a. compared to 3.6 mm p.a. on Waunafon 'D'. Average ground loss on Milfraen's vegetated profiles was 3.5 mm p.a. compared to 2.2 mm p.a. on Waunafon 'E'. So, the Waunafon slopes besides being half the age of those at Milfraen are also morphogenetically younger because the processes of slope evolution are working more slowly.

The pattern of erosion on the two Waunafon profiles is, however, quite similar. In both cases, slope foot incision is demonstrated by a high ground loss recorded from the erosion pins in the basal channel. Lateral under-cutting of the slope foot is also clearly demonstrated, especially on Waunafon 'D.', by the high retreat scores immediately upslope.

Away from the channel, both slopes again show a bimodal pattern of ground loss with peak erosion scores located at either extreme of the central rectilinear mainslope. However, on the more rapidly

8

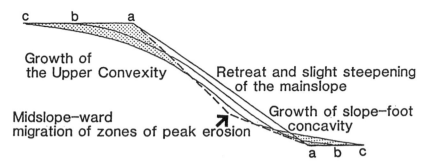

Fig. 5. Pattern of slope evolution: base level constant (cf: Milfraen).

evolving, unvegetated, profile, Waunafon
'D.', the two erosion peaks are very much
closer together.

7 ANALYSIS OF RESULTS

It is possible to generalise from these
three sets of experimental results.
Figure 5 attempts to summarize the results
from Milfraen, N. Gwent, as a single
diagrammatic model, which includes the
premise that the presence of vegetation on
the slope influences the rate of erosion
but not the overall patterns of slope
evolution. It charts slope evolution from
A-A, an initial recti-linear slope, to
C-C, roughly the condition of Milfraen
'E.' (Fig. 1). The critical features of
this process of transformation are: the
mid-slopeward migration of the two zones
of maximum erosion, the increase of the
radius of curvature of the upper convexity
and its extension into the recti-linear
mainslope, a slight steepening of the
recti-linear mainslope, the development of
the lower concavity, and a progressive
increase in its radius of curvature.
At Milfraen, there is a minimal
accumulation of sediment at the slope
foot. At Utica, the area available for
slope-foot deposition is severely
restricted and the lower-slope tends to
become buried beneath colluvial deposits.
Slope evolution at Utica is generalized
(see Fig. 6). The pattern remains similar
except that the lower slope zone of
maximum erosion is absent, and the rate of
increase in the radius of curvature of the
lower concavity is reduced. In addition,
both the life-span and rate of steepening
of the rectilinear mainslope is reduced.
   Figure 7 is based on the results
collected at Waunafon, Gwent. Here, the
local relief of the profile is not reduced
by deposition but, instead, is
substantially enhanced by basal incision.

Slope evolution proceeds by a progressive
increase in the radius of curvature of the
upper convexity and its extension into the
rectilinear mainslope element. This
becomes steeper, a tendency which is
progressively more marked down-slope and
which tends to lead to rounding.
Accumulation, of course, is inhibited at
the slope foot but episodic undercutting
may allow the development of a
channel-side bench. This feature affects
the energy relations of the slope and
encourages local deposition. The slope
inflexion so created is exaggerated by the
existence of the lower slope retreat
maximum immediately upslope (cf. Fig. 4).

8 DISCUSSION

There are, of course, major similarities
between the three models. There is a
general tendency for initially rectilinear
technogenetic slope profiles to evolve
towards a sigmoid shape. Peak erosion
scores seem to be adjacent to, and
associated with, disjunctions in the
profile.
   The processes which shape these slopes
are mainly due to rainwater. The slopes
are affected by soil creep, but its
impacts are much less obvious (Haigh,
1978). Many students of surface erosion
have noted that sediment loads increase
with increases in slope angle and slope
length. Zingg (1940) originally
formulated this relationship as:

$$A = c\ S^{1.4}\ L^{1.6}$$

where A is soil loss per unit area, S is
slope angle, L is slope length and c a
constant (cf. Meyer and Kramer, 1968).
   The relationship leads one to expect
that peak erosion will occur at the foot
of the steepest slope element. However,
the slope foot is a zone where deposition

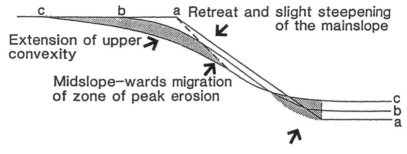

Extension of upper convexity

Retreat and slight steepening of the mainslope

Midslope-wards migration of zone of peak erosion

Growth of slope -foot concavity inhibited by burial

Fig. 6. Pattern of slope evolution: basal deposition (cf: Utica)

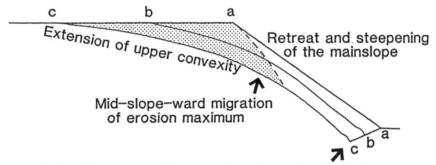

Extension of upper convexity

Retreat and steepening of the mainslope

Mid-slope-ward migration of erosion maximum

Episodic undercutting creates slope-foot bench

Fig. 7. Pattern of slope evolution: basal undercutting (cf: Waunafon)

occurs as a result of the loss of slope steepness. Inevitably, the peak of erosion must occur upslope of the parts of the profile affected by deposition, even the temporary deposition of sediment.

This is patently the case in the more mature slope profiles where peak erosion scores occur, at the steepest part of the midslope, at the point slope angles begin to decline away towards the slope foot. However, on the immature profiles, there are two zones of peak erosion, and the larger is usually upon the upper convexity.

This phenomenon may reflect patterns in the variation of the increase of slope angle and slope length down the profile of the immature slope. Ground retreat scores reflect the contribution of a unit area of slope to the total sediment load. An above average retreat score reflects an unusual increase in the capacity for sediment transport. This capacity can be increased in one of two ways. If the terms of Zingg's simple equation are accepted, either there is an increase in

average slope angle or an increase in slope length. Slope length increases uniformly downslope, but slope angle incréases abruptly across the upper convexity on an immature technogenetic profile. This creates a large initial addition to the sediment load. Subsequen' increments to the slope angle factor are small, so less is added to the sediment load and there is less erosion. The progressive increase in the slope length factor, however, continues to build sediment loads to a new peak above the lower concavity where the slope angle factor declines.

Table 1 provides an attempt to simulate this process for two simple slope profiles. Here, L is distance from the slope crest, S is average slope from the slope crest at point L. E is total erosion and G the ground loss increment between each station of L. The profile data in Part A, is based on Milfraen 'N.E.' This slope displays a bimodal pattern of erosion in actuality. The simulated results display a similar

10

Table 1: Simulation of erosion patterns employing the Zingg (1940) equation
Part 'A': Data based on the profile of Milfraen 'N.E.'
Part 'B': Data based on a simple 0°-34°-0° slope.

Average slope angle (Degrees) S	Cumulative slope length (Metres) L	Zingg exponents for soil loss estimates $S^{1.4}$	$L^{1.4}$	Sediment load $E=0.01S^{1.4} L^{1.6}$ (estimate per unit time) E	Ground loss (E increments per unit slope length) G
A 0	0	0	0	0	0
5	5	9.5	13.1	1.3	1.3
8	10	18.4	39.8	7.3	6.0
13	15	36.3	76.2	27.7	20.4
18	20	57.2	120.7	69.0	41.3
21	25	71.0	172.5	122.5	53.5
24	30	85.6	230.9	197.7	75.2
26	35	95.7	295.5	282.8	85.1
26	40	95.7	365.8	350.1	67.3
26	45	95.7	441.8	422.8	72.7
26	50	95.7	522.8	500.3	77.5
27	55	100.9	609.0	614.5	114.2
26	60	95.7	699.9	669.8	55.3
24	65	85.6	795.5	681.0	11.2
B 0	0	0	0	0.0	0.0
0	2	0	3.0	0.0	0.0
11	4	28.7	9.2	2.7	2.7
17	6	52.8	17.6	9.3	6.6
20	8	66.3	27.9	18.5	9.2
22	10	75.8	39.8	30.2	11.7
23	12	80.6	53.3	43.0	12.8
20	14	66.3	68.2	45.2	3.2
18	16	57.2	84.5	48.3	3.1

pattern. The second part of the table gives results from a simple recti-linear profile of two upper and lower 0° elements and five central elements at 34°. Here, there is only one erosion maximum, which is also the case for most mature, concave and rounded slope profiles.

9 CONCLUSION

The evolution of technogentic slopes is controlled by the migration of two zones of especially accelerated erosion which round the profile and coalesce at the midslope. The erosion of these slope profiles may be modelled in terms of their average slope angle and the distance from the slope crest. Further research is required to match observed values for ground loss with coefficients of slope angle and slope length under different conditions of soil and vegetation cover, -and to accomodate the impact of the basal control.

ACKNOWLEDGEMENTS

This paper revises and updates a presentation made originally to the Third British-Bulgarian Seminar on Applied Geography. The research has received grants at various times from the Natural Environment Research Council (1971-1974), University of Chicago (1975-1979) and Oxford Polytechnic (1980-). Thanks for assistance in field research go to K. Pearce, M.F. Haigh, and W.L. Wallace.

REFERENCES

Barr, D.J. and Rockaway J.D. 1980, How to decrease erosion by natural terrain sculpturing. Weeds, Trees and Turf, January 1980, 31-35.
Bowden, K. 1974, Illinois in R.P. Carter et al: Surface-Mined Land in the Midwest, p 281-95. Chicago, Illinois: Argonne National Laboratory ANL/ES-43: Coal Mining (VC-88)(N.T.I.S. PB-237830/S.

Chironis, N.P. 1978, Imaginative plans
make mined land better than ever. In
Chironis, N.P. (ed) Coal Age Operating
Handbook of Coal Surface Mining and
Reclamation, p278-281. N.J.: McGraw
Hill.

Clouston, J.B. 1974, British experience in
mined-land reclamation and pre-planning
of mineral workings. Research and
Applied Technology Symposium on Mined
Land Reclamation (National Coal
Association, U.S.A.) 2: 217-241.

Commission on Energy and the Environment.
1981, Coal and the Environment, London:
H.M.S.O.

Croxton, W.C. 1928, Revegetation of
Illinois coal stripped lands. Ecology
9: 155-175.

Doubleday, G.P. and Jones, M.A. 1977,
Soils of Reclamation. In Hackett, B.
(ed) Land Reclamation Practice p85-126.
Guildford: I.P.C. Science and Technology
Press.

Down, C.G. and Stocks, J. 1976,
Environmental Impact of Mining. London:
Applied Science Publishers.

Evangelou, V.P. and Karathanasis A.D.
1985, Reactions and mechanisms
controlling water quality in
surface-mined spoils. In Plass, W.T.
(ed). American Society for Surface
Mining and Reclamation, Symposium on the
reclamation of land disturbed by surface
mining, p213-247, London: Science
Reviews Ltd.

Goodman, J.M. and Haigh, M.J. 1981, Slope
evolution on abandoned spoil-banks in
eastern Oklahoma. Physical Geography 2
(2): 160-173.

Grandt, A.F. 1985, History of reclamation
research. In: Plass W.T. (ed) American
Society for Surface Mining and
Reclamation, Symposium on the
reclamation of lands disturbed by
surface mining, p164-187, Northwood,
Middx: Science Reviews Ltd.

Haigh, M.J. 1978, Evolution of slopes on
artificial landforms, Blaenavon, Gwent.
University of Chicago, Department of
Geography, Research Paper 183, 1-311.

Haigh, M.J. 1979a, Ground retreat and
slope evolution on plateau-type colliery
spoil mounds at Blaenavon, Gwent,
Transactions, Institute of British
Geographers N.S. 4 (3): 321-328.

Haigh, M.J. 1979b, Ground retreat and
slope development on regraded surface
mine dumps, Waunafon, Gwent. Earth
Surface Processes 4 (2): 183-190.

Haigh, M.J. 1984, Microerosion processes
and sediment mobilization in a road-bank
gully catchment in central Oklahoma. In
Burt, T.P. and Walling, D.E. (eds),
Catchment Experiments in Fluvial
Geomorphology, p.247-264. Norwich:
Geobooks.

Haigh, M.J. and Wallace, W.L. 1982,
Erosion of strip-mine dumps in La Salle
County, Illinois: preliminary results.
Earth Surface Processes and Landforms 7
(1): 79-84.

King, A.D. 1977, Preliminary Guidance for
Estimating Erosion on Areas disturbed by
Surface Mining Activities in the
Interior Western States: Interim Report.
United States Environmental Protection
Agency 980/4-77-005: 1-56.

Law, D.J. 1984, Mined-land rehabilitation.
New York: Van Nostrand Reinhold Co.

Mallary R. and Carlozzi, C.A. 1976, The
Aesthetics of Surface-Mine Reclamation:
an on-site survey in Appalachia.
Arstechnica, University of
Massachusetts, Amherst, Publication
R-76-5: 1-53.

Mann, C.E. 1979, Optimizing sediment
control systems. Surface Coal-mining
and Reclamation Symposium p.293-299.
New Jersey: McGraw Hill.

McKenzie, G.D. and Studlick, J.R.J. 1978,
Determination of spoil bank erosion
rates in Ohio by using interbank
sediment accumulations. Geology 6:
499-502.

Mayer, L.D. and Kramer, L.A. 1968,
Relation between land-slope shape and
soil erosion. American Society of
Agricultural Engineers Paper 68-749:
1-30.

Paone, J., Struthers, P. and Johnson, W.
1978, Extent of disturbed lands and
major reclamation problems in the United
States. In: Schaller, F.W. and Sutton,
P. (Eds) Reclamation of Drastically
Disturbed Lands, p.11-23. Wisconsin:
A.S.A. - C.S.S.A. - S.S.S.A.

Pole, M.W., Brown, T.H., and Zimmerman, L.
1980, Water movement in a reclaimed site
and associated soils at a mine site in
western North Dakota. North Dakota Farm
Research 37(6): 5-8,20

Shaefer, M., Elifritts, D. and Barr D.J.
1979, Sculpturing reclaimed land to
decrease erosion. Symposium on Surface
Mining Hydrology, Sedimentology and
Reclamation, (University of Kentucky BU
119) Proceedings: 99-110.

Smith, R.M. and Sobek, A.A. 1978, Physical
and chemical properties of overburdens,
spoils, wastes and new soils. In
Shaller, P.W. and Sutton P. (Eds).
Reclamation of Drastically Disturbed
Lands, p 149-172. Wisconsin: A.S.A. -
C.S.S.A. - S.S.S.A.

Taylor, R.K. and Spears, D.A. 1970, The
breakdown of British Coal Measure rocks.

International Journal of Rock Mechanics
and Mining Science 7, 481-501.
Thomas, W.G. 1981, Big Pit, Blaenafon.
Cardiff: National Museum of Wales.
Touysinhthiphonexay, K.C.N. and Gardner,
T.W. 1984, Threshold response of small
streams to surface coal mining,
bituminous coal fields, central
Pennsylvania. Earth Surface Processes
and Landforms. 9(1): 43-54.
Toy, T.J. 1984, Geomorphology of
surface-mined lands in the western
United States. In: Costa, J.E. and P.J.
Fleisher (eds), Developments and
applications of geomorphology, p133-170.
Berlin: Springer Verlag.
U.S. Bureau of Mines, 1979, Abandoned
Coal-Mined Lands: Nature, Extent and
Cost of Reclamation. United States,
Department of Commerce, Washington:
N.T.I.S. PB-299535: 1-29.
White, J.R. and Plass, W.T. 1974, Sediment
control using modified mining and
regrading systems and sediment control
structures. Research and Applied
Technology Symposium on Mined Land
Reclamation (National Coal Association,
U.S.A.) 2. 117-121.
Zingg, A.W. 1940, Degree and length of
land-slope as it affects soil loss in
runoff. Agricultural Engineering 21:
59-64.

Environmental Geotechnics and Problematic Soils and Rocks, Balasubramaniam et al. (eds)
© 1987 Balkema, Rotterdam. ISBN 90 6191 785 9

Fly ash utilization for soil improvement

R.C.Joshi
Department of Civil Engineering, University of Calgary, Canada
T.S.Nagaraj
Department of Civil Engineering, Indian Institute of Science, Bangalore

ABSTRACT: Fly ash, a by-product of pulverized coal combustion in suspension-fired furnaces used in electric power plants, is being produced in increasing quantities all over the world. This paper describes the physical, chemical and engineering properties of fly ash and its uses in construction.

1 INTRODUCTION

Fly ash is a by-product of the combustion of pulverized coal in thermal power plants which utilize suspension-fired furnaces. The ash from coal is largely from mineral matter or rock detritus which fills up the fissures in coal seams. Since the ash is in a finely powered state and has, in the dry state, a tendency to fly when disturbed, the term fly ash has come into vogue. Normally fly ash is removed by mechanical collectors or electrostatic precipitators as fine particle residue from hot flue gases before they are discharged into the atmosphere. Depending upon the type of furnace, varying amounts of fly ash, bottom ash or boiler slag are produced. In excess of 75 to 85% of the coal ash produced by suspension-fired furnaces is fly ash. The remainder is bottom ash or boiler slag. Generally, the fly ash is either stored in hoppers or sluiced to lagoons. Where power plants are restricted by environmental regulations, the fly ash is moistened with water and transported to distant places for disposal, unless it can be used for other purposes.

In Canada, over 3 million tonnes of fly ash is produced by power plants located in Nova Scotia, New Brunswick, Ontario, Manitoba, Saskatchewan and Alberta. The production, use and properties of fly ash in Canada were assessed and reviewed by Berry and Malhotra (1977).

The University of Calgary is actively involved in carrying out research on fly ash and studying its possible applications, both for general use and with special reference to ash produced in Alberta. Although there has been active interest in fly ash utilization both in the field of concrete technology and geotechnical engineering, this paper deals with the characterization of fly ash, and its properties and utilization in geotechnical engineering.

2 FLY ASH PROPERTIES

Since rock detritus or mineral matter in coal is likely to vary from one coal to another, variations are to be expected between ashes. Further variations are introduced by the actual firing process, the degree of pulverization the coal receives, the type of furnace, and the firing conditions.

2.1 Mineralogical and chemical properties

The majority of mineral matter consists of clays, pyrite and calcite. More than 85 percent of most fly ashes consist of glassy and crystalline compounds formed by the thermal treatment of minerals.

In considering the composition of fly ashes, Manz (1976) noted that a convenient division based on their chemical nature, could be made, notably between those derived from sub-bituminous and lignite coal sources and those derived from bituminous coal. Sub-bituminous and lignite coal ashes are characterized by a relatively high proportion of CaO and MgO to Fe_2O_3, as distinct from bituminous coal ash, which is usually low in CaO and MgO and relatively rich in Fe_2O_3. Some of the coal particles remain unburned and are collected with the fly ash as carbon particles. The amount of unburned coal is determined by such factors as the rate of combustion, the air/fuel ration, and the degree of pulverization of the coal.

Fly ashes from Alberta and Saskatchewan are derived from sub-bituminous coal and lignite coal respectively. Fly ashes produced in the Eastern provinces are derived from bituminous coals.

2.2 Physical properties

Fly ash varies in color from light cream to dark brown. The color is affected by the proportion of unburnt coal, the iron-rich particles and the moisture. The majority of fly ash particles are glassy and spherical.

The average specific gravity of fly ash particles varies from 1.9 to 2.9. Fly ash containing high amounts of unburnt carbon and hollow spherical particles, called cenospheres, has a low specific gravity. Larger proportions of iron-rich particles and other compounds impart a high specific gravity to the ash. The bulk density of dry fly ash is in the order of 50 to 60 lbs per foot.

2.3 Pozzolanic activity in fly ash

As early as 1930, it was discovered that fly ash possessed various physical and chemical properties characteristic of pozzolans. Pozzolans are natural and artificial materials which by themselves are not cementitious but which react with hydrated lime and water to form cementitious compounds, namely calcium silicate and aluminate hydrates of low solubility. The usefulness of fly ash as a pozzolanic material is very limited, and is determined by its highly variable physical and chemical properties. Such variability arises not only from varying amounts of mineral matter in the coal from which the fly ash originates, but also from the conditions of operation of the boiler, the fineness of the pulverized coal prior to combustion, the additives mixed with the coal for pollution control, and the method of collection. This variability of the ash has hindered the assessment of the causes of pozzolanic activity and the development of standard tests for pozzolanic activity. Conceivably pozzolanic activity could be studied by examining the possibility of separating the constituents of fly ash and allowing

them to react with lime individually. This separation of fly ash constituents was found to be practically impossible (Joshi 1979). Laboratory synthesization of fly ash from various mixtures of minerals associated with actual coal under conditions simulating those found in suspension-fired furnaces has been carried out successfully (Joshi and Rosauer 1973a). A study of the pozzolanic behavior of this type of fly ash revealed that the pozzolanic activity of a pure alumino siliceous synthetic fly ash is influenced more by the amount of strained glassy phase present than by the particle size (Joshi & Rosauer 1973b). It was further possible to infer that iron in an alumino-siliceous ash reduced pozzolanic activity, whereas calcium had the opposite effect. It was also recognized that the pozzolanic reaction was initially reaction-controlled, and after 8-10 weeks became diffusion-controlled.

In addition to pozzolanic activity, most sub-bituminous and lignite ashes also exhibit a self-hardening property i.e. when mixed with water and compacted, they harden due to cementation. In general, this self-hardening effect is related to the presence of free lime and/or the water soluble fraction in the fly ash. Most types of Alberta fly ash, being basically sub-bituminous, exhibit a self-hardening property.

3 FLY ASH UTILIZATION

For a number of years the potential uses of fly ash have been explored in civil engineering. Joshi, Duncan & McMaster (1975) provide a critical appraisal of the new and conventional engineering uses of fly ash. In concrete technology fly ash has been used in the production of cement and lightweight aggregates as a cementitious material along with lime or cement to stabilize aggregates or soil. Fly ash has been used in concrete for over five decades. In a cement mix, three different proportions of fly ash have been used, namely:
1. for partial replacement of cement;
2. as an additive to the fine aggregate; and
3. as a partial replacement of both cement and fine aggregate.

Fly ash has also been directly used to make bricks, building blocks and as a filler in bituminous concrete, rubber and plastics.

In geotechnical engineering, fly ash

has been used as a construction material to replace soil, as a stabilizing agent for surface soils, and for the treatment of deep-seated in situ soils. This paper deals with the uses of fly ash in geotechnical engineering. Fly ash has been used in practice specifically for the following purposes:

1. structural fill;
2. back fill;
3. embankment material;
4. stabilization of soils and aggregates for pavements;
5. in situ improvement of soils and sanitary land fills by lime-fly ash injection.

Fly ash consists of particulate materials similar to silty soils in a dry state. Before discussing its practical applications, it is necessary to describe the problems involved in handling such fine materials, and the geotechnical properties of fly ash notably its compaction, consolidation shear strength, and permeability characteristics.

3.1 Handling characteristics

a) Dry ashes, or ashes stored in hoppers: Fly ashes collected by electrostatic precipitators and stored in silos or hoppers generally contain over 80 percent particles with a mesh sieve number of over 200. Hopper collected ashes are much coarser and are generally of uniform gradation. [These ashes lack plasticity]. Paddle mixers and sprayers are used to moisten the ash before loading it in trucks for transportation.

b) Stockpiled fly ashes: The conditioned or moist fly ash in stockpiles generally gets agglomerated. Sub-bituminous fly ashes may even harden to shale-like consistency in stockpiles, depending on the degree of self-hardening. Such ashes may require special excavation equipment if they need to be reclaimed. Ashes with a low degree of self-hardening do not pose any problem since they do not get compacted, permitting easy transportation and reclamation from the stockpiles.

c) Lagoon or ponded fly ashes: Fly ashes are often sluiced in ponds and lagoons. Sedimentation of ash particles produces layers of variable particle size and specific gravity. Ponded ashes are, therefore, likely to be highly heterogeneous in particle size gradation. Reclaiming such ashes may be relatively easy if they are not self-hardening. However, they have a high moisture content and a very low density due to sedimentary deposition, and this may lead to liquifaction and handling problems. Some bituminous ashes have a high moisture holding capacity and do not dry easily. On the other hand, it might even be necessary to resort to blasting and other rock excavation techniques to mine the ponded self-hardened ashes produced from sub-bituminous and lignite coals.

d) Scrubber sludge and ashes: Fly ashes mixed and stored with calcium sulfite/sulfate sludge, generated by scrubbers installed to control sulfur dioxide emissions, generally never harden. The mixture is thixotropic and turns to a toothpaste like consistency with slight working due to its invariably high initial moisture contents. If calcium sulfite is oxidized to calcium sulfate, the sludge can be easily used for fill purposes. Otherwise the wet sludges, which are rich in calcium sulfite, are hard to handle and transport and are thus unsuitable for fill purposes.

4 GEOTECHNICAL ENGINEERING PROPERTIES

Fly ashes are particulate media akin to soils which interact with water in developing low to no plasticity characteristics but can develop cementation bonds either due to pozzolanic material or due to an inherent self-hardening property. Hence the mechanical properties of fly ashes, such as shear strength, compressibility and permeability, are of relevance in assessing their potential for utilization as a construction material.

4.1 Compaction characteristics

To realize the desired engineering performance using bituminous or non self-hardening fly ashes, either independently or when mixed with soil, densification under external compactive effort is necessary. Since fly ashes interact with water, the compaction characteristics are similar to those of fine grained soils and ashes, and therefore sensitive to moisture content. The compacted dry density can be increased by increasing the compactive effort. It has been observed that hopper or silo ashes as well as fine dredged ashes from ponds generally yield a compaction curve with a sharply defined peak. It has been observed that the compaction characteristics of stockpiled ashes are only slightly different from the fresh ashes.

The optimum moisture content and maximum dry density for Alberta ashes and some other western fly ashes are presented in Tables 1 and 2 (Joshi 1981).

Sub-bituminous and lignite ashes which also exhibit a high degree of self-hardening are difficult to compact. These ashes harden during transportation and cannot be compacted. Such ashes may be compacted if a retarder is used to delay the self-hardening of the ashes. However, hardened ashes do not disintegrate on innundation and saturation. Therefore, the compaction density is not particularly important. It is the problem of handling and finishing such ash fills which requires attention. Gypsum or any other retarder may be used to retard setting and improve the compaction and finishing of an ash fill.

Compressibility and volume change

It has been found that self-hardening ashes are not susceptible to settlement under light to moderate loads. From a practical standpoint, specifications are needed to allow the development of minimum strength. Moderately self-hardening ashes, when compacted, exhibit unconfined strengths of 6 to 10 ksf. Such ashes will deform very little even under structural loads of 6 to 10 ksf. On the other hand, it is very essential to assess the possible moisture absorption and the consequent strength reductions, if any. Results of laboratory tests on compacted sub-bituminous fly ash (Joshi 1978) indicated (Table 3) that samples compacted to 90 to 95% of their maximum density can still absorb 10 to 15% additional water. This is due to capillary water rise due to initial partial saturation. This increase in moisture content can cause a loss of strength, depending on age, particularly in non-self-hardening fly ashes.

Shear strength

Test data on different fly ashes indicate that compacted fly ashes develop moderate to very high strength depending upon the self-hardening characteristics. The compaction moisture content seems to have little effect on the strength. Even in the case of non-self-hardening bituminous ashes, the strength increases with time due to pozzolanic activity when used with cement and/or lime. The level of strength that can be obtained depends upon the density of compacted lime-fly ash or cement-fly ash. Even a fly ash with a low to moderate degree of self-hardening develops increasing strength with time. However, the degree of compaction is very important for strength development in ashes with a low to moderate degree of self-hardening. In a study by Joshi (1976), such fly ash samples compacted to about 80% of the maximum density, and, attained about 70% as much strength as those compacted to maximum density. Invariably, in practice the strength levels that can be achieved in most of the self-hardening ashes are adequate to meet the strength requirements for geotechnical purposes.

Permeability

The air void content of fly ashes at their maximum dry densities vary from 5 to 15 percent. Generally the permeability varies from 5×10^{-4} to 6×10^{-6} cm/sec. Experience has indicated that minor variations in compacted density do not seem to influence the permeability of the ashes very significantly. Since the permeability is low, the rate of infiltration is low. The permeability seems to reduce with time in compacted self-hardening ashes. Apparently, the voids between particles are partially filled with hydration products.

5 STRUCTURAL FILLS USING FLY ASH

Structural fills are controlled or engineered fills constructed to perform as desired. They have to satisfy strength and settlement criteria and should exhibit satisfactory performance under natural or man-made environmental conditions. Case histories on fly ash usage are presented below.

5.1 Fly ash structural fill under fuel story tank

A fuel storage tank of 1-million gal. capacity was constructed in the Missouri river flood plain by Kansas City Power and Light Company, at the Hawthorne power plant. The subsurface profile consisted of soft, highly plastic clays to a depth of about 20 ft. underlain by loose to medium dense sands to a depth of about 80 ft. To minimize excessive settlements, soft clays had to be replaced by properly compacted structural fill. The cost of transporting granular soil or low plastic silty clay fill was considered to be prohibitive. Hence the potential use of moistened fly ash from the Grand Avenue plant, stockpiled closely at a distance of 300 m, was explored.

On the basis of the laboratory test data, fly ash from the stockpile was selected and used as a structural fill (Joshi, Duncan & Durbin 1976). The moisture content of the fly ash from the stockpile varied from 18 to 38 percent. Experiments indicated that the large variation in moisture content did not pose significant compaction problems. Ninety-five percent relative compaction could be attained without any difficulty in most of the field situations. A nuclear moisture density gauge and hand driven samplers were used to verify the field moisture and density data.

The compacted ash generally attained high compressive strength, as was observed during foundation excavation for the ring wall. A flexible bottom steel tank was constructed on the top of the fly ash fill. The tank was water tested for 110 percent of the anticipated fuel oil loads. Settlement measurements on the four points on the periphery of the tank indicated a minimum of 12 mm and a maximum of 25 mm settlement. The tank has been in service for the past 12 years.

5.2 Fly ash as backfill material

Since fly ashes are particulate media with frictional characteristics as well as some permanent cohesion or cementation, they exhibit a higher angle of shearing resistance than soils with appreciable fines. Also the cementation due to self-hardening in a fly ash backfill is likely to reduce lateral pressures against a retaining structure. A mildly self-hardening fly ash was used as backfill behind a 28 ft. high 230 ft.

long rigid wall. The wall abounded the 3-storey parking garage constructed by Farmland Industries in Kansas City, Missouri. The parking structure measured 115 ft. by 230 ft. in plan dimensions and was built into the slope of a ravine (Fig. 1). The natural slope was cut back about 10 to 30 ft. for wall construction. A tied-back reinforced concrete cantilever wall was constructed. About 3500 tons of Grand Avenue fly ash was used as a backfill material. This ash exhibited self-hardening properties. When compacted, it developed strengths of 140 to 320 lbs/sq.in. within a period of 28 days. Within two to three days itself, the compacted ash hardened to such an extent that it resisted shovel excavation.

Six Glotzel cells were installed to measure the lateral and vertical pressures generated by the backfill Figure 2 shows typical vertical and lateral pressures measured with time. Theoretical at rest values were computed for an average bulk unit weight of 86 lbs/ft., and, a coefficient of lateral earth pressure at rest, k_o of 0.5. The construction details and the relevant data have been provided elsewhere (Joshi, Duncan & Durbin 1976). The lateral pressures were reduced appreciably due to self-hardening property of fly ash, compaction techniques, stage construction, and arching. Arching was also considered responsible for the low vertical pressures recorded in this project. Figure 3 shows the construction details.

5.3 Lime-fly ash slurry injection

Lime slurry pressure injection has been a method of treating soils in situ to reduce their volume change potential and in some instances to increase their bearing capacity. In soils of low clay content, lime injection alone will not produce the desired stabilization effect. Since fly ashes are composed of silica and/or alumina, it is an ideal artificial pozzolan for use with lime in the stabilization of soils. A unique characteristic of lime/fly ash mixtures is their inherent ability to heal or recement across cracks by a self-generating mechanism.

Lime fly ash slurry is injected into soil under pressures of 350 to 1380 kPa. Slurry is injected through 38 to 41 mm diameter hollow injection pipes fitted with nozzles. The injection pipes are pushed into the ground and lime slurry containing 30% solids is injected into the ground at spacings of 300 to 350 mm. Facilities capable of handling great depths of over 40 meters have been developed in recent years. Detailed laboratory and field experiments reveal (Joshi, Natt & Wright 1981) that lime migrates by diffusion and alters the strength properties of clay.

Six case histories, two on subgrade stabilization by a lime-fly ash slurry injection process to reduce settlement, one on earth-dike stabilization to reduce seepage and increase shear strength of the soil, and three on slope embankment stabilization to reduce settlement and increase slope stability (Joshi and Wright 1978) reflect the diversified use of this technique in achieving desired in situ improvement.

Total and differential settlements of the order of 38 mm and 25 mm due to heavy loads in an industrial building at Ohio could be totally arrested by lime-fly ash slurry injections. The subsurface soils consisted of about 10 feet (3 m) of soft to firm clays overlying dense sand. Settlement observations of probes and plates installed at various depths did not show any settlement under heavy loads.

In another situation the surface erosion and rotational and translational slides of the steep high plastic clay slopes could be arrested by slurry injection. The injection probes were located at a 4.5 ft. square grid and were staggered in two stages of injection. In each stage lime slurry was injected to about 10 ft. depth. After treatment, the slopes did not show any signs of further failure and longitudinal cracks.

Most recently (Sharma and Pariti 1985) a lime-fly ash injection was applied by Alberta Transportation in a localized area of the Highway 16 Twinning project near the village of Gainford, 60 miles west of Edmonton, Alberta. In July 1983, soon after construction, surface cracks appeared. Site investigation revealed that soft deposits and sea shells at approximately 36 ft. were responsible for observed surface movements and subsidence. Lime-fly ash slurry injection was resorted to up to 36 ft. depth under a pressure of 215 psf at approximately 6 to 9 ft. intervals in each hole as the drill rod was pushed downwards. This resulted in dramatic reductions in surface deformations. This portion of the highway, which is currently in operation is still under observation.

A recent new application of lime-fly ash slurry injection has been for the stabilization of land fills to support construction of buildings and parking lost, and for repair and renovation of existing buildings and parking lots (Joshi, 1983; Blacklock, Joshi and Wright, 1983). Because fly ash is an inexpensive by-product and is readily available in many parts of the world, it is an ideal material to use in this type of application where large volumes of cementitious fill material are required due to the large cavities found in land fills. The combination of lime and fly ash has proven to be extremely successful when used in the pressure injection grouting method, for strength, gas and liquid cutoff and structural stabilization of large landfills.

5.4 Geocrete

The strength requirements of placed or roller compacted concretes in structurally safe pavements and gravity dams up to 150 m high are relatively low compared to the strengths realized by the use of conventional mass concretes using portland cements. Fly ashes can be conveniently used to replace cements in mass concrete, to reduce the heat of hydration, and to effect economy. It is also possible to increase the proportions of non-cementitious material to cementitious material to achieve economy without sacrificing strength and durability requirements. The method of placement can be essentially by roller compaction, similar to that practiced in embankment construction. Geocretes are materials composed of earth materials (sand, clay and aggregates) mixed with low fractions of cementitious materials (cement and fly ashes) where the compaction characteristics are sensitive to the moisture content. Joshi and Natt (1983) report the engineering properties, and potential applications of roller compactible geocretes. Table 4 details the compressive strengths of various mixes containing lime-cement and fly ash as cementitious materials.

Roller compactible mixes exhibited good durability characteristics, as evaluated from dry shrinkage and freeze-thaw tests.

Geocretes have a potential usage for pavements, dams, retaining walls, coastal protections, canal lining and river protection. A few dams have been constructed in the U.S.A. and Japan using geocrete or roller compacted concrete. Such dams eliminate the need for an overflow or spillway structure and intake tower and are, therefore, less expensive than even earthfill or rockfill dams. More information about long term durability, dimensional stability and permeability is still needed to enhance its usage confidently in geotechnical engineering.

6 DISCUSSIONS AND COMMENTS

This paper provides a brief description of the potential uses of fly ash in soil improvement. Due to space constraints, several other applications have not been discussed.

As energy demands increase and the supply of oil diminishes and with public resistance to nuclear power, the use of coal in power generation is inevitable. This invariably will generate large volumes of fly ash which will require proper disposal or judicious utilization of the ash by the construction industry.

As a first step before exploitation, the physical, chemical and mechanical properties of fly ash need to be determined. There are several factors, such as the type, the amount of mineral matter in the coal, the coal processing, the furnace conditions, the collection methods, and the ash storage methods, which affect the properties of fly ash.

The following are the normal ranges of the engineering properties of fly ashes.

Optimum moisture content	18-34%
Maximum dry density	70-90 lbs/ft
Cohesion	4-6 psi
Angle of shearing resistance	32-35°
Unconfined compressive strength	4-6 KSF
Permeability k	4×10^{-4} to 6×10^{-6}
Compressibility	$C_c = 0.06$
	$C_r = 0.004$
Pozzolanic activity	Minimal to High

While fly ashes are used as structural fill, backfill or site fill, pneumatic rollers can be used after initial compaction with a tracked vehicle. Generally, vibratory rollers are not suitable. The compaction moisture content may vary between ± 6% of the optimum. The dry density may vary between ± 10% of the maximum. In general, compactibility should be the criterion rather than density or moisture content.

Experience has indicated that self-hardening fly ash fills possess a bearing capacity higher than natural soils. Generally, fly ash fills are relatively impermeable, and lateral and vertical pressures are less if fly ashes are of the self-hardening type. Some compacted ashes swell after curing. Such fly ashes may generate excessive lateral pressures when used as backfill.

Lime-fly ash slurry injection offers a promising method of increasing the bearing capacity of land fills with reduced settlement, and also of simultaneously rendering the volume of treated mass relatively impermeable. Lime-fly ash slurry fills seams and cavities and reacts with the adjacent soil forming strong and relatively impervious membranes locked into the soil mass. The effect of these membranes is to control the movement of moisture and to reinforce and confine the segmented portions of soil mass. With time, lime diffuses into the soil and hardens even the soil encased between the lime-fly ash slurry and the membrane created by it, due to the reaction with the adjoining soil.

In the not-so-distant future, a significant portion of the ash produced worldwide will be used as a construction material in many more diversified situations.

REFERENCES

Berry, E.E. and Malhotra, V.M. 1977. Production, Use and Properties of Fly Ash in Canada, presented at the Industrial Minerals, Dn. 79th Annual General Meeting of CIM, Ottawa, pp.16.
Blacklock, J.R., Joshi, R.C. and Wright, P.J. 1983. Pressure Injection Grouting of Landfills Using Lime and Fly Ash, Proceedings of Conference of "Old Landfill Stabilization & Restoration", Winnipeg, Canada, pp. 83-96.
Joshi, R.C. and Rosauer, E.A. 1973a. Pozzolanic Activity in Synthetic Fly Ashes 1 - Experimental Production and Characterization, The American Ceramic Soc. Bulletin, Vol. 52, No. 5, 456-459.
Joshi, R.C. and Rosauer, E.A. 1973b. Pozzolanic Activity in Synthetic Fly Ashes II, Pozzolanic Behavior, The American Ceramic Soc. Bulletin, Vol. 52, No. 5, pp. 459-463.
Joshi, R.C., Duncan, D.M. and McMaster,

H.M. 1975. New and Conventional Engineering Uses of Fly Ash, Transportation Engg. Journal, ASCE, Vol. 101, TE4, pp. 791-806.

Joshi, R.C., Duncan, D.M. and Durbin, W.L. 1976. Performance Record of Fly Ash as Construction Material, Proc. of Fourth Int. Ash Utilization Sym., St. Louis, pp. 300-320.

Joshi, R.C. 1978. Structural Fills Using Fly Ash, Proc. of Ash Management Conference, Texas, A & M University, College Station, pp. 186-213.

Joshi, R.C. and Wright 1978. In Situ Soil Improvement by Lime and Lime Fly Ash Slurry Injection Process, Proc. Symposium on Soil Reinforcing and Stabilizing Techniques in Engineering Practice, The New South Wales Institute of Technology, Sydney, Australia, October 16-19, 1978, pp. 545-558.

Joshi, R.C. 1979. Sources of Pozzolanic Activity in Fly Ashes, A Critical Review, Proc. Fifth, Int. Ash Utilization, Sym. U.S. Dept. of Engg., pp. 610-623.

Joshi, R.C. 1981. Geotechnical Aspects of Coal Ash Production Disposal and Utilization, Proceedings Coal Ash and Reclamation Workshop, Edmonton, April 29-30, 1981, Alberta Energy and Natural Resources, Edmonton, pp. 3-27.

Joshi, R.C., Natt, G.S. and Wright, P.J. 1981. Soil Improvement by Lime-Fly Ash Slurry Injection, Proc. Tenth Int. Conf. on Soil Mech. & Foundation Engg. Stockholm, 12/27, pp. 707-712.

Joshi, R.C. 1983. In Situ Stabilization for Structures, Proc. of Conf. an Old Landfill Stabilization and Restoration, Winnipeg, Canada, pp. 77-81.

Joshi, R.C. and Natt, G.S., 1983. Roller Compacted High Fly Ash Concrete (Geocrete), Proc. of the CANMET First Int. Conf. on the Use of Fly Ash, Silica Fume, Slag and Other Mineral By Products in Concrete, ACI sp 79, pp. 347-366.

Sharma, L. and Pariti, M. 1985, Lime Fly Ash Injection for Soil Stabilization, Presented at W.A.C.H.O. Conf., Edmonton, April 85.

Manz, O.E., 1976. Lignite Production and Utilization, Proc. Fourth Int. Ash Utilization Sym., St. Louis, pp. 39.

Table 1 Optimum Moisture Contents and Maximum Dry Densities of the Alberta Fly Ashes

	Forestburg		Sundance		Wabamum	
	25 Blows	10 Blows	25 Blows	10 Blows	25 Blows	10 Blows
Optimum Moisture Content (%)	18.0	20.4	19.6	22.0	16.2	18.0
Maximum Dry Density (kg/m³)	1350	1312	1280	1237	1526	1484

Table 2 Optimum Moisture Contents and Maximum Dry Densities of the Fly Ashes from the United States

	Kansas	Montana	Kansas Montrose	Kemmerer
Optimum Moisture Content (%)	17.0	24.0	17.0	21.0
Maximum Dry Density (kg/m³)		1825	1540	1501

22

TABLE 3

Average Strength and Compaction Characteristics of Sub-bituminous

Fly Ashes from Alberta, Canada (Joshi 1978)

Fly Ash No. 8

AGE	Compacted Dry Density	Compaction Moisture	Relative Compaction	Unconfined Compressive Strength		Water Absorption on Soaking
				Unsoaked	Soaked	
Days	pcf	%	%	ksf	ksf	%
28	83^a	19	98	39.0	-	-
28	80^a	19	95	25.5	-	-
28	74^a	15	88	7.1	-	-
18	82^b	20	97	16.5	15.5	20
18	82^b	20	97	36.5^c	37.0^d	
18	78^b	18	92	18.2	11.1	22
18	78^b	18	92	42.5^c	14.4^d	22

Note: a - Samples prepared by dynamic compaction.
         b - Samples prepared by static compaction.
         c - Samples oven dried and then tested for strength.
         d - Samples oven dried, soaked in water and then tested.

         All samples cured at 90% relative humidity to approximately
         simulate field conditions.

Table 4 Compressive Strengths of the Various Mixes Containing Lime-Cement-Fly Ash
Cementitious Material (.5:.5:8 lime:cement:fly ash) and Aggregate,
(Joshi & Natt, 1983)

Mix Nota- tions	Mix Ingredients					Wt. % of cementitious material of aggregate	Compressive Strength MPa			
	Cementitious Material			Aggregate			7 day	28 day	56 day	120 day
	cement	lime	fly ash	sand	aggregate					
AVe	1.11	1.11	17.78	80	0	25	3.8	9.6	11.45	14.15
AVi	1.11	1.11	17.78	40	40	25	4.85	10.15	12.95	14.5
BVe	1.11	1.11	17.78	80	0	25	4.8	14.2	15.4	17.5
BVi	1.11	1.11	17.78	40	40	25	5.2	16.0	21.6	22.9

A's - Wabamun fly ash
B's - Sundance fly ash

23

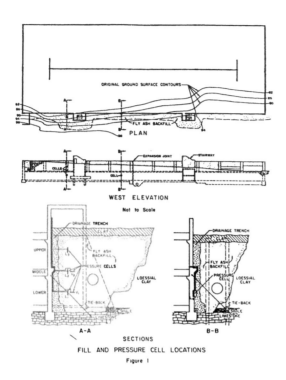

PLAN

WEST ELEVATION

Not to Scale

SECTIONS

FILL AND PRESSURE CELL LOCATIONS

Figure 1

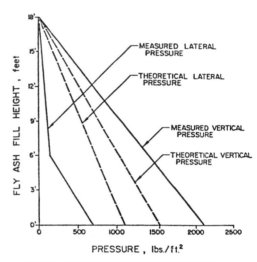

FIGURE 2: LATERAL & VERTICAL PRESSURE DISTRIBUTION

FIGURE 3: CONSTRUCTION DETAILS

24

Environmental Geotechnics and Problematic Soils and Rocks, Balasubramaniam et al. (eds)
© *1987 Balkema, Rotterdam. ISBN 90 6191 785 9*

Systems analysis in multi-dam reservoir systems including water quantity and quality controls

T.Kojiri
University of Gifu, Japan

ABSTRACT: The aim of this study is to establish the operational procedure for water quantity and quality control systems with multi-dam and multi-estimate points. That is to say, the computational complexity varies directly with the number of dam reservoirs and estimate points, because they add variables to the problem. Consequently, Dynamic Programming and the decomposition method are introduced and a formula is devised to gain the optimal solution and reduce the amount of the computer memory and the computation time. As for the water quantity and quality, multi-objective programming is used comparing the scalar and the vector optimization techniques. Moreover, the stochastic control for the real-time operation is discussed.

1 INTRODUCTION

Today there are many dam reservoirs in the world and some dam reservoirs are under construction or planned, because they are one of the most effective means for water supply, irrigation, electric power and flood prevention. Thus, many researchers have been investi gating the evaluation of storage facilities in the reservoir and the optimal policy of releasing discharge. Ripple(1883) estimated the safety rate of reservoir capacity through Mass Curve Analysis for the known inflow sequences. Introducing the queuing theory, Moran (1954) represented the storage state as the transition probability under conditions of the stationary inflow and fixed releasing rule. Lloyd and Anis (1975) and Nagao (1977) calculated the reliability of the intake discharge from a dam reservoir with Markov inputs.

On the other hand, Hall (1964) and Takasao and Seno (1970) formulated the optimization of dam control by Dynamic Programming, and LeVelle and et al. (1969) did it by Linear Programming. The increase of the number of dams, however, discreases the feasibility of computation in the digital computer. Hall and et al. (1969) and Heidari and et al. (1971) respectively modified the computation algorithm and proposed IDP (Incremental Dynamic Programming) and DDDP (Discrete Differential Dynamic Programming) by restricting the regions of state space. Takeuchi (1974) also proposed DCL (Dynamic Programming Coupled With Linear Programming) to optimize multi-reservoir system by LP in space and DP in time.

The operational problem of dam reservoirs involves three coordinates. The first denotes the time representing the statistical or probabilistic characteristc. This includes elements such as the short-term, the long-term, the design operation and the real-time operation. The second denotes the coordinate distinguishing between a single dam system, a multi dam system and a conjugative system with a dam, a big weir and a pumping well system. The third denotes a single objective, such as water quantity control, and a multi-objective, such as water quantity control and quality controls (temperature, suspended soil, or BOD). This is the important criterion for the consolidated operation of the short-term and the long-term operation or for environmental assessment. Figure 1 gives a cubic representation of the above systems.

In this paper, I will discuss the multi-dam and multi-objective operational system, and propose a stochastic control rule for the real-time operation considering the prediction procedure of inflow.

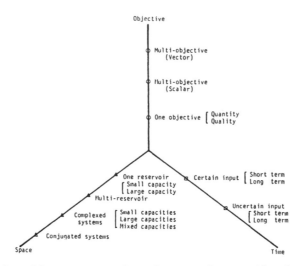

Figure 1. Cubic representation of reservoir operational systems.

2 OPTIMAL OPERATION OF DAM RESERVOIRS

2.1 Objective function of dam operation

It is important to define the objective function for the optimization of the control system. As for water qantity and quality controls, there are three objectives, namely, low flow, high flow (flood), and flow turbidity, which is one of the control criteria in water quality. Since t it is easy to understand the controlled effect and to apply any water management system, each objective can be defined not through the economic index but through the physical index:

low flow ; $P = \min (\overline{Q_{ml}/Q_{md}}) \longrightarrow \max$ (1)

turbidity ; $D = \max (C_{mmax}/C_{md}) \longrightarrow \min$ (2)

high flow ; $K = \max (Q_{mp}/Q_{md}) \longrightarrow \min$ (3)

Where, Q_{ml} is the lowest level of flow discharge after being controlled at estimated point m, Q_{md}, the allowable limit level of low flow, C_{mmax}, the worst level of turbidity, C_{md} the allowable turbidity level, Q_{mp}, the peak level of flood flow, and Q_{md}, the allowable flood level at defence point m.

2.2 Formulation of dam operation by Dynamic Programming

Dynamic Programming is a powerful approach to find the optimal solution based on the Principle of Optimality (Bellman, 1957). The control system is composed of four variables, namely the decision variable, the state variable, the evaluation variable and the disturbance variable. In the dam reservoir system, they correspond to the release discharge, $QO_n(t)$ (n is the number of dam from 0 to N; N is the tota number of dams and t is the control time stage), the storage level of dam n, $S_n(t)$ ($0 \leqq S_n(t) \leqq V_n$; V_n is the allowable storage capacity level of dam n), the flow discharge $Q_m(t)$ at estimate or defence m (m = 1,2,...., M; M is the total umber of estimate points), and the inflow, $QI_n(t)$ and tributary flow discharge $q_n(t)$. If the evaporation from the reservoir surface is negligible, the continuity equation at dam n is

$$S_n(t)=S_n(t-1)+QI_n(t)-QO_n(t) \qquad (4)$$

and the constraint equation is as follow

$$0 \leqq QO_n(t) \leqq \min(S_n(t-1)+QI_n(t),V_n(t),GCA_n) \qquad (5)$$

where GCA_n is the maximum release discharge from dam n. With a set such that the initial condition C_n is $S_n(0)$, the recursive equation of forword DP is formulated as follows:

Table 1. Computed results of optimal control (A-3 type).

Inflow																					S(0)=0, V=10	
t	0	1	2	3	4	5	6	7	8	9	10	11	12	13	14	15	16	17	18	19	20	$Q_{nl}/\bar{Q}_{md}$
$QI(t)$		4	2	3	6	2	4	7	3	5	4	2	4	3	5	6	2	4	2	4	5	
$q_1(t)$		0	0	0	2	3	2	0	0	1	2	2	0	0	0	2	0	2	0	2	2	
$q_2(t)$		2	1	2	0	0	2	2	1	0	1	1	0	2	0	1	0	2	3	1	3	
When $\bar{Q}_{1d}=1$ and $\bar{Q}_{2d}=1$,																						
$S(t)$	10	9	6	4	7	7	8	10	8	9	10	9	8	6	6	9	6	8	6	7	10	
$Q0(t)$		5	5	5	3	2	3	5	5	4	3	3	5	5	5	3	5	2	4	3	2	
$Q_1(t)$		5	5	5	5	5	5	5	5	5	5	5	5	5	5	5	5	5	4	4	5	4	4
$Q_2(t)$		7	6	6	5	5	7	7	6	5	6	6	5	7	5	6	5	6	7	6	7	5
When $\bar{Q}_{1d}=1$ and $\bar{Q}_{2d}=4$,																						
$S(t)$	10	9	6	4	6	5	7	10	8	8	9	8	6	5	4	7	3	5	4	6	10	
$Q0(t)$		5	5	5	4	3	2	4	5	5	3	3	6	4	6	3	6	2	3	2	1	
$Q_1(t)$		5	5	5	6	6	4	4	5	6	5	5	6	4	6	5	6	4	3	4	3	3
$Q_2(t)$		7	6	6	6	6	6	6	6	6	6	6	6	6	6	6	6	6	6	5	6	1.25

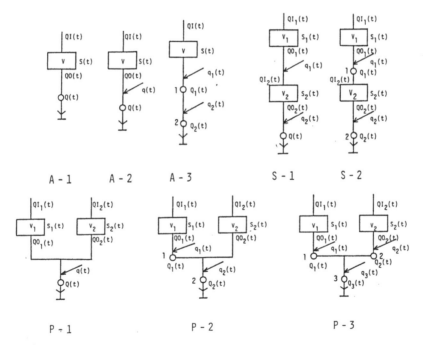

A - 1   A - 2   A - 3   S - 1   S - 2

P - 1   P - 2   P - 3

Figure 2. Basic patterns composing of complex system with milti-reservoir and multi-defence points.

$$f_t(S_1(t),...,S_N(t)) = \min_{(O_n(t))} \left(\sum_{m=1}^{M} D_m(Q_m(t)) + f_{t-1}(S_1(t-1),...,S_N(t-1)) \right) \quad (6)$$

subject to

$$f_1(S_1(1),...,S_N(1)) = \sum_{m=1}^{M} D_m(Q_m(1)) \quad (7)$$

where, $D_m(.)$ is the estimate function at point m, and $f_t(.)$ is the accumulated value of the controlled estimation function from time stage 1 to t. Table 1 shows the controlled sequences of release discharge and storage level in the case of an A-3 type shown in Figure 2 and using the following estimation function according to the low flow control.

$$D_m(Q_m(t))=((RV+1)M)^{**}(-a_mQ_m(t)+b) \qquad (8)$$

$$RV = \max_{(t,n)} (\min (VA_{nt}, t-1))$$

where, VA_{nt} is the range of the state variable at time stage t, a_m, positive integer (a_mQ_{md}=const.), and b, constant, respectively. If m(n) and m(h) denote the number of the release discharge and the tributary flow discharge directly flowing at estimated point m, the flow discharge $Q_m(t)$ is expressed as follows:

$$Q_m(t)= \sum_{n=1}^{m(n)} QO_n(t)+\sum_{h=1}^{m(h)} q_h(t) \qquad (9)$$

2.3 Approximative method for computation

In the performance of dynamic programming, as given in eqs. 6 and 7, there appear several problems, notably the overflow error and the limited memory capacity attributed to multi-dimension or multi-stage. These are solved by the following methods. The first method is to use the following equation as a substitute for the recursive equation:

$$f_t(S_1(t),...,S_N(t))= \max_{(O_n(t))} (\min(Q_m(t)/\bar{Q}_{md},$$
$$f_{t-1}(S_1(t-1),$$
$$...,S_N(t-1)))) \qquad (10)$$

The computed sequences may differ from those solved by using eq. 6 from the viewpoint of the duration periods of the lowest flow.

The second method is to decrease the dimensions of the system. For example, a parallel reservoirs system (P-1 type) having only one estimate point can be aggregated as follows:

$$V_0 = V_1 + V_2 \qquad (11)$$

$$QI_0(t) = QI_1(t) + QI_2(t) \qquad (12)$$

The release discharge of the composite reservoir is decided in accordance with the following space ratio deduced from Havard Water Program (Mass and et al., 1962):

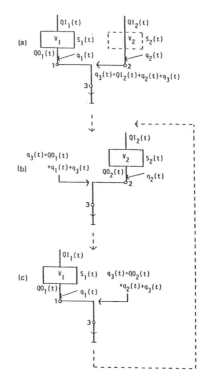

Figure 3. Computational procedure of iteratively approximative method.

$$\frac{V_n-S_n(t)}{\sum_{i=1}^{2}(V_i-S_i(t))} = \frac{QI_n(t)}{\sum_{i=1}^{2}QI_i(t)} \qquad (13) \quad (n=1,2)$$

If the value of $QO_n(t)$ or $S_n(t)$ is negative, it must be set to zero. This procedure can also be applied to the serial reservoirs system (S-1 type).

In the multi-estimate points system, it is useful to approach the optimal solution by computing the approximative solution on each reservoir, iteratively. At the first step, only one dam is controlled regarding the release discharge from another dam as one of the tributary inflows. At the second step, the next dam is controlled, fixing the release discharge from another dam conversely. Figure 3 shows these iterative procedures. The convergence process in computation is proved by using eq. 8 (Takasao and et al., 1975).

For the multi-time stages problem, the control period is divided into several periods and the iterative computation is performed among two adjacent periods as shown in Figure 4 (Takasao and et al., 1975). Assuming the storage level $S(T_{II})$,

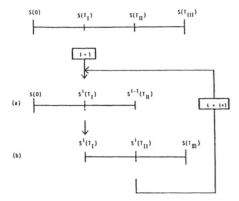

Figure 4.Computational procedure of approximative method with divided control periods.

in the first sub-period and the second sub-period the reservoir is separately optimized by forward DP and backward DP. The optimal storage level $S(T_I)$ is decided as the minimum value of the summations of accumulative functions $f_I(S(T_I))$ and $f_{II}(-S(T_I))$. Then, after optimizing the second sub-period and the third sub-period, $S(T_{II})$ is decided. The experimental results show the convergence with a few of iterations, and the consistency with optimal solution.

The last method is to use DDDP in which the width of the corridor equals VA_{nt} and the number of the required memories can be reduced considerably.

3 OPTIMAL OPERATION IN MULTI-DAM AND MULTI-ESTIMATE POINTSSYSTEM

3.1 Introduction to the decomposition method

Decomposing the whole system into N subsystems, according to the dam formation, the system is represented by Linear Programming as follows (Buras, 1975):

a)The constraints in the whole system:

$$\sum_{n=1}^{N} A_n((S_n(t)),(QO_n(t)),Q_{m(n)}(t)))=b_0 \quad (14)$$

b)The constraint in the sub-system n:

$$B_n((S_n(t)),(QO_n(t)),(Q_{m(n)}(t)))=b_n \atop (n=1,2,\ldots,N) \quad (15)$$

c)The objective function:

$$z = \sum_{n=1}^{N} C'_n ((S_n(t)),(QO_n(t)),(Q_{m(n)}(t)))$$
$$\longrightarrow \text{max(or min)} \quad (16)$$

Where, A_n and B_n denote the coefficient matrixes, b_0, the column vector representing the whole system, b_n, the column vector representing the sub-system and ', the row vector, respectively. With a setting $X_n =((S_n(t)),(QO_n(t)),(Q_{m(n)}(t)))$, the new Linear Programming where u_n^j is the decision variable is formulated as follows:

$$\sum_{n=1}^{N} \sum_{j=1}^{l_n} A_n u_n^j X_n^j = b_0$$

$$\sum_{j=1}^{l_n} u_n^j = 1 \quad (17)$$

$$z^0 = \sum_{n=1}^{N} \sum_{j=1}^{l_n} C'_n u_n^j X_n^j$$
$$\longrightarrow \text{max(or min)}$$

Where, l_n is the total number of used vertexes. Equation 17 is called the restricted master program and the parameter X_n is found by solving the following subprogram on each reservoir:

$$B_n X_n = b_n$$
$$z_n^0 =(C'_n-RM'_0 A_n) X_n \quad (18)$$
$$\longrightarrow \text{max(or min)}$$

Computing eqs. 17 and 18, iteratively, the grobal optimum will be performed when $z_n^0 - RM_n \neq 0$, where, RM'_n and RM_n are the simplex multipliers according to eq. 14 and u_n^j respectively.

3.2 Application to dam reservoir systems

In parallel reservoirs as shown in Figure 5, the sub-problem is related to each reservoir and a neighbouring estimate point. The restricted master program is composed of the confluence conditions and an estimate point after confluence. In serial reservoirs, each reservoir cannot compose the sub-problem because of its dependence on the upstream reservoir. Therefore the operation of the downstream

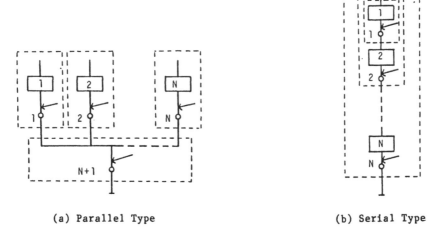

(a) Parallel Type   (b) Serial Type

Figure 5. Decomposition procedure of multi-reservoir systems.

reservoir is optimized through the restricted master program. Dam 2 is operated in accordance with the master program, including the result of dam 1. Then, dam N is similarly calculated, including the results of upstream dams. In the subprogram, dam reservoirs may be operated by Dynamic Programming (Takasao and et al., 1977).

4 DAM OPERATION CONSIDERING FLOW ROUTING MECHANISM

4.1 Formulation considering flow routing mechanism

The flow routing may be expressed by the Storage Function Method. If $SC_m(t)$ is the storage level in river channel m and $CQ_m(t)$ is the outflow from the channel m, the Storage Function is represented as follows:

$$SC_m(t+TL_m^*)=AL_m \ (CQ_m(t+TL_m^*))**BL_m$$

$$QO_n(t)-CQ_m(t+TL_m^*)=(SC_m(t+TL_m^*)$$

$$-SC_m(t+TL_m^*-1))/ \ dt \quad (19)$$

$$Q_m(t)=QC_m(t)$$

where AL_m and BL_m are parameters of the Storage Function, TL_m^* is the lag time, and dt is the interval of control time stage. Using the approximative estimate function (the first method in the approximative

method for computation) for flood control, the recursive equation of DP is formulated as follows:

$$f_t(S_1(t),...,S_N(t),SC_1(t+TL_m^*),$$

$$...,SC_M(t+TL_m^*))$$

$$= \min_{(QO_n(t))} (\ \max(Q_m(t+TL_m^*)/Q_{md},f_{t-1}(S_1(t-1),$$

$$...,S_N(t-1),SC_1(t+TL_1^*-1),$$

$$...,SC_M(t+TL_M^*-1))) \quad (20)$$

After computing the accumulated function till the last control stage, the optimal objective function f_{opt} is determined as follows:

$$f_{opt} = \min_{0 \lessgtr SC_m(T+TL_m^*) \lesseqgtr SC_m max \ (m=1,2,...,M)} (\ f_T(S_1(T),...,S_N(T),$$

$$SC_1(T+TL_1^*),...,SC_M(T+TL_M^*)) \quad (21)$$

where, $SC_m max$ is the maximun level of storage capacity in the river channel m. Thus, the last storage level in the river channel is found by eq. 21, and the optimal sequence of release discharge is also found by deducing the release discharge forward from time stage t to 1.

4.2 Simplified method in computation

As the state and decision variables are treated as the integer number in the traditional process of DP, the continuity in

30

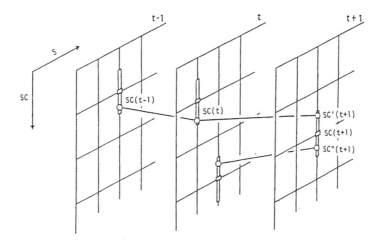

Figure 6. Schematic representation of DP algorithm where continuity in channel flow is held.

channel flow cannot be held because of its non-linearity. Therefore, I propose an approximative algorithm in which the computational grid $SC_m(.)$ of channel storage level represents the real number of state variables from $SC_m(.)-0.5$ to $SC_m(.)+0.5$. If the grid includes the several storage levels, the optimal pass satisfying the objective function at that time is selected among all of feasible passes, as shown in Figure 6.

For a reduction of the computer memory and the computational time, the state variable is represented onlyt by the reservoir storage level. Remembering the channel storage as a function of the reservoir storage levels, the recursive function is expressed as follows:

$$f_t(S_1(t),...,S_N(t)) = \min (\max (Q_m(t (QO_n(t))$$

$$+TL_m^*)/Q_{md},...,f_{t-1}(S_1(t-1),$$

$$...,S_N(t-1)))) \qquad (22)$$

  subject to

$$SC_m(t+TL_m^*) = g_m(S_1(t),...,S_{wn}(t)) \qquad (23)$$

here, $g_w(.)$ is the function of channel storage level represented by the storage levels of reservoirs 1 to wn. Table 2 shows a comparison of the computed results by eq. 20 (the exact method) and eq. 22 (the approximative method). Though the value of the objective function in the latter case is a little worse than that in the former case, the peak flow is low enough. The approximative method tends to

store the flood water in the earlier control stage because the decision of the storage level in the channel does not consider the influence of storing water and decreasing the feasible volume for dam control in a future.

5 MULTI-OBJECTIVE CONTROL IN DAM RESERVOIR

5.1 Scalar optimization

During long periods of no-rainfall, there are two objectives, namely low flow and turbidity controls. Taking tne reciprocal of eq. 2, both objectives are solved as the maximum value problem within the same domain. When the same weights are assigned to P and D (reciprocal objective of turbidity), the synthesized scalar objective is expressed as follows:

$$J = \min(Q_{m1}/\bar{Q}_{md}, C_{md}/C_{mmax}) \longrightarrow \max \qquad (24)$$

In the performance of Dynamic Programming, the release discharge may be determined by satisfying the above objective and the recursive equation can be formulated as:

$$f_t(S_1(t),...,S_N(t),CS_1(t),...,CS_n(t))$$

$$= \max(\min (Q_m(t)/\bar{Q}_{md}, C_{md}/C_m(t),$$

$$f_{t-1}(S_1(t-1),...,S_N(t-1),CS_1(t-1),$$

$$...,CS_N(t-1)))). \qquad (25)$$

where, $CS_n(t)$ is the turbidity level in the reservoir n at time stage t. Although

Table 2. Computed results in the case where the concept of lattice point in DP is changed.

Inflow	S(0)=0, S(10)=14, V=19, SCmax=29, SC(0)=10, Q_d=20, AL=5, BL=0.5 TL*=0										
t	0	1	2	3	4	5	6	7	8	9	10
QI(t)		3	6	9	12	15	10	7	5	3	2
q(t)		1	3	3	6	8	15	10	6	3	1
Integer type											
S(t)	0	1	4	0	2	9	18	19	19	12	14
QO(t)		2	3	13	10	8	1	6	5	10	0
SC(t)	10	11	13	11	11	11	11	11	8	9	1
Q(t)		4	4	5	5	5	5	5	3	4	0

$Q_p/\overline{Q}_d$=0.25

Real type											
S(t)	0	0	0	0	0	7	17	19	19	13	14
QO(t)		3	6	9	12	8	0	5	9	5	1
SC(t)	10.0	10.0	12.6	15.3	18.9	19.6	19.5	19.4	19.4	16.5	12.4
Q(t)		4.0	6.4	9.3	14.3	15.3	15.1	15.1	15.0	10.9	6.1

$Q_p/\overline{Q}_d$=0.765

the inflow turbidity has a complicated behavior in the reservoir, in this paper its behavior is treated simply by the complete mixing model.

5.2 Vector optimization

Vector optimization is the method used to perform a social goal with conflicting objectives. In a problem involving two objectives, the constraint method is easily applied as follows:

Setting the additional constraint for one objective (turbidity):

$$C_{mmax} \leqq C_{md}' \qquad (26)$$

another objective (low water) is optimized by using the following recursive equation:

$$f_t(S_1(t),\ldots,S_N(t),CS_1(t),\ldots,CS_N(t))$$

$$= \max(\min(Q_m(t)/\overline{Q}_{md}, f_{t-1}(S_1(t-1),$$

$$\ldots,S_N(t-1), CS_1(t-1),$$

$$\ldots,CS_N(t-1)))) \qquad (27)$$

Reducing the value of C_{md}', many couples of P^i and D^i are found until it becomes impossible to compute the optimization and the Transformation Curve (TC) (Haimes et al., 1975) is drawn as shown in Figure 7. The goal of low water, which denotes the optimum condition of the system, is defined as the average flow discharge for the control time periods. On the other hand, the goal of turbidity is defined as the turbidity concentration in the depletion period after floods. Then the Indifference Curve (IC), which means the same as the estimation level or satisfaction level, is also drawn to determine the grobal optimum. Assuming that the limit points on TC (P=1 and D=1) are estimated as the same value, point A and C are on the same IC. Thus if $|AG| \neq |CG|$, the scale of turbidity coodinates is multiplied by:

$$dw = (((a_1-g_1)^{**}2-(c_1-g_1)^{**}2)/(-(a_2-g_2)^{**}2$$

$$+(c_2-g_2)^{**}2)))^{**}0.5. \qquad (28)$$

where, a_i, c_i and g_i denote the elements of points A, C and G respectively. A modified IC is drawn as a concentric circle in the tramsformed coordinate system, where $|AG| = |CG|$. The contact point B' of TC and IC becomes the optimum point and $|B'G'|$, i.e. is the minimum radius of IC, is the deviation from the goal. Therefore the ratio between $|B'G'|$ and $|E'G'|$, which is the maximum radius:

$$F = |B'G'|/|E'G'| \qquad (29)$$

denotes the control effect of vector optimization.

32

Table 3. Input conditions in the case of multi-objective control.

					S(0)=25, S(10)=29, V=30, $\overline{Q}_{1d}$=7, $\overline{Q}_{2d}$=9, C_{1d}=15, C_{2d}=15					
t	1	2	3	4	5	6	7	8	9	10
QI(t)	13	19	21	17	14	12	7	5	4	7
q_1(t)	2	4	5	7	9	6	4	2	5	7
q_2(t)	3	2	3	4	5	7	8	7	4	5
CI(t)	3	5	7	10	6	3	2	1	2	3
Cq_1(t)	4	6	8	10	12	14	15	18	12	15
Cq_2(t)	3	5	6	7	8	10	12	13	15	16

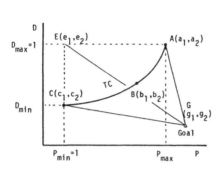

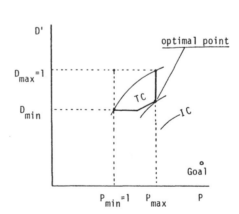

(a) Original Graph   (b) Modified Graph

Figure 7. Tranformation curve and indifference curve.

5.3 An example of optimization

As an example, one dam and two estimate point system, A-3 type, is applied to compare the results of two optimization algorithms. Table 3 shows the input conditions. The value of the objective function in scalar optimization is 1.49. Figure 8 shows the modified TC, and the coefficient dw is 1.87. For the goal, the value of the low water, which is the minimum among the ratios
between the average water and the allowable flow discharge, is 3.11. The value of turbidity, which is the maximun among the ratios between the lowest concentration of non-controlled flow for the control time stages and the allowable value, is 0.21. The TC is not a smooth curve because the reservoir is operated by a discrete level. The control effect F is 0.59, and a similar estimate value of scalor optimization is 0.62. A more accurate analysis of the reservoir, such as a one-dimensional

Figure 8. An example of optimal point in multi-objective control.

model (Iwasa and et al., 1978), and a more detailed control, such as the selected intake operation (Takasao and et al., 1979), can be applied in the same way.

6 APPLICATION TO REAL-TIME OPERATION OF LOW FLOW AND TURBIDITY CONTROL

6.1 Fundamental procedureof real-time operation

In the period of low flow between floods, the water stored in the reservoir is utilized for municipal, industrial and agricultural water demands. The fundamental procedure for the real-time operation consists of seven steps, and depends on the iterative method in the process of time as follows:
1) set the present states of storage and turbidity inreservoirsasinitial values
2) decide whether or not low flow and turbidity controls at the present time should be changed to flood control ,based on meteorological and forecasting information at the present and in the future;
3) when changing to flood control, the stochastic flood control procedure may be introduced with a flood objective, and a unit time and forecasting method inherent to flood control;
4) if low flow and turbidity controls are being continued, the period of no-rainfall in the future will be estimated in accordance with the observed data and the Baysian inference method, and is set up as for a period of low flow and turbidity con trols;
5) the time series of input data during the period (hydrograph and turbidity graph) may be estimated an accordance with the observed data and the Kalman filtering theory;
6) as for the estimated input data, the optimal solution technique based on Dynamic programming may be implemented and the optimal release may be determined;
7) with the exception that the special release rule may be applied because of the intense depression of low flow and water quality, the present low flow and turbidity controls will be continued and the control time be moved to the next time, and the procedure will be returned to step 1).
The total system is divided into three sub-systems, i) prediction and observation systems of no-rainfall duration and rainfall volume; ii) prediction system of water quantity and turbidity in no-rainfall duration time; and iii) decision system of release discharge. The last two sub-systems will be discussed in the following sections because the first sub-system has already been published in another paper (Ikebuchi and et al., 1981).

6.2 Prediction of water quantity and turbidity

The recession curve of low flow is assumed between inflows at time t and $t+TA$ in logarithmic type,

$$\ln(QI_n(t+TA))=\ln(QI_n(t))-CE_n \, TA \qquad (30)$$

where CE_n is the recession parameter. $Y_{qn}(TA|t)$ $(=\ln(QI_n(t-TA))-\ln(QI_n(t)))$ and $X_{qn}(TA|t)$ is described as the observation variable and state variable at time t for parameter CE_n. The system state equation and observation or measurement equation are expressed as follows:

$$X_{qn}(TA|t)=PG_{qn,t|t-TA}X_{qn}(TA|t-TA)+v_t \qquad (31)$$

$$Y_{qn}(TA|t)=HA_{qn,t}X_{qn}(TA|t)+w_t \qquad (32)$$

in which $PG_{qn,t|t-TA}$ is the recession coefficient of lag time TA, $HA_{qn,t}$, the coefficient corresponding to eq. 30 and v_t, w_t, the system state and measurement errors, respectively. When v_t and w_t are regarded as Gauss white noise, Kalman filtering theory can be applied to the identification of parameter $CE_n(t)$, and the inflow hydrograph may be predicted with the parameter identified from time to time.
On the other hand, the turbidity into the reservoir is approximatively estimated in condition with the inflow hydrograph.

$$CI_n(t)=AD_n(QI_n(t))**BD_n \qquad (33)$$

Expressing the observation variable as $Y_{cn}(t)$ $(= \ln(CI_n(t)))$, and the observation vector $HA_{cn,t}$ $(= (1 \quad \ln(QI_n(t))))$, and the state vector as $(\ln(AD_n(t) \quad BD_n(t)))$, the dynamic system equation is formulated as follows:

$$\begin{bmatrix} \ln(AD_n(t)) \\ BD_n(t) \end{bmatrix} = \begin{bmatrix} 1 & 0 \\ 0 & 1 \end{bmatrix} \begin{bmatrix} \ln(AD_n(t-1)) \\ BD_n(t-1) \end{bmatrix} + v'_t \qquad (34)$$

$$Y_{cn}(t) = (1 \quad \ln(QI_n(t))) \begin{bmatrix} \ln(AD_n(t)) \\ BD_n(t) \end{bmatrix} + w'_t \qquad (35)$$

The procedure for the identification of parameter $AD_n(t)$ and $BD_n(t)$ is the same as the procedure described in the prediction of the inflow hydrograph.

6.3 Decision of release discharge

The release discharge must be decided with consideration of not only the predicted inflow series during the control period but also the secular balance of the water supply. When $w_{qn,t}$ has mean 0 and variance $W_{qn,t}$, the predicted variance of the recession paremeter is expressed as follows:

$$ET_{qnt} = FI_{qn,t|t-TA}PA(TA|t-TA)$$

$$FI_{qn,t|t-TA}^!+W_{qn,t} \qquad (36)$$

where, $PA(TA\ t-TA)$ is a covariance of the error. Thus the predicted inflow obeys the following log normal distribution function:

$$g_{QI}(QI_n(t+TA))=1/((2\ 3.14\ ET_{qnt})**0.5$$

$$QI_n(t+TA))\exp((-0.5)(\ln(QI_n(t+TA)$$

$$-\ln(QI_n(t+TA)))**2\ /ET_{qnt} \quad (37)$$

$$QI_n(t+TA)=QI_n(t)\exp(-Y_{qn}(TA|t))$$

The inflow turbidity also obeys the log normal distribution around the estimated state value, similar to eq. 37. Under these circumstances, stochastic control on multi-reservoir system (Kojiri et al, 1985) can be applied. The control procedure in a system composed of a single reservoir and a single estimate point has needs to be considered.

The control objective is to express the reliability of controlled results according to the form of the probability density function (pdf). Therefore the stochastic control objective is defined as the criteria by which the most reliable sequence of the storage volume is extracted among the sequences satisfying the necessary controlled effect.

$$PK=\max(P^K(S_1(t),\ldots,S_N(t),CS_1(t),\ldots,$$

$$CS_N(t))) \longrightarrow \min. \quad (38)$$

$$P^K(S_1(t),\ldots,S_N(t),CS_1(t),\ldots,CS_N(t))$$

$$=\max(P_m^K(S_n(1),CS_n(1)\ S_n(0),$$

$$CS_n(0)),\ldots,P_m^K(S_n(t),$$

$$CS_n(t)|S_n(t),CS_n(t))) \quad (39)$$

where the super script "K" means the stochastic controlled criteria similar to eq.24, and is given as the necessary performance value. In a simple river basin

system composed of one reservoir and one estimate point, each value is as follows:

$$P^K(S(t),CS(t)|S(t-1),CS(t-1))$$

$$=1-\int_{QImin(t)}^{QImax(t)}g_{QI}(QI(t))dQI\cdot\int_{CImin(t)}^{CImax(t)}g_{CI}(CI(t))dCI$$

$$(40)$$

$$QImax(t)=S(t)-s(t-1)+J\ Q_d \quad (\gtreqless 0) \quad (41)$$
$$QImin(t)=S(t)-S(t) \qquad\qquad (\gtreqless 0)$$

$$CImax(t)=h^{-1}(C_d/J) \quad (42)$$
$$CImin(t)= 0$$

where J is the performance value and $h^1(.)$ is the function representing inflow turbidity $CI(t)$ by a couple of storage levels $(S(t), s(t-1))$, a couple of turbidity levels in the reservoir $(CS(t), CS(t-1))$, and the mean inflow $QI(t)$. The recursive function of DP is:

$$f_t^K(S(t),CS(t))= \max\ (\min(P^K(S(t),$$
$$(S(t-1))$$

$$CS(t)\ S(t-1),CS(t-1)),f_{t-1}^K(S(t-1),$$

$$CS(t-1))). \quad (43)$$

Thus the expected release discharge $o(t)$ is as follows:

$$O(t)= \int_{QImin(t)}^{QImax(t)} (S(t-1)-S(t)+QI(t))\ g_{QI}(QI(t))dQI$$

$$/\ \int_{QImin(t)}^{QImax(t)} g_{QI}(QI(t))dQI \quad (44)$$

As the controlled results keep the performance value J of the stochastic control objective, this method is known as the stochastic control based on reliability criteria (SCRC). By changing the value of J, the optimal control policy is found with regard to the reliability for low water and turbidity. It is useful to apply the above procedure to a multi-reservoir system by combining the Shift Operation Method (Kojiri et al., 1985).

6.4 An example of real-time operation

Figure 9 shows the predicted inflow sequences at the predicted time stage 2, 10, and 21. The parameter is optimized through the daily observed data from 1975 to 1976. Though the flood peak is not exactly predicted, the inflow hydrograph

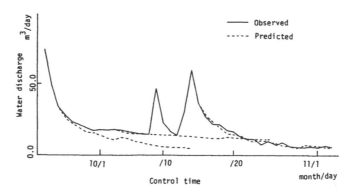

Figure 9. Comparison of water discharge between predicted inflow and observed inflow.

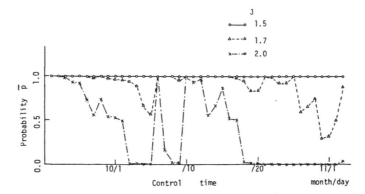

Figure 10. Comparison of controlled probability between three performance values.

in the recession period is represented with high accuracy. The inflow turbidity is also predicted with the same accuracy. Figure 10 shows the controlled sequences of reliability on the fixed performance value. When $J = 2.0$, the reliability suddenly decreases at time stage 11 and 28. When $J = 1.5$, the reservoir can be operated with high reliability (= 1.0). Therefore, in the most hopeful operational policy where the reliability is greater than 0.9, J is 2.0 from the time stage 1 to 5, 1.7 from 6 to 12, 1.5 from 13 to 15, 1.7 from 16 to 19, 2.0 from 20 to 22, 1.7 from 23 to 28, 1.5 from 29 to 30, 1.7 from 31 to 35, and 1.5 from 36 to 40.

7 CONCLUSIONS

In this paper, the optimal operartion in a system composed of multi-dam reservoirs and multi-estimate points and the real-time operation of a single dam reservoir with stochastic control has been discus-

sed. To summarize, the following results were obtained:
i) The formulation of the optimal operation of dam reservoir systems:

The optimal operation of dam reservoirs is formulated by Dynamic Programming. Moreover, for a complex system consisted of multi-reservoirs, approximative methods depending on the dam sites is proposed, or the decomposition method is introduced to synthesize the divided subsystems after optimization in each subsystem.
ii)Evaluation of flow routing mechanism:

Introducing the storage function method to represent the flow routing mechanism, the optimal release discharge is determined by DP. It is especially important to evaluate the controlled effects for flood prevention and the powerful technologies in the computer is also proposed.
iii)Formulation of multi-objective control:

Multi-objective control problems on the

low flow and turbidity is formulated by scalar and vector optimization techniques. Dam reservoirs can be operated to satisfy the control goal considering the turbidity of release discharge and the storage water.

iv)Decision of the real-time operational policy:

Combining the filtering theory and the stochastic control, the real-time operation of a dam reservoir is proposed. These approaches can be applied to the basin-wide system with multi-reservoirs by using the Shift Operation Method and will become more effective and more practical.

REFERENCES

Aya, S., Iwasa, Y. & Matsuo, N. 1978. On the Transport Processes of Turbidity. Proc. 26th Japanese Conference on Hydraulics, Japan, pp.131-138.

Bellman, R. E. 1957. Dynamic Programming. Princeton University Press, Princeton, New Jersey, USA.

Haimes, Y.Y., Hall, W. A. & Freedman, H.T. 1975. Multiobjective Optimization in Water Resources Systems - The Surrogate Worth Trade-off Method -. Elsevier Scientific Publishing Company.

Hall, W. A. 1964. Optimum Design of a Multi-Purpose Reservoir. Jour. of Hydraulic Division, Vol.90, pp.141-149.

Hall, W. A., Harboe, R. C., Yeh, W. W. G. & Askew, A. J. 1969. Optimum Firm Power Output From a Two Reservoir System by Incremental Dynamic Programming. Water Resources Center Contribution, No.130, Univ. of California, USA.

Heidari, M., Chow, V. T., Kototovic, P. V. & Merdith, D. D. 1971. Discrete Differential Dynamic Programming for Approach to Water Resources Systems Optimization. Water Resources Research, Vol.7, No.2, pp.273-282.

Ikebuchi, S., Takasao, T. & Kojiri T. 1981. Real-time Operation of Reservoir Systems Including Flood, Low Flow and Turbidity Controls. Experience in Operation of Hydrosystems, Water Resources Pub., Littleton, Colorado, USA, pp.25-46.

Kojiri, T., Ikebuchi, S. & Hori. T. 1985. Stochastic Control of Dam Reservoirs Considering Flood Inundation Probability on Whole River Basin. 4th Int., Hydrology Sympo. - Multivariate Analysis of Hydrologic Process -, Fort Collins, Colorado, USA, In Publication.

Lloyd, E. H. & Anis, A. A. 1975. An Outline of the State of the Art as Understood by Applied Probabilists, IIAA Research Report, Austria.

Moran, P. A. P. 1954. A Probability of Dams and Storage Systems, Australian Jour. of Applied Science, Vol.5, Australia, pp.116-124.

Nagao, M. 1977. Application of Markov Chain to a Stochastic Reservoir Theory for the Estomation of Reliability on Water Supply. Proc. 21th Japanese Conference on Hydraulics, Japan, pp.133-141.

ReVelle, C., Joeres, E. & Kirby, W. 1969. The Linear Decision Rule in Reservoir Management and Design. 1, Development of the Stochastic Model. Water Resources Research, Vol.5, No.4, pp.767-777.

Ripple, W. 1883. The Capacity of Storage Reservoirs for Water-Supply, Proc. of ICE. Vol.71, pp.270-283.

Takasao, T., Ikebuchi, S. & Kojiri, T. 1975. Dynamic Programming for Multi-reservoir System Design based on Water Quality Control. Proc. JSCE, No.241, Japan, pp.39-50.

Takasao, T., Ikebuchi, S. & Kojiri, T. 1978. A STUDY for the Optimal Control of Multi-reservoir and Multi-defence Point System. Annuals, Disas. Prev. Res. Inst., Kyoto Univ., No.21 B-2, Japan, pp.193-206.

Takasao, T., Ikebuchi, S. & Kojiri, T. 1979. A Study for the Water Quantity and Turbidity Control by Multi-dam Reservoirs System. Annuals, Disas. Prev. Res. Inst., Kyoto Univ., No.22 B-2, Japan, pp.167-178.

Takasao, T. & Seno, K. 1970. A Study on Flood Control by Group of Dam Reservoirs (I). Annuals, Disas. Prev. Res. Inst., Kyoto Univ., No.13-B, Japan, pp.83-103.

Takeuchi, K. 1974. Optimal Control of Water Resources Systems Using Marginal Loss Function of Remaining Reservoir Storages. Proc. JSCE, No.222, Japan, pp.93-104.

Environmental Geotechnics and Problematic Soils and Rocks, Balasubramaniam et al. (eds)
© *1987 Balkema, Rotterdam. ISBN 90 6191 785 9*

Development energy and the environment: India

K.S.Murty
University Department of Geology, Nagpur, India

ABSTRACT: The development of a country has become energy intensive in the twentieth century. India's achievements in food production and industrial growth along planned lines in the last three decades have been in no small measure due to an increased input of oil, coal and hydropower energy in different forms. However, 70 per cent of all available water in India is said to be polluted about 73 million workdays are estimated to have been lost due to water-related diseases. Levels of sulphur and other particulate matter in several Indian cities exceed permissible limits. Over one million hectares of forests are cut every year and some 0.15 million hectares of forests are lost to development projects annually, while a significant portion of the 15,000 plant species and 75,000 animal species found in India are threatened by the pressure of human activity on land and forests. Steps are being taken towards soil conservation, afforestation and wild life preservation. Clearance from the Department of Environment of the Government of India is needed to proceed with any development project now. It is realised that development has to be achieved, but not at the expense of the environment.

1 INTRODUCTION

"If we desire to fight successfully the scourge of poverty and want from which 90% of our countrymen are suffering, if we wish to remodel our society and renew the springs of our civilsation and culture, and lay the foundations of a strong and progressive national life, we must make the fullest use of the power which our knowledge has given us. We must rebuild our economic system by utilising the resources of our land, harvesting the energy of our rivers, prospecting for the riches hidden under the bowels of the earth, reclaiming deserts and swamps, conquering the barriers of distance".(Meghnad Saha, 1938).

When Meghnad Saha made the above statement, India's population was about half of its present 690 million. The value of the total mineral production was Rs.440 million, to which Burma's contribution was approximately Rs.100 million. The production of coal was around 28 million tons (Dey, 1955),of petroleum around 5,500 barrels a day (Kuriyan, 1980), and no nuclear energy was produced at all. Hydropower generation was mainly from the Western Ghats and to a smaller extent from the Himalayas, totalling hardly a million kW. The production figures for 1983-84 were: 138.39 million tonnes of coal, 30 million tonnes of oil, 14,000 MW of hydel energy, 16,000 MW of thermal power, and 640 MW of nuclear energy. Several dams were built to provide water for irrigation and power generation, and the overall mineral production rose to a value of Rs.53,000 million. But a price had to be paid for this growth in terms of environmental degradation, in loss of forest and soil cover, in floods, and in health problems for the people.

2 ENERGY INPUTS

Energy inputs largely determine the development of any country, and India's effort in this direction, though fair, lags behind several countries in energy use per capita.(Table 1)

Table 1. Present energy use per capita.

Country	Energy use per capita (BOE/year)
U.S.A.	51.7
Germany	26.9
USSR	22.4
China	2.2
India	0.9
World average	6.0

1 BOE = Barrel of Oil Equivalent
= 5.8 million BTUs
= 158.7 M^3 of natural gas
= 0.88 Tonnes of Coal Replacement value
= 0.27 Tonnes of Coal Equivalent
= 1700 MW Hrs. of Electricity

(Malhotra, 1981)

Fuelwood, agricultural waste and animal dung form the main sources of non-commercial energy in India. At present coal accounts for nearly 60 per cent of the total commercial energy. About 1/8th of the hydel potential in the country has been developed and another 1/12th is under development. The southern region has developed 48 per cent of its potential (Table 2). India has a hydroelectric potential of 41,155 MW at 60 per cent load factor (Fuel Policy Committee, 1974) (Table 3). In the last three decades, the country invested over Rs.100,000 million and built over 600 storage dams of various sizes, accounting for a storage capacity of 16 million hectare metres for irrigation, flood control and power generation. Since 1951, the irrigation potential has been raised from 22.6 million hectares to about 65 million hectares. The total irrigation potential has been assessed at 113 million hectares which is hoped to be achieved by 2000

A.D. The total water resources of India are assessed at 176.8 million hectare metres.

Table 2. Hydro-electric potential and status of development(April'82)

Region	Annual Energy potential Twh	Annual Energy potential developed Twh	Annual Energy potential under development Twh
Northern	147.39	14.88	11.83
Western	36.95	6.36	1.65
Southern	68.31	21.10	11.45
Eastern	37.81	3.07	4.64
North-eastern	105.86	0.73	1.56
Total	396.32	46.14	31.13

(Inst. of Engineers, 1981)

Table 3. Estimated hydro-electric potential of India.

Region	MW at 60% load factor	Billion kwh	Per cent of total potential
Eastern	2,694	14.2	6.5
Northern	10,731	56.4	26.1
Western	7,189	37.8	17.4
Southern	8,097	42.6	19.7
North-eastern	12,464	65.4	30.3
Total	41,155	216.4	100.0

Energy consumption in India from the commercial primary energy sources has risen manyfold in the last three decades, during which six five-year plans were executed (Table 4). Also, projections of the energy consumtion up to 2000-2001 from both commercial and non-commercial sources reveal that the share of the former might rise to 80 per cent by 2001 A.D. from its present 46 per cent (Malhotra, 1981) as shown in Table 5.

3 ENVIRONMENTAL PROBLEMS

The exploitation of natural resources by man has become inevitable

Table 4. Trends in consumption of commercial primary energy (energy consumption in original units).

Year	Coal (Million tonnes)	Lignite	Oil (MT)	Natural gas (NM3) (million m^3)	Hydropower Twh. (trillion units)	Nuclear power Twh.
1953-54	34.10	–	3.50	–	2.90	–
1960-61	49.90	0.05	6.70	–	7.80	–
1965-66	62.30	2.57	10.49	–	15.22	–
1970-71	71.23	3.37	14.97	704	25.25	2.42
1973-74	77.68	3.32	18.22	866	28.97	2.40
1975-76	92.22	3.01	18.68	1286	33.30	2.63
1980-81	109.42	4.99	26.01	1566	46.53	3.00
1981-82	117.66	6.30	26.75	2311	49.53	3.03
1982-83	124.22	6.40	30.60	2735	48.25	2.02

Table 5. Energy consumption trends and projections 1953-2001 (in million BOE per day).

Year	Coal	Oil	Electricity	Total commercial	Fire wood	Agricultural waste	Cow dung	Total non-commercial
53-54	0.0888	0.0746	0.0235	0.1869	0.2544	0.0777	0.0575	0.3896
60-61	0.1251	0.1359	0.0523	0.3132	0.2928	0.0900	0.0674	0.4502
65-66	0.1603	0.2000	0.0947	0.4550	0.3213	0.0987	0.0739	0.4940
70-71	0.1591	0.3009	0.1507	0.6107	0.3467	0.1065	0.0798	0.5330
75-76	0.2198	0.3582	0.2043	0.7823	0.3916	0.1204	0.0904	0.6024
78-79	0.2130	0.4368	0.2613	0.9111	–	–	–	0.6171
82-83	0.3000	0.5111	0.3972	1.2080	–	–	–	0.6318
87-88	0.4071	0.6721	0.5919	1.6711	–	–	–	0.6278
92-93	0.5776	0.8996	0.8730	2.3503	–	–	–	0.6061
2001	0.9535	1.4931	1.4581	3.9047	–	–	–	0.5061

in order to achieve a decent standard of living. Of the four basic elements - air, water, fire and earth - the most exploited are perhaps water and earth (Varma, 1984). Man is taking out minerals from the earth at a much faster rate than before, and is bringing more areas under cultivation to grow more food, using both surface and underground water. Electric power is generated from thermal, hydro and nuclear sources. These activities have created environmental problems.

3.1 Mining and environment

Considerable damage has been attributed to mining in the Jharia and Raniganj coal fields, and from Goa, Sandur, Dehra Dun and Singhbhum.

Extensive damage has been reported to houses in villages which are in the neighbourhodd of blasting operations in Goa. Washings from overburden dumps, low grade friable ores and powdery ores have damaged tracts of agricultural land, besides causing pollution of streams and water bodies (Sharma, 1982). Though the damage to agricultural land is not thought to be very extensive, reclaiming such land so as to bring under cultivation again has become a problem (Bhargava et al., 1982). In the Kudremukh iron ore project, extensive pollution of rivers and streams due to soil washoff in the rainy season and escape of tailings from a tailing dam has been noticed. Soil erosion has caused destabilisation of the hill slopes.

In the Jharia coal field, 32.26 km^2 of land out of 223 km^2 of land has been affected by subsidence, causing damage to railway tracks, river beds and roads. Besides, fires in the coal fields have damaged the land and caused atmospheric pollution. In the industrial belt of Dhanbad-Jharia region, studies were made during 1979-80 on a range of concentrations of pollutants like suspended dust, SO_2 and NO_2, and of lead, manganese, arsenic, chromium and cadmium and also of the incidence of respiratory and cardiovascular diseases in different seasons, and they showed increases (Ghosh et al., 1982). In the Mussourie-Dehra Dun hills, quarrying limestone has gone to such an extent as to cause serious environmental disasters like landslides, siltation of rivers, and destruction of fertile land. On a writ petition filed against limestone quarrying in that area, the Supreme Court passed an order on March 12, 1985 directing several lessees to stop mining operations in that region, thus creating a watershed in Indian mining history (MGMI, 1985). Similar environmental consequences have been reported from the lead-zinc mines of Zawar.

3.2 Dams and the environment

Equally serious are ecological problems caused by dams constructed on some of the rivers of India. The submergence of a site by a reservoir could cause loss of forests, shifting of human settlements, and the disappearance of wild life, besides disturbing the existing hydrological conditions and processes in the region. The Idukki reservoir region in Kerala has lost 50 per cent of the forest cover and the width of the active river channel of the Periyar river has shrunk from 80 metres to less than two metres, even during the monsoons. The soil erosion resulting from the absence of trees has attained dangerous proportions. The rate of deforestation in the western ghats ranges from 60 to 113 km^2 a year. Indeed, between 1951-52 and 1975-76 deforestation for various purposes claimed about 4.14 million hectares in the country: agriculture - 2.51, river valley

projects - 0.48, establishment of industry - 0.13, road construction - 0.06, and others - 0.96. According to unofficial estimates, over one million hectares of forests are cut every year and some 0.15 million hectares of forests are lost to development projects annually. The cost of soil and forests in India have been put at billions of rupees, Rs.7000,000 million and Rs.1150,000 million respectively (Ranganathan, 1981). Only eight States in the Union of India have forest cover above the stipulated average of 33.3 per cent of the total area. The all-India average is indeed only 22.75 per cent. Large-scale deforestation has been held responsible for loss of soil cover and floods in the Himalayan rivers, and it was one of such floods in the Alaknanda in 1970 which swept away several villages, bus-loads of tourists, bridges, cattle and agricultural lands in its wake. Now there is strong opposition to the location of dams on the Bhagirathi and Alaknanda rivers in the Garhwal Himalayas. Public awareness of the ecological changes has led to the 'Chipko movement' in the Himalayan region by the people. This reaction has spread to other parts in the country and has resulted in dropping at least one project - the Silent Valley project in Kerala. Ecologists felt that the construction of the dam would submerge India's last near-virgin tropical rain forest that contains a multitude of plant and animal species that evolved in tranquility in that valley for six million years. The hydel project would have generated 120 MW of power, but it could cause major changes in rainfall over major agricultural zones, not only in the western ghats, but over the entire continent. A joint Central-State committee made an in-depth study of this project and recommended against the construction of the dam and the view was accepted by the government. This has encouraged ecologists to raise their voice against any project that may cause serious environmental disorder or disaster. It has now become essential to study the environmental impact of projects or developmental activities. The preservation of environment in the catchment area will be made an integral part of the river valley projects, and clearance from the Department of Environment is mandatory before a project is put into effect, whether it is to generate power, irrigate land, mine for a mineral or start an industry. It is now development in relation to energy requirements and the environment.

> "They knew the art of living and dying, and of racial survival,
> They had a silent compact with the earth's gradations of consciousness
> And inferred intimate filiations with bird, plant, tree, animal."
> (Australia Helix)

REFERENCES

Bhargava, D.N. & S.Chand 1982. Eco-development Plan for Iron Ore-bearing Areas of Western Ghat Region. National Seminar on Minerals & Ecology, Indian School of Mines, 1982, III/1:1-4.

Blitz Special 1984. Whither Conservation?

Dey, A.K. & J.C. Brown 1955. India's Mineral Wealth. Bombay, Oxford University Press.

Fuel Policy Committee Report 1974. Government of India.

Ghosh, S.K., J.K. Sinha & S.P. Banerjee 1982. Air Pollution in Dhanbad and its impact on human health. National Seminar on Minerals & Ecology, V/1:1-5.

Journal of Institution of Engineers (India) 1981. 62 Pt EL 3, 109.

Kuriyan, G. 1980. India: A General Survey. p.201. New Delhi: National Book Trust.

Malhotra, N.K. 1981. The Search for Energy. Third Jagasia Memorial Lecture. Bombay.

Newsletter 1985. Environmental concerns in Mining of Minerals. Vol. 10, No.3. Mining, Geological and Metallurgical Institute of India. Calcutta.

Ranganathan, S. 1981. Population: the neglected factor, Relevant Forestry. Ion Exchange. Bombay.

Saha, Meghnad 1938. The Problems of Indian rivers, Presidential address, National Institute of Sciences of India.

Sharma, R.N. 1982. Keynote address, National Seminar on Minerals & Ecology. p.XIII-XVIII.

Varma, C.V.J. 1984. Dams and Environment. Indian & Foreign Review, 15 September 1984, p.14.

Environmental Geotechnics and Problematic Soils and Rocks, Balasubramaniam et al. (eds)
© 1987 Balkema, Rotterdam. ISBN 90 6191 785 9

Divagation of stabilized pavement materials

T.Ramamurthy & G.Venkatappa Rao
Civil Engineering Department, Indian Institute of Technology, Delhi
C.P.Nag
Civil Engineering Department, College of Engineering, Jodhpur, India

ABSTRACT: Stabilized soils are being increasingly used in the construction of roads. Characterization of these materials in terms of strength and stress-strain response have been extensively made. With the increasing demand for higher wheel loads, more appropriate characterization of these materials is called for. Fracture energy of the material appears to be the most promising and seems to reflect the realistic response of the material under any desired stress field subjected to cyclic loading, including the stage of material divagation.

In this paper properties of cement-stabilized silts which are extensively used in India are presented. Using energy concept, remaining life of pavements has been expressed in a mathematical form. This equation can be used to estimate the remaining life of pavements from the data obtained from the laboratory tests. Illustrations have been provided for direct use by practising engineers.

1 INTRODUCTION

Early pavement design methods were empirical or semi-empirical and required material characterization of pavement components on a qualitative basis. Today there is a need for the quantitative evaluation of elastic properties under simulated traffic conditions. Though there is considerable understanding of the cement stabilization technique and thousands of kilometers of cement stabilized roads have been constructed by now, the engineering behaviour of soil-cement still needs attention in the light of the present day design requirements. While laying stress on the need to evaluate elastic constants appropriately, if elastic theory or modified elastic theory is to be used for pavement design, Pell and Brown (1972) have pointed out how the stresses acting on an element undergo changes under the moving load. A generalised stress condition of normal and shear stresses on an element exists (Fig.1) and these . stresses undergo cyclic variations.

As such even the conventional triaxial tests need modifications to simulate field conditions.

The life of paving materials is normally defined on the basis of a failure criterion based on fatigue tests, since fatigue is the most commonly observed mode of failure. Field observations of existing pavements have indicated that properties of paving materials undergo changes like enhanced compaction, degradation, weathering etc, under the effect of traffic and environment with the lapse of time, (Foster, 1972). The change in material property under the effect of repetitive loading has been described as material divagation (Stauffer and Strauss 1977) to represent the damage caused.

The basic concept underlying the theory of material divagation can be presented by a simple experiment considering two specimens that are physically and chemically identical in energy. One of the specimens is subjected to a cyclic deformation process so that after the process is complete, the specimen is in the

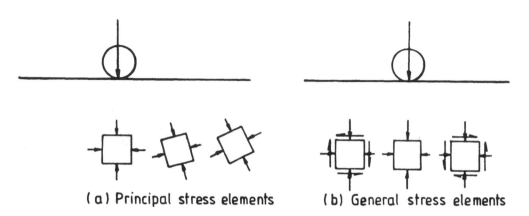

(a) Principal stress elements   (b) General stress elements

Figure 1(a). In situ stresses beneath a rolling wheel

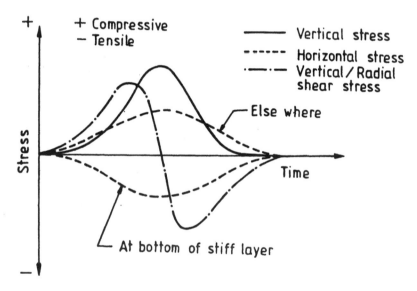

Figure 1(b). Cyclic variation of stresses on a pavement element
            due to passage on a rolling wheel load (after Brown
            and Pell,1972)

original geometric configuration but with an altered microstructure. The observed response of this test specimen would in general be different than the one tested directly without load cycling, as shown in Fig.2. The divagated material may either show increased brittleness or ductility associated with a change in fracture energy. This difference reflects the change in the material properties induced by the cyclic load history applied to the first test specimen. The process is like a "before and after" study of load cycling.

In order to evaluate the pavement at any time of its use, it is necessary to define the pavement materials in their divagated form. Although relationships between number of load cycles and divagated values of material characteristics, e.g. modulus of elasticity and permanent deformation, have been suggested by Yamanouchi 1973 and Lentz and

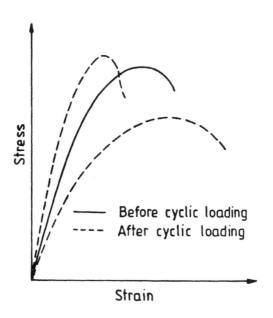

Figure 2. Stress-strain curves before and after load cycling-effect of material divagation

Balachi 1980, physical explanation to these relations on the fatigue phenomenon is not yet available. Yamanouchi's observations for modulus of elasticity of model pavements prepared from cement-stabilized gravelly sand and volcanic ash have led to Equation 1,

$$E_N/E_0 = a + b \ \log_{10}N \qquad (1)$$

where E_0,E_N = Modulus of elasticity at the initial and N cycles of loading respectively

      N = Number of load cycles, and

  a,b = experimental constants.

Further review of literature reveals that results of constant stress and constant strain type of fatigue tests yielded different results. Similarly the material characteristics from repeated triaxial test, viz. modulus of resilient deformation (M_R) has been found to vary largely with change in test conditions.

In view of this, there is a need to isolate a parameter to represent the material status after it has undergone repeated loading due to traffic. One such parameter that appears to be promising is the energy of the specimen, which can be brought into picture because pavement acts as a media to transfer the energy imparted by traffic and environment to the underlying subgrade. Some researchers (e.g. Majidzadeh,et al 1971) have considered dissipated energy as the factor to predict the number of cycles to failure. But such a concept is beset with the limitations that in a pavement under service it is impossible to assess the energy dissipated or in other words, the traffic that has actually passed over the pavement. Due to this reason, a concept of remaining energy has been developed and described herein.

Laboratory simulation of field conditions, as they exist in pavements under traffic is a complex problem. Under a moving wheel load, an element in a pavement is subjected to generalised conditions of stresses comprising of both normal and shear stresses, which undergo cyclic variations both in magnitude and direction. Many researchers have considered repeated load test in axisymmetric triaxial conditions

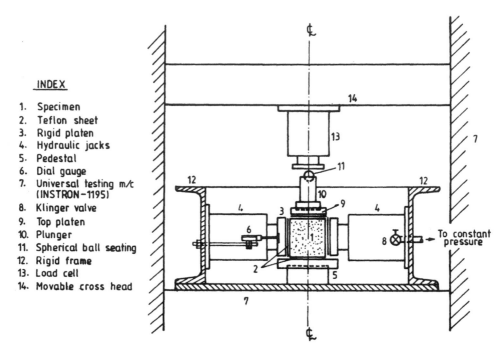

INDEX

1. Specimen
2. Teflon sheet
3. Rigid platen
4. Hydraulic jacks
5. Pedestal
6. Dial gauge
7. Universal testing m/c (INSTRON-1195)
8. Klinger valve
9. Top platen
10. Plunger
11. Spherical ball seating
12. Rigid frame
13. Load cell
14. Movable cross head

Figure 3. Sectional elevation of Universal Triaxial Apparatus-II, placed in Instron-1195, UTM.

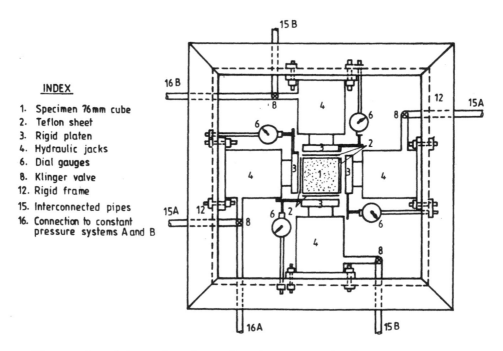

INDEX

1. Specimen 76mm cube
2. Teflon sheet
3. Rigid platen
4. Hydraulic jacks
6. Dial gauges
8. Klinger valve
12. Rigid frame
15. Interconnected pipes
16. Connection to constant pressure systems A and B

Figure 4. Schematic plan of Universal Triaxial Apparatus-II

as an approximation to the problem. For a realistic evaluation of pavement material behaviour, it is imperative that characteristics be determined under generalised stress field. With this in view, an apparatus designed Universal Triaxial Apparatus-II (UTA-II) shown in Figs.3 and 4 was developed at IIT Delhi wherein normal stresses in three mutually perpendicular directions can be applied on 76 mm cuboidal specimens (Ramamurthy,et al 1979). Thus an equivalent principal stress system could be generated. Using UTA-II, large amount of experimental data has been obtained concerning the mechanical behaviour of a silty soil and soil-cement under static and cyclic loading in different stress fields. This data has been employed to explore the possibility of using remaining energy as a useful material characteristic in pavement evaluation.

2 REVIEW OF PREVIOUS WORK

For many years soil-cement has been in use as base and sub-base layers of flexible pavements. In general the performance of these layers has been judged as excellent. AASHO Road Test, Highway Research Board (1964) and Nussbaum and Larsen(1965) have reported that properly designed and constructed soil-cement base courses have longer and better service as compared to untreated gravel base courses. Many researchers (Felt and Abrams 1957,Reinhold 1955) in the past have reported work on the strength properties of soil-cement in terms of unconfined compressive strength or results from CBR or Plate load tests. The limitations of these tests in evaluating individual material characteristics are well known.

Strength and durability increased with increasing cement content in the case of soil-cement prepared with granular soils (Abrams 1959). Soil-cement specimens prepared with A-1-b, A-2-4, A-6 and A-4 soils, when tested under static and repeated loads in compression and flexure (Felt and Abrams 1957) have shown that the static modulus in compression as calculated at 33% of ultimate load was approximately equal to 60 to 75% of the dynamic

modulus; the dynamic modulus and modulus of rupture were almost equal for all soil-cement specimens. Poisson's ratio was found to lie between 0.22 and 0.36.

Delhi silt (sand 46%, clay 12%, w_l=24.5%, w_p=19%)was mixed with ordinary portland cement in 2,4, 6 and 8 percent by weight of soil to form soil-cement mixes. Cylindrical specimens of 38mm diameter and 76mm height were prepared at optimum compaction conditions and cured under compacting moisture conditions for 1, 7, 14 and 28 days before testing. These specimens were subjected to unconfined and axisymmetrical triaxial compression. The stress-strain response of these specimens were obtained and material characteristics viz., modulus of elasticity (E) and Poission's ratio (μ) along with peak compressive strength and shear strength parameters,c and φ were evaluated (Venkatappa Rao, Ramamurthy and Nag 1978, Nag, Ramamurthy and Venkatappa Rao(1979). These studies indicated that

1. Unconfined compressive strength and modulus of elasticity increase with increasing compactive effort, cement content and curing period. The Poisson's ratio decreases on addition of cement. Variation of content does not seem to effect this value.

2. The effect of confining pressure is to further increase the strength parameters of soil-cement. An increase of 25% was recorded due to a confining pressure of 1.05kg/ cm^2. Hence the magnitude of confining pressure should be given due consideration in determining the strength properties of soil-cement for design of pavements. The effect of low confining pressures at longer curing periods was found to be insignificant on strength.

3. The cohesion intercept c in terms of total stresses has also been found to increase significantly with cement content and curing period. This is of importance while considering the immediate stability of pavements.

4. The angle of shearing resistance, φ, increases significantly by the addition of cement in small proportions (upto 2%). Further increase in cement content and increased curing period caused little variation.

Table 1. Influence of volume of test specimen on strength and modulus

Properties		Cylindrical, of diameter		Cubical
		38 mm	100mm	76 mm side
UCS (kg/cm^2)	Proctor	2.30	2.32	2.28
	M.AASHO	2.65	2.58	2.60
E(kg/cm^2)	Proctor	43.4	41.6	40.8
	M.AASHO	55.0	50.2	52.0

5. A linear relationship between unconfined compressive strength and log of volume of specimen was found to exist for cylindrical specimens, with different l/d ratios. The constant of proportionality is a function of curing period,Table 1.

6. The modulus of elasticity was found to increase linearly with cement content at all confing pressures. (Fig.5).

3. FRACTURE ENERGY

Phenomenologically any load-structure system undergoes an energy transfer. The energy imparted by the loading system (in any form-potential, kinetic, thermal, chemical, etc.) is stored in the structural element in the form of strain energy. As a consequence of the energy transfer, the structural element undergoes deformations which are both elastic and plastic in nature. On removal of the applied load, the stored energy is not totally recovered, rather, a part of it is dissipated or permanently adsorbed by the structural system as irrecoverable or dissipated energy. Some energy is also list in the form of heat in the system. The capacity to adsorb or store energy is a material property known as toughness.

The area under the stress-strain curve upto failure or fracture for a material tested in axial compression or tension represents the fracture toughness or fracture energy absorbed per unit volume of the

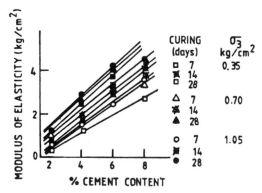

Fig.5. Variation of E with cement content (Triaxial test)

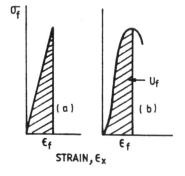

Fig.6.(a) Idealised stress-strain curve
(b) Normal stress-strain curve

specimen and is equal to the work done by the external load under ideal conditions (assuming no dis sipation of energy as heat loss).

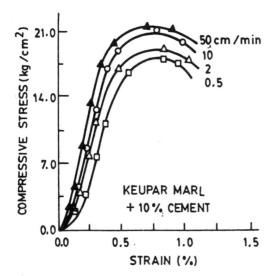

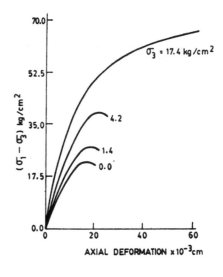

Fig.7. Effect of rate of strain on the stress-strain curves of soil-cement(After Dunn and Obi,1969).

Fig.8. Effect of confining stress, σ_3 on stress-strain curves of soil (After Nash et al, 1966)

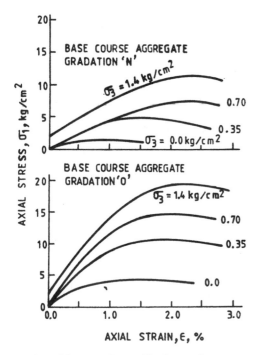

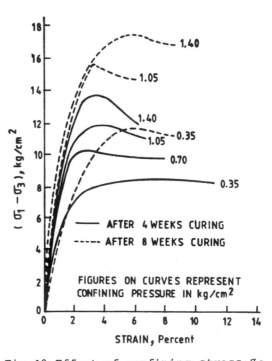

Fig.9. Effect of confining stress,σ_3 on stress-strain curves of base coarse aggregates(After Sebastyan,1967)

Fig.10. Effect of confining stress,σ_3 on stress-strain curves of lime-flyash stabilized Delhi silt(After Rekhi,1979).

51

Assuming idealised conditions of isotropy, homogeneity and elasticity, the following mathematical formula holds good for fracture energy calculations (Fig.6.)

$$U_f = \frac{1}{2} \sigma_{1f} \varepsilon_{1f} \qquad (2)$$

Fracture energy is a scalar quantity and is essentially a product of stress and strain, when calculated over a unit volume. For a particular material under identical conditions of testing, it almost assumes a constant value and does not undergo large change like other material characteristics, e.g. with increased rate of loading most materials indicate an increase in strength and modulus of elasticity and a decrease in strain; the fracture energy being a product of an increasing stress and a decreasing strain is not expected to show much variation. As an example, reference is made to Fig.7 reproducing stress-strain curves (Dunn and Obi 1969) for soil-cement specimens tested at a rate of compression ranging between 0.5 to 50 cm/min. When the rate of deformation increased from 0.5 to 50 cm/min, the axial strength at failure increased by about 20 per cent and modulus of elasticity increased by about 33 per cent, whereas the fracture energy was found to reduce only by less than 3 percent.

Fracture energy characterizes the material resilience and as such is more relevant to the fatigue conditions to which pavements are subjected. A review of the literature has revealed that laboratory fatigue test results obtained from constant stress and constant strain conditions yield different results in regard to fatigue life, but have confirmed that the total dissipated energy in either case remains same. Van Dijk et.al (1972) have also opined that failure criterion based on fracture energy has better applications to fatigue conditions.

Apart from the above observations a study of Table 2 (wherein various material characteristics from axisymmetric triaxial tests on different materials ranging from subgrade soils to base course aggregates from Nash et al. 1966, Sebastyan 1957, Rekhi 1979, Figs.8, 9, 10

respectively, are compared) reveals that fracture energy normally follows a uniform trend with variations in confining pressure (σ_3) and generally showed higher sensitivty to the increase in confinement. In comparison to strength, fracture energy has been found to be more sensitive. Strain has not been found to follow any uniform trend with σ_3. Modulus of elasticity represents the material state only at one point, and may assume different values for the same material depending upon the manner of its measurement, either as tangent or secant modulus.

4 EXPERIMENTAL INVESTIGATIONS

Delhi silt and ordinary Portland cement with 4 and 8 percent cement content were used to form soil and soil-cement specimens. Specimens statically compacted at optimum proctor conditions and cured at 100 percent relative humidity for 28 days were tested in UTA-II under both static and cyclic loading. In order to investigate the extent of material change due to load cycling, a new test procedure representing "before and after load cycling" to study the material properties was adopted. A number of identical specimens were prepared and cured. One of these specimens was first tested monotonically to failure, at a constant rate of 0.55 mm/min. The remaining were subjected to constant stress cycles of 100, 1000 and 10,000 cycles, at a constant rate of strain of 20 mm/min. After the designated number of load cycles were applied, the specimen was brought back to the initial stress-field (as adopted for the first specimen) and tested monotonically to failure under a constant strain rate of 0.5 mm/min. The effect of material change is revealed when the stress-strain diagram of the first specimen is compared with those subjected to load cycling. The investigations reported here pertain to σ_3 values ranging between 0.5 to 1.5 kg/cm^2, lateral stress ratio (σ_2/σ_3) between 1 to 6 and cyclic deviatoric stress (σ_{d1cy}) ranging between 20 to 80 percent of the deviatoric stress at failure under static conditions. In all,

Table 2. Comparison of material characteristic for different paving materials

Sl. No.	Material	Test conditions	Confining pressure σ_3 kg/cm²	Failure stress σ_{1f} kg/cm²	Change %	Failure strain ε_{1f} %	Change %	Modulus of Elasticity E kg/cm²	Change %	Fracture energy U_f kg/cm²	Change %
1.	Delhi silt + 4% lime (Fig.10)	4 weeks cured	0.70	9.40	–	2.30	–	820	–	0.228	–
			1.05	10.68	14	3.80	65	860	5	0.352	55
			1.58	15.68	67	3.70	61	940	14	0.442	94
		8 weeks cured	0.35	12.50	–	5.90	–	310	–	0.420	–
			1.05	16.90	35	3.30	44	613	99	0.720	72
			1.58	18.50	48	5.90	0	680	120	0.883	110
2.	Soil + 4% cement (Fig.8)	16 days cured	0.00	21.40	–	1.90	–	2000	–	0.260	–
			1.40	27.33	28	1.94	2	2400	20	0.83	28
			4.20	41.30	93	2.20	16	2600	30	0.518	99
3.	Base course		0.35	4.75	–	1.75	–	428	–	0.063	–
	Aggregate Gradation-N		0.70	7.25	48	2.25	28	460	9	0.106	68
			1.40	11.25	130	4.25	142	578	35	0.172	163
	Gradation-C (Fig.9)		0.35	10.75	–	2.25	–	1000	–	0.181	–
			0.70	14.70	38	2.00	– 15	1190	20	0.200	11
			1.40	19.50	90	2.25	0	1280	28	0.306	69

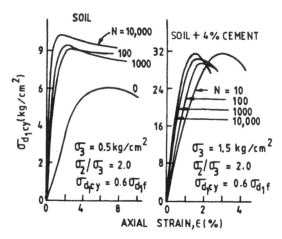

Fig.11. Influence of load cycles on
stress-strain curves for
(a)soil and (b)soil-cement

about 150 specimens of soil and
soil-cement were tested (Nag 1982).

5 RESULTS

Fig.11 represents typical stress-
strain curves obtained in static
tests after N cycles for soil and
soil-cement specimens for $\sigma_2/\sigma_3 = 2$
and a σ_{d1cy} of 60 percent deviatoric
stress at failure. It is clear that
increase in N considerably changes
the stress-strain behaviour. The
area under the stress-strain curve
upto failure, when calculated for
unit volume is designated as the
fracture energy.

Data from such tests were analysed
in terms of deviatoric stress at
failure (σ_{d1fN}), modulus of elastic
deformation (E_{stN}), and fracture
energy (U_{fN}). Figs.12 and 13 present typical variations of σ_{d1fN},
E_{stN} and U_{fN} with N for soil and
soil-cement. From these figures it
is clear that no definite trend can
be isolated in the variation of
σ_{d1fN}, E_{stN}; but fracture energy U_{fN}
seems to follow a near linear varia-
tion with respect to log N.

An overall review of the various
test results indicated that fracture
energy increased with increase in
confining pressure, lateral stress
ratio, cement content, compactive
effort, and curing period - the
factors which attempt to stabilise
the material. It decreased with

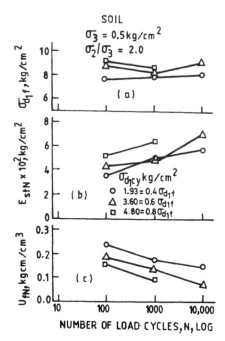

Fig.12. Influence of load cylces and
cyclic deviator stress on
strength,elastic modulus and
fracture energy for soil

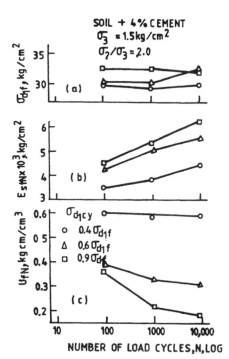

Fig.13. Influence of load cycles and
cyclic deviator stress on
strength,elastic modulus and
fracture energy for soil-cement

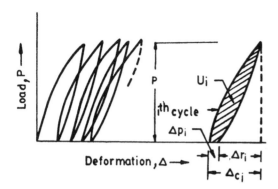

Fig.14.Load deformation plot under constant stress cyclic test

increase in cyclic stress level and number of cycles, as expected. However the same could not be stated in respect of some of the other material characteristics, specifically, strength.

A linear multiple regression analysis was carried out to establish a relationship between fracture energy and other material and test variables which yielded the relation,

$$U_{fN} = b_0 + b_1(c) + b_2(\sigma_3) + b_3(\sigma_2/\sigma_3)$$

$$+ b_4(\sigma_{dlcy}/\sigma_{dlf}) + b_5(\log N)$$

where   (3)

$b_0, b_1, b_2, b_3, b_4, b_5$ are the coefficients of the regression equation, as in Table 3,
c = cement content,
σ_3 = minor principal stress,
σ_2/σ_3 = lateral stress ratio, and
$\sigma_{dlct}/\sigma_{dlf}$ = cyclic stress level.

Unlike Modulus of elasticity, fracture energy represents the material in totality in the form of the area under the stress-strain curve upto failure.

Further Miner's law of cumulative effect is also applicable to the fracture energy and as such this parameter can be utilised to represent the overall response of a pavement structure at any time of its use. The advantage of the applicability of Miner's law to fracture energy lies in the algebraic accumulation of the effects of the

various factors like the random application and the magnitude of wheel loads, and environment.

From the above discussion it can be inferred that fracture energy is a superior material characteristic and defines the material in a more distinct way as compared to other conventional characteristics.

6 FRACTURE ENERGY RELATED TO REMAINING LIFE

For a better understanding it is first essential to evaluate the effect of load cycling on the fracture energy. Let us consider a specimen under ith cycle of constant stress cyclic load test. From load deformation diagram, Fig.14, we get

Work done in the loading cycle,

$$W_1 = \frac{1}{2} P.\Delta c_i$$

Work recovered during unloading cycle,

$$W_v = \frac{1}{2} P.\Delta r_i$$

Net work done in ith cycle,

$$W_i = W_1 - W_v = \frac{1}{2} P(\Delta_{ci} - \Delta_{ri}) \quad (4)$$

Also under idealised conditions, $W_i = u_i$, the dissipated energy. Therefore, $u_i = \frac{1}{2} P.\Delta_{pi} = \frac{1}{2}\sigma.\varepsilon_{pi}$ x volume of the specimen.
Then,
Dissipated energy per unit volume during ith cycle,

$$u_i = \frac{1}{2} \sigma.\varepsilon_{pi} \quad (5)$$

Applying Miner's law of cumulative effect, the total dissipated energy per unit volume at the end of N cycles, for constant stress condition,

$$U_N = \int_{N=0}^{N} u_i = \frac{1}{2} \sigma \int_{N=0}^{N} \varepsilon_{pi}$$

or $U_N = \frac{1}{2}\sigma.\varepsilon_{pN} \quad (6)$

where σ = constant cyclic stress, and ε_{pN} = total permanent or plastic at the end of N cycles.

Table 3. Multiple regression analysis for fracture energy

Sl. No.	Variable	Range	Mean	Standard deviation	Coefficient of regression equation	Standard error or coefficient
1.	Cement content, c	4-8 percent	0.437	1.165	$b_1 = 0.852 \times 10^{-2}$	0.668×10^{-2}
2.	Confining pressure, σ_3	0.5-1.5 kg/cm^2	0.968	0.433	$b_2 = 0.126$	0.171×10^{-2}
3.	Lateral stress ratio σ_2/σ_3	1-6	2.347	1.538	$b_3 = 0.180 \times 10^{-1}$	0.484×10^{-2}
4.	Cyclic stress level, $\sigma_{dlcy}/\sigma_{dlf}$	20-80 percent	0.657	0.286	$b_4 = -0.285$	0.351×10^{-1}
5.	Log number of loadcycles, log N	0-4	2.074	1.514	$b_5 = -0.854 \times 10^{-1}$	0.686×10^{-2}
6.	Fracture energy U_{fN}	0.155-0.700 (kg cm/cm^3)	0.381	0.226	–	–
	Regression equation constant				$b_0 = 0.544$	0.517×10^{-1}

Now assuming fracture energy of a specimen (U_f) under certain conditions of testing and environment as constant, and that the specimen is considered to have failed at the end of N_f cycles when its fracture energy is totally dissipated, then,

$$U_f = U_{nf}$$

Also total permanent strains for $N = N_f$ will be equal to ε_{pNf}

Hence, $\quad U_f = U_{Nf} = \frac{1}{2}\sigma \cdot \varepsilon_{pNf} \qquad (7)$

Now, if the fracture energy of the the specimen tested after N load cycles of constant stress is designated as U_{fN}, then by law of summation,

$$U_{fN} = (U_f - U_N)$$

Substituting values of N_N and U_f, from Eqns. 6 and 7, we get

$$U_{fN} = \frac{1}{2}\sigma(\varepsilon_{pNf} - \varepsilon_{pN}) \qquad (8)$$

For most of the materials, permanent strain has been found to follow a linear variation with log number of load cycles, and therefore

$$\varepsilon_{pN} = a + b\,\log_{10}N$$

and $\varepsilon_{pNf} = a + b\,\log_{10}N_f \qquad (9a)$

Substituting for ε_{pN} and ε_{pNf} in Eq.8, we get,

$$U_{fN} = \frac{1}{2}\sigma \cdot b\,\log_{10}\frac{N_f}{N}$$

or $\log_e \dfrac{N_f}{N} = 2.303\left(\dfrac{2U_{fN}}{\sigma \cdot b}\right)$

$= K.U_{fN}$ (putting $\dfrac{2.303 \times 2}{\sigma \cdot b} = K$)

or $\dfrac{N_f}{N} = e^{K.U_{fN}} \qquad (9b)$

or $\dfrac{(N_f - N)}{N_f} = \left(1 - \dfrac{1}{e^{K.U_{fN}}}\right)$

Now $N_f - N = N_R$, remaining life, hence,

$$N_r = \left(1 - \frac{1}{e^{K.U_{fN}}}\right).N_f \qquad (10)$$

For any specimen under certain conditions of testing and environment both N_f, the life to failure and K are constant terms and can be estimated from laboratory fatigue tests (constant stress system). Equation 10 therefore, suggests that remaining life of the specimen (N_R) is a function of fracture energy at the end of N cycles, i.e. U_{fN} only.

The relationship derived above can be used for materials in each component layer of the pavement and algebraically added to observe the net effect. The process of summation is justified, as the parameter involved (i.e. U_{fN}) is a scalar quantity unaffected by the direction of applied stress. Since fracture energy is affected by mechanical and chemical changes which may occur with lapse of time on account of compaction or degradation under traffic or on account of ageing, the fracture energy of the existing pavement is expected to undertake these effects automatically.

7 APPLICATION OF FRACTURE ENERGY TO PAVEMENT EVALUATION

The preceding discussions and the mathematical relationship between fracture energy and the remaining life of a specimen have emphatically brought out the fact that fracture energy has potential for a realistic characterisation of pavement material at any time of its use. The remaining life of an existing layered pavement will also depend on number of other factors like layer compositions and their inter-relationship. To ascertain effects of these factors on remaining life, long term field and laboratory investigations are required to be carried out. However, to make the outcome of this paper more useful to the profession, a step by step method is presented hereunder to estimate the remaining life of an existing pavement, using the concepts of fracture energy and material divagation.

In order to eliminate various parameters associated with multilayered pavements, a single layer cement stabilized pavement is considered for evaluation.

To evolve various variables of Equation 9 laboratory testing has to be conducted in two stages as given below.

Stage-1

(i) Sampling:- Undisturbed sample blocks may be cut out or cylindrical cores be obtained from the freshly laid pavement, after compaction and curing if any, but definitely before any traffic is allowed on the pavement. A minimum of 3 samples be obtained from different places using standard sampling techniques.

(ii) Specimen preparation: If samples are not sufficiently cured, laboratory curing shall be done before testing. Cylindrical or cubical specimens shall be prepared in conventional manner from the cores or blocks.

(iii) Determination of N_f and K: A constant stress cyclic load test (fatigue test) in any of the stress-fields (as per apparatus available) is conducted on these samples. The stress range may be appropriately chosen to suit the design values. In case the design life of the pavement is not already assigned, the test may be conducted till failure, to assess the design life (N_f). Where N_f is already known as a an assigned value (e.g. many pavement design methods adopt 10^6 applications of design wheel load as the life of the pavement) the test may be conducted to ascertain predetermined number of load cycles, (say 10,000). Permanent strain (ε_{pN}) values be observed for 3 or 4 intermediate values of N.

From a plot of ε_{pN} v/s $\log_{10} N$, (Fig.15), the variable b is determined as the slope of the line represented by the following equation,

$$\varepsilon_{pN} = a + b \log_{10} N$$

The constant K may then be computed

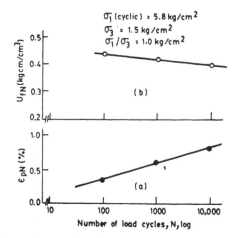

Fig.15. Variation of (a) Permanent strain ε_{pN}, (b) fracture energy U_{fN} after N cycles of constant stress on soil + 4% cement cubical specimen in general stress system

as

$$K = \frac{2.303 \times 2}{\sigma . b}$$

where σ = applied cyclic stress.

The values of N_f and K as well as and the stress field are recorded for future use.

Stage-2

(i) Tests at the time of pavement evaluation: When it is desired to estimate the remaining life of pavement, samples shall be obtained from the existing pavement and laboratory specimens prepared following the procedure outlined under Stage 1. These specimens are then subjected to static test to failure under the same stressfield as adopted under Stage 1 above. The area under the stress-strain curve upto maximum stress is measured as fracture energy (U_{fN}).

(ii) Estimation of remaining life: substituting the values of U_{fN}, K and N_f in Equation 10, the remaining life N_R is estimated. This value can also be computed as percentage of the design life as

58

Remaining life $= (1 - \dfrac{1}{e^{K} U_{fN}}) \times 100$
(per cent)

The procedure described above is illustrated as an example given below:

Example

(Adopting the data from the laboratory fatigue tests (Nag,1982) on soil + 4 per cent cement cubical specimens tested in general stress conditions, σ_1(cyclic) = 5.8 kg/cm^2 σ_2/σ_3 = 1.0 and σ_3 = 1.5 kg/cm^2.)

Let us assume that pavement layer consists of soil + 4 per cent cement and that thickness of the pavement is designed for a design life (N_f) of 10^6 applications of design wheel load (σ_1=5.8 kg/cm^2).

Assuming lateral stresses as $\sigma_2 = \sigma_3$ = 1.5 kg/cm^2. From Fig.15(a), the slope of ε_{pN} v/s $\log_{10}N$ plot, b = 0.275. Then

$$K = \frac{2.303 \times 2}{0.275 \times 5.8} = 2.879$$

Adopting U_{fN}= 0.416 kg cm/cm^3 from Fig.15(b) for an assumed value of N = 10,000 repetitions of design wheel load, substituting in Equation 10, we get

$$N_R = (1 - \frac{1}{e^{2.879 \times 0.416}}) \ 10^6$$

$$= (0.818) \ 10^6 \text{ repetitions.}$$

$$= 81.8 \text{ per cent of design life.}$$

Since the fracture energy is a scalar quantity, and the law of algebraic summation is applicable for its use, it is possible to extend the suggested procedure of remaining life estimation to multi-layered pavements also.

8 LIMITATIONS

The method discussed above has considered only a single layer cement stabilised pavement, from understanding point of view. However,in practice a flexible pavement is a multi-layered system of both bound and unbound materials. Under these conditions, the application of fracture energy concepts to pavement evaluation raises two major problems, viz. (a) obtaining reasonably undisturbed core sample of the multilayered pavement and (b) the applicability of the law of cumulative effect on heterogeneously composed pavement layers. Further experimental investigations both in the laboratory and in-situ may be necessary to search out the solutions to these problems.

It is possible to sample out a core specimen from paving layers of hard bound materials like cement concrete, bituminous concrete and stabilised soils, but the same may be difficult in the case of granular and loosely bound base and subbase materials. Sampling techniques for obtaining an undisturbed core from coarse and unbound materials need development amd modifications.

As regards the applicability of the method to the multi-layered pavements, the concepts of fracture energy are more fundamental and basically phenomenological is as compared to other conventional material characteristics, e.g.stress, strain or elasticity. Being a scalar quantity fracture energy is subjected to Miner's law of cumulative effect and therefore the method presented here is applicable to multilayered pavement system. However, further research in this direction is suggested to evolve suitable modifying factors that may be necessary.

9 CONCLUSIONS

A realistic representation of pavement and pavement materials at any time of its use is essential for rational evaluation and overlay design. In this Paper, the potential of fracture energy as material characteristics has been explored for estimation of remaining life of in-service (existing) pavements. The following are the major points brought out:

(1) The concept of fracture energy as a material characteristic has been developed in the context of pavement performance and evaluation.

(2) A comparison of the various

material characteristics (viz. strength, strain, modulus of elasticity, fracture energy etc.) of a wide range of paving materials revealed that fracture energy characterizes the material more distinctly and represents its overall state at any stage of its use.

(3) A mathematical expression derived on the basis of theory of material divagation, relates remaining life to the fracture energy of the paving material at any time of its use.

(4) A step by step procedure to estimate remaining life of pavement is hypothesised for a single layer pavement. This method is capable of further extension to multilayered pavements, as the various factors affecting pavement response can be easily taken into consideration, and as fracture energy is a scalar quantity and Miner's Law of cumulative damage is applicable to it.

10 REFERENCES

Abrams, M.S.1969. Laboratory and field tests of granular soil-cement mixtures for base courses. ASTM, STP254, pp.229-243.

Dunn,C.S. and Obi, B.C.A. 1969.Some engineering properties of cement stabilized keupar marl with particular reference to rapid and repeated compression tests. The Quarterly Journal of Engineering Geology, The Geological Society of London, Vol.2, No.1, pp.25-48.

Federal Aviation Agency.1964. Thickness design procedures for airfields containing stabilized pavement components. U.S. Dept. of Commerce, Office of Technical Services, Washington.

Felt, E.J. and Abrams, M.S. 1957. Strength and elastic properties of compacted soil-cement mixtures. ASTM STP 206, pp.152-178.

Highway Research Board.1962. The AASHO road test; Part 5. Pavement Research, Spl.Rep.No.61E,Washington D.C

Lentz, R.W. and Baladi,G.Y.1980. Prediction of permanent strain in sand subjected to cyclic loading. Transportation Research Record No.749.

Majidzadeh,K., Kauffman,E.M. and Ramasamooji,D.V. 1971. Application of fracture mechanics in analysis of pavement fatigue. Proc. The Association of Asphalt Paving Technologists, Vol.40,pp.227-246.

Nag, C.P. 1982. Soil-cement characterization under cyclic loading in general stress system. Unpublished Ph.D thesis, Department of Civil Engineering,Dellhi.

Nag, C.P., Ramamurthy,T. and Venkatappa Rao, G.1979. Engineering characteristics of stabilized soil. Proc.VI Asian Reg.Con. on SM & FE, Singapore,Vol.II,pp.159-162.

Nash,J.K.T.L., Jardine,F.M. and Humphreys, J.D. 1966. The economic and physical feasibility of soil-cement dams. Proc.6th Int. Conf. on SMFE, Vol.2, pp.517.

Nussbaum, P.J. and Larsen,T.J.1965. Load deflection characteristics of soil-cement parameters.H.R.B. Bulletin No.342.

Pell,P.S. and Brown,S.F.1972. The characteristic of materials for design of flexible pavement structures. Proc. 3rd Int.Conf.SDAP, London, pp.326-342.

Ramamurthy,T., Venkatappa Rao,G. and Nag, C.P.1979. Development of universal triaxial test apparatus for evaluation of pavememt material characteristics. Proc. Int. Symp. Pavement Evaluation and Design, Rio-de-Janerio, Brazil, Vol.3, Paper 9.

Rekhi,T.S.1979. Physico-chemical mechanisms governing the engineering behaviour of lime and lime fly-ash stabilized silt. Unpublished Ph.D thesis, Department of Civil Engineering, I.I.T. Delhi.

Reinhold,F.1955. Elastic behaviour of soil-cement mixtures. H.R.B. Bull. No.108.

Sebastyan,G.Y.1967. Flexible airport pavement design and performance. Proc.2nd Int.Conf. on SDAP, University of Michigan,Michigan, pp.71-87.

Stauffer, D.C. and Strauss,A.M.1979. The concept of material divagation and its application. Advances

in Research on Strength and Fracture, Ed., DMR Taplin,Vol.3A, 4th
Int. Conf. on Fracture,Waterloo.

Timoshenko, S.P. and Goodier,J.N.
1979. Theory of elasticity. Mc-
Graw Hill Kogakusha, pp.244-248.

Van Dijk, W., More and H.Quedeville,
A. and Uge, P.1972. The fatigue
of bitumen and bituminous materials. Proc. 3rd,Int.Conf. on SDAP,
Michigan,Vol.1,pp.354-366.

Van Dijk,W. and Visser,W.1977. The
energy approach to fatigue for
pavement design. Proc. The Association of Asphalt Paving Technologists, Vol.46, pp.1-40.

Venkatappa Rao,G. and Rekhi, T.S.
1978. Undrained strength behaviour
of lime-fly ash stabilized silt.
Proc.Symp. on Soil Reinforcing and
Stabilizing Techniques, Sydney,
Australia, pp.433-450.

Yamanouchi,T.1972. Some studies on
the cracking of soil-cement. HRB,
HRR 442.

Venkatappa Rao,G., Ramamurthy,T.
and Nag, C.P.1978. Strength and
deformation behaviour of cement-
stabilized silt. Proc.Symp. on
Soil Reinforcing and Stabilizing
Techniques, Sydney,Australia,
pp.417-432.

Environmental Geotechnics and Problematic Soils and Rocks, Balasubramaniam et al. (eds)
© 1987 Balkema, Rotterdam. ISBN 90 6191 785 9

Engineered clay liners: A short review

R.M.Quigley & F.Fernandez
Geotechnical Research Centre, Faculty of Engineering Science, University of Western Ontario, London, Canada
V.E.Crooks
I.T. Corporation, Irvine, Calif., USA

ABSTRACT: Chemical flux predictions and clay/leachate compatibility with respect to liner performance are briefly reviewed. Typical flux-time plots are presented for a range of advective velocities and diffusion coefficients. Proof of the importance of migration by <u>diffusion</u> is demonstrated by profiles for Na^+, K^+, Mg^{++}, Ca^{++}, Cl^- and DOC through clay below a domestic landfill. Fifteen year migration to ~ 1-3 m is compared to heavy metal migration to ~ 0.1 m.

Clay leachate compatibility is discussed using hydraulic conductivity as the assessment tool. Cation retardation is demonstrated and the complex and dangerous role of low dielectric organic liquids is demonstrated by permeation experiments.

1 INTRODUCTION

Clay barriers below waste disposal sites are required to control the rate of migration of noxious leachate solutes into the groundwater environment. Generally speaking, natural, undisturbed, unfissured clay deposits are preferred over compacted clay liners. However, the latter are often required, especially if use is to be made of existing excavations such as large abandoned gravel pits.

This brief position paper discusses five major aspects of liner design, namely:

1. The purpose and predicted performance of clay liners, with emphasis on the important role of diffusion.
2. A case history of a sanitary landfill at Sarnia confirming the significance of migration by diffusion.
3. Problems of liner compaction and measurement of hydraulic conductivity.
4. Soil-leachate compatibility (domestic waste).
5. Soil-leachate compatibility (liquid hydrocarbons).

2 PURPOSE AND PREDICTED PERFORMANCE

Some of the typical requirements for a clay liner may be summarized as follows:
1. Liners should be 1.0 to 1.5 m thick for domestic waste and 3 to 4 m thick for industrial waste.
2. The hydraulic conductivity, k, with respect to water should be 10^{-7} to 10^{-8} cm/s.
3. The hydraulic conductivity shall NOT increase when permeated with the leachate being contained.
4. Normally a limit of some kind is placed on the chemical flux exiting from the liner, and often Cl^- is used as the conservative (non-reacting) control solute. No contamination beyond the boundaries of a site is an increasingly common approval requirement.
5. Depending on the political jurisdiction and the design position of the leachate level within the waste, a downward gradient may be required to prevent surface seepage from the toe of the cover clays. Alternatively, removal of leachate by pumping and trucking to a treatment facility may be required.

Test techniques are rarely specified in detail by environmental authorities and the consultant is left with the difficult task of performing laboratory measurements which frequently require such extensive chemical control that they appear more like research projects than practical tests for design.

The time rate of solute migration through a liner may be estimated using coupled

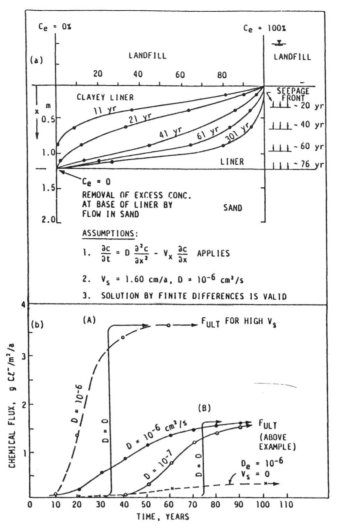

Figure 1. Migration of chemical solutes through a clay liner by coupled diffusion and advection: (a) C/C_0 vs. depth; (b) Flux vs. time plots

advective flow and diffusion, and plots of concentration versus depth may be obtained as shown on Figure 1a. It is important to note that for the input parameters employed, conservative species such as chloride are calculated to arrive at the base of the 1.3 m liner within 20 years even though the seepage front (assuming an average linearized seepage velocity) has only advanced half way through the liner. For this particular system, it was assumed that the base of the liner would be washed by salt-free ground water, whereas in many

situations the concentration would continue to increase with depth, even through sands.

The chemical flux which would derive from the above example is illustrated in Figure 1b (Curves B). For the situation of advective transport alone (i.e. $D_e = 0$), abrupt arrival of the chemical flux would occur in 75 years. If diffusion is also considered ($D_e = 10^{-6}$ cm^2/s), a small chemical flux is calculated to arrive in 15 years. For the case of zero gradient

($V_s = 0$), very slow arrival and a very small chemical flux are indicated. This particular situation is often referred to as a secure landfill. The penalty for having a high gradient or a high hydraulic conductivity is illustrated by the large flux and early breakthrough time shown by curves (A).

3 CASE HISTORY (SARNIA)

The Confederation Road landfill near Sarnia serves as an excellent field example demonstrating the importance of diffusion (Goodall and Quigley, 1977; Crooks and Quigley, 1984; and Quigley, Crooks and Yanful, 1984). Since the underlying clays are over 30 m thick, short boreholes for field investigations offer no environmental risks.

3.1 Site description

The silty clay till at Sarnia, Ontario is believed to be a waterlain till of late Wisconsin age, and has been extensively studied by Quigley and Ogunbadejo (1976). The soil in the 3 m zone directly below the refuse is a grey, carbonate-rich, clayey silt containing 40% of <2 μm sizes, and having the following composition: ~ 34% carbonate, ~ 25% illite, ~ 24% chlorite, ~ 15% quartz and feldspar and ~ 2% smectite. The effective cation exchange capacity is approximately 12 meq/100 g measured on the <74 μm fraction.

A slight downward flow gradient exists in the general area. Piezometric observations indicate a small groundwater mound in the refuse and a steady downward average linearized porewater velocity of approximately 0.24 cm/a in the till. In situ measurements of till hydraulic conductivity (Goodall and Quigley, 1977) indicated values of approximately 1.5 x 10^{-8} cm/s at an in situ water content of 20-25%.

Prior to its use as a landfill facility, the site was a clay borrow pit for construction of a nearby highway embankment. The borrow pit area readily impounded precipitation and runoff, and the initial lifts of refuse were probably placed in standing water. The refuse probably attained full saturation shortly after placement, allowing leachate migration through the clay to begin within a short time period. At the time of the 1983 investigation, migration from the refuse

had been in progress for approximately 15 years.

3.2 Cation (Na^+, K^+, Mg^{++}, Ca^{++}) Migration

Typical results for porewater cations and exchangeable adsorbed cations are presented on Figure 2. The results obtained to date for this and several other boreholes indicate that after 15 years, migration has occurred to depths of 0.7 m to 1.5 m below the refuse/soil interface. Spatial variability in the composition of the refuse produces somewhat different measured migration profiles at various locations within the site. This may result in significantly different maximum cation concentration values at the refuse/soil interface and slightly different estimates of the maximum depth of migration at a particular location.

Calculated sodium adsorption ratios (S.A.R.) often correspond to the adsorbed sodium on the soil clay minerals. In the example borehole (Figure 2) the calculated S.A.R. values decrease from 6 at the refuse/soil interface to 2 at a depth of 0.5 m, suggesting that over this distance up to 6% of the clay exchange sites should be occupied by Na^+. These and subsequent analyses indicate that calcium is predominantly adsorbed. This absence of large cation transfer between the leachate and clay minerals is related to a Na^+ concentration of the leachate which is not high enough to promote exchange. In addition, we find that the silver thiourea exchange technique employed (Chhabra et al., 1975) tends to solubilize Ca^{++} from carbonate. Since the average linearized porewater velocity in the till is very small ($V_s = 0.24$ cm/a) seepage over a period of 15 years would only have progressed to a depth of ~ 3.6 cm. It is apparent then, that in such low-flow situations, diffusion becomes the dominant mechanism of contaminant transport. (See also Desaulniers et al., 1980).

3.3 Chloride migration

A typical chloride migration profile is shown on Figure 3 for the same samples presented on Figure 2. Chloride migration has apparently advanced to a depth of ~ 1 m, although measured chloride results show the same variability as discussed previously with respect to the cation profiles.

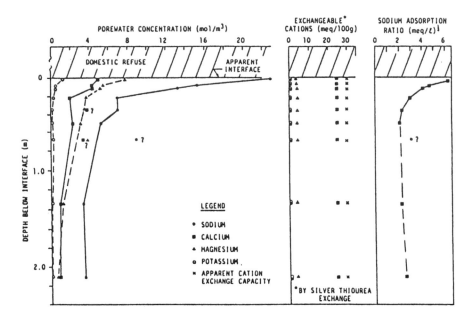

Figure 2. Porewater cation concentration, exchangeable cations and sodium adsorption ratio (S.A.R.), BH83-2 (Adapted from Quigley, Crooks and Yanful, 1984)

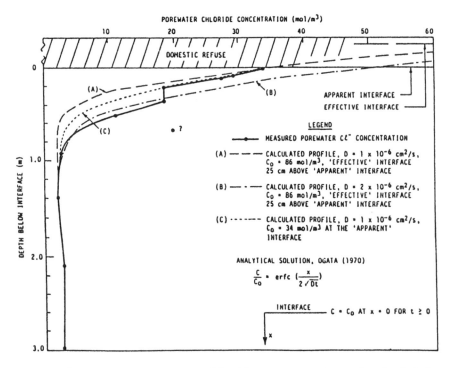

Figure 3. Measured porewater chloride concentration and calculated profiles, BH83-2 (Adapted from Quigley, Crooks and Yanful, 1984)

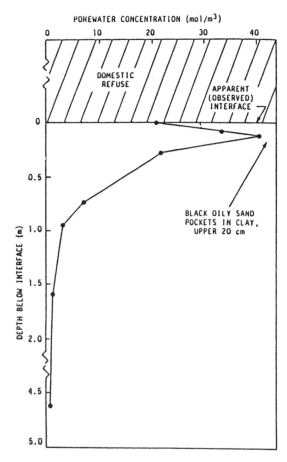

POREWATER CONCENTRATION (mol/m³)

DOMESTIC REFUSE

APPARENT (OBSERVED) INTERFACE

BLACK OILY SAND POCKETS IN CLAY, UPPER 20 cm

DEPTH BELOW INTERFACE (m)

Fig. 4. Porewater dissolved organic carbon, BH83-14 and BH83-3 (Directly from Quigley, Crooks and Yanful, 1984)

Since chloride is a conservative tracer (i.e. does not participate in geochemical reactions) the migration profile may be used to obtain "best-fit" values of effective diffusion coefficient D. Ogata (1970) has presented a number of analytical solutions for dispersive and advective transport of contaminants which may be readily used with a hand calculator to estimate potential migration profiles for preliminary work. The analytical solution used to analyze the results was derived for the case of one-dimensional diffusion (without advection) from a constant source, and is applicable in our situation due to the negligible advection component (V_s = 0.24 cm/a) of transport from the landfill. A "best-fit" value of D = 2 x

10^{-6} cm²/s has been obtained for these data (see Quigley, Crooks and Yanful, 1984, for a more complete description). Profiles calculated using the analytical solution are shown on Figure 2 and illustrate that "agreement" between calculated and measured profiles is somewhat dependent on the position of the effective soil/refuse interface and on the chosen value of the leachate concentration.

An alternative procedure which assumes an interface partition and a very low diffusion coefficient within the interface has recently been published by Rowe and Booker (1985) and Rowe et al. (1985). This method applied to the Sarnia site and thin barriers was recently reviewed by Quigley and Rowe (1985).

3.4 Dissolved organic carbon

A preliminary indication of the extent of migration of organic substances from the landfill may be obtained by measuring dissolved organic carbon (DOC) concentrations in the soil pore water. The measured profile is shown on Figure 4 and indicates that migration of organics has reached a depth of approximately 0.9 m directly below an area of the interface overlain by a black oily substance. Rapid diffusion of dissolved organic carbon is implied by these data and agrees with comments by Griffin et al. (1976).

3.5 Heavy metal migration

Preliminary results for Cu, Zn, Fe and Pb are presented on Figure 5, and represent the total acid extractable heavy metals including those adsorbed and contained within the crystal structure of the soil constituents. The results indicate that the heavy metals have migrated to a maximum depth of approximately 15 cm below the apparent interface. At the present time, it is believed that the "above background" heavy metals are present as a complex mixture of carbonates, sulphides and solid organics (Yanful, 1986).

One of the most important conclusions from the field study is that diffusion is an important short-term transport mechanism. Furthermore, a small chemical flux will indeed arrive at the base of a 1.0 m thick liner within 15 years or so as suggested by Figure 1b.

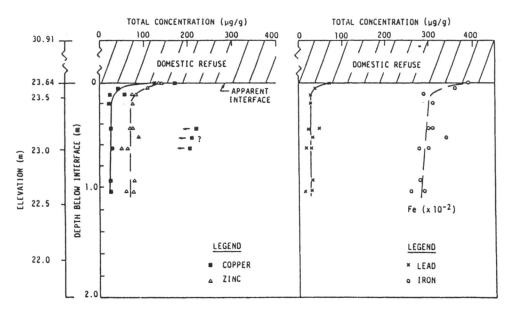

Figure 5. Total concentration of heavy metals (combined analysis U.W.O. and X-ray Assay Laboratories Ltd.), BH83-7 (Directly from Quigley, Crooks and Yanful, 1984)

4 LINER COMPACTION

Generally accepted practice calls for liner compaction wet of Proctor optimum using kneading compaction such as obtained from a sheepsfoot roller. The greatly decreased hydraulic conductivity which results from wet of optimum compaction is illustrated in Figure 6 adapted from Mitchell et al. (1965). Although high energy compaction (heavy equipment) at lower values of water content (curves C compared to curves A) should produce even lower values of hydraulic conductivity, it is apparently much more difficult to destroy the macropores between the hard soil clumps and a lower k may not be achieved, especially in the field. Such problems were recently discussed by Daniel (1984). Since destruction of the macropores is the dominant purpose of field compaction, extra passes which produce extra remoulding may be well worth the time and money even though there is no increase in field density.

Reproducible laboratory values of hydraulic conductivity are also made very difficult by the problem of macropores. As shown in Figure 7, k was shown to vary from 10^{-5} to 10^{-8} cm/s at constant dry unit weight (17.0 kN/m^3) over a moisture

range of 19 to 20.5% (Mitchell et al., 1965).

Along with the obvious benefits of wet of optimum compaction, there seem to be a few problems. Thixotropic flocculation effects may increase k by a factor of 3 or more just due to aging. Also, significant shrinkage on drying may occur. Although this shrinkage by drying is normally prevented by a sand blanket, it does suggest that chemical shrinkage could also be a problem for clay liners containing certain clay minerals in contact with certain types of leachate. This is a problem of soil-leachate compatibility which is discussed next.

5 SOIL-LEACHATE COMPATIBILITY
   (DOMESTIC LEACHATE)

Most inactive soils whose clay minerals consist of illites and chlorites are fairly insensitive to leachate from domestic waste and the hydraulic conductivity will normally actually decrease (Griffin et al., 1976). Such a situation is shown on Figure 8a for a Sarnia soil containing ~ 35% illite and chlorite plus ~ 5% smectite. The sanitary landfill leachate was fresh, unoxidized and unsterilized and had

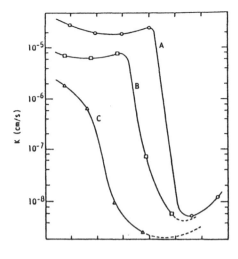

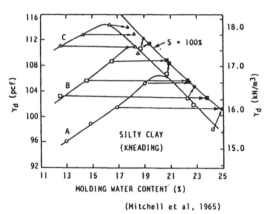

(Mitchell et al, 1965)

Figure 6. Hydraulic conductivity vs. moulding water content for different energies of kneading compaction (Mitchell et al., 1965)

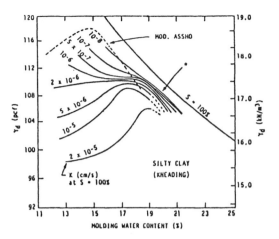

Figure 7. Variability in laboratory measured hydraulic conductivity vs. dry unit weight and moulding water content (Adapted from Mitchell et al., 1965)

a salinity of ~ 5 g/L. The k value with water was 2×10^{-9} cm/s and it can be seen that k actually decreased slightly during the passage of some 12 pore volumes of leachate. The reasons for this decrease are probably related to double layer expansion caused by the presence of Na^+. However, such factors as biological plugging are also important in tests that require several months.

The accompanying effluent concentration versus time curves required for chemical control (Figure 8b) are especially interesting. Chloride passes rapidly through the sample, the effluent reaching

the influent (leachate) concentration of 1.600 g/L after only 2 pore volumes. Na^+ also passes through rapidly, the effluent reaching influent concentration after only 3 pore volumes, although some retardation relative to $C\ell^-$ is apparent. Ca^{++} and Mg^{++} show large excess amounts relative to the influent after one pore volume, a feature probably mostly due to cation exchange and commonly referred to as a hardness halo. The most interesting curve is for K^+ which demonstrates very marked retardation for the entire 12 pore volumes of leachate and was nearly non-existent for the first 4 pore volumes.

It is possible that K^+ adsorption and fixation may be an important factor in the chemical shrinkage of liners containing significant amounts of the clay mineral vermiculite. For example, the solid particles of a soil containing 20% vermiculite would contract about 5.5% as the mineral contracts from 1.4 nm to 1.0 nm. This would be accompanied by reduction in charge deficiency on the clay particles and accompanying reductions in double layer size and cation exchange capacity. All these factors together could produce an increase in the size of the soil pores available for flow. The hydraulic conductivity might therefore increase with time as K^+ diffuses into the vermiculite structure.

More recent testing on samples compacted

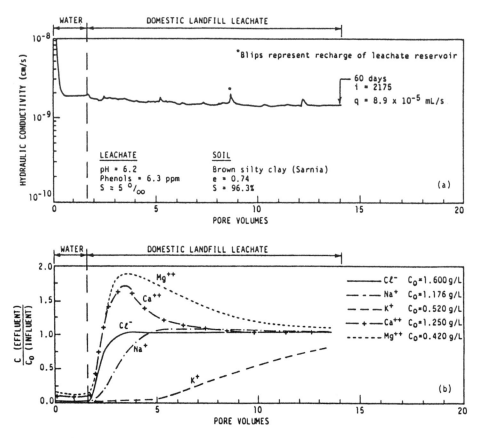

Figure 8. Soil-leachate compatibility; (a) hydraulic conductivity and (b) effluent chemistry vs. pore volumes of leachate passed through

at lower water contents to void ratios of about 0.5 invariably showed early arrival of leachate indicating compaction-induced cracks in the soils. This is a very serious problem in both the laboratory and the field.

6 SOIL-LEACHATE COMPATIBILITY (LIQUID HYDROCARBONS)

One of the major factors in the control of double layer size around clay particles is the dielectric constant of the fluid medium. Figure 9 graphically illustrates how the double layer contracts as water ($\varepsilon = 80$) is replaced by alcohols ($\varepsilon = 20$–30) and then by a liquid hydrocarbon such as benzene and its simple derivatives ($\varepsilon \approx 2$). If the soil mass remains constant in volume, then this double layer contraction represents an enormous

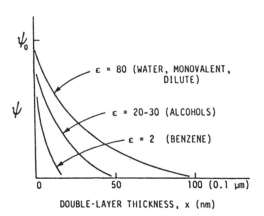

Figure 9. Electric potential, ψ, versus distance from clay particle for various values of dielectric constant, ε

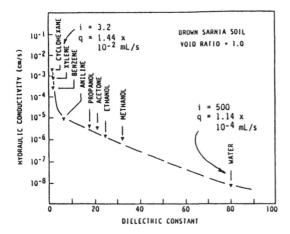

Figure 10. Measured hydraulic conductivity vs. dielectric constant (soil moulded with the fluid indicated)

increase in "free" pore space available for advective flow. The large values of hydraulic conductivity associated with liquid hydrocarbons have been discussed by such authors as Michaels and Lin (1954), Mesri and Olson (1971). Recent work at The University of Western Ontario is plotted on Figure 10. The hydraulic conductivity was measured in a sealed constant flow permeameter on soils mixed with the liquids indicated. Values of $k = 10^{-3}$ cm/s for the benzene compounds are typical of values reported in the literature.

In practice clay liners are moulded and compacted with water. Subsequent flow of liquid hydrocarbons through water-wet clays requires displacement of at least some of the pore water. For liquids of intermediate dielectric constant (ε = 20-35) such as acetone, glycol and alcohol, this displacement is relatively easy since they are mutually soluble with

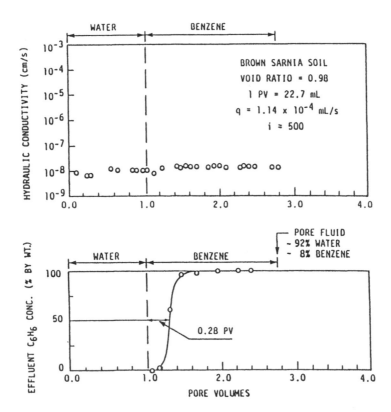

Figure 11. Hydraulic conductivity and effluent concentration versus pore volumes passed through soil sample (Fernandez and Quigley, 1985)

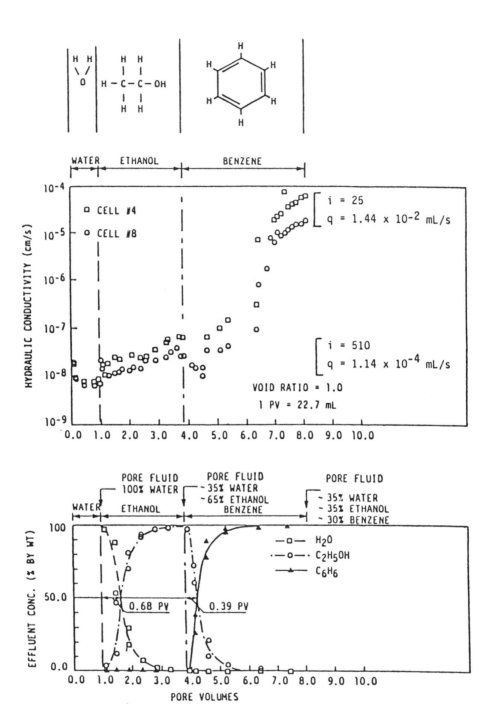

Figure 12. Hydraulic conductivity and effluent concentrations versus pore volumes for sequential permeation by water, ethanol and benzene (Fernandez and Quigley, 1985)

water. Rapid increases in k occur when these liquids are permeated through soil as discussed, for example, by Anderson and Brown (1981, 1983).

For insoluble, non-polar liquids such as benzene, aniline and xylene (ε = 2) displacement of the water is much more difficult and our work indicates that rapid increases in k do not occur even when the effluent from the soil consists of 100% benzene as shown on Figure 11. For this particular test, the pore fluid remaining in the soil sample after testing contained 92% water and only 8% benzene. Flow is believed to have occurred through macropores or channels in the soil with no increase in hydraulic conductivity.

If, however, an "association fluid" such as ethanol is first passed through the water saturated specimen, much of the water is easily replaced as shown on Figure 12. For the test illustrated, the pore fluid remaining after 3 pore volumes of ethanol were passed through consisted of 35% water and 65% ethanol, the effluent consisted of nearly 100% ethanol, and the hydraulic conductivity had increased from 1×10^{-8} cm/s to about 5×10^{-8} cm/s. Subsequent permeation with benzene produced a dramatic increase in k to about 5×10^{-5} cm/s. At the end of testing the effluent consisted of 100% benzene whereas the pore fluid remaining in the soil contained only 30% benzene, which had replaced only ethanol which was reduced to ~ 35%.

In summary, it appears that industrial liquids of low dielectric constant may produce dramatic increases in the hydraulic conductivity of a water saturated clay liner if mutually soluble "association liquids" such as acetone, glycol or alcohol are present. The accidental dumping of such liquid waste over a 1 m thick clay liner could conceivably create a significant, highly conductive "hole" in the liner.

7 CONCLUSIONS

A few conclusions that might be drawn from this brief overview of engineered clay barriers are:
1. The problem of chemical migration is very complex: the chemistry is complex, the control of hydraulic conductivity is complex, decisions concerning compaction are complex and the interrelationships are very difficult to comprehend.

2. In practice it is very difficult to separate a practical test from a research test because in most cases it seems necessary to have good chemical control.
3. Finally, for a sanitary landfill liner of only 1 to 2 m thickness, the accidental dumping of industrial hydrocarbon liquids may be the greatest danger to the integrity of the liner. An increase in hydraulic conductivity from 10^{-8} cm/s to 10^{-3} cm/s over even a small area of the liner might be considered equivalent to a large hole, thus allowing contamination of any underlying ground water.

ACKNOWLEDGEMENTS

This work has been largely funded by the Natural Sciences and Engineering Research Council of Canada. This particular contribution is an updated version of a position paper presented at a Working Seminar on the Design and Construction of Municipal and Industrial Waste Disposal Facilities, sponsored by the Canadian Geotechnical Society and the Consulting Engineers of Ontario, Toronto, June 1984.

REFERENCES

Anderson, D. and Brown, K.W. 1981. Organic leachate effects on the permeability of clay liners. Soil and Crop Sciences Department, Texas A & M University.

Brown, K.W. and Anderson, D.C. 1983. Effects of organic solvents on the permeability of clay soils (EPA-600/2-83-016). U.S. EPA Municipal Environmental Research Laboratory, Cincinnati, Ohio (National Technical Information Service No. PB83-179978).

Chhabra, R., Pleysier, J. and Cremers, A. 1975. The measurement of the cation exchange capacity and exchangeable cations in soils: a new method. Proc. Int'l Clay Conference, Applied Publishing Ltd., Wilmette, Illinois 60091, U.S.A.: 439-499.

Crooks, V.E. and Quigley, R.M. 1984. Saline leachate migration through clay: a comparative laboratory and field investigation. Can. Geotech. Journ., 21: 349-362.

Daniel, D.E. 1984. Predicting hydraulic conductivity of clay liners. Journ. Geotech. Engrg., A.S.C.E., 110(2): 285-300.

Desaulniers, D.E., Cherry, J.A. and Fritz, P. 1981. Origin, age and movement of pore water in argillaceous Quaternary deposits at four sites in southwestern Ontario. Journ. Hydrology, 50: 231–257.

Fernandez, F. and Quigley, R.M. 1985. Hydraulic conductivity of natural clays permeated with simple liquid hydrocarbons. Can. Geotech. Journ., 22: 205–214.

Goodall, D.E. and Quigley, R.M. 1977. Pollutant migration from two sanitary landfill sites near Sarnia, Ontario. Can. Geotech. Journ., 14: 223–236.

Griffin, R.A., Cartwright, K., Shimp, N.F., Steel, J.D., Ruch, R.R., White, W.A., Hughes, G.M. and Gilkeson, R.H. 1976. Attenuation of pollutants in municipal landfill leachate by clay minerals: Part I – Column leaching and field verification. Illinois State Geol. Survey, Environmental Geology Notes No. 78, 34 p.

Mesri, G. and Olson, R.E. 1971. Mechanisms controlling the permeability of clays. Clays and Clay Minerals, 19: 151–158.

Michaels, A.S. and Lin, C.S. 1954. Permeability of kaolinite. Industrial and Engineering Chemistry, 46(6): 1239–1246.

Mitchell, J.E., Hooper, D.R. and Campanella, R.G. 1965. Permeability of compacted clay. Journ. Soil Mechanics and Foundations Div., A.S.C.E., 91(4): 41–65.

Quigley, R.M. and Rowe, R.K. In press. Leachate migration through clay below a domestic waste landfill, Sarnia, Ontario, Canada: Chemical interpretation and modelling philosophies. Proc. 5th Int'l Symposium on Industrial and Hazardous Waste, Alexandria, Egypt, June 1985 (ASTM).

Quigley, R.M., Crooks, V.E. and Yanful, E. 1984. Contaminant migration through clay below a domestic waste landfill site, Sarnia, Ontario, Canada. Proc. 1984 Int'l Symposium on Groundwater Resources Utilization and Contaminant Hydrogeology, Montreal, Quebec, Vol. II: 499–506.

Quigley, R.M. and Ogunbadejo, T.A. 1976. Till geology, mineralogy and geotechnical behaviour, Sarnia, Ontario. In: Glacial Till, an interdisciplinary study. Ed. R.F. Legget. Royal Society of Canada Special Publication No. 12: 336–345.

Rowe, R.K. and Booker, J.R. 1985. 1-D pollutant migration in soils of finite depth. Journ. Geotech. Engrg. Div., Amer. Soc. Civil Engineers, 111(4): 479–499.

Rowe, R.K., Caers, C.J., Booker, J.R. and Crooks, V.E. 1985. Pollutant migration through clayey soils. Proc. XI Int'l Conf. on Soil Mech. and Fdn. Engrg., San Francisco, Vol. 3: 1293–1298.

Yanful, E. and Quigley, R.M. 1986. Heavy metal deposition at the clay/waste interface of a landfill site, Sarnia, Ontario. Proc. Third Canadian Hydrogeological Conference, Saskatoon, April 1986, 35–42. Also Univ. of Western Ontario, Faculty of Engineering Science Research Report GEOT-7-86, 14 p.

Environmental Geotechnics and Problematic Soils and Rocks, Balasubramaniam et al. (eds)
© *1987 Balkema, Rotterdam. ISBN 90 6191 785 9*

Environmental protection of human settlements from technological hazards
Report on the state-of-art in the development of industrial towns in India

R.C.Gupta
School of Planning and Architecture, New Delhi, India

ABSTRACT: Environmental protection of human settlements against technological hazards associated with major development projects, particularly in areas of mining and power generation undertaken as part of national economic development programmes in developing countries, is a matter of deep concern for physical planners and planning administrators alike. It is a new area wherein research and technological innovation should go side by side with an integrated approach to physical planning to enable a pattern of development of human settlements which would provide a safe living environment on the one hand, and minimize areas of environmental conflict on the other. The paper discusses some of the key issues involved, and outlines the present-day efforts for such an integrated approach to planning and development of human settlements in the context of two major mining-cum-power development projects undertaken in India, namely Singraulli in Central India and Neyveli in the South, the former still largely in the planning stage, and the latter already established but contemplating major expansion.

INTRODUCTION

Land is the platform of all human activities, and wherever activities take place, people congregate, establishing the process of urbanisation and giving rise to problems usually associated with unplanned urban growth and environmental degradation. This is a phenomenon encountered irrespective of the nature of the activity, and gets more pronounced in a developing economy wherein large-scale industrial development projects are established as part of the national social and economic development endeavour, leading to proliferation of both the activities and the people through a series of chain reactions.

The intensity and scale of operation create not only various forms of geotechnic stresses but also various environmental disorders threatening the human settlement system, which is concommitent to the activity itself and is required for its sustanance. The two are intimately tied up as 'cause and effect' in a two-way process of industrial growth and physical development. The concern for environmental protection of such settlements in its macro-and-micro-dimensions assumes greater significance in

large-scale efforts for natural resources exploitation and resultant industrial activities. Coal-mining and thermal power generation are two key inter related activities which are basic to the national industrial development programme of a developing country like India, and both are more or less location-tied, the former for obvious geological reasons and the latter due to the economics of production costs. Ancillary, associated and auxilliary industries also get located close by, drawing upon the availability of power and industrial infrastructure creating eventually a coal mine thermal power-based industrial complex over a period of time. This is a scene commonly withnessed in India where the first public sector steel plants which were located in the proximity of the mineral-rich coal-belt of the Chota Nagpur plateau eventually led to the establishment of a multitude of other industries in the region, now known as the "Ruhr of India". There are other examples as well of such industrial agglomerations around a coal power-based nucleus.

Neyveli, near Madras in Tamil Nadu state of India, has been on the national energy scene for the last 25 years as a major lignite mining-cum-power generation complex. Singrauli, in Mirzapur district of Madhya Pradesh, where a series of super-thermal power projects have been planned to draw upon the massive coal resources of the Singrauli coal-fields, is fast emerging as the energy capital of the country.

The technological hazards of such massive mining-cum-industrial activities in these two areas are too pronounced in their impact on human settlements (the former already existing and the latter in the planning and development stage) to be covered in all their ramifications. At the macro-level intensive coal mining activities, have major polluting effects, viz. air, water and noise pollution, problems of mine drainage, acid rain, land subsidence, adverse effecs on ground-water, besides visual disfigurement of the landscape, adverse impacts on cropping pattern and ecology, and dislocation of villages, to name just a few of the human problems. They have also created problems with regard to effective handling of the huge over-burden consequent to mining operations. Power development programmes also lead to major pollutants, resulting in air, water, and noise pollution, problems connected with ash-dumping and disposal, fly-ash menace, and visual disorder consequent to transmission corridor networks, railway sidings and merry-go-rounds (MGR) required for haulage of coal to the plants, etc.

How far human settlements, both the existing ones and the new ones which are envisaged for these areas, could be effectively protected from such technological hazards would depend on the in-built protection measures taken in the plant designs, location and siting of the various operational uses and requirements, vis-a-vis the plant township and other settlements, and the protection measures taken in the planning and development of the settlements themselves.

Based on studies carried out by the author during recent years for the two areas as part of overall planning consultancy, and drawing upon other studies, research notes, and documents made available by concerned agencies, it has been possible to prepare this document on the 'state-of-the-art' of the approaches towards environmental protection of industrial towns against technological hazards as an integral component of urban development planning.

I. INTEGRATED PLANNING FOR SINGRAULI-ENERGY CAPITAL OF INDIA

Development brief

The Singrauli Area, falling partly in the Mirzapur district of UP (Uttar Pradesh) and partly in the Sidhi district of MP (Madhya Pradesh) states in India, is experiencing major development activity in the field of power generation, coal mining, and related development. Some important development projects have been programmed for this region, which is emerging as the energy capital of the country. A Special Area Development Authority (SADA) was established for the MP area of the region, and similar action by the UP State Government is also contemplated. The NTPC (National Thermal Power Corporation), CCL (Central Coalfields Ltd.) and UP SEB (UP State Electricity Board) are major partners in this development programme. With a view to devising an appropriate planning framework for this area, a Planning Co-ordination Committee was set up in the Ministry of Energy, with the task of analysing the various development and infrastructure requirements and proposing a plan framework for integrated developemnt. The Central Town and Country Planning Organisation of the Government of India (TCPO), which was represented on this Committee, was entrusted to carry out various planning studies at the regional and local levels and to suggest the overall spatial planning strategy and formulate a structural development plan for the Singrauli Area, keeping both the regional and local requirements in view. It was proposed that the committed and cleared projects should comprise the immediate plan requirements, and the long-term plan should take note of the overall requirements of expansion, both in the mining and power generation sectors.

According to current development programmes, the CCL have targeted a coal production of 76 million tonnes by the year 2001 from their 11 mines prospected in the Singrauli Area, whereas the NTPC have proposed establishing eight super-thermal power projects with a combined capacity of 20,000 MW> In a general study of the development programme, it has been observed that a string of project town-ships would need to be developed in close proximity to the projects to cater for housing and ancilliary needs of the manpower employed in the mines and the power plants. the Special Area Development Authority in Madhya Pradesh has further envisaged developing a Service

City around the existing Waidhan town to cater for the ancilliary growth which will take place consequent to these major development activities, and also to accommodate the supporting population and services in this area. It needs hardly be pointed out that such a massive coal production and power generation programme along with its associated operational requirements would have considerable environmental and ecological implications, and also would pose physical constraints for the orderly development of the townships complex and infrastructure. To overcome this, it has been rightly considered essential to devise a spatial planning strategy for the Singrauli Area in which various activities and conflicting land uses could be accomodated appropriately, with adequate environmental safe-guards.

Need for re-organising the operational network

The projected programmes for NTPC projects, apart from identifying eight plant sites, indicated a network of wide transmission corridors, and extensive zones for ash dumps. Ash dumping along Rihand reservoir was also proposed. Despite the care which may be taken by NTPC by constructing suitable embankments and retaining walls, it is feared that large-scale dumping of ash along the banks of the Rihand reservoir, an impounded reservoir which presently is the only source of water in the area, may eventually lead to silting, and may also pose environmental and ecological hazards. Similarly the huge over-burden from the coal mines, which requires as much as 55.50 km^2 of land for dumping, is also going to be a serious environmental problem for large-scale habitation in the area. On a broad estimate of the man-power requirements, and also the supporting population for the projects and the SADA's Service City, it is estimated that a population of nearly one million would be living in this area amidst a scatter of such operational zones. While technological innovations are imperative for proper utilisation of the thermal-ash, it would be essential to reorganise the operational network of the project activities suitably, so that environmentally safer zones are identified and reserved for townships and infrastructure. This factor has assumed key significance in the planning for the Singrauli Area.

Approach to planning

In order to analyse the environmental impacts of the various projects and operational activities of the NTPC and CCL in this area, and to cater for the various land requirements, both for operational needs and for the township's complex and infrastructure, consistent with desirable standards for a living environment, it has been necessary to organise studies at two levels (a) at the macro level; and (b) at the micro level. The former would lead to a strategic plan for the overall region, the latter to the structural development plan for the comprehensive planning area. For purposes of studies at the macro level, a planning region has been tentatively delineated by the TCPO team, encompassing the entire Singrauli coal belt and contiguous areas covering Sidhi district in MP and Dudhi and Roberts - gunj tehsils in UP. This would be the area for regional studies, both physical and socio economic, which are intended, to suggest guidelines for future development activities and for an environmental protection programme for urban and rural settlements. Studies at the micro level will be more intensive and related to the area under comprehensive planning, and will cover the part of the coal belt which has been prospected and is being operated, project areas of the NTPC, the developments which are likely to gain momentum consequent to such activities (particularly on the UP side), and the area requirements of the various supporting activities for the immediate development programme of the NTPC and CCL. Tentatively this intensive study area has covered approximately 1800 km^2 (see figure 1).

Coal & power development programme

Coal potential and programme

The Singrauli coalfields are estimated to contain about 7,500 m. tonnes of coal of which 2,700 m. tonnes is estimated to be quarriable. There are two main coal horizons in the major parts of the coal field - Purewa and Turra. The total thickness of exploitable reserves in the major sector of the coalfields varies from 24 to 46 meters. The gross calorific value of the worst coal in this field is 3300 koal/kg (Purewa top/bottom) and that of the best coal 5300 koal/kg (Turra seam in Dudhi-chua block). the ash content varies from 17 to 46 percent, and the moisture content from 60 to 10 percent.

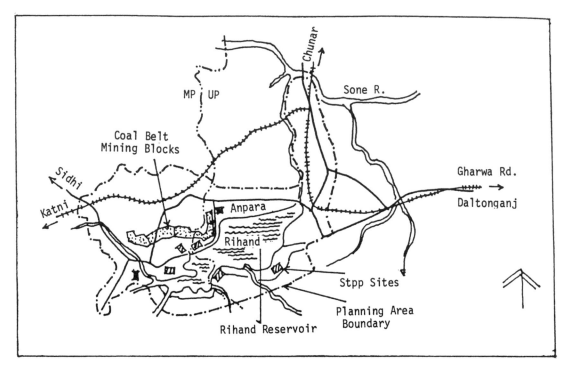

Fig. 1. Singrauli Complex: General Location

The low calorific value and high ash and volatile matter content makes this coal unsuitable for industrial use, other than power generation.

For the present, the ratio of coal to over-burden ranges from 1:2.8 to 1:3.9. The coal authorities visualise that ultimately the entire Singrauli field may have to be quarried with a coal to over-burden ratio of 1:6. The coalfield has been divided into 11 blocks with a combined annual capacity of 86.5 million tonnes per annum over a lifespan varying from, 16 to 40 years.

The entire field drains practically into the Rihand reservoir. During the process of mining these over-burden dumps, natural processes are likely to interfere with normal drainage flow patterns or carry silt into the Rihand reservoir. As it is, even though Rihand reservoir has been functioning for nearly 20 years, there has not been any siltation. Large-scale mining in this area is likely to cause problems which have to be taken care of in the planning of the mining operations, dumping of over-burden, and associated activities.

In addition to coal mining operations, a coal gasification plant and a coal-based fertilizer plant have been projected for this area which will also give rise to additional township and infrastructure requirements, besides the requirements of the mines. Other coal-based industrial plants are also a future possibility for this coal-rich area.

Power development potential

Planning and development of the coal mines are expected to match the demand of coal for power. With this new source of energy, it is possible to sustain about 25,000 to 30,000 MW of power in the area for a substantial period to meet the immediate demand of the country in the coming 25 to 30 years. This takes into consideration the availability of water for thermal power from Rihand reservoir to be supplemented possibly by the construction of other reservoirs in the nearby areas.

The total projected capacities of thermal power development projects so far constructed or under construction and under active consideration for clearance in this area are as follows:

Sl.	Name of Station	Total capacity MW
1.	Renusagar privately owned by Birlas	310
2.	Obra of UPSEB	1500
3.	Shaktinagar STP Singrauli I of NTPC	1900
4.	Waidhan STP of NTPC	3260
5.	Bijpur STP Ssingrauli II of NTPC	3000
6.	Anpara STP- A,B,C	3100

Apart from these existing or under-construction and contemplated plants, there are six to seven new sites identified for future plants. the power stations to be constructed in this region are expected to serve the demand of the northern and western regions of the country.

Rihand waters & utilisation

In this coal-rich region, UPSEB in the early sixties created Govind Ballabh Pant Sagar (Rihand reservoir) by constructing the Rihand Dam, primarily for the development of hydro power. The reservoir was originally designed for an annual yield of 6.92 m acre feet. It has a catchment area of 5148 square miles, and a maximum water spread of 180 square miles.

According to studies carried out by the CEA (Central Electricity Authority) and the NTPC, it has been estimated that thermal power stations with a total capacity of about 20,000 MW could be installed in the Singrauli coalfield area from considerations of water availability, leaving a balance of 1,202 million acre feet. per annum of water for utilisation for other purposes, as required, including exisiting and future industries.

The NTPC have also estimated, on the basis of preliminary studies, that several of their power stations would have to utilise closed cyle cooling with cooling towers. This means that Rihand waters are proposed not only to serve as the source of water for the NTPC projects, but to constitute the only source of water supply for all other requirements, including human habitation. Conservation of this reservoir lake and its protection from any polluting effects is an essential step to be taken.

General profile of the Singraulia area

The area is rugged with elevations varying up to about 300 metres above mean sea level. Towards the Rihand dam, i.e. the downstream end of the reservoir, the area is extremely rugged, and virtually no level space is available for any large-scale construction. It is only towards the upstream end of the reservoir, again between the coal-bearing plateau area and the reservoir, that we get patches of level area which have been identified by the NTPC as proposed sites for the pwoer plants and related development.

Soil and geology

The soil in the area consists of sandy silt with a soft vareity of sand rock. At places shales and flaky limestones are also available. Near the Rihand reservoir, a good quarry site for granite exists. The soil and sub soil is light, brown in colour on the top and yellow immediately below. The thickness of the soil and subsoil horizons vary from nearly 2 metres to about 10 to 12 metres. While the upper layer of the soil is sticky when wet, the lower layers are sandy. Sandstones being the predominant rock type in the area, the derived soil is mostly sandy. The soil zone is intermixed with calcareous nodules resembling "Kankar", the size of which, though mostly around a few millimeters, may be as big as 2.5 cm in places. Practically the entire area is sedimentary formation. Sandstone is the predominating rock type, and is encountered practically in the entire region. The underlying rock, though predominantly compact fine-grained sandstone with argillaceous matrix, is generally decomposed up to 10 to 13 metres below the soil cover. The area is practically free from any faulting or geological distrubance.

Seismosity

The Singrauli area falls in zone III of the seismic zoning map of India, prepared by the Indian Standards Institution, and bears the code IS-1893-1970.

Meteorology

The climate is typical with a severe summer. The temperature in summer goes up as high as 48° C in May-June months, and the average minimum summer temperature is 21° C. In winter, the average maximum and minimum temperatures are 21° C and 4° C from November to February. The rainy season is generally from July to September. The average annual rainfall is about 1000 mm, out of which the rain season contributes 95%. The predominant wind directions are from west to east and south to north.

Existing infrastructure

The infrastructural facidlities in the shape of access by rail, road or air is far too inadequate. The nearest air-field is at Varanasi about 225 km away from the centre of the coalfields. A railway line, even though it reaches up to Ranukut, is not yet connected by fast trains. A railway loop line is currently under construction from Kerala road, skirting the mines, for evacuating the coal from the area. The highway is absolutely inadequate to catering for the movement of the coal even, not to speak of the movement required for large-scale industrial goods of several organisations like the NTPC, the UPSEB, UP Irrigation, the MPED the Central Coalfields, and other private organisations. To make matters worse, on either side of the existing road, haphazard growth has started creating traffic bottle-necks and problems for planned development in the future. The area has no telegraph office and the nearest telegraph office is about 50 km away. Telephone, telex and postal facilities are very inadequate. The social infrastructure, viz. health and educational facilities have yet to be developed to the minimum acceptable standards.

Socio-economic profile

The planning region as delineated comprises an area of 17,687 km^2, spreading over 5 tehsils, Gopadhanas, Deosar and Singrauli in the Sidhi district of MP and Robertsganj and Dudhi tehsils of Mirzapur district in UP. It has a population of roughly 17,63,833 living in nine towns and 3306 villages. The nine towns collectively have a population of 1,13,812 according to the 1981 census.

Some of the towns are potential industrial centres, and a town-wise development potential study will need to be carried out to identify their future development potential and pattern in the context of large-scale development activities being considered for the planning area.

The planning area as delineated covers approximately 1800 km^2 of which approximaetly 840 km^2 is on the UP side and 960 km^2 on the MP side. Of the area on the UP side, a substantial component is

the Rihand reservoir. The planning area as delineated on the MP side, comprising 960 km^2, has 133 villages with a population of 67,940 according to the 1981 census. Many of these villages would fall within the urban development zone, and while some of them will have to be relocated, several others could be integrated within the township complex along with requisite environmental improvement measures. With regard to villages which fall outside the urbanisable zone, specific environmental improvement and infrastructure development measures will also have to be devised, as these villages will come within the direct influence of the large-scale urban development programme. As most of the urban development programme falls within the SADA (MP) area, the villages on the MP side will have closer socio economic interactions with the future urban-industrial complex.

Industries

Amongst the industries that are already functioning in the region are the Hindustan Aluminium Corporation Limited (HINDALCO), Kanoria Chemicals, Churk, Chunar, Kajarhat and Dalla Cement factories. With the mineral resources available both in MP and UP, there are possibilities for further large-scale industrial development in this area. Heavy water projects, explosive factories, coal based fertilizers, and coal gasification plants are some of the other possible industries which may be established in this area.

Agriculture

Agricultural activities in the area are very limited and are primarily on the fringes of the lakes and tributaries. There is no irrigation, and produce consists primarily of maize and a small quantity of wheat of inferior quality, and is consumed locally. Cultivation of cash crops, like vegetables or fruits, are not much in evidence.

Rehabilitation of tribals

Excepting for the towns of Waidhan and Robertsganj, the polulation in the area consists primarily of Adivasi (tribal) people. Unfortunately some of these people have already faced eviction twice-once during the construction of the Rihand dam in the late fifties, and secondly during the construction of NTPC's

Singrauli Project and UPSEB's Anpara
Project and Colliery development work.
With further development envisaged in coal
and power sectors in the area, a third
shifting of these people cannot be
completely ruled out. It will, therefore,
be necessary to provide planned model
villages for final settlement of these
people in an integrated manner. In
addition, a certain amount of eviction and
resettlement will also be involved for the
development of mines in this area.

Analysis of land requirements

On the basis of individual manpower
requirements indicated by the NTPC and CCL
for their respective projects, and making
allowance for 33% as the service/
supporting population, it has been
analysed that ultimately the NTPC and CCL
projects would account for a population of
778,000 to be accommodated in the
Singrauli Area. Besides, the central
Service City to be developed by SADA (MP
Government) to provide higher level city
functions and to cater for ancillary and
auxillary industrial growth taking place
in this area, would on a broad estimate
account for a population component of
200,000. The overall population to be
accommodated in the township complex would
thus be approximately 986,100 or say 1.00
million. The township complex and
infrastructure will have to be conceived
therefore on a metropolitan scale.

Individual project land requirements for
both the NTPC and CCL would also cover
specific area requirements for operations
uses, land for which has to be allocated
suitably in the planning area. These
include besides main plant, areas for ash
dumps, railway MGR, miscellaneous uses in
respect of the NTPC projects, and
extensive zones for mine infrastructure
and over-burden/top soil dumps for coal
activities.

For the development of individual
townships, an overall density of 75
persons per acre has been considered
appropriate, keeping in view the blastic
impact hazards and the network of power
corridors, which would limit most of the
construction to two storeys only.
Accumulative requirements of land for the
township complex work out to approximately
53.43 km^2, a small portion of which would
account for smaller township complexes for
power projects to be located across the
Rihand reservoir on its south bank. The
main township complex would thus cover
between 45 to 50 km^2, besides reservation

for land for ancillary industries and
associated industrial plants, viz. coal
gasification and coal-based fertilizer
plant. According to the development
concept, the township complex will consist
of a string of individual project
townships to be developed in phases in an
integrated manner.

Provision for the service population has
to be kept within individual township
requirements and each project township
will have to be developed to cater for the
respective service components along with
the project requirements. Besides, the
balance service component, part of which
would obviusly be drawn from the villages
falling within the township zone, would be
catered by the Service City to be
developed by SADA, which will accommodate
larger city level functions and the
concommitant population. The idea is that
the total complex should be developed on
an integrated basis with all trunk
services and land development
responsibilities to be assigned to SADA
(MP), and individual project management
will develop its township on land
earmarked/allotted for a particular
project in the overall plan, according to
specified development guidelines.

Environmental pollution factors and
safeguards

The large number of construction
activities associated with mining
operations, removal and dumping of
overburden, actual mining operations and
transportation of coal, coupled with the
establishment of several Super thermal
power stations is bound to cause serious
problems and hazards due to environmental
pollution of the atmosphere and water, and
the consequential effect on vegetation and
animal life in the area. The operation of
such magnitude will also cause
considerable ecological impacts and
implications in the area. This
environmental pollution is likely to be
caused due to:

a) Rejection of the heat load into the
   Rihand lake, especially from the
   stations which are cooled by a lake
   surface cooling system.
b) Atmospheric pollution caused by flue
   gas emission from the chimney.
c) Pollution caused by other obnoxious
   plant effluents.
d) Pollution caused by operation and
   maintenance of townships.
e) Pollution caused by decanted and
   seepage ash water.

f) Pollution caused by dust raised from the mining activity in the coal mines.

g) Pollution of the lake caused by overburden and the consequent siltation of the lake.

h) Noise pollution.

The source of pollution would need to be examined in detail item by item and preventive measures taken to limit their consequential effects, even though it may not be possible to completely eliminate them. The extent to which the pollution effects from the above mentioned sources can be controlled and eliminated would have a great bearing on the quality of life despite the best planning efforts with regard to development of townships in this area. With regard to the thermal pollution emanating from the power projects, the NTPC propose to ensure that all the stations are not put on lake cooling, which incidentally will help in the reduction of the thermal pollution of the lake. Adverse effects on the marine life are also expected to be reduced by minimising the heat effects on the lake water. However, further detailed studies on these aspects are necessary.

So far as atmospheric pollution consequent to emission from the plants is concerned, the NTPC propose to adopt multiflue construction, and by installation of effective electrostatic precipitator systems in their projects, which have been found to be quite efficient expect to substrantially ontrol atmospheric pollution. Further improvements could also be considered to keep the dispersion concentration to a minimum. The effeuent frodm the individual main plant will be disposed of after neutralisation.

With regard to mining activity, this is likely to bring in a huge quantum of coal dust in the area, causing serious pollution problems in the inhabited zones. It is imperative to undertake dumping of coal over burden systematically, otherwise it may also affect the natural drainage system. The coal dust could be confined through occasional spray and the plantation of bushy trees. Protection of the Rihand lake from the monsoon run off passing over loose soil of the dumping areas is also necessary, and adequate safeguards will need to be taken by the coal authorities to check the flow of silt into the Rihand reservoir.

Similarly proper dumping and utilisation of ash from the power plants will have to be done on a scientific basis so that the monsoon run off does not carry the ash into the reservoir, creating not only silt but affecting the quality of water. The location of ash disposal sites also has to be properly thought of as much away from the reservoir bank as possible. Research in the field of utilisation of the ash for construction activities, or other similar uses will have to be undertaken, as the accumulation of ash otherwise would cause a formidable problem. Already the use of fly-ash in building activities has been established and some more investigation would be necessary to use the fly-ash to the maximum possible extent. Transportation of the fly-ash however is a remote possiblity unless it is compacted.

Enviornmental pollution is also cused by noise, some of which is occasional, viz. steam blow off during commissioning and testing periods or puffing of the boiler. As these are occasional, they may be tolerated. However other types of noise created by equipment like diesel generating sets, compressors, fans, cooling towers etc. would need to be minimised. Besides other measures which are in built in the latest technical know-how for installation and operation of such activities, suitable buffer plantation around such noise sources will substantially reduce the noise effects outside the plants.

Whereas every individual power station is expected to have adequate pollution control to limit the emission of pollutants within the prescribed standard, it would also be necessary to ensure planned location and development of power stations in the area as well as mining activities, thereby minimising the accumulation of pollutants to the extent possible. Co ordinated study of the pollution aspects is necessary. The development of a major urban-industrial complex which in actual fact will be in the form of a cluster of several townships integrated with each other through transportation and communication networks, will also be a source of environmental pollution unless proper programmes are undertaken for the disposal and management of solid waste.

Besides the above major industrial activities, there would be a possibility of other industries also getting established in this area, some of which could be directly related to ancillary industrial growth. There is a possibility for development of a coal gasification plant, a coal-based fertilizer plant and brick manufacturing industry in the area. The availability of rock phosphate and silica sand deposits in the Bundelkhand region are the main resources which may attract such industries. In addition animal husbandry, dairy farming and massive plantations of a particular type of long fibre-yielding tree may be found suitable in the region. With these plantations, there are also prospects of developing paper industry. These ancillary industries may also create environmental pollution, and proper safeguards would be necessary, particularly as these will be within the fold of the integerated complex.

Locational imperatives for operational uses & urban form

Both the CCL and NTPC have estimated extensive land requirements for respective opertional uses and activities, and owing to certain technical considerations, the location of such uses and activities has to be considered in close relation to the mines and the projected and committed power plant sites. While there may be certain possibilities for rationalising the location of such operational uses and activities, particularly in so far as the power generation programme is concerned, viz. ash dumps and network of transmission corridors, the operational activities in respect of coal development might not permit scope of major adjustments owing to their generally fixed nature. These have constituted operational constraints to basic locational decisions for urban development planning in the area. These are indentified as follows:

1) CCL's coal overburden and mine infrastructure requirements would entail huge land reservations close to the mining blocks. Consequent to the latest thinking, however, the CCL had indicated some reduction in land requirements for coal overburden as a result of the improved mining technology now being adopted.
2) The staggered location of presently committed and projected super thermal power plants around the periphery of Rihand lake and vast areas coming under each plant and their operational requirements.
3) The location of extensive ash disposal areas for each STPP.
4) A network of power transmission corridors ranging from 1/2-2kms in width and its configuration emanating from each STPP, and connecting to the regional power grids.
5) A network of MGR system required for coal transportation from mining areas to the individual plant sites.

6) The alignment of broad-guage railway line for the safe movement of passenger traffic as also to cater to extensive coal movement, requiring segragation of railway facilities for pasenger and coal traffic.
7) Railway siding facilities for each plant.
8) Existing village settlements and their future development and rehabilitation of the oustees from villages acquired for development of CCL and NTPC projects is yet another factor to be considered in formulation of an integrated development plan. This is a serious socio economic problem, besides having physical and environmental implications related to the development of human settlements.
9) Environmental and ecological problems consequent to pollution hazards are likely to be caused by coal extraction and haulage, pollution of Rihand waters owing to ash water draining into the reservoir, atmospheric pollution due to dust and foul gasses, denudation of forest, mine drainage, acid rain and several other factors yet unknown in terms of the intensity of their environmental impact.
10) Project requirements of township location in close proximity to plants requiring sufficient reservation of developable land to cater for expansion.

While the above list is not comprehensive, these major constraints while working in relation to each other have posed several problems and have had a profound bearing on the development and infrastructure facilities.

The predominant wind directions being from west to east and also from south to north, within the locational constraints of the operational activities outlined above, an environmentally safe zone has to be so identified for purposes of the townships as to be located close to the respective work centres and also enable a properly integrated urban form for the total complex. A generally linear form alighed north-east to south-west and taking within its fold the existing Shakti Nagar and Jayant townships appears to be a logical choice.

Such a linear urban form should be amenable to a non-polluting electric trolley mass transit system connecting all parts of the township complex with the city centres and to the work centres whether they are the industrial plants, coal mines or the power projects. The linear form would also enable phased development of the individual townships by the project authorities and their integration with the Service City to be developed by SADA.

The Basic Plan Frame

The TCPO Planning team during 1982-83 carried out in-depth planning studies along with its associates after analysing the operational and infrastructure requirements of the individual development programmes, environmental problems likely to be created by the coal and power development activities, safeguards necessary for creating a healthy environment for development of individual components and their linkages, through several site visits, consultantions and discussions with all concerned agencies. As a result of this an overall development plan framework was evolved, discussed and debated at various forums thereby resolving several issues relating to integrated development. This is illustrated in figure 2.

The plan provides a development framework for township and infrastructure development, establishing landuses for township, plant and associated activities of both CCL and NTPC taking note of the commitments already made and the discussions held regarding land requirements for various activities. An alternative development framework was also evolved keeping in view latest thinking of the CCL in regard to reduced land requirements for over-burden dumps, thereby releasing more land area near the mining infrastructure zone for township development. This is also illustrated in figure 3, although from overall environment considerations, such close proximity of the townships to the coal mining and operational zone, is not considered to be conducive to a safe living environment. Both the conceptual proposals are presently under consideration of the Coordination Committee for integrated development of Singrauli Area. Centain policy issues have emerged and are highlighted here.

Control on development

The primary issues on which immediate action would be necessary relate to exercising development control both in the context of macro and micro level areas, their enforcement and management, and undertaking land and infrastructure development programmes in the planning area to enable integrated development of the urban complex and its components in a systematic and phased manner. A suitable mechanism with statutory powers covering the entire area and both private and public developments would need to be devised immediately.

Environmental protection measures

In regard to the enviormental protection and control measures, while the NTPC projects will get clearance from the Department of Environment and shall be governed by the safeguards to be observed by them, proper afforestation, development of green buffers and green belts wherever indicated on the development plan should be the joint responsibility of SADA, CCL and NTPC. In regard to specific enviormental problems likely to be experienced in the area as a result of coal overburden dumping and ash disposal, and how to minimise their adverse impacts, some recommendations have been put forth by the planners and outlined in the following paragraphs:

Mining and overburden

The over-burden dumps in most of the mine projects are left unattended and contribute significantly to the problem of soil erosion. Most of the dumps have 30° angle. No terracing or protective embankments are built. As a consequence serious aesthetic and environmental degradation problems are created. Over-burden dumps can be successfully vegetated for environmental conservation. The borehole and mine drainage water could be used for growing vegetation on over-burden dumps. A Master Plan for conservation and plantation of trees on over-burden dumps with specific funds earmarked for this purpose is most desirable alongwith funds for all inputs required to facilitate growing of plants.

Acid mine drainage would require immediate attention. Acid mine drainage of Gorbi mine has PH 2 i.e. highly acidic and creates problems of pollution as it is discharged like in any other mine drainage. The volume of Gorbi acid mine drainage is approximately 5000g/min. A monitoring programme on mine drainage water needs to be initiated to evaluate its physical and chemical properties.

Serious problems of atmospheric pollution have already arisen in the area and evident in Singrauli where discharge from Renusagar Power Plant remains

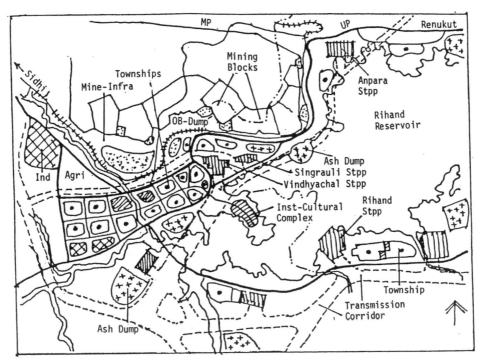

Fig. 2.

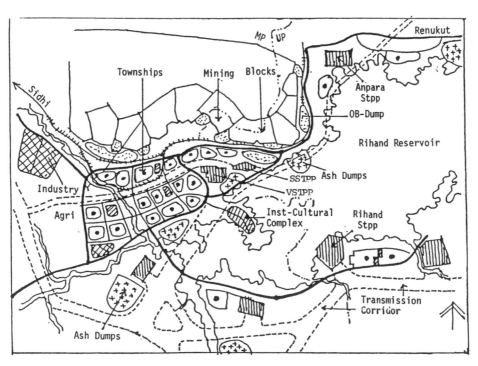

Fig. 3.
Singrauli Basic Plan Framework: Alternative Concepts

confined in this area which is surrounded by medium size hills on all sides. The coal transport by roads can also be a cause of serious road hazard. These problems need to be tackled on a priority basis to the extent possible.

Ash disposal sites

It may be mentioned that ash disposal, reclamation of ash disposal sites and other related aspects should be given as much importance by NTPC as any other aspect concerned with power generation. It may be added that reclamation of an existing abandoned site would provide valuable experience in developing plans and proposals for reclaiming ash disposal sites in future. A timely relcamation of the abandoned site would set an example for other projects in this high industrial activity area, to minimise environmental disruption.

In regard to the current thinking for locating the ash disposal area on the shallow waters of Rihand reservoir, the planners have observed that there are considerable disadvantages which may cause serious environmental degradation and disruption of the ecology and fisheries potential of the Govind Sagar lake. Ecologically the lake shore is most vulnerable and highly fragile. Various authorities have warned against the reclamation of the lake shores for construction of ash pond or for any other purpose. For example, UNEP Industry and Environment Guidelines Series Vol. I: Guidelines for Assessing Industrial Environmental Impact and Environmental Criteria for siting of Industry, UNEP 1981, has indicated that "wet lands, estuaries and coastal zones are environmentally sensitive areas". Also the Biologist's Manual for the Evaluation of Impact of Coalfired Power Plants on Fish, Wild Life, and their Habitats, US, Department of the Interior states "The close the ash and sludge disposal site is to wet lands, the less suitable the site, particularly if the wet lands are down the gradient of the proposed pond area".

In view of the flourishing fisheries resources of the Govind Sagar lake, it will be a serious mistake to reclaim any part for ash pond or for any other facility. It has been opinioned that the Govind Sagar lake is going to be under progressive stress on more than one account, because of the various developments contemplated in the Singrauli area for coal and power production. Unless very strict measures are taken, the lake and its enormous fishery resources are likely to get spoiled irretrieveably. It may also be recalled that there are going to be a large number of thermal power plants all along the periphery of the lake and if the lake margins are to be reclaimed for ash disposal of all the power plants envisaged in the area, the lake will suffer a major disaster apart from disfiguring the landscape on a permanent basis. Inview of this the planners have strongly recommended that ash disposal sites should be located appropriately away from the lake area for all future plants.

It would also be desirable that the disposal of sedimented ash in the new ash ponds should be carried out in a sectionwise manner to enable reclamation of silted sections progressively without waiting till the end of the life of the ash pond. This would require careful thinking and appropriate designing in advance. It would also be very desirable to recycle the ash transport water to minimise the discharge of liquid effluents into the Govind Sagar lake.

It has also been suggested by the planners that land for ash dumping should not be acquired in one go covering the requirement for the entire life span of the project. It may be expected that some new technology may be developed in the future to utilise fly ash for extensive industrial construction use with the eventuality that the ash once dumped would be evacuated and dumping capacity of a site would be recreated. Technological innovations are already on in this regard, at the governmental level and are being pursued in connected institutions.

Prospects for planned development

The Basic Plan is presently in the process of further refinement through mutual consultation at various committees and forums before it is taken as the final development framework acceptable to all. During the course of finalisation there may be changes consequent to refinement of projections, new and modified requirements of NTPC and CCL primarily relating to their operational requirements, and policy decision at government levels pertaining to investment pattern, sharing of responsibilities for common infrastructural services, institutional arrangements for urban development etc. and allied factors. Whatever be the shape of final things to emerge, it is imperative that a futuristic view be taken of the urban scenario, the short term and long term implications of the several key environmental factors, the need for pooling of resources for creating common infrastructure network particularly in regard to transport, utilities and services, and providing for sustaining investment to cater for the needs of non-plant non-company employment, formal and informal, that will get converged to this area. This can be achieved only through a spirit of sharing of responsibilities, financial and institutional, by the major development authorities functioning in the area.

In this regard rehabilitation of displaced population appropriately, both social and occupational, would require to be considered objectively and dealt on priority basis jointly. It could hardly be expected to be left to the development authority (SADA) to cope with this problem effectively owing to common constraints.

It is also imperative that the planning issues be discussed and resolved objectively and not as a result of expendiency, as every issue reflected as a component of the Basic Plan, is inter-linked as part of the whole. For this purpose, the implication of the operational requirements, viz. area and location of ash disposal sites, network and width of transmission corridors, area and sites for over-burden dumps, and mine infrastructure, as well as conventional constraints for mining townships to be close to pit-head, should be ressessed, giving due consideration to environmental factors. After all it is not a single project area, but a multitude of power and coal development projects where the environmental implications get multiplied manifold. With such a heavy projected investment on the development projects by NTPC and CCL, it is imperative that the hitherto conventional approach to opertional requirements should be discarded and new technology practised elsewhere, be adopted and given a trial atleast, which will no doubt entail technological research in environmentally sensitive areas of power development and coal mining activities. If piecemeal isolated decisions are taken as a matter of expediency, despite all best intentions and despite any amount of integrated planning precepts, the results would be disastastrous, and Singrauli may well turn out to be an environmental blunder and a challenging opportunity for achieving planned development lost to planners and planning administrators.

II NEYVELI LIGNITE COMPLEX – ENVIRONMENTAL PROTECTION MEASURES & ISSUES

General description of Neyveli

Neyveli Lignite Corporation (NLC) integrated mining cum industrial complex is situated 200 kms. south of Madras and occupies an area of approximately 17000 acres (6800 hectares). It is approached by the Cuddalore-Viridhachalam State Highway which passes south of the First Mine and continues to function as the main transportation artery of this region along with the railway line running almost parallel to it. The Neyveli Lignite Corporation's initial temporary colony came up along-side of this road when excavation of First Mine in the NLC Area started in the late fifties. The integrated Complex of Neyveli was inaugurated in May, 1957. The temporary colony of NLC Complex is still under active use. The permanent township was developed on a site towards the north of the First Mine and in the northern sector of NLC Area, outside the lignite belt which extends south-wards even beyond the area presently under occupation of the NLC.

Neyveli Lignite Corporation has a huge area under its jurisdiction. Generally a flat terrain, the NLC land has a natural gradient from north and north-east towards south and south-west, a factor which has come in handy in providing a natural protection to the township zone. Although the lignite belt extends to an area of 260 sq.kms. The area under open cast under the First Mine is 15 sq.kms, the existing area under occupation of NLC covers the Second Mine area also. However, with the opening of Second Mine, its expension, and the Third Mine alongwith related Thermal Power Plant, NLC's activities shall inevitably extend over a much larger area. But all such increase would account for the operative uses of NLC's mining and industrial activity. The integrated complex at Neyveli has the following component units:

1) A lignite mine producing 6.5 million tonnes of lignite per annum which is the only fossil fuel (energy source) in this region;

2) The biggest lignite-fed pit-head power station (600MW capacity) in south-east Asia meeting around 35 per cent of the needs of Tamil Nadu;

3) A fertiliser factory producing urea (46 per cent nitrogen) with an

installed capacity of 1,52,000 tonnes per annum;

4) A low temperature carbonisation factory producing lighite coke known commerically as 'Leco' a most economical and smokelss fuel and a few valuable base chemicals. The lignite coke has now become a popular industrial fuel for industries such as electro-chemical, electro-metallurgical, cement, carbide, tea, etc. replacing scarce furnance oil, coal, coke and charcoal; and

5) A clay beneficiation plant producing washed clay and pulverised clay.

The configuration of mining block (First Mine) under operation, extensive coverage of land under overburden, development of a major factory area with associated industrial infrastructure and operational requirements particularly ash dumps of the Thermal Power Station (TPS-I) and growth of hutment colonies at various strategic places have given rise to certain constraints which have come in the way of coordinated development of the whole area and created wide ranging impacts on the New Town's environment.

Impressed by its performance and as lignite is the only fossil fuel energy source in this region, Government of India, have sanctioned a Second Mine (First stage 4.7 MT per annum, and second stage 15.0 MT per annum of lignite) and Second Thermal Power Station, TPS-II (First stage 630 MW and second stage 1470 MW). The Second Mine development is in an advanced stage and lignite production for Second Thermal Power Sation is in progress. Investigation is on for the Third Mine, a 1500 MW Power Station, expansion of First Mine, expanding First Thermal Power Station, addtional coke plant etc.

New town and infrastructure

The New Town located on the northern end of NLC boundary has been developed as a garden city comprising 31 residential blocks which have been developed on an exceedingly low density (average 3 houses per acre). Each block has been conceived as a self-contained residential area with all basic amenities according to generally contemporary standards. Housing is invariably in the form of single storey detached houses on large size plots laid out strictly in accordance with garden city layout with wide avenues. It is only recently (1982 onwards) that new housing development has been conceived in two and three storeyed residential blocks, more or less on a group housing basic. The road side trees are of a wide variety, shady and mature. The town is fully serviced in regard to protected water supply from bore-wells, underground sewerage and surface drainage, and electricity, with a profusion of a wide variety of trees both in private and public spaces. This coupled with wide tree-lined avenues and low rise deeply set-back small and big dwelling units impart it the character of a garden suburb.

In addition to the New Town catering to almost 1,00,000 population in its 11000 single storeyed detached, and 808 flat type units, hutments have come up in fairly organised manner in three major pockets, accommodating almost 1500 families. The hutments of course, have been allowed to come up by the NLC virtually on "Site and Services" basis to accommodate its low paid employees who have not been provided accommodation in the permanent township. (Refer Fig. 4).

Environmental pollution associated with mining activities of Neyveli

The surface mining produces sharp changes in the landscape due to excavation. First, the over-burden has to be excavated

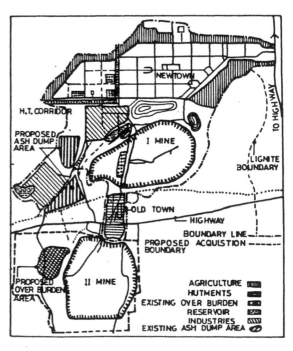

Fig. 4 Neyveli Lignite Complex Activity Areas.

in order to reach the ore body. The amount of overburden is substantial in volume. Surface mining leads to rapid erosion of land because of denudation of vegetation. Mining also results in despoilation of land which is associated with the problem of solid waste disposal. The waste dumps which are created as a result of surface mining could, if not stablised and treated properly, lead to sliding or erosion.

Mining, whether it is by open cast or underground methods, has adverse ecological and enviornmental impacts. Only their magnitude and significance vary in different cases. It depends on the method of mining and beneficiation scale and concentration of mining activity in conjunction with the geological and geomorphological setting of the area, the nature of mineral deposits, the landuse pattern before commencement of mining operations, and the natural resources existing in the area. The general environmental problems of serious nature in order of priority in mining settlements like Neyveli and for which appropriate remedial measures have to be devised, could thus be summarised as under:

1) Land degradation by open cast mining operation;
2) Deforestation during mining operation in the forested areas;
3) Soil erosion and land slides;
4) Subsidence of land due to underground mining operations;
5) Disposal of tailings, slime and effluents in case of beneficiation plants;
6) Wash-off overburden wastes and fine mineral particles into the nearby water bodies and agricultural land;
7) Acid mine drainage;
8) Dust generation in the open cast mines, on hand operations and in beneficiation plants;
9) Pollution of air due to noxious fumes released into the atmosphere by blasting and coke making operation and mine fires;
10) Disruption of water regime due to mining; and
11) Noise and ground vibration resulting from heavy blasting and operation of heavy machinery.

Industrial pollutants

Other pollutants which have impacts on the macro-environment emanate from the associated industrial plants and are liable to cause air, water and noise pollution. In the case of Neyveli, the main industrial pollutants emanate from the Thermal Power Station. Apart from emission of smoke from its chimneys, the hydro-handling of the fly-ash while preventing air-pollution to some extent, requires extensive land for ash-ponds. The ash-slurry water being let out for irrigation could also be a source of pollution, unless treated before being let out to the fields. The effluents from the Fertiliser, Briqueting and Carbonisation plants which have toxic matter, are also important sources of water pollution and require proper treatment.

Another sourse of environmental pollution is the fine dry lignite dust in conveyors and stamp-pits, which is liable to be wind-blown into the atmosphere unless preventive measures are taken. The high tension transmission corridors associated with power plants are also sources of visual pollution particularly when they pass through the New Town area. Low height plantations under the corridor can minimise the adverse impacts.

Air, water and noise pollution

An important factor for air pollution in mining areas is the mining activity itself. The direction of the wind contributes to the level of air pollution. The meteorology of air pollution in this context becomes most important and a critical parameter for the environmental issues in mining settlements. Complete dust prevention in such operation is however difficult.

Dust, grit and gases especially sulphur-dioxide are amongst the significant pollutants of air at Neyveli. The gases originate due to use of lignite as fuel in the Thermal Power Station and use of furnance oil or gasification for production of urea. An assessment of the environment dust levels (dust of particle size 5 microns and below lead to health hazards) in the mines was made by the Regional Laboratory Institute, Madras, in collaboration with the Central Laboratory of Neyveli Lignite Corporation Limited, and the dust concentration at various points in the mines assessed. Based on free silica content of the samples of overburden and red soil, the threshold limit (TLV) for 8 hours exposure was 8.8 MPPCE and 16.00 MPPCE. The dust concentrations are found to be within TLV's in most of the places in the mines at Neyveli.

Mine operations also contribute towards water pollution through the discharge of acid mine water into the streams and water bodies and by adding toxic active elements to the water bodies. Water pollution in mining areas also results from sttlements, mining activities, water temperature, trash and debris in water etc. Acid mine water is a unique pollutant because acid generation and discharge continue to occur even after mining has ceased. Some of the toxic substances dissolve in water and are carried into the water course, thus polluting them. Emphasis therefore is necessarily placed on prevention in NLC's mining operation.

Noise, particularly in mining activities, is ubiquitious in nature, due to the general din and hum of mining activities and through air and surface transportation of mined materials. But, as a part of the occupational health and safety measures, certain safeguards are usually incorporated in the mining activities to mitigate noise pollution in working environment. The prevalent techniques commonly used are:

1) Preventing the noise from generation;
2) Confining it by isolating the sources;
3) Modulating the noise by air silencer, etc;
4) Isolating the whole machine or plant from people e.g. remote control operations;
5) Isolating people from the main sources of noise; and
6) Reducing the time that workers are exposed to high level noise.

Measures taken for ecological and environmental care in Neyveli

Preservation of ecological balance

NLC is observed to be quite conscious of preserving ecological balance especially where vast stretches of land had to be acquired to lay the township and the industrial complex. Of late the speedy and massive compaign for tree plantation had led to planting over 33 million saplings in the last two years. Neyveli complex has an average of 200 trees per family. Each house has been allowed a spacious garden. Residents take care of the entire stretch which is made greener and greener as years roll by. Employees enjoying gardens also benefit by the usufructs.

Eucalyptus trees are found to contain noise and dust pollution and organisational efforts have led to the planting of 1.39 million eucalyptus trees in the last two years. Plants are reared

up also along the heaps of quarried stretches called spoil banks of over-burden dumps, and when the growth becomes abundant there will be a gridling mountainous garden around the First Mine site. Along the ongoing project sites saplings have been planted for shade and to enable pollution control.

Controlling air pollution

As regards pollutants from the power house, effective control measures were thought of at the plant design stage and implemented to satisfaction. Fuelled by lignite, the power house heaves out fly ash but its hydrohandling prevents ash pollution in the neighbourhood. Yet to send the columns of smoke to higher altitude, chimneys in the new units of TPS-1 and the Second and Third Power Houses are conceived as taller structures (120m. 170m. and 200m.) than in the First Power House which had a chimney height of only 60m. Further, the electrostatic precipitators in operation help to achieve dust-control at 99 percent efficiency. The flue gases also escape through tall chimneys to higher altitudes so as not to affect the residents and immediate surroundings.

In the Fertiliser plant the pollutant Ammonium Carbonate in liquid form is led into a protected lagoon where it evaporates without very harmful effects and gaseous pollutants are burnt atop flare stacks.

The fine dry lignite dust in the conveyors and stamp pits are extracted by dedusting blowers where dust is sludged with water in order to minimise dust pollution in the atmosphere. The sedimented lignite dust in the sludge pits is also sold out.

Controlling water pollution

The effluent from the power house is the supernatant water from ash dumps, which is being chemically treated to be of no malign influence anywhere on its passage. Further, there are seperate effluent treatment plants in the Fertiliser plant and the Leco producing, Briqueting and Carbonisation plants. Samples are being collected periodically from these units and analysed in the R&D Wing to check effectiveness of pollution control methods. Liquid effluents of the Leco plant such as nitrates, chlorides, sulphates etc. are kept within the limits of tolerability. The toxic substances are removed in the Biological Treatment Plant.

At the Process Steam Plant which is servicing the Briqueting and Carbonisation Plant and the Fertiliser factory, a Deminerlisation Plant is in operation. Before being led into effluent channels the regenerated effluent is neutralised to be alkaline.

Controlling noise pollution

In noise pollution the tolerable limit in decibels is 85 decibels. In most workshops in Neyveli it is just less than that except of course, at driveheads and tail ends of the conveyors where the level may go up by two more decibels. While ear protection devices are given to employees, measures have been taken up to muffle noise pollution with silencers. Greasing and lubrication of machinery is being attended to regularly. The trees are planted in large numbers to lessen, if not totally abate noise level. However, noise of the worksports does not affect residential areas. The R&D Wing of the Corporation Centre for Applied Research and Development, is actively engaged in monitoring various types of pollution by collecting samples and processing them for suggesting remedial measures. The R&D experts have noted that the noise level is within the tolerable range.

Dust formation cannot be dispensed with altogether in the open cast mining system, but is kept under check through needed modification to machinery, and transit equipment in material handling. The roads within the mines are drenched off and on, to prevent raising of dust upon plying of vehicles. Those who work in the Mine and other Units are being periodically examined to prevent them getting proned to occupational diseases because of dust pollution. The experience gained in the First Mine is taken note of to bring under control noise and dust pollution in the Second Mine site. Recycling of ash water for use in TPS-II has also been contemplated which will minimise environmental hazards related to ash disposal in slurry form.

Overall enviornmental assessment

External environment

Environmental studies carried out by the author for Neyveli have revealed that in so for as macro level environmental impact consequent to Lignite Mining, Power generation, Fertiliser production and Briqueting and Clay washing plants are concerned, the emissions and effluents do not pose any discernible environmental hazards to the township area and nearby settlements at present, owing to inbuilt safeguards in respective projects to control environmental pollution.

Overburden dumping and consolidation which is a vital source of atmospheric and visual pollution is by and large, being scientifically done, and already tree plantation programmes to provide a green cover has been in progress to consolidate the overburden dumps of the First Mine. This programme has yielded successful results. Gradually, the specific variety of trees, identified through an ecological and plant material study, is taking root in the overburden soil and good growth can be expected in subsequent years. In fact, the township residential blocks closer to the over-burden dupm area would be able to have a visually satisfying land form as a landscape feature in an otherwise flat country, indicating the prospects of the over-burden dumps, an offshoot of open cast mining, providing eventually an eco-cover to the New Town. The planting operation is also expected to effectively check adverse etfects of rain and wind on overburden dumps.

Soil vegetative quality in the New Town development area and immediately around it including the adjoining open lands does not seem to have been depleted, as is evident from the profusion of trees and plantations which have come up creating a garden sub-urban environment around and within the New Town area.

Owing to appreciable distance (open land generally being 300 m to 800 m) between the toe of the overburden and town development zone, noise nuisance is not noticeable. Noise nuisance and consequent pollution in the New Town area resulting from mining and industrial activity has been effectively curbed, as considerable care to muffle the noise at source plant has been taken, which along with deep green planted buffer has successfully created adequate sound-insulation to the habitable area. Lignite from the First Mine having been excavated, the quarried mine has been backfilled by overburden and this has further increased the distance from the current mining activities, (expansion of First Mine) from the township zone for the better, in so far as environmental and ecological impacts are concerned.

Disposition of ash disposal ponds which usually are sources of water pollution and environmental hazard, also indicate that

their location and also construction takes
care of the adverse impact. In the ash
pond for First Thermal Plant expansion now
under construction, the embankments of the
trough-shaped ash pond is already being
vegetated to create visual harmony with
the green environment.

The First TPS-1 has a wet disposal
system of ash in the form of a slurry,
which is collected in the ash pond from
where after the ash has settled, the water
gets drained out for irrigation purposes.
There may be a possibility of this water
to seap through the soil and contaminate
the ground water, or to drain into the
surface water channel creating water
pollution. In view of this inherent
hazard, the NLC have envisaged to re-cycle
the ash slurry water in the TPS-II now
under construction, which would be welcome
as an added environmental improvement
measure.

The collection and disposal of ash in
dry form and its industrial use in cement
manufacture is also being practised for
some years at Neyveli, which would reduce
the burden of ash disposal in slurry form
and consequent reservtion of extensive
land area for future ash ponds. The NLC
management has a programme of continued
research and development of the use of fly
ash in building construction which could
be considered a pace setter for other
Thermal Power Plants in the country.

Summing up the foregoing analysis it is
observed that the New Town, by virtue of
its particular location on the highest
part of the NLC area along the leeward
side of prevailing winds, and seperated by
a deep buffer zone from the mining and
industrial areas, is favourably placed in
regard to general environmental protection
from adverse technological impacts. These
factors are further reinforced by various
environmental safety measures adopted in
mining and industrial plants to curb air,
water, dust and noise pollution. (Refer
Fig. 5).

Built environment

As regards Built environment it is
observed that the new town maintains a low
profile in the context of the man-made
environment created as a result of massive
mining and industrial activities in the
NLC Area. This is primarily due to a
sprawling garden-suburb pattern of town's
development at an exceedingly low
residential density of less than 3-4
houses per acre, and the low-rise

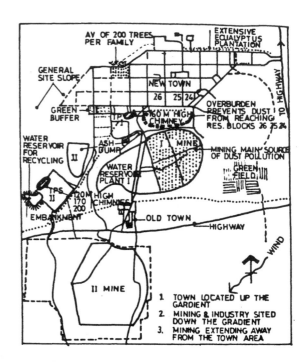

Fig. 5   Neyveli Lignite Complex
Environmental Safeguards.

character of the buildings. Consequently,
exceedingly large private and public open
spaces infiltrate the development.
Studded with trees of a wide variety,
these voids insulate more or less
completely the built components from the
immediate environment.

As the mining and consequent industrial
activities shall be extending in south and
south-west directions, the sources of
pollution, would be moving away from the
town area. While it is imperative to
restore vegetation on the worked out
mining tracts systematically, it should be
possible for NLC to even undertake
appropriate re-densification of the
developed residential blocks to cater to
its expansion needs, without apprehending
any deterioration in the quality of built
environment consequent to reduction in
excessive green spaces within the town.
It would also help in imparting an urban
character to the town, optimising use of
land designated for town development in
the NLC complex, and improving upon the
efficiency and economic functioning of the
urban subsystems and services.

Rehabilitation programme

An important environmental issue which both directly and indirectly affects a New Industrial Town like Neyveli, relates to rehabilitation of up-rooted village population as a result of land acquisition. This causes serious environmental problems, not only physical but also social, in the NLC area, and particularly the New Town.

NLC is a major mining and industrial complex of the state of Tamil Nadu and has an ambitious expansion programme, considering the massive lignite reserves. While the Second Mine has just been opened, the Third Mine is already contemplated with a Super Thermal Power Plant as an adjunct to the Mine. Rehabilitation problem would thus continue to be faced by the NLC management and may be of an increasing order. At the sametime occupational shifts from primary to secondary and tertiary activities among the up-rooted rural population particularly, consequent to additional direct and indirect jobs which would be created in the NLC Area, may become a socially accepted pattern of life.

It points to the need at the level of the State Government, to think in terms of developing a self-contained service town in close proximity to the NLC complex, preferably along the Cuddalore-Virdhachalam State Highway, which would sustain orgainsed rehabilitation programmes and servicing functions related to a large urban industrial complex like Neyveli. Such a step, would keep the growth of hutment colonies in the Neyveli New Town area within manageable limits, provide protection to the quality of built environment, and enable provision of environmentally safe infrastructure to rehabilitated population in an organised manner.

ACKNOWLEDGEMENTS

The author wishes to acknowledge the valuable assistance received from National Thermal Power Corporation (NTPC), Central Coal-Mines Planning and Design Institute Ltd.. (CMPDIL), Central Coalfields Ltd. (CCL) and Neyveli Lignite Corporation (NLC) in the form of reports bulletins and documents which provided the basic material for this article. Acknowledgements are also due to the Town and Country Planning. Organisation, Government of India, with whom the author had long association, for referance to planning studies, reports and documents relating to Integrated Planning for Singrauli Area, and to the Department of Environment, Government of India, who had funded the Environmental Assesment Study of Neyveli carried out by the author and referred in this article.

Environmental Geotechnics and Problematic Soils and Rocks, Balasubramaniam et al. (eds)
© *1987 Balkema, Rotterdam. ISBN 90 6191 785 9*

Improvement of geotechnical properties of residual sand with rice husk ash

Md. Anisur Rahman
Department of Civil Engineering, University of Ife, Ile-Ife, Nigeria

ABSTRACT: This paper presents the results of a series of laboratory tests carried out on residual sand from Ile-Ife (Nigeria) with various percentages of rice husk ash. It has been found that rice husk ash can potentially improve the various geotechnical properties such as unconfined compressive strength, California bearing ratio, cohesion and shear strength of residual sand. Moreover, the results of this study show that rice husk ash can be substitute for cement, lime or bitumen in soil stabilization. It recommends that 16% rice husk ash is optimum binder for this residual sand investigated and this stabilized soil can be used for sub-base and base materials for highway construction.

1 INTRODUCTION

Many materials have been used in different parts of the world to improve the geotechnical properties of soil for various civil engineering construction works. Some of these materials are difficult to get in some parts and some are expensive to produce for local construction purposes. Thus, those materials sometimes become very uneconomical. This is why many researchers in different parts of the world are carrying out intensive research work with abundant local materials. Cement, lime, bitumen, steel slags, fly ash, chemical compounds, etc. are very common materials used in soil stabilization all over the world. There is a continuous increase in the cost of the conventional items of constructional materials, particularly cement. This, therefore, calls for an urgent investigation into the possibility of using local materials as substitutes. One type of such local materials that justifies urgent attention is rice husk ash.

Rice husks are abundant all over the tropical regions of the world. Each country has the problem of utilization or disposal of this low-value by-product. Use or disposal has frequently proved difficult because of the tough, woody, abrasive nature of the husks, their low nutritive properties, resistance to weathering, great bulk and ash content. The knowledge of the practical usefulness of rice husk ash will not only benefit the highway construction but also other engineering works such as constuction of airfield, earth dam, low-cost housing, etc.

The purpose of this research work is to determine the effect of rice husk ash on the geotechnical properties of residual sand (A-1-b group) and to find out the optimum amount of rice husk ash required to stabilize the soil economically for sub-base and base materials in highway construction.

2 PREVIOUS WORKS

The chemical composition of rice husk ash was given by Korisa (1958), and Lazaro and Moh (1970) shown in Table 1.

The properties of the ash depend greatly on whether the husks had undergone complete destructive distillation or had only been partially burnt in the presence of adequate air. This fact was reported by Houstin (1972). He had also classified ash into: (a) high-carbon char, (b) low-carbon (gray) ash and carbon-free (pink or white) ash.

Rice husks and rice husk ash can be used in various civil engineering applications, although none of these uses has yet developed a commercial scale. Grist (1965) reported that rice husks had been used in building materials in India. These included light-weight concrete briquettes made partly from husks. Insulating bricks were also reported to have been made with cement and rice husk ash which resisted very high temperatures and were suitable for use in furnace. Also Korisa (1958) remarked that treated rice husks could be inert and suitable aggregate in high quality cement tiles and cement blocks.

Table 1. Chemical composition of rice husk ash given by Korisa (1958) and Lazaro & Moh (1970).

Composition (%)	Korisa		Lazaro & Moh
	1	2	
SiO_2	94.5	93.5	88.66
CaO	0.25	2.28	0.75
MgO	0.23	-	3.53
NaO	0.78	-	-
K_2O	1.10	3.15	-
Fe_2O_3	Trace	1.01	0.36
P_2O_5	0.53	-	-
Al_2O_3	Trace	Trace	1.48
MnO_2	Trace	Trace	-
CO_2	-	-	0.51
Loss on ignition	-	-	3.80

Lime-rice husk ash mixtures were used in the stabilization of deltaic clays by Lazaro and Moh (1970). He concluded that effective improvement of soil could be achieved by rice husk ash. In the recent past, investigations have been carried out with some Nigerian soils in order to determine the usefulness of soil stabilization in the building industry and highway consturction. Some useful results have been obtained. Mesida (1978) has established that soils in Okitipupa area of Ondo State need only 10-12% cement for stabilization to become reliable for building purposes in that area. Ola has done some research work in soil stabilization with Nigerian soil. He (1974) reported that lateritic soils can be effectively stabilized with less than 50% of the cement that is required for subgrade works in which temperate zone soils are used. He also used bitumen, lime and cement as stabilizers. Ola (1978) reported that lateritic soils stabilized with cement, lime and bitumen can be used for base and sub-base materials.

3 MATERIALS AND METHODS

Materials used in this study were rice husk ash (RHA) and residual sands.

3.1 Description of soil samples

All soil samples used in this investigation were collected from Ile-Ife (Nigeria). Large deposits of these residual sands were found under the ground surface and collected from pits. All soil samples were whitish gray in colour. General properties of original soils were determined in accordance with British Standard Specifications (1975). Uniformity coefficient and co-efficient of curvature of this soil were Cu = 4.5 and Cc = 1.1 respectively. These soil samples were almost well graded sands and these were classified under A-1-b group soil according to American Association of State Highway Officials.

3.2 Description of rice husk ash (RHA)

Rice husks used in this work were collected from Ekpoma, Bendel State, Nigeria. The moisture content of rice husks was 6% on a dry weight basis at 105°C. After burning at 800°C, the percentage of ash was approximately 19.3%. Rice husks burnt at 800°C showed that the remaining organic content was less than 3%. Whitish carbon-free ash was obtained. Specific gravity of ash was determined and it was about 2.35.

3.3 Laboratory tests and their procedures

A seires of laboratory tests was carried out on A-1-b group soil with various percentages of rice husk ash. The percentages of ash were 0, 4, 8, 12, 16 and 20. Laboratory tests conducted comprised Atterberg limits, standard Proctor compaction, unconfined compression, California bearing ratio and unconsolidated undrained triaxial compression. These tests were performed in accordance with British Standards (1975).

All soil-ash samples used in unconfined compression, California bearing ratio and unconsolidated undrained triaxial compression tests were compacted at optimum moisture contents. Different optimum moisture contents were determined for different percentages of ash with the help of standard Proctor compaction test.

A length/diameter ratio of 2.0 was utilized for all soil specimens in unconfined compression tests. Samples were sheared under strain-controlled test and the rate of strain was 1.14 mm/min.

Penetration testing in CBR tests was carried out with the help of a compression machine and the rate of strain was 1.27 mm/min for every specimen. CBR value was calculated at 2.54 mm penetration, since the load ratio was always less at 5.08 mm penetration.

The dimensions of every specimen in unconsolidated undrained triaxial test were 78 mm in length and 39 mm in diameter. All specimens were sheared under a strain-controlled test and the rate of strain was 1.14 mm/min.

4 RESULTS AND DISCUSSION

4.1 Standard Proctor compaction tests

The results of a series of standard Proctor compaction tests on residual sand with various percentages of rice husk ash are shown in Table 3. The changes of maximum dry densities and optimum moisture contents are presented in Fig. 1. The maximum dry density changes from 1.915 t/m³ at 0% ash to 1.637 t/m³ at 20% ash. It is the opinion of the author that the decrease in maximum dry density is due to two reasons. Firstly, the specific gravity of rice husk ash is lower (2.35) compared to soil grains. Secondly, ash raises air bubbles when mixed with soil.

Table 2. General properties of residual sand investigated.

Tests	Results
Natural moisture content, %	4.41
Liquid limit, %	Non-plastic
Plastic limit, %	Non-plastic
Plasticity index, %	Non-plastic
Specific gravity	2.72
% passing No. 200 BS sieve	0.06
Group index	0.0

Table 3. Effects of RHA on maximum dry density and optimum moisture content of residual sand.

RHA %	Maximum dry density, t/m³	Optimum moisture content, %
0	1.915	9.20
4	1.895	10.40
8	1.854	11.80
12	1.760	13.80
16	1.710	14.40
20	1.637	17.00

The optimum moisture content increases linearly from 9.2% at 0% ash to 17.0% at 20% ash content. The increase in optimum moisture content is due to the pozzolanic reaction of the ash with the soil constituents.

4.2 Unconfined compression tests

The results of unconfined compression tests of the soil with various percentages of ash contents are shown in Table 4. The trend of changes of unconfined compressive (UC) strength is also presented in Fig. 1. Compressive strength increases linearly from 10.71 KN/m² at 0% ash to 70.39 KN/m² at 12% ash and then it remains fairly constant. It indicates that the cohesion of the soil increases with increase in rice husk ash. The values of compressive strength of the stabilized soils are low. This is because the unconfined compression test is not suitable for sandy soils. Ola (1978) also obtained relatively low compressive strength values for lime stabilized soil for A-1-a group as compared to the more cohesive A-7-6 and A-2-4 group soils.

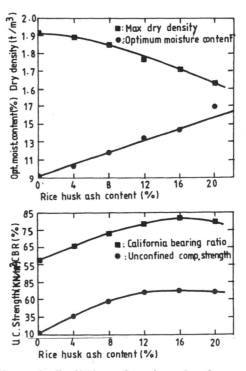

Figure 1. Variation of maximum dry density, optimum moisture content, unconfined compressive strength and California bearing ratio of residual sand with RHA contents.

Table 4. Effects of RHA on unconfined compressive strength, California bearing ratio, Cohesion and angle of internal friction of residual sand.

RHA (%)	UC strength (KN/m^2)	CBR (%)	Cohesion (KN/m^2)	$\emptyset$ (degree)
0	10.71	57.43	0	37.5
4	35.20	65.30	17	36.0
8	56.80	72.50	32	34.5
12	70.39	78.32	64	32.5
16	73.09	82.67	116	30.5
20	71.06	80.40	156	29.0

4.3 California bearing ratio (CBR) tests

Results of California bearing ratio tests on A–1–b soil group with various percentages of rice husk ash are summarized in Table 4. The nature of changes of CBR values with ash contents is also presented in Fig. 1. The value of CBR increases almost linearly from 57.43% at 0% ash to 82.67% at 16% ash. After reaching 16% ash, CBR value starts to decrease. These test results indicate a considerable improvement of the soil with rice husk ash contents. It is noted that this soil can be stabilized economically with 8 to 16% RHA for sub–base and base materials for highway construction.

4.4 Unconsolidated undrained triaxial compression tests

The results of a series of unconsolidated undrained triaxial tests with various percentages of rice husk ash are shown in Table 4. The trend of changes of cohesion and angle of internal friction with ash contents is presented in Fig. 2. Cohesion of soil increases from 0 KN/m^2 at 0% ash to 156 KN/m^2 at 20% ash. On the other hand, the angle of internal friction decreases linearly with increase in rice husk ash up to 16% ash content. Then, the trend of decrease of angle of internal friction tends to decrease. These results indicate that the soil can be stabilized economically at 16% rice husk ash for highway materials.

5 CONCLUSION

The effects of rice husk ash on various geotechnical properties of residual sand

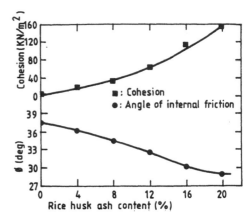

Figure 2. Variation of cohesion and angle of internal friction of residual sand with rice husk ash contents.

have been studied in this paper and the following conclusions can be drawn from the results obtained:
1) The maximum dry density decreases with increase in rice husk ash. But, the optimum moisture content increases linearly with increase in rice husk ash.
2) The unconfined compressive strength of the tested soils increases linearly with increase in ash content. After reaching 16% ash content, the compressive strength fairly becomes constant.
3) The California bearing ratio also increases linearly with increase in ash content. After reaching 16% ash contents, the California bearing ratio starts to decrease.
4) Angle of internal friction of residual sand decreases linearly as the percentage of ash increases. It remains fairly constant after 16% ash content. On the other hand, cohesion of the soil increases with increase in rice husk ash.

From the points of view of above results, the tested residual sand can be stabilized economically with 16% rice husk ash for base and sub–base materials for highway construction. Since rice husks are in abundant supply all over the tropical countries of the world, rice husk ash can be utilized as a substitute for cement, lime, bitumen, fly ash, etc. in soil stabilization in order to reduce the construction cost particularly in the rural areas of the less developed countries. Further research work relating to the improvement of various soils with rice husk ash are needed for implementing a large scale use of it.

REFERENCES

British Standards. 1975. Methods for test-
ing soils for civil engineering purposes.
BS 1377. British Standards Institution.
London.

Grist, D.H. 1965. Rice. London: Longman.

Houstin, D.F. 1972. Rice chemistry and technology. American Association of Cereal
Chemists. Minnesota: 301-340.

Korisa, J. 1958. Rice and its by-products.
London: Edward Arnold Ltd.

Lazaro, R.C. & Moh, Z.C. 1970. Stabilization of deltaic clays with lime-rice
husk ash mixtures. Proc. 2nd Southeast
Asian Conf. Soil Eng., Singapore: 215-
223.

Mesida, E.A. 1978. Soil stabilization for
housing in Okitipupa area. Occasional
research papers. Department of Geology.
University of Ife, Ile-Ife. Oyo State.
Nigeria.

Ola, S.A. 1974. Need for estimated cement requirements for stabilizing lateritic soils. Proc. ASCE. TE2. 100:
379-388.

Ola, S.A. 1978. Geotechnical properties
and behaviour of some stabilized Nigerian
lateritic soils. Quarterly Journal of
Engineering Geology. 11: 145-160.

Environmental Geotechnics and Problematic Soils and Rocks, Balasubramaniam et al. (eds)
© 1987 Balkema, Rotterdam. ISBN 90 6191 785 9

Some geotechnical aspects of seismic response

P.J.Moore
University of Melbourne, Australia

ABSTRACT: Solutions for the natural frequency of the ground have been developed for homogeneous sites in which the shear modulus of the soil overlying rock is constant with depth and for non-homogeneous sites in which the shear modulus increases linearly with depth. It was confirmed that the natural frequency increased as the ground rigidity increased. It was also found that an increase in the soil layer thickness has the effect of decreasing the natural frequency. Using a simple expression for the natural frequencies of buildings some remarks are made about undesirable building heights when avoidance of coincidence between the natural frequencies of the building and the ground is necessary.

1 INTRODUCTION

One of the geotechnical aspects of earthquake design of buildings that should be taken into account relates to the effects on seismic response of rigidities and thicknesses of soil and rock layers at the site. From a study of many earthquakes Duke (1958) found that the majority of structures were least damaged on firm ground. Soft ground was nearly always associated with highest damage. In the Alaskan earthquake of 1964, small rigid buildings in Anchorage were relatively undamaged whereas tall buildings were significantly damaged in most cases. This same type of experience has been mentioned by Seed and and Idriss (1969) for earthquakes in Mexico City. Here amplification of ground motions of 2 to 2.5 seconds period occurs because of the natural frequency of the bowl of soft soil on which the city is founded. Seed and Idriss further comment that for sites underlain by stiff soils, peak amplitudes tend to develop at low values of the fundamental period (0.4 to 0.5 sec), indicating that maximum accelerations would be induced in relatively stiff structures 5 or 6 stories high. For sites underlain by deep deposits of soft soils the fundamental period is higher (1.5 to 2.5 sec), for which the most susceptible buildings would be 20 to 30 stories high.

The preceding discussion demonstrates the desirability of avoiding coincidence between the natural frequency of the building and that of the ground. This is allowed for empirically in many building codes that permit the pseudo-static method of building design. In the American Uniform Building Code (1982) for example, the equation for base shear contains a "site-structure resonance" term which has a maximum value when the natural frequencies of the building and the ground coincide. The Code makes use of an equivalent single layer method or a multi-layer method based upon equations for the homogeneous profile (see section 4 below) for the evaluation of the characteristic or fundamental period of vibration for the site.

The use of a one-dimensional ascending shear wave model for the evaluation of the natural frequency of a site is an oversimplification of a very complex three-dimensional problem. Nevertheless it has wide appeal (see for example, Ohsaki et al (1984)) and in many cases it is known to give very good results. In this paper natural frequencies of the ground have been evaluated for homogeneous profiles and for non-homogeneous profiles in which the shear modulus increases linearly with depth.

2 NATURAL FREQUENCIES OF BUILDINGS

A number of simple empirical formulae have been proposed for the estimation of the natural frequency of a building for the

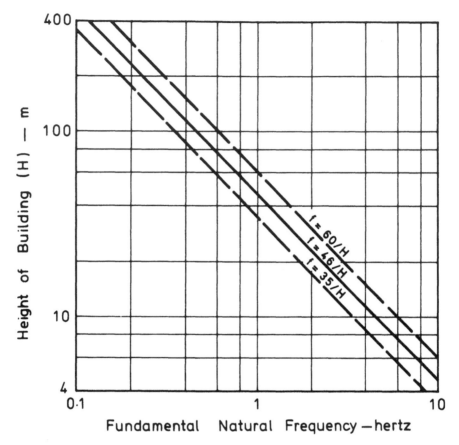

Figure 1. Natural frequencies of buildings

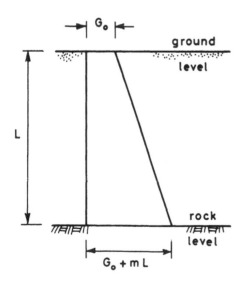

Figure 2. Shear modulus variation with depth

fundamental transverse mode (mode 1) of vibration. Anderson et al (1952) have examined the popular formula for the fundamental period (T) of vibration

$$T = K \ H/D^{\frac{1}{2}} \ sec. \qquad (1)$$

where H = building height (m), and
D = depth of building (m) parallel to the direction of vibration

Based upon a large number of observations they found that the constant K should be 0.091. They also found that the calculated natural frequency for a building could be in error by more than 50%. Salvadori and Heer (1960) proposed that the constant K should be 0.109 but still acknowledged that the formula was not very precise. Housner and Brady (1963) examined a number of formulae, among the simplest being

$$T = 0.1 \ N \ sec. \qquad (2)$$

102

for rigid frame buildings, and

$$T = 0.5 \, N^{\frac{1}{2}} - 0.4 \text{ sec.} \tag{3}$$

for steel frame buildings, where N = number of floors in the buildings.

More recently, Ellis (1980) has examined the accuracy of some of these simple formulae and proposed that a slightly better expression for natural frequency (f_n) is

$$f_n = 46/H \text{ hertz} \tag{4}$$

While errors of ± 50% were not uncommon in the prediction of natural frequency, he found that simple formulae, such as equation (4), were likely to be as accurate at computer based predictions. Equation (4) has been plotted in Fig. 1. The two dashed lines, f = 60/H and f = 35/H enclose the majority of the observations used by Ellis.

3. NATURAL FREQUENCY OF THE GROUND

The natural frequency of the ground has been the subject of some confusion in years gone by. It was apparently considered in some quarters that the natural frequency of the ground was dictated by the type of soil at the ground surface. Reynolds (1954), for example, produced a tabulation of natural frequencies of soils. These values were based upon earlier measurements at various sites by means of an oscillator on the ground surface. Apparently it was not fully appreciated that the natural frequencies measured were dictated as much by the oscillator characteristics as by the nature of the ground. This is now understood and appreciated by those involved with machine foundation design. It is now known that the natural frequency of the machine-foundation-soil system can be widely varied by changing the foundation characteristics such as mass and size.

The natural frequency of the ground in the present context is quite different from that discussed in the preceding paragraph. Here the natural frequency is determined by the characteristics of the various soil strata lying between the ground surface and the underlying basal rock. The characteristics of major importance are the thickness and the rigidity of each soil layer. Excitation of the soil is introduced at the soil-rock interface by shear stresses produced by rock motions. The natural frequency of the ground may be defined as the frequency of horizontal vibration in the basal rock at which there is such amplification through the soil strata (as the stress wave propagates

upwards) to yield theoretically infinite vibration amplitude at the ground surface.

Soil motions resulting from stress wave propagation vertically through soil strata may be evaluated by the technique described by Heierli (1962) and by Streeter et al (1974) using the method of characteristics. The latter named authors have also developed a computer program CHARSOIL to assist in the evaluation of soil motions.

Exact mathematical solutions for soil motions may be developed for the case where the soil is considered to be a non-homogeneous elastic solid with shear modulus linearly increasing with depth as illustrated in Fig. 2. The procedures followed to obtain solutions for this case are described briefly below.

4 CONSTANT SHEAR MODULUS CASE

A solution will first be developed for the case of a soil layer with a constant density and constant value of shear modulus throughout the layer thickness. Referring to Fig. 2, this means that the value of m is zero and the shear modulus of the soil layer is G_0 throughout the layer thickness, L. The horizontal motion at the soil-rock interface will be assumed to be of the form.

$$u = u_0 \sin \omega t \tag{5}$$

where u_0 = displacement amplitude.

For convenience the origin (z = 0) will be located at the soil rock interface, so that z = +L at the ground surface. It can be shown that the equation of motion is

$$\partial^2 u/\partial t^2 = v_s^2 \, \partial^2 u/\partial z^2 \tag{6}$$

where $v_s = (G_0/\rho)^{\frac{1}{2}}$ = shear wave velocity, and ρ = mass density of the soil.

The solution to equation (6) yields the variation of displacement amplitude (u_z) as a function of z

$$u_z = u_0 \cos \theta + u_0 \tan (\omega L/v_s) \sin \theta \tag{7}$$

where $\theta = \omega z/v_s$ and at the ground surface the displacement amplitude (u_L) becomes

$$u_L = u_0 / \cos(\omega L/v_s) \tag{8}$$

If the amplification ratio (AR) is defined as the ratio of the displacement amplitude at the ground surface to that at the soil-rock interface, then

$$AR = 1/\cos(\omega L/v_s) \tag{9}$$

From equation (9) it is clear that the fundamental (mode 1) natural frequency is

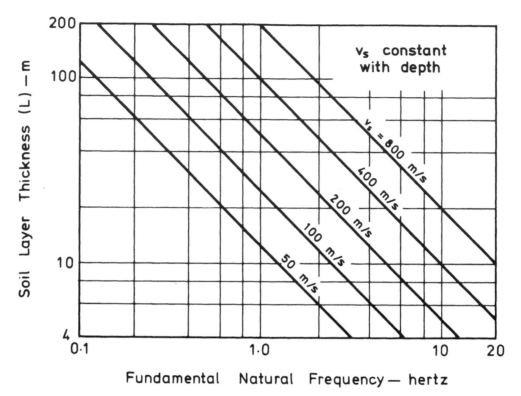

Figure 3. Natural frequency for constant shear modulus case

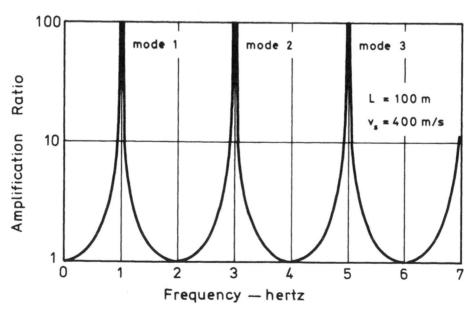

Figure 4. Various modes of Vibration - Constant shear modulus case

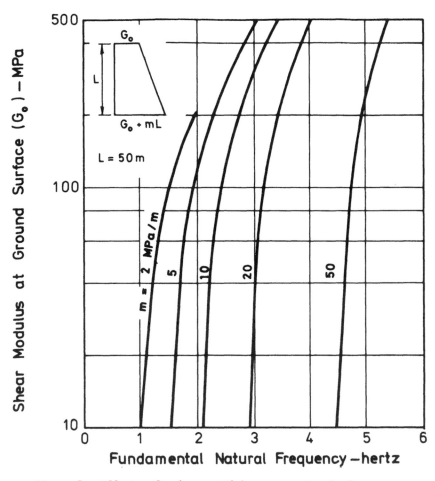

Figure 5. Effect of shear modulus on natural frequency

given by

$$\omega = 2\pi f = v_s \pi / 2L \qquad (10)$$

and the natural frequency is plotted in Fig. 3 for various values of the soil stratum thickness and the shear wave velocity. The higher mode natural frequencies, which are illustrated in Fig. 4 for particular values of L and v_s, are given by

$$\omega L / v_s = 3\pi/2, \ 5\pi/2, \text{ etc.} \qquad (11)$$

for mode 2, mode 3, etc. respectively.

5 NON-HOMOGENEOUS CASE

For this case, in which the shear modulus increases linearly with depth, the origin is again located at the soil-rock interface and the horizontal motion at this location is also assumed to be given by equation (5). The equation of motion for this case may be shown to be

$$\rho \partial^2 u / \partial t^2 - G_z \ \partial^2 u / \partial z^2 + m \ \partial u / \partial z = 0 \qquad (12)$$

where ρ = mass density of soil,
$\quad G_z$ = shear modulus at depth z,
$\quad\quad$ = G_o at ground surface
$\quad\quad$ = $G_o + mL$ at the soil-rock interface, and
$\quad m$ = rate of increase of shear modulus with depth.

By separation of variables the following solution to equation (12) may be obtained

$$u_z = A \ J_o(x) + B \ Y_o(x) \qquad (13)$$

where $J_o(x)$ = Bessel function of the first kind of order zero,
$\quad Y_o(x)$ = Bessel function of the second

105

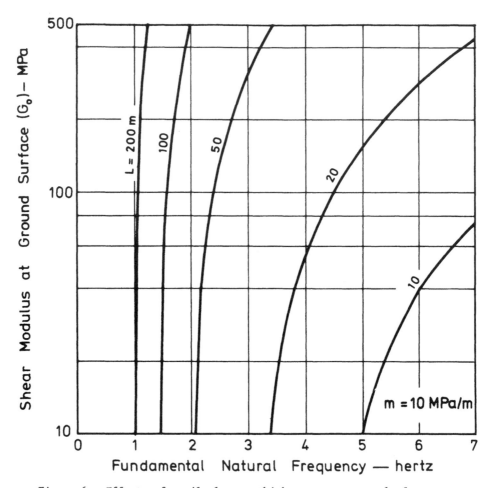

Figure 6. Effect of soil layer thickness on natural frequency

kind of order zero, and

$$x = 2\omega(\rho G_z)^{1/2}/m$$

The ratio of the parameters A and B may be evaluated from boundary conditions to yield

$$A/B = - Y_1(x_o)/J_1(x_o) \qquad (14)$$

where $Y_1(x_o)$ = Bessel function of the second kind of order unity,

$J_1(x_o)$ = Bessel function of the first kind of order unity,

and $\quad x_o = 2\omega(\rho G_o)^{1/2}/m$

From equations (13) and (14) the amplification ratio (AR) may be expressed as follows

$$AR = \frac{(A/B)J_o(x_o) + Y_o(x_o)}{(A/B)J_o(X_L) + Y_o(x_L)} \qquad (15)$$

where $x_L = 2\omega[\rho(G_o + mL)]^{1/2}/m$.

The amplification ratio has been evaluated for various values of L, G_o and m but for a single value of ρ of $1800 kg/m^3$. From these calculations the natural frequencies for the first three modes of vibration have been extracted.

The effect of shear modulus on the fundamental (mode 1) natural frequency is presented in Fig. 5 for m values varying from 2 to 50MPa/m. This shows that the natural frequency increases as shear modulus increases in general agreement with the conclusions drawn from Fig. 3. Fig. 5 was prepared for a single value of the soil layer thickness (L = 50m) and in Fig. 6 the effect of thickness of the soil layer is shown. An increase in the soil layer thickness has the effect of decreasing the natural frequency. At large values

106

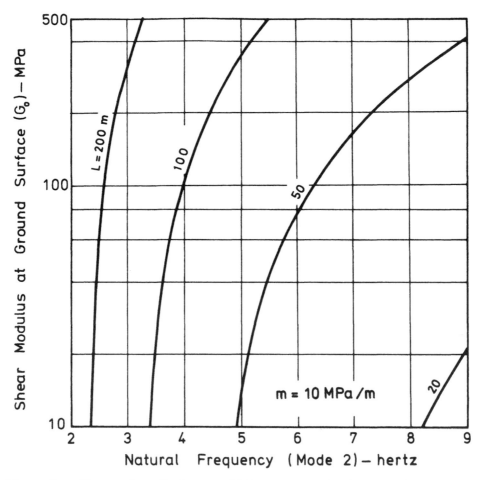

Figure 7. Effect of soil layer thickness on mode 2 natural frequency

of the soil layer thickness (L > 100m) the natural frequency tends to become almost independent of the ground surface value of the shear modulus (G_o). The effects of soil layer thickness on the mode 2 and mode 3 natural frequencies are illustrated in Figs. 7 and 8 respectively. Figs. 6,7 and 8 show that the trends of the curves are similar, the major effect being that the curves are displaced in the direction of higher natural frequency for higher modes of vibration.

6 EQUIVALENT HOMOGENEOUS CASE

It has already been observed that the trends regarding variations in natural frequency for the non-homogeneous case in which the shear modulus increases linearly with depth are in general agreement with those for the homogeneous case in which the shear modulus is constant with depth. It is a simple matter to calculate (using equation (10))the magnitude of an equivalent shear modulus (G_{eq}) for a homogeneous case which would yield the same value of the fundamental natural frequency as that for the non-homogeneous case. The equivalent depth (L_{eq}) corresponding to this equivalent shear modulus was evaluated for each of the non-homogeneous cases examined. The equivalent depth, expressed as a percentage of the soil layer thickness (L) is shown in Fig. 9 in terms of the ratio of shear moduli (G_L/G_o), where G_L is the shear modulus at the soil-rock interface. Although there is a scatter in the plotted points, Fig. 9 indicates that the equivalent depth varies from about 0.69L, for G_L/G_o = 1, to less than 0.60L for G_L/G_o greater than about 20. This enables

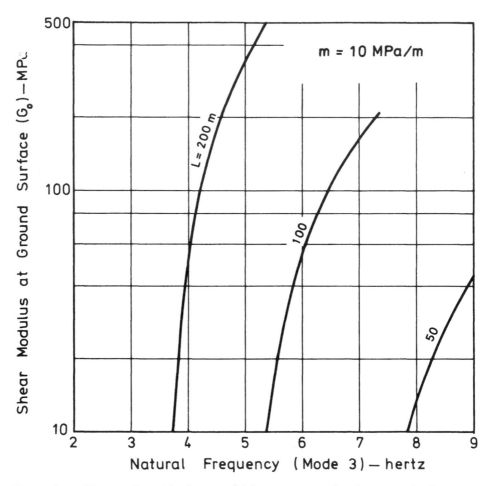

Figure 8. Effect of soil layer thickness on mode 3 natural frequency

the equivalent shear modulus to be estimated for a case in which shear modulus increases with depth. This then makes it possible for a simple calculation procedure (using equation (10)) to be followed for the evaluation of the fundamental natural frequency.

It appears that this equivalent shear modulus cannot be used for the calculation of natural frequencies for higher modes of vibration. This point is illustrated in Fig. 10, which shows the response curve for a non-homogeneous case for particular values of G_0, m and L. The equivalent shear modulus is 165.9MPa yielding the same magnitude of fundamental natural frequency. It is clear from the figure that the higher mode natural frequencies do not coincide.

From section 4 it is seen that the ratio of the mode 2 natural frequency to the

mode 1 (fundamental) natural frequency is 3.00 for the homogeneous case. For the non-homogeneous cases examined this ratio varied from 2.30 to 2.79 with an average of 2.53.

7 UNDESIRABLE BUILDING HEIGHTS

One of the applications of the analyses described above is in the identification of those building heights which may be considered as potentially hazardous. The hazard that is being examined here is that associated with the increase in amplification of vibratory motion of the building, which occurs when the natural frequency of the building coincides with one of the natural frequencies of the ground. This can best be illustrated by means of an example.

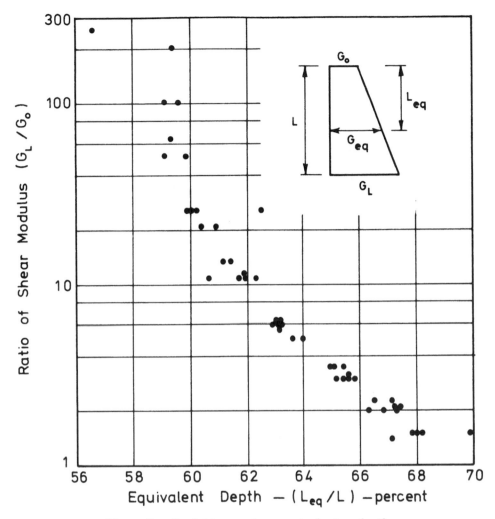

Figure 9. Variations   in   equivalent   depth

Consider a site where the soil layer thickness (L) is 100m, G_O is 40MPa and m is 10MPa/m. From Figs. 6,7 and 8, the first, second and third mode natural frequencies of the ground are respectively 1.5 hertz, 3.6 hertz and 5.8 hertz. From equation (4) the building heights corresponding to these natural frequencies are 31m, 13m and 8m respectively. To avoid resonance effects these building heights should be avoided at this site. Probably a better way to quantify the hazard associated with certain building heights would be to make allowance for the scatter of observed points from which equation (4) was derived. This can be done by making use of the two other equations for the enveloping lines in Fig. 1 and quoting

ranges of hazardous building heights. This leads to the following ranges of building heights that should be avoided : 23m to 40m (mode 1 resonance), 10m to 17m (mode 2 resonance) and 6m to 10m (mode 3 resonance). That is, the building heights that should be considered for design should be within the range of 17m to 23m or greater than 40m at this site.

The information described above is most easily presented in graphical form such as in Fig. 11. The shaded areas indicate the zones to be avoided in order to minimise resonance effects. Fig. 11(b) illustrates the effect of higher shear modulus at the ground surface. It is seen that the high shear modulus has the effect of displacing the zones of undesirable building heights

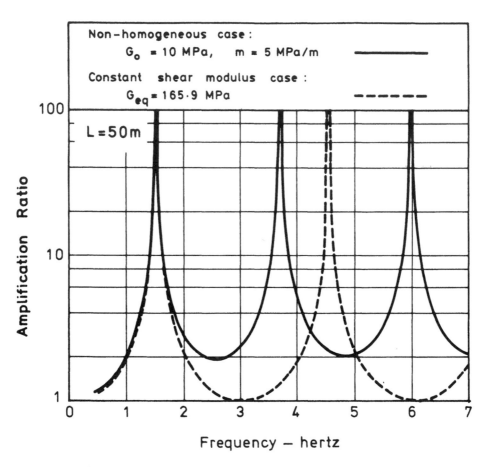

Figure 10.   Comparison   of   dynamic   response   curves

in the direction of decreasing height of building.

For the homogenous case (constant shear modulus over the thickness of the soil layer), Figs. 12(a) and 12(b) illustrate the ways in which the zones of undesirable building heights are displaced in comparison with the positions of these zones for the hon-homogeneous case.

It needs to be emphasised that the preceding comments are based on elastic analyses in which damping has not been considered. The effects of damping which have yet to be evaluated are likely to produce significant reductions in the amplification ratios in the regions of the natural frequencies. This will be of major importance for the higher modes of vibration if the ground is assumed to be subjected to viscous damping.

8   CONCLUDING REMARKS

Natural frequencies of the ground have been evaluated for homogeneous and non-homogeneous cases by means of a one-dimensional undamped wave propagation model, in terms of the shear modulus and its variation with depth and the layer thickness of the soil overlying rock. It was found that the natural frequency increases as the shear modulus increases and it decreases as the soil layer thickness increases. Based on the analyses described, zones of undesirable building heights have been identified when it is necessary to avoid coincidence between the natural frequencies of the building and the ground.

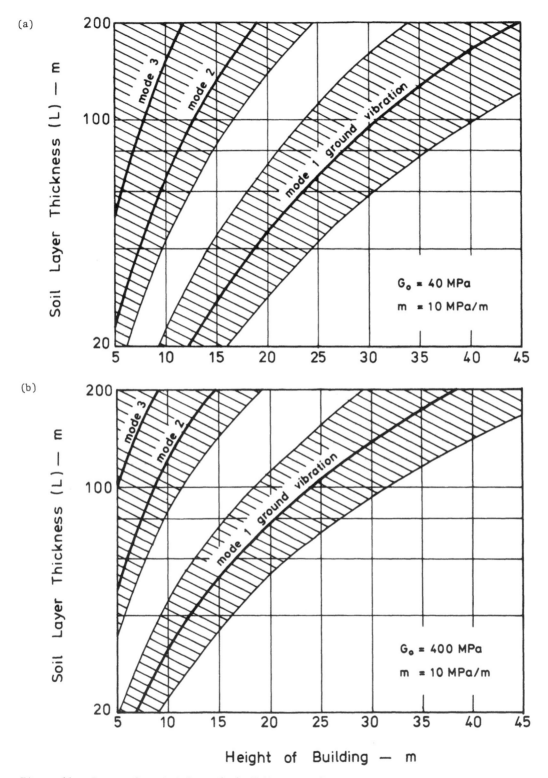

(a)

(b)

Figure 11. Zones for height of building –non-homogeneous case

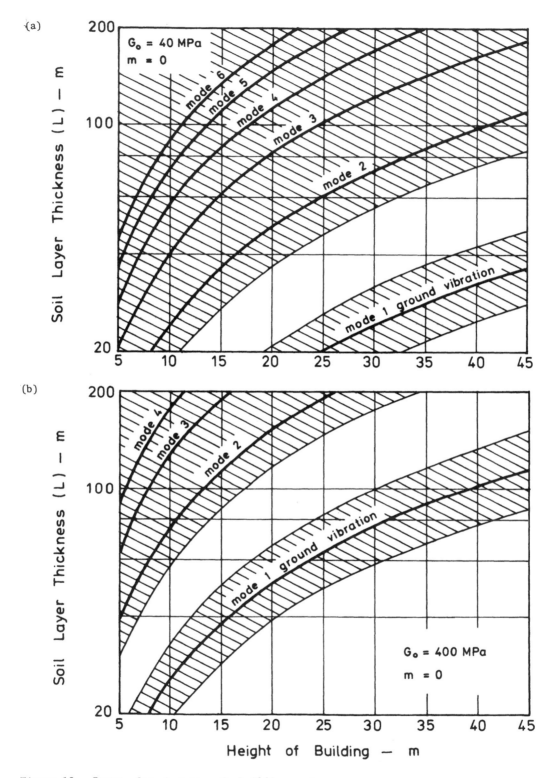

Figure 12. Zones for height of building – homogeneous case

112

REFERENCES

Anderson, A.W. et al 1952. Lateral forces of earthquake and wind. Trans. ASCE, 117 : 716-780.

Duke, C.M. 1958. Effects of ground on destructiveness of large earthquakes. Jnl. Soil Mech. and Found. Div. ASCE, 84 (SM3) : 1730-1-1730-23.

Ellis, B.R. 1980. An assessment of the accuracy of predicting the fundamental natural frequencies of buildings and the implications concerning the dynamic analysis of structures. Proc. Instn. Civ. Engrs, 70 (Part 2) : 763-776.

Heierli, W. 1962. Inelastic wave propagation in soil columns. Jnl. Soil Mech. and Found. Div. ASCE, 88 (SM6) : 33-63.

Housner, G.H.and Brady, A.G. 1963. Natural periods of vibration of buildings, Proc. ASCE. Eng. Mechanics Div. 89 (EM4) : 31-65.

Ohsaki, Y., Watabe, M., Tohdo, M. and Ohkawa, I. 1984. Characteristics of surface ground motions considering the various property combinations of subsoils and earthquakes. Proc. 8th World Conf. on Earthquake Eng. San Francisco, 2 : 801-808.

Reynolds, H.R. 1954. Notes on the effects of vibration on soils. Civil Eng. and Public Works Rev. 49 (Part 1) : 835-837.

Salvadori, M.G. and Heer, E. 1960. Periods of framed buildings for earthquake analysis. Proc. ASCE. Struct. Div. 86 (ST12) : 59-71.

Seed, H.B. and Idriss, I.M. 1969. Influence of soil conditions on ground motions during earthquakes. Jnl. Soil Mech. and Found. Div. ASCE, 95(SM1) : 99-137.

Streeter, V.L., Wiley, E.B., and Richart, F.E. 1974. Soil motion computations by characteristics method. Jnl. of the Geot.Eng., Div. ASCE, 100 (GT3): 247-263.

NOTATION

AR	(u_L/u_o)
D	depth of building parallel to direction of vibration
f	frequency of vibration (Hertz)
f_n	natural frequency (Hertz)
G_{eq}	equivalent shear modulus
G_L	shear modulus at soil-rock interface
G_o	shear modulus at ground surface
G_z	shear modulus at depth z
H	building height (metres)
$J_o(x)$	Bessel function of the first kind of order zero
$J_1(x)$	Bessel function of the first kind of order unity
L	soil layer thickness
L_{eq}	depth below ground surface at which G_{eq} occurs
m	rate of increase of shear modulus with depth
N	number of floors in building
T	fundamental period of vibration t time
u	horizontal ground displacement
u_L	displacement amplitude at ground surface
u_o	displacement amplitude at soil-rock interface
u_z	displacement amplitude at depth z
v_s	shear wave velocity
x_L	$(2\omega[\rho(G_{o_1} + mL)]^{\frac{1}{2}}/m)$
x_o	$(2\omega(\rho G_o)^{\frac{1}{2}}/m)$
$Y_o(x)$	Bessel function of the second kind of order zero
$Y_1(x)$	Bessel function of the second kind of order unity
z	depth measurement from origin at soil-rock interface
ρ	mass density of soil
θ	$(\omega z/v_s)$
ω	frequency of vibration (rad/sec)

Environmental Geotechnics and Problematic Soils and Rocks, Balasubramaniam et al. (eds)
© 1987 Balkema, Rotterdam. ISBN 90 6191 785 9

Road damage due to ground freezing in the Republic of Korea and a method to predict frost depth for pavement design

Daesuk Han
Korea Institute of Energy and Resources, Seoul, South Korea

ABSTRACT: Some of the roads in the Republic of Korea suffer serious damages caused by the frost heaving during the winter and the thaw weakening preceded by noticeable or unnoticeable heaving during the spring melting period. In order to eliminate such damage the depth of frost penetration below the pavement should be considered for design purposes. The frost depth for pavement design can be theoretically determined based on the modified Berggren equation. To facilitate use of the equation, two varieties of maps were produced showing the distribution of design freezing index values and the zonal distribution of freezing periods. A diagram also was constructed illustrating the relationship between freezing index and thermal ratio. Using the above information, it is possible to estimate the depth to which frost penetrates anywhere in the Republic of Korea.

1 INTRODUCTION

The cold regions of the world centering around the poles are divided into two types of frozen ground: (1) seasonally frozen ground and (2) perennially frozen ground, the so-called permafrost. In the Northern Hemisphere, ground freezing is estimated to occur over about half (48 per cent) of the land mass.

The seasonally frozen ground includes the freeze-thaw layer of the temperate regions and the layer above the permafrost which is subject to freezing in the winter and thawing in the summer. Its area constitutes about 26 per cent of the land area in the Northern Hemisphere.

The southern limit of the seasonally frozen ground of the temperate regions is characterized by frost penetration to a depth of about 30.5 cm (1 ft) at least once in 10 years (Bates & Bilello 1966). On this basis, approximately 98 per cent of the land area of the Republic of Korea, which lies between latitude 33°30'N and 38°30'N, is affected by seasonal freezing.

The general problems to be considered for engineering design and construction in areas of seasonally frozen ground are frost heaving in the soil during the winter and the loss of strength of the frozen soil during the spring melting period. These phenomena may cause serious damage to roads, airfield runways, building foundations and so on.

2 ROAD DAMAGE

Road damage by freezing in the Republic of Korea is due to two kinds of action, frost heaving and thaw weakening, which can be commonly observed in the temperate regions of other countries. The total cost of maintenance of damaged roads throughout the Republic of Korea has not been reported, but the annual cost of repair of frost-related damage in the city of Seoul alone amounts to about US $2,300,000.00 according to the city authority.

When the pore water in the soil converts into ice, it increases its volume by almost 10 per cent. Ground heaving may result from this volume change in the ground where deep frost penetration occurs. This heaving may, however, be much less than that attributed to the ice lenses which form and grow in the presence of the following three factors: (1) a frost-susceptible soil; (2) freezing temperature in the soil; and (3) the availability of water to the freezing front. As any subgrade soil with the above three factors is subject to freezing, ice lenses form and grow in it approximately in parallel with the pavement surface, generating expansive forces which may

Table 1. Results of a hand auger boring below the pavement shown in Figure 2.

Number of layer	Depth, cm	Description of material	Physical properties	Thermal properties			Remarks
				Thermal conductivity, cal/cm·s·°C	Volumetric heat capacity, cal/cm³·°C	Latent heat of fusion, cal/cm³	
1	0–8	Bituminous concrete	–	3.30×10^{-3}	0.449	0	
2	8–19	Gravel	–	5.69×10^{-3}	0.453	10.30	Base course
3	19–43	Mixture of gravel (20%), sand(35%), silt(30%) & clay (15%)	W_{cf}: 23.0% W_{cu}: 11.0% ρ_d : 1.8 g/cm³	5.61×10^{-3}	0.473	15.84	Subbase
4	43–80	Silty sand, fill (SC): gravel 3%, sand 48%, silt 35% & clay 14%	W_{cf}: 45.0% W_{cu}: 22.8% ρ_d : 1.52 g/cm³ W_L : 30.5% W_P : 21.7%	5.34×10^{-3}	0.534	27.72	Subgrade
5	80–	Clayey silt, alluvial soil(ML): gravel 1%, sand 31%, silt 50% & clay 18%	W_{cu}: 28.1% ρ_d : 1.34 g/cm³ W_L : 29.5% W_P : 23.5%	3.35×10^{-3}	0.524	24.64	

Notes: W_{cf} = water content for frozen state.
W_{cu} = water content for unfrozen state.
ρ_d = dry density for unfrozen state.
W_L = liquid limit.
W_P = plastic limit.

cause surface deformation. If the soil profile, drainage pattern, and the soil thermal properties are not uniform, differential upward displacement occurs, producing the undesirable effects seen in Figure 1. The heave amount of the pavement surface in the figure was measured at about 8 cm.

The pavement surfaces in the Republic of Korea often suffer severe cracking caused by differential frost heaving. This crack- ing is well illustrated in Figure 2. The heaving in the figure amounted to as much as 10 cm in the ground frozen to a depth of about 80 cm. At the beginning of March 1984 an investigation was conducted into what gave rise to such damage with a hand auger boring to a depth of 3 m and laboratory testing. Based on the results of this investigation (Table 1), the subgrade and subbase materials were identified as frost-

Figure 1. Uplift of pavement surface due to ground freezing, Choch'iwon, 1984. Note that the arrow marked portion of the road middle line is convex.

Figure 2. Damage to bituminous concrete pavement caused by frost heaving, Seoul, 1984.

susceptible. The obtained samples contained a large number of thin ice lenses ranging in thickness from 1 to 2 mm with their particle size characteristics also indicating frost susceptibility. The alluvial soil overlain by the subgrade must have been a source of water to feed the ice growth.

The loss of soil-bearing strength by the thaw weakening during the melting period is preceded by noticeable or unnoticeable frost heaving. As the ice lenses in the upper layers of subgrade melt during the spring, the melt-water cannot drain downward into the lower layers which are still frozen, inducing a very high moisture content. When traffic loads are applied to the pavements directly above the saturated layers, these pavements fail as can be seen in Figures 3 and 4. The water content of the subgrade soil collected from the damaged area shown in Figure 3 was determined

Figure 3. Disruption of pavement due to spring thaw, Seoul, 1984.

Figure 4. Road damage caused by thaw weakening, Ch'ungju, 1984.

at as much as 46 per cent, almost twice as high as that of the nearby subgrade soil before it was frozen.

3 PREDICTION OF FROST-PENETRATION DEPTH

The degree of frost damage primarily depends upon the depth of frost penetration.

Since it may be impractical to obtain a direct measurement of the depth to which frost penetrates multilayers such as pavement structures, the use of the modified Berggren equation (Aldrich & Paynter 1953) is suggested for the purpose of calculating frost depth. The equation is written as follows:

$$X = \lambda \left(\frac{172800 \cdot F}{(L/k)_{eff}}\right)^{\frac{1}{2}}$$

where
X = frost penetration, cm
F = freezing index, $^{\circ}$C·days

$$(L/k)_{eff} = \frac{2}{X_1^2}\left[\frac{d_1}{k_1}\left(\frac{L_1 d_1}{2} + L_2 d_2 + \cdots + L_n d_n\right)\right.$$

$$+ \frac{d_2}{k_2}\left(\frac{L_2 d_2}{2} + L_3 d_3 + \cdots + L_n d_n\right)$$

$$\left. + \cdots + \frac{d_n}{k_n} \cdot \frac{L_n d_n}{2}\right]$$

$X_1 = d_1 + d_2 + \cdots + d_n$: estimated trial penetration, cm
$d_1, \cdots, d_n$: thickness of each layer within the above estimated depth, cm
$k_1, \cdots, k_n$: thermal conductivity of each layer, cal/cm·s·$^{\circ}$C
$L_1, \cdots, L_n$: latent heat of fusion of each layer, cal/cm^3
$\lambda = (u, \tau)$: correction coefficient
$u = C_{wt} \cdot F/L_{wt} \cdot t$: fusion parameter
$\tau = T_0 \cdot t/F$: thermal ratio
T_0 = average annual air temperatures, $^{\circ}$C
t = freezing period, days
$C_{wt} = (C_1 d_1 + \cdots + C_n d_n)/X_1$: weighted value of volumetric heat capacity, cal/cm$^3 \cdot ^{\circ}$C
$L_{wt} = (L_1 d_1 + \cdots + L_n d_n)/X_1$: weighted value of latent heat of fusion, cal/cm^3

In order to utilize the above equation, two maps (Figures 5 & 6) and a diagram (Figure 7) were prepared for the Republic of Korea. As Figure 5 illustrates, the map is composed of the seven isolines of the design freezing index values, say 150, 250, 350, 450, 550, 650, and 850, all in $^{\circ}$C·days. These isolines were drawn on the basis of all the design freezing index values of 87 weather stations located throughout the

Republic of Korea. Each of the design freezing index values were acquired by summing the differences between the average daily air temperatures and 0°C for the coldest winter in the years from 1974 and 1984. All the average daily air temperatures used are the mean values of the four temperatures observed at 03:00, 09:00, 15:00, and 21:00 o'clock. Figure 6, which consists of the six zones of the freezing periods, 41-50, 51-60, 61-70, 71-80, 81-90, over 100 days, was also drawn based on the freezing periods of 87 weather stations for the coldest winter in the years given above. Figure 7 was derived from the design freezing indexes of twenty-two selected weather stations and their thermal ratios.

Utilizing the above three kinds of information and the modified Berggren equation, it is possible to estimate the depth of frost penetration into the uniform or multilayer soils for anywhere in the Republic of Korea, provided that the thermal and physical properties of the soils are known.

The depth of frost penetration below the bituminous concrete pavement described in Table 1 was found to be 83 cm using the modified Berggren equation. This theoretical value is about 3 cm greater than that actually measured.

4 CONCLUSION

In general, the following two problems should be considered for pavement design and construction in areas of the seasonal freezing and thawing: (1) what is the depth of frost penetration below the pavement? and (2) how can inadequate subgrade soils (frost-susceptible soils) be treated? For the Republic of Korea, the depth of frost penetration can be ascertained by using the modified Berggren equation along with the two maps and one diagram described in the preceding paragraph. One of the methods of treating the frost susceptible soils is to replace them with a non-frost susceptible material for complete protection. The replacement thickness can also be calculated by the above method. Since the design freezing index values for the Republic of Korea do not appear so high as to require a thickness in excess of economic constraints, the replacement method may be adopted throughout the country.

The method described above for predicting the depth of frost penetration can also be applied to such structures as water pipes and building foundations, which are generally protected against the effects of ground freezing by being placed below the level of frost penetration.

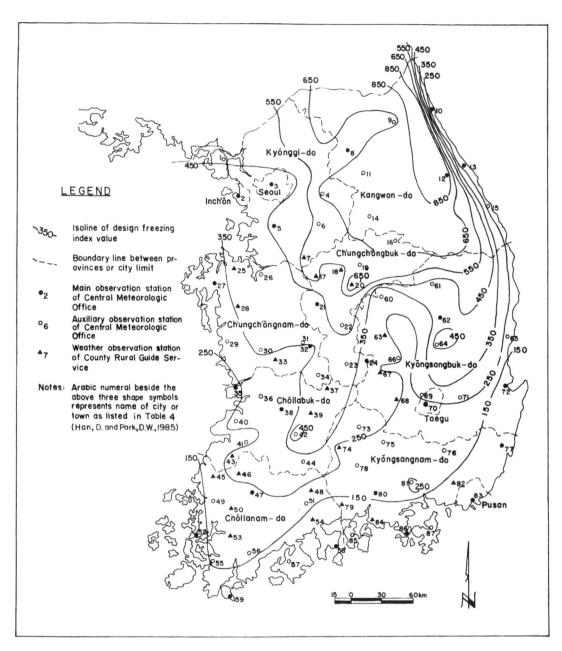

Figure 5. Distribution of design freezing index values for the Republic of Korea.

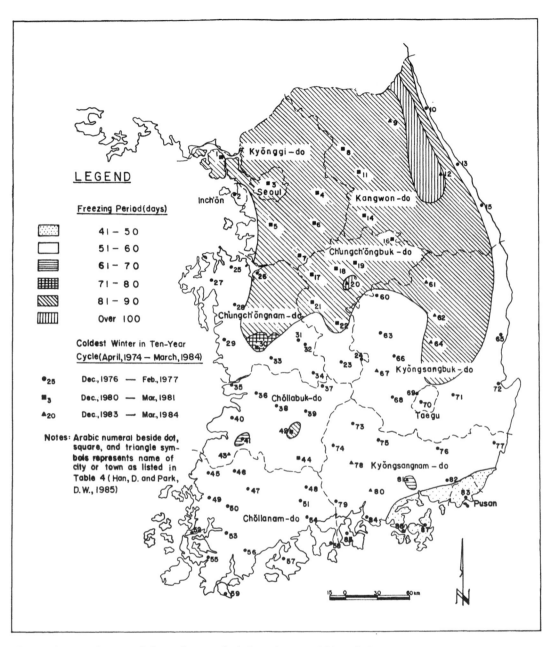

Figure 6. Zonal map of freezing period for the Republic of Korea.

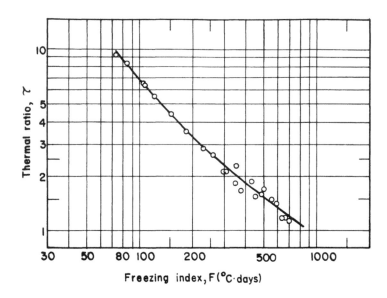

Figure 7. Relationship between freezing index and thermal
ratio for the Republic of Korea.

REFERENCES

Aldrich, H.P. & H.M.Paynter 1953. Analyti-
cal studies of freezing and thawing in
soils. US Army Corps Eng. Constr. Frost
Eff. Lab. First Interim Rep., Boston,
Mass.
Bates, R.E. & M.A.Bilello 1966. Defining
the cold regions of the Northern Hemisphere. US Army Cold Regions Res. and Eng.
Lab. Technical Rept. 178.
Burdick, J.L., E.F.Rice & A.Phukan 1983.
Cold regions: descriptive and geotechnical
aspects. In Andersland, O.B. & D.M.Ander-
son (eds.), Geotechnical engineering for
the cold regions, p.1-18. New York:
McGraw-Hill.
Corte, A.E. 1969. Geocryology and engineer-
ing. In Varnes, D.J. & G.Kiersch (eds.),

Reviews in engineering geology vol.2,
p.119-185. Boulder: The Geological Society
of America.
Gillott, J.E. 1968. Clay in engineering
geology. Amsterdam: Elsevier.
Han, D. & D.W. Park 1985. Distribution of
design freezing index values for the Re-
public of Korea and its application.
Korea Institute of Energy and Resources
Research Report 85-20: 39-88.
Hunt, C.B. 1972. Geology of soils. San
Francisco: Freeman.
Phukan, A. & O.B.Andersland 1983. Founda-
tions for cold regions. In Andersland, O.
B. & D.M.Anderson, Geotechnical engineer-
ing for the cold regions, p.276-282 &
p. 348-351. New York: McGraw-Hill.
Washburn, A.L. 1979. Geocryology. London:
Edward Arnold.

Environmental Geotechnics and Problematic Soils and Rocks, Balasubramaniam et al. (eds)
© *1987 Balkema, Rotterdam. ISBN 90 6191 785 9*

Soil-cement technology: A key instrument for rural housing development

V.Poopath
Kasetsart University, Bangkok, Thailand

ABSTRACT: Housing development in the rural sector of Thailand has been greatly disrupted by the lack of a key material resource, i.e. wood, which has been depleted at a very fast rate during the last three decades. There has been an attempt to search for wood substituting materials. Among them, soil-cement, a mixture of soil and a low percentage of cement with an additional amount of water under pressure, is found to have very high potential for wood substitution. The development of soil-cement technology, has been pursued. The paper is a review of development results, emphasizing the technical aspects of the technology, the background of the country's needs for rural housing development and finally the appropriateness of the technology for the rural housing development.

1 INTRODUCTION

Thailand, like many other developing or less developed countries has a dual social structure. Its population is high in the rural sector, about 8 in 10 of the total population. Even with this majority advantage the rural sector had been neglected for a long period of time. Efforts in development have been encouraged in the past three decades of accelerated development of the country. Physical development in the rural sector has imposed heavily upon the development of infrastructures and public utilities. Housing (or shelter) development has been left in the hands of the local people. In the days before the accelerating development period, wood which had been the main local resource for housing development, was still plentiful. Housing development was a traditional activity of the rural people. But with the country's development the resource has been quickly depleted leaving a major constraint on rural housing development and in other sectors.

With this sudden local resource shortages efforts for the development of substitution materials have been encouraged. It has been discovered recently that the lateritic soils in Thailand have high potential for use in the making of soil-cement by mixing them with 13 percent by weight of cement and a certain amount of water. The technology of soil-cement has been developed and strongly promoted for housing development in many areas of the country.

This paper is intended to encourage an awareness of the need for technologies for rural development. These technologies should be appropriate to the rural conditions and environment. But although the technologies may be simple in the technical sense, they may not be so simple in terms of their matching with the required conditions.

2 SOIL-CEMENT AND SOME OF ITS PROPERTIES

"Soil-cement" is a mixture of selected soils (mainly lateritic soils) with about 13 percent by weight of cement and water at optimum moisture content. In its moist state it can be used to make brick by moulding at a pressure of about 10 MN/m^2. Soil-cement can also be mixed to a wet state and used to cast floor slabs, mat foundation or even columns. However, it is still not recommended for elevated floors or beams, since deflection control is still a problem.

2.1 Mixture of Soil-cement

a) Proportions

The proportions of soil, cement and water should be laboratory established. After obtaining the amount of CaO in the soil, the amount of cement used can be calculated from:-

$$C = \frac{G - F}{E - F} \times 100 \ \%$$

when	C	=	percent by weight of cement,
	F	=	percent by weight of CaO in the soil,
	E	=	percent by weight of CaO in the cement,
and	G	=	percent by weight of CaO in the mixture.

(The percent of cement, for practical purposes, in the range of 10-15% by weight.)

The proportion of water is equal to the amount of water at optimum moisture content of the soil. The OMC of the soil is obtained from the well-known Standard AASHTO Test Method.

b) The mixture

The mixture should be uniform and homogeneous. Soil and cement should be throughly mixed before adding the water. Immediately after the water is added, mixing should begin to get a uniform and homogeneous mixture.

2.2 Soil for Soil-cement

In Thailand, where lateritic soils are plentiful, especially in the north-eastern and northern regions of the country, the soils in many areas have been proved by TISTR[1] to be applicable for soil-cement making.

Soils that are appropriate for soilcement making should have the following properties:-

a) Chemical properties:-

Fe_2O_3	ranges	1.5-3.0%
Al_2O_3	"	8-12%
SiO_2	"	75-85%
CaO+other matter	"	1.5-3.5%

b) Physical properties:-
- Loss on ignition less than 5%
- Fire shrinkage 2-8%

(1)   "Physical Development of Rural Settlements in Thailand", TISTR, Sept. 1981.

- Particles should be mostly silt with some clay (all soils should be sieved through a No. 4 sieve before use)

2.3 Soil-cement Blocks

Soil-cement blocks are the result of pressing soil-cement mixture in a mould under pressure (about 10 MN/m^2). The pressing process can be partially or fully mechanized. The partial process (normally used in rural housing programs) developed elsewhere and adapted Thai application is called "CINVARAM".

The shape and size of the soil-cement blocks can vary but the usual ones are shown in the figure 1. The first form is normally used with wet-mix soil-cement joint plaster to form a wall while the second can be used without the joint plaster. The second form is a recent development and has been proved to save construction time and labor as well as costing less than the first.

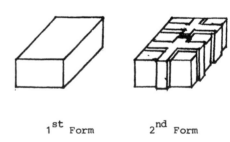

1st Form   2nd Form

Figure 1: Typical Forms of Soil Cement Blocks

2.4 Some Properties of Soil Cement

a) Strength of Soil-cement

The strength of soil-cement very much depends on soil properties, cement content and curing time, very much like cement concrete. Its average variation range is from 30 ksc to 60 ksc for 12% cement content and 21 days curing time.

Even with the large variations in strength (rather low in many cases), the strength properties do not constrain rural housing utilization, since most such houses are very low rise structures requiring very low material strength.

b) Modulus of Elasticity (E)

This property of soil-cement is normally very low in comparsion with concrete. The E-value as tested in the unconfined compressive laboratory test is normally in the range of about 1,000 ksc to 4,000 ksc, where the E-value of concrete can go as high as 400,000 ksc. The low E-value of soil-cement does significantly constrain its application for house construction since the elevated steel reinforced floor beams or slabs cannot be constructed without large deflection. With this limitation, wood substitation is still not achieved, since the elevated flow and the roof frame of the soil-cement house still depends on wood. Technically, reinforced concrete floor or roof beams can also be used in place of wood. But in economic terms, this is unsafisfactory if the beams have to be imported from urban areas.

c) Durability

Tests for durability were mainly conducted by the use of wet and dry processes in the laboratory. TISTR experience is that even at six rounds of wetting-drying, the soil-cement blocks were still in good shape without scratches and with very low strength reductions (not more than 10%).

d) Moisture Absorbtion

Moisture absorbtion is normally found not to exceed 25% in lateritic soil. This is a rather high figure (to prevent the effect of moisture in house construction, the exposed surfaces should be plastic painted). But rural houses are normally very open in design, so the moisture effect should be very low.

3 BASIC COUNTRY BACKGROUND

Thailand, like other developing countries, has a dual social structure, urban and rural. The population majority is in the rural sector (80%) but development in the sector is still very low. The major constraint for rural development is probably inadequate human resource development in the sector. Human development for a large population takes time while the need for physical development proceeds apace. It is necessary that the task of physical development must be approached with an awareness of the human resource problems of the rural sector.

In this regard, other development constraints which have to be considered in the development process are as follows:

(1) the dispersed nature of the rural areas causes low cost effectiveness for rural investment,

(2) the economic conditions of the rural people are, or average, much lower than those the urban, thus generating much lower perchasing power,

(3) the rural people have lower management capabilities and a lower capacity to acquire and use the new technologies. However, it is found that the rural people do have many skills, including skill in constructing simple structures.

The limited management capability of the rural people brings problems speeding development, even though the cooperative system has been effective in the sector for quite some times.

(4) resource constraints are another importion aspect in rural development. As formerly mentioned, the wood resources of the country have been depleted, which increases the need for other sources of materials to substitute for wood.

4 THE NEED FOR RURAL HOUSE-IMPROVEMENT

Field studies of 15 representative provinces in the north, northeast and south regions of the country were conducted by TISTR in 1981 to analyse the needs and demand for shelter development. The study was carried out in two ways. One was to observe the physical conditions and features of the rural houses, the other to interview house-owners using prepared questionaires to seek the perceived needs of the people.

The main study results summarised here.

4.1 Housing Condition

(1) Age of the houses

Even though the majority of the rural houses were found to be less than 5 years of age, about 7-20% were older than 20 years. The age of the house does not necessarily dictate its physical condition. Additional factors are the materials used for the construction, which linked to the age of the house can have a significant effect on its condition.

(2) Building Materials

A very high percentage of wood was used as building material (see Table 2). About 10% of the materials were found to be bamboo, especially in the upper floors and wall parts.

Table 1 Age Distribution of Rural Houses

(sample percent)

Age Range(yrs)	North	Northeast	South
1-5	53.21	48.25	26.49
6-10	21.15	20.10	24.5
11-15	12.18	6.7	11.92
16-20	4.49	2.58	11.92
21-25	5.13	5.16	9.93
26-30	1.28	5.64	7.64
31-50	1.28	-	7.28
Over 50	0.64	1.03	0.66
Not decer	-		2.65

Table 2 Building Materials

(percent)

Materials	Super-Structure	Under-Structure	Upper Floor	Wall
Concrete & Brick	-	15.57	-	-
Wood	92.98	83.29	87.27	58.71
Bamboo	7.02	1.14	12.73	15.63

Table 3 Living Space

Areas	North	North-east	South	Recommended min.space
Sleeping Area	18.5	25.3	16.3	8.64
Multiper-pose Area	31.17	34.18	29.3	12.96
Kitchen	10.30	11.1	11.2	7.2
Terrace	11.5	12.1	10.8	5.4
Total	71.41	82.68	67.6	34.2

Table 4 Percent Distribution of Rural Houses According to Physical Condition

Physical Condition	North	Northeast	South
Very Good	3	3	5
Good	25	16	28
Fair	62	67	44
Poor	10	14	23

The appearance of the houses was mainly good with only about 15-20% having a rather poor appearance.

(3) Living Space

The comparison of the average living space (as presented in Table 3) and the recommended minimum space (for a single family) indicates sufficient space for living if the family is a single one. But for enlarged families, which are still a common part of family patterns in rural areas, the space may or may not be sufficient depending on family size. But ample living space in rural houses is a tradition of Thai houses which reflects the plentiful local material resources of former days and the living habits of the rural people, which favor the enlarged family pattern.

The survey results of the existing living space and the living habits of the rural people reflect a moderate need for housing expansion or new houses for large family sizes or too many famlies in the same houses.

(4) Physical Condition

As presented in Table 4, the physical condition of the rural houses was an average fair to good. About 10-23% were in poor condition.

From the overall survey results on the physical condition of rural houses, it can be established that the need for housing development in any form involves on average about 10-15% of households. These estimated needs should be checked with owner perceptions, to be presented in the next paragraph.

4.2 Owner Perceptions

From the results of interviews with the house owners, it is indicated that a large percentage are not satisfied with heir houses. The house owners also suggest major needs for house improvement in three catagories, (1) extension and expansion, (2) new houses and (3) replacing some compenents of house. There were also stated needs for house repairs.

4.3 Need for House Improvements

Combining the two study results with other relevant information acquired in the study

Table 5 Percentage Distribution of House Improvement Needs			
Needs for Improvement	N	NE	S
Rebuild	22.63	20.22	19.72
Change Floors & Roofs	10.22	8.74	9.86
Change Floors & Walls	8.76	6.01	4.93
Extention and Expansion	21.9	25.69	38.73
Repair	7.3	11.48	11.27
Modification	1.46	2.73	-
No Answer	27.74	25.14	15.49

Table 6

Type of House		Cash Flow from Rural to Urban
One-Storey House	Wood	90.3%
	Concrete Block	74.4%
	Soil-cement	66.4%
Two-Storey House	Wood	93.7%
	Concrete Block	
	Soil-cement	81.4%

project, the need for rural house improvement is estimated to include about 200,000 housing units per year during 1982. The need is also forecast to become a demand if the situation does not significantly change. And to respond to the demand with minimum impact on the wood resources of the country, a new material substituting wood needs to be developed.

5 HOW APPROPRIATE IS SOIL-CEMENT IN SOLVING RURAL HOUSING DEVELOPMENT PROBLEMS

To answer this question criteria were set to handle the word "appropriateness". This paper proposes criteria for checking appropriacy.

Anything including technology will be appropriate to any society or country if it satisfies the following conditions:

a) Economic Conditions

Economic condition should be considered at 2 levels:-

(1) Country level

The general economic indicators at this level should be the global benefit cost ratio, the sharing of capital investment and finally, and very important, the cash flow out of the country.

Housing development is a part of social development which will give social benefit. The benefit is justified by the decrease in material disparity between rural and urban areas. The benefit cost ratio indicator then automatically justified.

As for the capital investment indicator, it can be estimated directly from the housing need, which is about 200,000 housing units per year or about 6,000 million

baht per year. This sum of money is only 3 percent of the agro-product value.

The most important point that should be considered at this level is the cash flow out of the country. Soil-cement technology can justify this indicator in that its requirement for imports is only for some materials used in the cement production. The fraction of such materials is only about 5 percent of the total cost, which is very low. The cash flow out is there very low.

(2) Sectoral Level

The economic indicator used at this level should be the cash flow out of the rural sector. The figures are estimated and presented in Table 6 where wood, block and soil-cement houses are compared. It can be seen that soil-cement house can reduce the rural-urban out flow by about 33% in comparison to 10% for a wooden house.

(3) Individual level

At this level the main economic indicator is house cost in comparison to the purchasing power of the rural people.

At this point a comparative estimation is made for the house costs of wooden and soil-cement houses. It is found that for the same design, a soil-cement house is 15-30% cheaper than wooden house. An attempt has also been made to design soil-cement houses with bamboo and thatch roof and it is found that with zero labor costs the cost of a soil-cement house can be as low as 5,000 baht for a small single-family housing unit.

b) Social Conditions

To satisty social conditions the technology must suit average social knowledge,

Table 7

Type of House		Skill Labor	Labor
One-Storey House	Wood	51.1%	48.9%
	Concrete Block	51.1%	48.9%
	Soil cement	39.9%	60.7%
Two-Storey House	Wood	53.5%	46.5%
	Concrete Block		
	Soil cement	29.4%	70.6%

skills and attitudes, and must also not drastically change the life-style and customs of the people.

This aspect is evaluated for soil-cement technology and it is found that even though soil-cement technology may not completely suit rural knowledge and skills, a short training program can be sufficient to upgrade knowledge and skills as required. There are good attitudes towards soil-cement houses. And most of the technological developments have been pursued with proper condideration for the preservation of the people's life-stye and customs.

c) Resource Condition

This condition will involve two components of resources, the human and the material. The condition will be satisfied only if the technology uses available resources at proper rates of consumption.

Soil-cement technology has been designed to be highly labor-intensive in response to high rural labor waste during the off-season period. The soil-cement house has been found to use a higher proportion of labor to skilled labor as presented in Table 7, which responds very well to the manpower structure in the rural sector.

For material resource utilization, the amount of soil used to meet housing demand is comparatively low compared with the use of soil in road construction.

A close study of this condition reveals that the main advantage of soil-cement technology is here, where available human resources will be used to make a contribution to social development, conserving a very important natural resource, that is wood, and finally reducing the gap between urban and rural areas in physical terms.

Environmental Geotechnics and Problematic Soils and Rocks, Balasubramaniam et al. (eds)
© *1987 Balkema, Rotterdam. ISBN 90 6191 785 9*

Modeling of gradual earth-fill dam erosion

Vijay P.Singh
Department of Civil Engineering, Louisiana State University, Baton Rouge, USA
Panagiotis D.Scarlatos
Louisiana Water Resources Research Institute, Louisiana State University, Baton Rouge, USA

ABSTRACT: Prediction of the flash flood hydrograph emanating from a breached earthen dam is important for contingency evacuation planning and decision-making for dam safety. Gradual failure of a dam is a complicated, unsteady, nonlinear, multi-phase, three-dimensional phenomenon. Mathematical modeling of this phenomenon requires an integrated approach to the two main phases involved, water and soil, through physical aspects from various disciplines such as hydrology, hydrodynamics, sediment transport and geotechnology. A successful model should simulate both the outflow discharge hydrograph and the breach enlargement processes. In this paper the state-of-the-art of dam breaching is discussed and a review of the existing numerical models given. A new physically-based computer model (Breach Erosion of Earthfill Dams – BEED) is presented and applied for the failure of Teton dam in Idaho, USA. Suggestions for further research are made.

1 INTRODUCTION

Failure of a dam is a major event which may have very serious consequences. During historical times, more than 2,000 dam failures were recorded around the world. As a result of those failures, thousands of lives were lost and tremendous financial losses suffered. Today it has been estimated that there are approximately 150,000 dams worldwide which impose a risk of serious damage in case of failure (Gruner, 1967). In spite of such statistics, knowledge of triggering and control mechanisms of dam breaching and failure is very limited. The large majority of dams are either man-made earth-fill dams or embankments resulting from land-slides. It has been reported that earth dams break during a time period ranging from a few minutes to several hours (Singh and Snorrason, 1982). Failure can be caused by unpredictably large inflows; differential settlement; landslides; earthquakes; poor design, construction or operation; or acts of war. The initial stages of dam deterioration occur in the form of: (a) internal erosion (piping), (b) seepage and progressive erosion of the downstream face, or (c) overtopping of the crest and erosive enlargement of the breach. Once a breach has developed, erosion and sloughing contribute to the destructive processes of dam failure. The ability to simulate these processes is essential for: (a) flood forecasting, (b) contingency evacuation planning, and (c) operational and management decisions for dam safety. Indeed, duration and shape of the flash flood hydrograph is in direct relation to the pattern of dam failure.

The dam-break problem can be divided for simulation purposes into two parts: dam failure description and flood routing. Solution of the first part is essential because it provides the necessary initial condition for solution of the second part. However, equal attention has not been paid to the study of the two problems. There is a very extensive literature on aspects of flood routing physical description, modeling and solutions. The processes of dam-breaking have recently been given proper attention, and there are only few simulation techniques. In most past applications, the dam was considered as breaking instantaneously.

The problem of earth dam breaching and failure is a three-dimensional, unsteady, nonhomogeneous and nonlinear combination of hydrologic elements, hydrodynamics, sediment transport processes and geotechnical aspects. At this stage of theoretical knowledge, technical expertise and practical experience, mathematical simulation of earth dam failure is possible only under certain simplifying assumptions and simplifications.

The purpose of this paper is to present the physical characteristics of earth dam erosion and to provide a theoretical basis for mathematical simulation. The state-of-the-art is analyzed through the discussion of existing models. A model recently developed by the authors is also included (Singh and Scarlatos, 1985). Emphasis is given to the validity of the various modeling approximations, and suggestions for further research are provided.

2 PHYSICAL CHARACTERISTICS OF DAM BREACHING

Breach morphology at its initial stage and during the erosion processes is the main characteristic of the failure of a dam. There is a large quantity of data regarding breach shapes and dimensions, but predictive correlations are very limited. Through analysis of existing data, Singh and Snorrason (1982) concluded that the width of the breach is usually two to five times the height of the dam, and that with a fifty percent probability the overtopping hydraulic head at the moment of initiation of breaching is less than 0.5 m. Ponce (1982) derived an approximate relationship between breach Froude number F and shape factor S_F given as

$$F = 0.20 \ S_F^{-0.39} \tag{1}$$

The two quantities are defined as

$$F = \frac{Q_p}{B(gd^3)^{1/2}} \tag{2}$$

and

$$S_F = \frac{Bd}{B_D Z_o} \tag{3}$$

where Q_p is the peak outflow discharge, B is the top width of the breach, d is the depth of the breach, B_D is the top width of the dam, Z_o is the maximum height of the dam, and g is the acceleration of gravity. Another interesting compilation of data is provided by MacDonald and Langridge-Monopolis (1984). They also suggested an empirical methodology for the prediction of the size, shape and failure duration time for an earthfill dam. In their study the breach shape was assumed to be triangular with 2V:1H side slopes; after it reached the bottom of the dam, the breach shape became trapezoidal. Based on the same set of data, Houston (1984) proposed a trapezoidal-shaped breach with base width equal to depth and 1V:1H side slopes.

Although all these studies provide valuable information on the physical characteristics of the breach, their applicability is rather limited. The scattering of data points and lack of theoretical explanation imply that the results should be used with caution and judgement, and that additional basic research is needed.

3 REVIEW OF EXISTING MATHEMATICAL MODELS

Attempts at mathematical simulation of gradual failure of earthfill dams started in the mid 1960's. Since then, a few models have been developed based on the principles of hydraulics, hydrodynamics and mechanics of sediment transport. The level of simplicity and the assumptions included in each of the existing models vary. A general discussion of these models is given here.

3.1 Christofano's Model

This is probably the first model ever developed for simulation of earthfill dam erosion processes (Cristofano, 1965). By equating the force of water flowing through the breach to the resistant friction force acting on the wetted perimeter, Cristofano obtained an analytical expression which reads

$$\frac{Q_s}{Q_b} = K_c \exp \left(- \frac{\ell \tan \phi}{h}\right) \tag{4}$$

where Q_s is the sediment discharge, Q_b is the water discharge through the breach, K_c is an empirical coefficient, ℓ is the length of the breach bottom in the direction of flow, ϕ is the developed angle of repose and h is the hydraulic head. The model incorporates a trapezoidal shaped breach of constant width, and side slopes equal to the developed angle of internal friction. The proportionality coefficient K_c is not related to any physical quantity and is therefore difficult to estimate.

3.2 Harris and Wagner's Model

Harris and Wagner (1967) treated the problem by assuming a parabolic breach subjected to erosive action. The water flow was approximated by spillway overflow hydraulics, sediment transport by the Schoklitsch bed-load formula. Tailwater effects and sloughing of the breach sides were neglected. The model required specification of the geometrical features of the breach in addition to sediment grain size and threshold value of discharge for initiation of erosion.

3.3 BRDAM Model

The Bureau of Reclamation computer model BRDAM, presented by Brown and Rogers (1977, 1981), is based on the work of Harris and Wagner. The model was able to simulate erosion from overtop flow or from piping.

3.4 DAMBRK Model

The National Weather Service computer model DAMBRK, developed by Fread (1977) is a parametric model which can handle rectangular, triangular or trapezoidal breach shapes. The breach was assumed to grow at a predetermined rate while the reservoir water depletion is described by a simple continuity equation. For its application the model required definition of the terminal breach geometry and the failure time duration. These requirements reduce the model to a means of identifying a range of possible flood events but not the one most likely to occur.

3.5 Lou's Model

In Lou's (1981) model, the hydrodynamic field was simulated by the St. Venant system of equations from which the inertia terms were neglected. The solution was based on Priessmann's finite difference scheme. The sediment transport mechanisms were approximated by a simplified equation as follows:

$$M = e_i \, t_d \, u^4 \qquad (5)$$

where M is the mass of eroded soil, e_i is an empirical coefficient, t_d is the failure duration time and u is the mean water velocity. Uncertainty in the estimation of the parameters involved restricts the applicability of the model.

3.6 Ponce and Tsivoglou's Model

Ponce and Tsivoglou (1981) extended Lou's work. In their model they treated the hydrodynamic field the same way as Lou did but improved the simulation of sediment transport processes by utilizing the Meyer-Peter and Muller bed-load formula. For the breach morphology, they introduced a regime type relation between rate of flow and top width of the breach. This relation was applied from the initiation of erosion until the occurrence of peak outflow discharge, after which the breach was kept constant. The weakness of this model is the determination of the rate of growth of the top width of the breach.

3.7 BREACH Model

This model was presented by Fread (1984) and was an important contribution to earth dam failure simulation techniques. The model is an iterative one based on a quasi-steady state uniform flow discharging over a broad-crested weir. Tailwater effects were included for correction of the outflow discharge. Sediment transport was estimated by the Meyer-Peter and Muller formula. Besides the erosion effects, the breach was allowed to grow from sloughing due to side instability. The geotechnical analysis was based, however, on dry soil conditions. The bottom of the breach was assumed always to be parallel to the downstream face of the dam. The main limitation of the model is the difficulty of estimating the values of critical shear stress for erosion, and terminal breach width, which are both required as input data.

3.8 Evaluation of the Existing State-of-the-Art

All of the existing models were applied to historical dam failure cases and they all exhibited a satisfactory degree of accuracy. Proper adjustment during calibration of the parameters involved improved the performance of the models considerably. Since most of these parameters, unfortunately, cannot be estimated from field measurements, the models cannot be utilized successfully for predictive purposes. In spite of the significant thrust of these models in the methodology of earth dam failure simulation, there are still many aspects that can be improved. Improvements can be made mainly by the inclusion of more detailed and realistic descriptions of the physical processes, and by using mathematical expressions with well-defined, measurable quantities.

4 THEORETICAL PRESENTATION OF EARTH DAM EROSION PROCESSES

Gradual failure of an earth dam is intuitively regarded as a two-phase, water-soil interacting system. Water discharging from the reservoir through the breach, causes enlargement of the breach by erosion and/or sloughing. This process continues until either the reservoir is emptied or the breach resists any further erosive action. It is on the basis of this general qualitative description that the various physical aspects and their mathematical formulation will be analyzed in the following sections.

4.1 Reservoir Water Mass Balance

The motion of the water from the reservoir towards the breach is dynamic and is essentially described by both continuity and momentum equations. However, due to the comparatively small velocities and to the locality of dynamic effects, the water depletion within the

reservoir can be described by a single water mass balance equation as

$$A_s(H) \frac{dH}{dt} = I_o - Q_b - Q_o \qquad (6)$$

where A_s is the surface area of the water within the reservoir, H is the reservoir water level, I_o is the inflow discharge, Q_o is the outflow discharge through spillway and/or power house and t is the time. A schematic representation of the variables is presented in Fig. 1.

Inflow discharge I_o can be predicted through statistical analysis of all possible water sources such as riverine water, overland runoff, direct precipitation or groundwater flow. The outflow Q_o is a predetermined function of both water level H and time. Also the relation $A_s = A_s(H)$ is a function known from the topography of the undulated valley. However, Eq. 6 cannot be solved until Q_b is expressed in terms of the reservoir water level H.

4.2 Hydraulics of Breach Outflow

The hydraulics of breach outflow is similar to that of flow over a broad-crested weir. Therefore, for the case of trapezoidal breach, outflow can be estimated as

$$Q_b = (C_1 b + C_2 h \tan \theta) h^{3/2} \qquad (7)$$

where C_1, C_2 are constant dimensional coefficients and θ is the angle between breach side and the vertical. The hydraulic head h is a function of time given as

$$h = H - Z \qquad (8)$$

where Z is the elevation of the breach bottom. Flow over the downstream face of the dam can be approximated by quasi-steady state conditions for each time step. Thus, from critical conditions on the breach at the crest of dam, the flow will reach supercritical conditions and normal depth by forming an S-2 profile. The water profile can be computed from the momentum equation

$$\frac{d}{dx}\left(\frac{Q_b^2}{2g A_b^2}\right) + \frac{dy}{dx} + (S_f - S_o) = 0 \qquad (9)$$

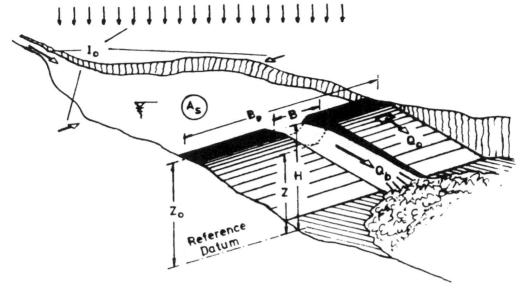

Figure 1. Failure of earthfill dam.

where A_b is the wet cross-section area
of the breach, y is the water depth, S_o
is the slope of the breach at the downstream face of the dam and S_f is the
frictional energy dissipation as it is
given by Manning's or Chezy's formula.
Eq. 9 can be easily written in the form

$$\frac{dy}{dx} = F(y) \tag{10}$$

and then integrated numerically (Prasad,
1970) In most of the cases however, the
flow reaches its normal depth over a
short distance and can be assumed to be
under normal conditions on the entire
downstream face of the dam.

In case the breach outflow is submerged, tailwater effects should be
included by reducing the coefficients
C_1, C_2 according to the relation

$$C_{1,2} = C_{1,2} \, [1.0 - 27.8 \, (\frac{y_o - Z}{H - Z})$$
$$- 0.67)^3] \tag{11}$$

where y_o is the water depth at the tailwater section.

Combination of Eq. 6, 7 and 8 results in
an ordinary differential equation with
two unknowns, H and Z. For the com-
pleteness of the problem, another equation must be introduced that will

describe the behavior of the variable Z.
This can be achieved by the analysis of
erosive processes.

4.3 Mechanics of Breach Erosion

The mechanics of erosive processes under
extremely dynamic conditions such as
those of dam failure have not been
investigated as yet. This limits the
choice for the description of sediment
transport to the conventional formulae
derived for alluvial channels. The most
widely accepted of these expressions is
the Einstein-Brown bed load formula
(Simons and Senturk, 1976). The advantage of this method is that it does not
require the determination of the critical shear stress for erosion, and that
it matches a wider range of experimental
results. Einstein-Brown formula reads

$$\Phi = f \, (\frac{1}{\Psi}) \tag{12}$$

where Φ is the sediment transport rate
function and Ψ is the inverse of Shields
shear stress. Explicitly, the
quantities Φ and Ψ are given as

$$\Phi = \frac{q_{bw}}{\gamma_s \, k_E \, [g \, (\frac{\gamma_s}{\gamma} - 1) \, D_s^3]^{1/2}} \tag{13}$$

and

133

$$\frac{1}{\Psi} = \frac{\tau}{(\gamma_s - \gamma) \, D_s} \qquad (14)$$

where q_{bw} is the bed-load discharge in weight per unit width, γ_s is the specific weight of soil, γ is the specific weight of water, D_s is a representative size of sediment grains, τ is the bed shear stress, and $k_E = K_E(D_s, \gamma, \gamma_s)$ is a known constant parameter. Since the functional relationship of Eq. 12 has been established only for a limited range of $1/\Psi$ for high shear stresses, an extrapolation is required.

Once the bed-load discharge q_{bw} is calculated, the rate of erosion can be estimated accordingly as

$$\frac{\Delta Z}{\Delta t} = \frac{q_{bw}}{(1 - p) \, \gamma_s \, \ell} \qquad (15)$$

where p is the porosity and Δ is the finite difference symbol. Eq. 15 describes the rate of change of breach bottom elevation Z.

4.4 Geotechnical Stability of Breach Slopes

Stability of breach slopes is a very complicated process. For simplification it will be assumed that failure occurs along a linear shearing plane that passes through the toe of the slope. Utilizing the "contour" method (Chugaev, 1964) the stability can be analyzed as the result of gravity, hydrodynamic seepage forces, cohesion and soil friction. Combination of all these forces yields the stability relation

$$F_H + G \tan (\zeta - \phi) \leq C \, x_p \, [1$$
$$+ \zeta \tan (\zeta - \phi)] \qquad (16)$$

where F_H is the horizontal component of the seepage force, G is the weight of the sliding wedge, ζ is the angle between the shearing plane and the horizontal, x_p is the horizontal projection of the shearing plane and C is the cohesion. A schematic representation of the problem, along with information for computation of the F_H and weight G, are given in Fig. 2.

5 BREACH EROSION OF EARTHFILL DAM (BEED) NUMERICAL MODEL

The Breach Erosion of Earthfill Dam (BEED) model is a numerical model developed for the prediction of flash flood hydrograph produced by gradual failure of a dam. The model is suitable for both personal and main-frame computer. Simulation of the failure processes is based on the theoretical information previously given.

5.1 Physical Description

The model is able to simulate gradual failure of a homogeneous earth dam by utilizing rectangular, triangular or trapezoidal shaped breach. The size and shape of the initial breach should be provided as initial conditions. The selection of these conditions is based entirely on engineering justification. Once the simulation starts, the flowing water causes erosion on both the crest and downstream face of the dam. How-

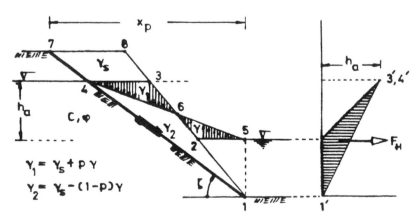

Figure 2. Stability of breach sides.

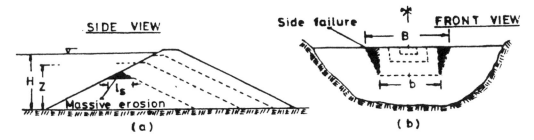

Figure 3. Erosion pattern simulated by BEED model

ever, erosion at the crest occurs at a much lower rate than the one at the downstream face. Schematically, the erosion processes are represented in Fig. 3a. Whenever the downstream face meets the upstream face, a mass erosion of the top cone is assumed to occur, so that the crest always maintains a specific horizontal length ℓ_s (Fig. 3a). The lateral erosion of the breach is assumed to occur in such a way that the depth over the bottom width ratio remains constant. Sloughing however can alter the initial breach shape. Schematically, the situation is presented in Fig. 3b. The top width of the downstream face breach is always set equal to the top width of the breach at the crest. When tailwater effects are present, erosion is considered only on the crest and not on the downstream face. The program terminates when the hydraulic head h becomes zero.

5.2 Description of Solution Algorithm

Since the variables H and Z are interdependent, the solution algorithm of the model is iterative. The computational steps are represented by the flow chart in Fig. 4. The scheme is very efficient and converges after very few iterations. A detailed analysis is given elsewhere (Singh and Scarlatos, 1985).

6 APPLICATION-RESULTS

The BEED model was applied to simulate the outflow discharge produced by gradual failure of the man-made Teton dam in Idaho, USA. On June 3, 1976, field engineers detected leakage at the toe of the 100 m high dam. A rapid rate of piping exacerbated the situation and

by noon on June 6, 1976, the crest of the dam was breached. Duration of failure was approximately 4 hours and a maximum discharge of 68,000 m^3/s was recorded. The final breach was about 170 m wide and 87 m deep. Embankment material used for the construction of Teton dam consisted of clay, silt, sand and rock fragments obtained from excavations and borrow areas in the Teton river canyon.

The BEED model started the simulation from the minute that an initial rectangular breach of 1.0 m width and 0.6 m depth developed on the crest of the dam. The material properties of the breach was assumed as follows: $\phi = 40°$, C = 5 t/m, γ_s = 2.5 t/m^3, D_s = 0.005 mm. The simulated terminal breach was 100 m deep and 175 m wide. The computed outflow hydrograph is presented in Fig. 5. From this it is evident that the timing, shape and magnitude of the actual hydrograph closely resembles the hydrograph derived from the BEED model.

7 CONCLUSIONS

The existing state-of-the-art of gradual dam-breaking modeling is continuously improving. An effort towards this goal was the development of the BEED numerical model. This model utilizes broad-crested weir flow for breach hydraulics and quasi-steady conditions for flow over the downstream face of dam. The Einstein-Brown sediment transport formula estimates the erosion processes while the 'contour' method is applied for the slope stability under the action of seepage forces. The model is based on more realistic physical aspects than any other model and requires fewer

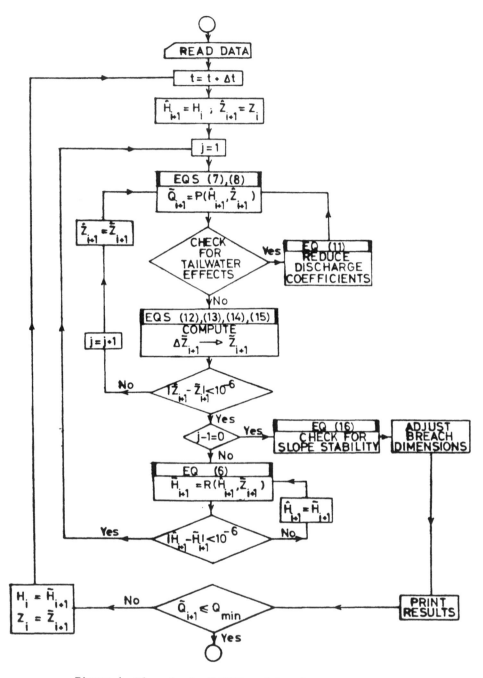

Figure 4. Flow chart of BEED model solution algorithm.

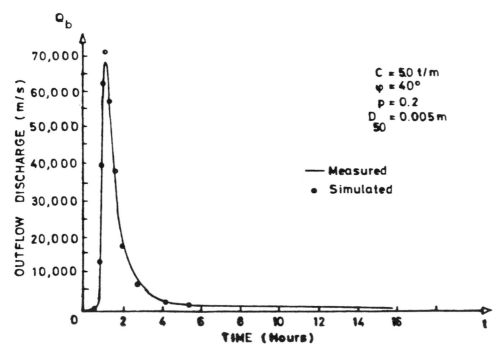

Figure 5. Outflow discharge from Teton Dam, Idaho, USA.

empirical parameters. The iterative
solution algorithm is very efficient and
suitable for both micro and main-frame
computers. Application of the model
demonstrated a very close agreement
between actual data and simulation
results. Further research on the subject should be concentrated on experimental work on erosive processes under
high dynamic conditions.

8 ACKNOWLEDGMENTS

This study is part of the project "Sen-
sitivity of Flood Wave Parameters by Dam
Breach Erosion Modeling Procedures,"
which was funded by the Environmental
Laboratory, Waterways Experiment
Station, U.S. Army Corps of Engineers.

REFERENCES

Brown, R.J. and Rogers, D.C. (1977), "A
    simulation of the Hydraulic Events
    During and Following the Teton Dam
    Failure," Proceedings, Dam-Break Flood
    Routing Model Workshop, Bethesda,
    Maryland, October 18-20, pp. 131-163.
Brown, R.J. and Rogers, D.C. (1981),
    "Users Manual for Program BRDAM,"
Engineering and Research Center,
    Bureau of Reclamation, Denver,
    Colorado.
Chugaev, R.R. (1964), "Stability Analy-
    sis of Earth Slopes," Israel Program
    for Scientific Translations, Jerusa-
    lem.
Cristofano, E.A. (1965), "Method of Com-
    puting Rate of Failure of Earth Fill
    Dams," Bureau of REclamation, Denver,
    Colorado.
Fread, D.L. (1977), "The Development and
    Testing of a Dam-Break Flood Fore-
    casting Model," Proceedings, Dam-Break
    Flood Routing Model Workshop, Bethesda,
    Maryland, October 18-20, pp. 164-197.
Fread, D.L. (1984), "A Breach Erosion
    Model for Earthen Dams," NWS Report,
    NOAA, Silver Spring, Maryland.
Gruner, E. (1967), "The Mechanism of Dam
    Failure," Neuvieme Congres des Grands
    Barrages, CIGB, Q. 34, R. 12,
    Istanbul, pp. 197-206.
Harris, G.W. and Wagner, D.A. (1967),
    "Outflow from Breached Earth Dams,"
    B.S. Thesis, University of Utah, Salt
    Lake City, Utah.
Houston, M. (1984), "Breaching Charac-
    teristics of Dam Failures, Discus-
    sion," Journal of Hydraulic Engi-
    neering, ASCE, Vol. 110, No. 11,
    pp. 1125-1129.

Lou, W.C. (1981), "Mathematical Modeling of Earth Dam Breaches," Ph.D. Dissertation, Colorado State University, Fort Collins, Colorado.

MacDonald, T.C. and Langridge-Monopolis, J. (1984), "Breaching Characteristics of Dam Failures," Journal of Hydraulic Engineering, ASCE, Vol. 110, No. 5, pp. 567-586.

Ponce, V.M. and Tsivoglou, A.J. (1981), "Modeling Gradual Dam Breaches," Journal of the Hydraulic Division, ASCE, Vol. 107, No. HY7, Proc. Paper 16372, pp. 829-838.

Ponce, V.M. (1982), "Documented Cases of Earth Dam Breaches," San Diego State University Series No. 82149, San Diego, California.

Prasad, R. (1970), "Numerical Method of Computing Flow Profiles," Journal of the Hydraulic Division, ASCE, Vol. 96, No. HY1, pp. 75-85.

Simons, D.B. and Senturk, F. (1976), "Sediment Transport Technology," Water Resources Publications, Fort Collins, Colorado.

Singh, K.P. and Snorrason, A. (1982), "Sensitivity of Outflow Peaks and Flood Stages to the Selection of Dam Breach Parameters and Simulation Models," State Water Survey Division Report 289, Surface Water Section at the University of Illinois.

Singh, V.P. and Scarlatos, P.D. (1985), "Breach Erosion of Earthfill Dams and Flood Routing: BEED Model," Report to U.S. Army Corps of Engineers, Environmental Lab., WES, Vicksburg, Mississippi.

Environmental Geotechnics and Problematic Soils and Rocks, Balasubramaniam et al. (eds)
© *1987 Balkema, Rotterdam. ISBN 90 6191 785 9*

A study on the use of sea water in lime and cement stabilization

Zaki A.Baghdadi & Ahmed M.Khan
King Abdulaziz University, Jeddah, Saudi Arabia

ABSTRACT: There may be occasions when potable water may not be easily available for construction and subsoil saline water or sea water may be economical to use. This paper presents results of investigation on the use of sea water in lime and cement stabilization of a clay soil. Compressive strength and durability characteristics have been studied in order to compare the effect of using sea water in soil stabilization. The results indicate that sea water may be used successfully as mixing water for soil lime stabilization. The compressive strength increases upon curing and is higher than that obtained by using distilled water. With cement, strength increase is also observed with sea water upon curing but compressive strength values are lower as compared to distilled water. Similarly, the sea water lime treated soil is found to be more durable as compared to distilled water lime treated soil. In the case of cement stabilization, the distilled water shows higher durability.

1 INTRODUCTION

In many areas, geotechnical projects may involve usage of poor soils which necessitates their improvement by physical or chemical methods depending on type of soil and the nature of the project. Mixing water affects the properties of improved soil, for example, as discussed by Kezdi(1979) and others. Near sea coasts, where considerable geotechnical construction activity often takes place, the sea water is easily available; and in many land forms, subsoil water is saline. Such water is cheap and available in abundance. Desalination plants provide amounts of water that hardly meet the drinking water demands. The present research on usage of sea water was initiated with the objective of studying the behaviour of improved soils using sea water. Results are compared by taking distilled water as reference.

The study was conducted on a problem soil requiring improvement before being employed in construction. Various percentages of stabilizing agents were added and compaction, compressive strength and durability characteristics were investigated.

The results are encouraging, with both lime/and, to a certain extent, cement and are presented in the following sections.

2 MATERIALS

The materials used in this study are described here.

2.1 The soil

The soil utilized was taken from Akol near the city of Madinah, Saudi Arabia. Classification of this soil according to the AASHTO system shows that it belongs to the (A-7-6) group with 57% clay content. The soils of this group are considered poor construction materials which often require improvement. Tests to characterize the material reveal the following:

a) X-ray diffraction traces (Figures 1 & 2) show that the clay present in the soil is mainly kaolinite with some illite and montmorillonite.
b) Specific gravity, G_s = 2.56
c) Atterberg limits LL = 49%, PL = 20% and PI = 29%. Since PI is high, the soil requires improvement.
d) Average soluble solids = 3.55%
e) pH = 8.8; this indicates that the soil is basic and thus favourable to lime and cement stabilization.

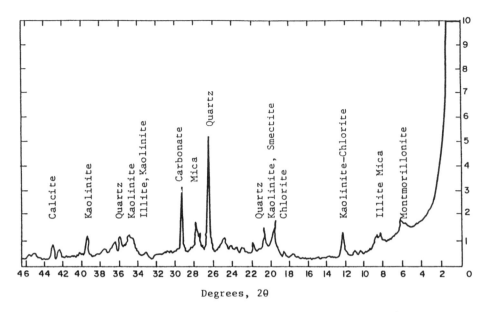

Fig. 1 X-ray diffraction of Akol Clay (random sample)

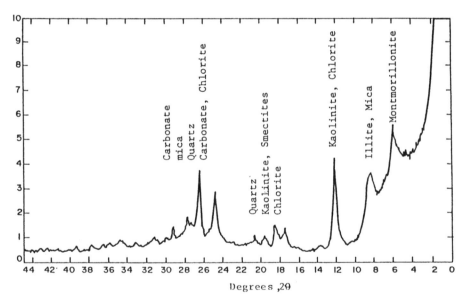

Fig. 2 X-ray diffraction of Akol Clay(oriented sample)

2.2 Cement(Type I) normal portland cement, manufactured by Rolaco (Greece).

2..3 Lime: Hydrated lime, Ca (OH)$_2$ manufactured by Riyadh Cement Co., Suaūdi Arabia.

2.4 Mixing Water:
   a) distilled water,
   b) sea water; Red sea water was utilized with the following properties:

- amount of total solids = 34.55g/l
- amount of sulphates = 2092 mg/l
- pH = 8.3

3 EXPERIMENTAL WORK

The experimental work in this research was aimed at the characterization of materials used and determination of compaction, compressive strength and durability characteristics. Most of the tests were conducted according to available ASTM standards (1977) and these included Atterberg limits, specific gravity, sedimentation, pH, modified AASHTO compaction, unconfined compression and durability (wetting and drying). In addition, X-ray diffraction analysis was carried out on random and oriented soil specimens and the amounts of total solids and sulfates in sea water were also determined.

The compaction tests were conducted on natural and stabilized soil with distilled and sea water respectively. With both distilled and sea water, 10% of cement and 10% of lime, by weight of dry soil, was used for comparison.

In the unconfined compression tests on stabilized specimens, a series of tests was run using various percentages of cement and lime for both types of mixing water. Tests were done on 5cm dia and 10cm high cylinders stabilized by 0%, 6%, 8%, 10% and 12% of cement and lime by weight of dry soil. The waxed cylindrical specimens were tested after 7 days, 28 days and 60 days of curing at 73°F.

The durability tests (wetting and drying) were done on soil-cement and soil-lime mixes using both distilled and sea water. Cement contents were 8%, 10%, 12% and lime contents were 10%, 12% and 14% respectively. The durability study specimens were cured for 28 days before the testing was started.

4 DISCUSSION OF RESULTS

The results of this investigation give an insight to the practicing engineer as to what to expect when he decides to use sea water in soil-cement and soil-lime mixes. There are clear trends in the results that will help in drawing conclusions from the study.

Examination of the compaction curves of the soil (Figure 3) reveals that the compaction characteristics are more or less the same for both distilled water and sea water cases. However, when the soil is treated with 10% cement, improvement is noticed in both cases. The soil-cement distilled water mix is slightly better than the soil-cement-sea water mix. The greater bonding in the former as evidenced by its higher strength may be the reason for its better compaction results (as may be seen later).

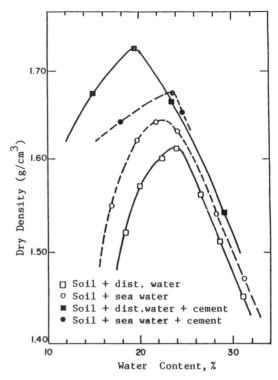

Fig.3 Dry Density water content relationships of natural soil and soil stabilized with 10% cement using distilled and sea water.

The stress-strain curves of specimens cured of 7 days, 28 days and 60 days for both cases are shown in Figures 4 and 5, using 10% cement.

As expected, the addition of cement results in a marked improvement of compressive strength, as shown in Figures 6 and 7. However, soil-cement-sea water gave lower strengths than the soil-cement-distilled water by as much as 35% for the 7-day curing, 40% for the 28-day curing and 18% for the 60-day curing. Even for the no-cement soil specimens, the soil with distilled water showed higher strength, although in most cases the difference was relatively small (3% - 4%).

Sea water definitely affects the mechanism of the treatment. It is most probable that the salt content (sulphates and chlorides) retarded both the primary and secondary reactions within the soil-cement system. It should be noted that rates of strength increase for each group, i.e. the rates of increase in strength of sea water mixed specimens from 7 days to

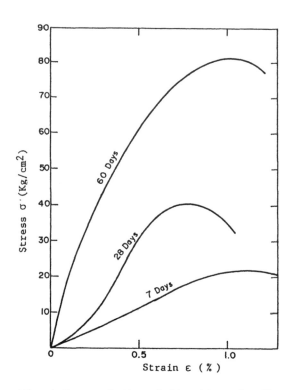

Fig. 4 Stress-strain relationships of soil
with 10% cement and distilled water.

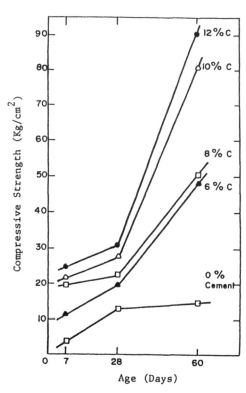

Fig. 6 Compressive strength of soil-cement
with distilled water

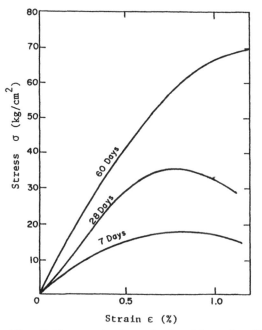

Fig. 5 Stress-strain relationships of soil
with 10% cement and sea water

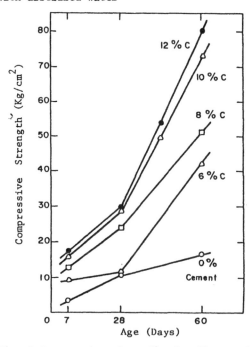

Fig. 7 Compressive strength of soil-cement
with sea-water

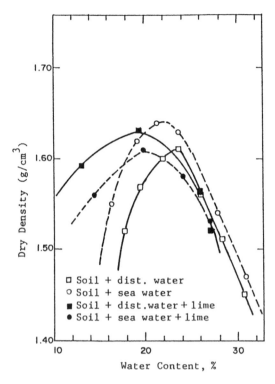

Fig. 8 Dry density water content relationships of natural soil stabilized with 10% lime, using distilled and sea water.

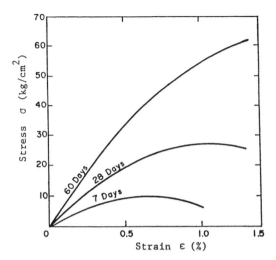

Fig. 9 Stress-strain relationships of soil with 10% lime and distilled water.

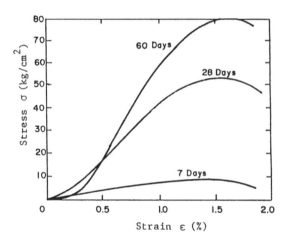

Fig. 10 Stress-strain relationships of soil with 10% lime and sea water.

28 days curing and from 28 days to 60 days are essentially the same.

The wet-dry cycling of soil-sea water and soil-distilled water mixed with 12% cement shows the greater durability of the soil-cement-distilled water mix; 18.5% versus 20.8% weight loss was observed.

Stabilization of the same soil used in this study with lime mixed with distilled water and sea water shows no improvement in the compaction characteristics although the optimum water contents are slightly less than those of the unstabilized soil, as shown in Figure 8.

Figures 9 and 10 show the stress-strain relationships for distilled water and sea water mixed specimens cured for 7, 28 and 60 days.

The addition of lime results in an increase in the compressive strength, as illustrated in Figures 11 and 12. However, the trend here is reversed as compared to cement-stabilized results. Soil-lime-sea water specimens gave higher compressive strength values. This trend is of course, subject to chemical reactions that took place, which will be studied in more detail

in the near future.

Also, wet-dry cycles revealed higher durability of the soil-lime-sea water mixes. Plasticity index results show that mixing with sea water has reduced the P I value of the soil, from 29 to 11. But, the addition of 10% cement and lime to the distilled water-soil mixtures reduced the P I to about the same value (from 29 to 11). Also the addition of 10% cement and lime to the sea water-soil mixtures reduces the P I to the same value (from 11 to 8). This indicates that the sea water has compounded the effect of the hydrated lime in reducing the plasticity index.

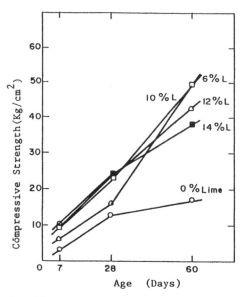

Fig. 11 Compressive strength of soil-lime with distilled water.

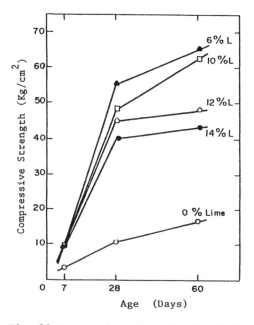

Fig. 12 Compressive strength of soil-lime with sea water.

5 CONCLUSIONS

In the light of this research, the following conclusions are drawn:

1. Soil dynamic compaction using distilled water or sea water yields nearly similar results in terms of maximum dry density and optimum water content. Hence sea water may be used in field compaction.
2. Sea water may be used as mixing water in soil-lime stabilization. With curing, the compressive strength of soil-lime-sea water mixes increases.
3. At various percentages of lime-content, the sea water mixes give higher strengths than distilled water mixes.
4. In the case of stabilized mixes using ordinary portland cement, the distilled water gives higher strengths as compared to sea water. Durability tests (wetting and drying) on soil-cement mixes using distilled water show higher durability as compared to soil-cement mixes using sea water. Extreme caution should be observed if sea water is to be utilized in soil-cement. It should be expected that the strength will be lower, perhaps by as much as 30%.
5. Sea water is found to reduce substantially the soil plasticity index.

ACKNOWLEDGEMENTS

The authors wish to express their gratitude to the staff and students who extended cooperation and assistance in the completion of this study.

REFERENCES

Kezdi, A. 1979. Stabilized earth roads p.108-173. Amsterdam: Elsevier Scientific Publishing Company.
Al-Tatari, M.M. & Hamadah, Z.H. 1984. A study on the use of Red sea water in cement and lime stabilized Soil. p.198. Jeddah: King Abdulaziz University.
A.S.T.M. 1977. Annual book of ASTM Standards p. 494. Philadelphia, Pa. American Society for Testing and Materials.

Environmental Geotechnics and Problematic Soils and Rocks, Balasubramaniam et al. (eds)
© 1987 Balkema, Rotterdam. ISBN 90 6191 785 9

Statistical analysis on road slope collapse

K.Makiuchi
Nihon University, Funabashi, Japan

S.Hayashi
Kyushu University, Fukuoka, Japan

S.Mino
Sumitomo Construction Co., Tokyo, Japan

ABSTRACT: In order to investigate various factors of damage to road slopes, a statistical analysis using multi-variate theory was applied to the slope collapses occurring due to local rain. In this study, rainfall condition, type of slope (cut, embankment and natural slopes), gradient of slope, type of earth material, type of protective method, were examined by quantification method type-III. It was shown that the collapse of the igneous rock slope occurred through both severe local and preceding rainfalls, and the critical rainfall curve is presented. The quantification analysis reveals that the form of collapse is strongly related to the type of soil, and can be classified into three types.

1 INTRODUCTION

Among natural hazards which occur suddenly, a road disaster not only affects people's lives directly but also affects restoration work after damage and social life by the interruption of traffic. In addition, generally, road disasters occur at many places simultaneously, and consequently much labor and cost are required for restoring the communication.

In this study, various factors of road slope are examined statistically using the investigation data from road disasters resulting from severe local rain in Fukuoka and Saga prefectures, Kyushu, in Japan.

2 CONDITIONS OF RAIN

2.1 Major causes of road slope collapse

At the end of August 1980, a record heavy rain was caused by the seasonal rain front and low atmospheric pressure over the northern area of Kyushu district. Collapses of land and foundations, river fooding, agriculture and forestry damage, etc. occurred in various places. The damage caused by these disasters was serious, and 66 people were either killed injured (mainly by landslips). The rainfall in this district had contined intermittently from July until the severe local rain described above came in.

The ground condition over this area became extremely unstable because of an increase in the water content and the wet density of the ground and a reduction of its shear strength. In addition to the primary causes of slope collapses such as configuration of ground, type of soil and rock, and slope structure, this long rain must have also been a factor (exciting cause) of the serious damage. While the safety factor of the slope was thus lowered, the heavy rain on 29th and 30th August caused a sudden rise in the water table in slopes and an increase of the pore water pressure, the ground surface water flow was increased and the water level of rivers adjacent to the roads was raised. Those were the direct causes of the road slope collapses in various parts of the region.

2.2 Distribution of rainfall

The rainfall of 29th and 30th August and the integrated rainfall before that time, or the preceding rainfall, were taken as the index of the precipitation which was the cause of the road slope collapses. Although the preceding 16-days rainfall in the northern Kyushu district was less than in the eastern area, the rainfall over the two prefectures registered 400 to 700 mm, four to five times as much as in a normal year.

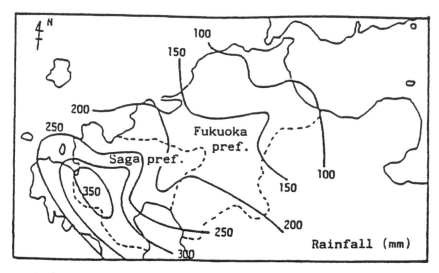

(a) Rainfall distribution on 29th August

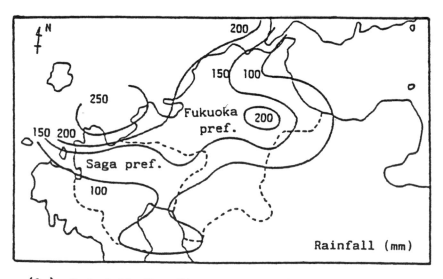

(b) Rainfall distribution on 30th August

Fig. 1 Rainfall Distribution on 29th and 30th August, 1980.

On the other hand, as shown in Fig. 1, the rainfall on 29th August was heavier in Saga preecture, especially in its western area, and much damage occurred in this area on the afternoon of the 29th. On the next day, the severe rain moved towards the east, and the disaster in Fukuoka prefecture occurred mainly before noon on the 30th.

3 OCCURRENCE CONDITIONS OF DISASTERS

3.1 Distribution of disasters

As shown in Fig. 2, the road disasters occurred over both prefectures, although they were not distributed evenly. For purposes of research into the disasters in Saga perfecture, the disasters under the

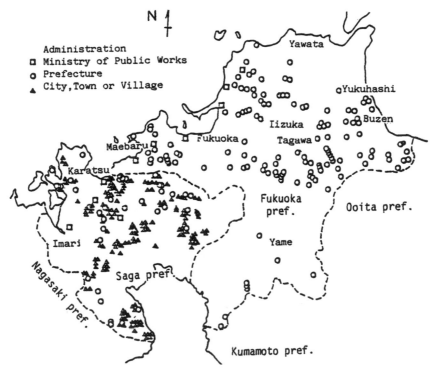

Fig. 2   Locations of road slope collapse

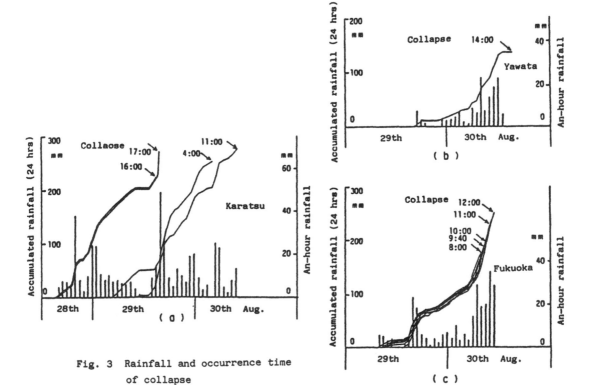

Fig. 3   Rainfall and occurrence time
of collapse

control of the prefectural administrative office were separated from those of the city, town and village offices. The number of disasters which occurred in both prefectures a mounted to about 4,000. The reason why they occurred over so wide an area is that the heavy rain came after the long rain caused by the seasonal rain front According to the road slope collapses with an accurate time of occurrence, occurrences were at the places where the integrated rainfall for the past 24 hours had exceeded approximately 150 mm, as shown in Fig. 3. The data also show that the rain came in waves and the slope collapses occurred when the rain intensity was relatively high.

3.2 Scale of damage

Restoration expenditure proved that the scale of damage varies from one area to another. After Karatsu district, the most small-scale disasters occurred within the area controlled by the public work office of Maebaru district. The number of disasters which occurred in the area controlled by the city, town and villages were about 2.8 times as high as that of the number of disasters in the area controlled by the prefectural offices, and the trunk roads suffered less damages. It should be noted that the number of small-scale disasters, which required restoration expenses of at most 2,000,000 yen was about 74% of the number of all disasters.

3.3 Differences of road slope collapse between districts

After investigating the disasters requiring restoration expenses it was found that Maebaru, Karatsu, and Imari districts differed from one another, although there had been much rainfall over those three districts. The number of disasters was extremely high in Maebaru and Karatsu, where the ground was granite and its composed soil (Masa-do). On the other hand, the disasters in Imari district were concentrated in the Sasebo and Ashiya strata of the tertiary formation. This implies that ground conditions are related to the occurrence of disasters.

3.4 Relationships between rainfall and density of disasters

In order to investigate the occurrence conditions of road slope collapses caused by

rain, the relationships between rainfall and density of co-lapse were investigated. Generally, the number of collapses increases in proportion to the rainfall, and the rain conditions at this time can be divided into two, the severe local rainfall on 29th and 30th, when many disasters occurred, and the preceding 16-days rainfall, which is an additional factor. On the other hand, as the index of the occurrence of disasters can be employed. This index is obtained by totaling the number of disasters in each city, town and village of both prefectures and dividing it by the total road length of the administrative district and multiplying by 10.

In Figs. 4 and 5, the horizontal axis indicates the severe local rainfall and the vertical indicates the rainfall for 16-days. The density of disasters is indicated in the figures, classified into the following four levels.

■ : slope collapse occurring easily (top 25%)

▨ : slope collapse occurring relatively easily (second 25%)

▥ : slope collapse not occurring relatively easily (third 25%)

□ : slope collapse not occurring easily (bottom 25%)

Figs. 4 and 5 indicate respectively the density of disasters in the igneous rock area and sedimentary rock area based on the geological survey map. In Fig. 4, a relatively clear critical rainfall curve can be drawn and it can be seen that the combination of the two types of rainfall about that curve could easily cause a disaster. It can be assumed that a disaster does not occur easily if the severe local rainfall is below 220 mm, even if the preceding rainfall is very heavy.

As shown in Fig. 5, the tendensy of disasters in the sedimentary rock area is different from that in the igneous rock area, and no clear critical rainfall curve can be drawn. The reason for this seems to be that the seepage mechanism of the rain water before the occurrence of a slope collapse of the igneous rock is different from that of the sedimentary rock.

3.5 Application of quantification theory

The disasters at this time had various causes. The district cause was the rain condition, and the primary causes are type of slope (cut slope, embankment slope, natural slope, etc.), gradient of slope type of earth material, etc. Many of these causes are qualitative variates which cannot normally quantified. However, they

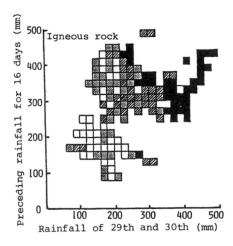

Fig. 4 Relation between rainfall
and density of disasters
for igneous rock

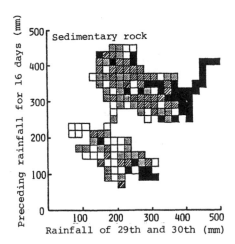

Fig. 5 Relation between rainfall
and density of disasters
for sedimentary rock

were analyzed statistically by quantification method Type III, which is a statistical method of multivariates analysis to
find out how such causes affect slope
collapses. For this analysis, the complete
disaster data in 380 places were used.

3.6 Quantification of variates

As the main factors in the disasters, the
ten items shown in Table 1 were selected,
with the variate of each item calssified
into several categories and quantified.
For the calculation, the HAYASHI III program in SPSS statistical package was used.
The outputs were the eigen values, correlation coefficients cumulative ratio of each
eigen value to all eigen values, distance
matrix among categories, case points, etc.
The five types of collapse in Table 1 are
defined as follows;

Surface slip: A large rill (or gully
erosion) on a slope caused
by the surface water (in-
cluding ground spring wa-
ter), or a fall-off from
the thin surface layer of
slope.

Fall collapse: A relatively deep scoop of
natural ground.

Landslide: A slide failure caused by
rotational slip to which
stability analysis can be
applied.

Flow: A flow failure of earth
slope.

River erosion: A collapse of road slope
adjacent to a river caused
by seepage water.

3.7 Meaning of axis

The graphs of the eigen values are given
below for the correlations among the categories and the properties of each case
from plotting the case score on the graphs
of eigen values.
The primary eigen value is indicated by
the vertical axis and the secondary is indicated by the horizontal axis. Since the
variates of rainfall increase as they go
up the vertical axis, this axis indicates
not only the rainfall but also the occurring order of the collapse from bottom to
top.
The horizontal axis in the positive indicates the bank slope and that in the negative direction indicates the cut slope.
This axis also denotes soil types since
the soil particle diameter decreases as
the positive value increases and becomes
rock type materials as the negative value
decreases.

3.8 Form of collapse

Quantification method Type III was analyzed
using the seven variates; the form of collapse, cut and bank slopes, type of soil,
method of protection work, gradient, severe
local rainfall, and preceding rainfall over
16-days.

Table 1 Quantification of variates

No.	Item	Quanti-fication	Category	No.	Item	Quanti-fication	Category
1	Form of collapse	1	Surface slip	5		6	61 – 80 m
		2	Fall collapse			7	81 – 120
		3	Landslide			8	121 –
		4	Flow	6	Height of collapse	1	0 – 5.0 m
		5	River erosion			2	5.1 – 10.0
2	Cut/Bank	1	Cut slope			3	10.1 – 15.0
		2	Bank slope			4	15.1 –
		3	Natural slope	7	Gradient	1	0 – 0.7
3	Type of soil	1	Clay			2	0.8 – 1.2
		2	Cohesive soil			3	1.3 –
		3	Sandy soil	8	Restora-tion expenses	1	0 – 2.5 million Yen
		4	Decpompoded granite			2	2.6 – 4.0
		5	Granite			3	4.1 – 9.0
		6	Sandstone			4	9.1 – 13.0
		7	Shale			5	13.1 – 20.0
		8	Gravelly soil			6	20.1 – 30.0
4	Protection method	1	Block			7	30.1 –
		2	Masonry	9	Rainfall of 29th and 30th	1	200 – 250 mm
		3	Co. retaining wall			2	251 – 300
		4	Co. mortar spray			3	301 – 350
		5	Planting			4	351 – 400
		6	No protection			5	401 – 450
5	Length of collapse	1	0 – 20 m			6	451 – 500
		2	21 – 30	10	Preceding rainfall for 16 days	1	250 – 300 mm
		3	31 – 40			2	301 – 350
		4	41 – 50			3	351 – 400
		5	51 – 60			4	401 – 450

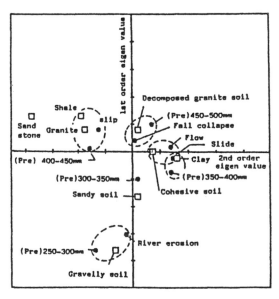

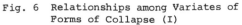

Fig. 6 Relationships among Variates of Forms of Collapse (I)

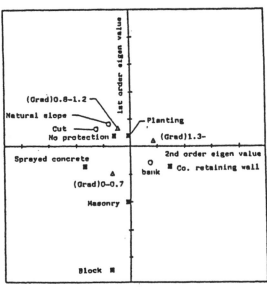

Fig. 7 Relationships among Variates of Forms of Collapse (II)

Table 2 Classification of slope collapse

Preceding rainfall for 16 days	Form of collapse	Type of soil
250 - 300	River erosion	Gravelly soil
300 - 350	Surface slip	Shale, Granite
350 - 400	Landslide	Clay
400 - 450	Fall collapse	Decomposed granite
	Flow	Cohesive soil

Fig. 6 shows only the form of collapse, type of soil, and preceding 16-days rainfall. It is found that the form of collapse and type of soil are strongly related to each other and correspond as shown in Table 2. The forms of collapse can be roughly divided into river erosion and others.

Generally the bank slopes adjacent to rivers are protected, thus fewer disasters occur on these slopes. However, some disasters occur while the rainfall is relatively light. The reason must be that the collapses on these slopes are caused by river water rather than rainfall.

The other forms of collapse (non-rivererosion) are divided by the horizontal axis into three types; the surface slip strongly related to slope and natural slope, flow and landslide related to the bank slope, and fall collapse related to both slopes.

To investigate the effects of slope structure on a disaster, the categories of gradient of slope, cut and bank slopes and methods of protective works were selected, as shown in Fig. 7. It is a general rule that when scores x_1 and y_1 are given to each item in quantification method Type III, the categories near the average values of each score are standard or important. Since the program used in this analysis outputs the scores of categories so that the average will be zero, the categories near zero are indicated as important to the disasters.

Fig. 7 shows that planting treatment and no-protective works are located near the origin and it can be seen that planting treatment gives little resistance to disasters. For gradient of slope, it is found from the distance from the origin that the effects of gradient on disasters are very small, but it is shown that gradient 1.2 (one in two slope, 1:1.2) is related to collapses on cut slopes and that gentler gradients are related to collapses on bank slopes.

Considering this phenomenon using the horizontal axis, which indicates the diameter of particles, it can be seen that rocks with large particles in the negative area collapse into surface slip when the slope is steep, but that clay or cohesive soil slope collapses in the form of surface slip or flow even if the slope is gentle.

3.9 Scale of collapse

The scale of collapse was analyzed and classified on the basis of the restoration expenses, and the results are shown in Fig. 8. The restoration expenses increase in the following order: river erosion, fail collapse, surface slip, landslide, and flow.

3.10 Type of slope

The cases of disaster were divided by type of slope: bank slope, cut slope, and natural slope as shown in Table 3. The gradient was then divided into three classes so that frequency of disaster would be almost the same. Then, the case score of each output case was plotted on the eigen value graph to examine the differences between slope structures.

In Fig. 9, the bank slopes are plotted a little right of center in many cases compared with cut slope. This tendency becomes more evident as the gradient of slope gets steeper. This indicates that the collapse of the bank slope is due to the fine soil, the bank slope failing in the form of fall collapse or landslide. It especially clear that when the gradient is steeper than 1.3 (Fig. 9(c)) failure occurs in the form of landslide or flow caused by the clay or cohesive soil. In the case of the bank slope, there is an obvious difference in the distribution of points in the vertical direction (a) and (b) in Fig.9.

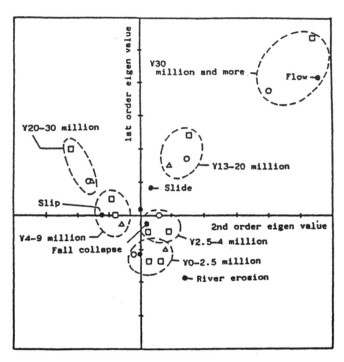

Fig. 8   Relation between Variates on the Scale of Collapse.

Table 3   Number of slope collapse

Slope	Gradient of slope		
	− 1:0.7	1:0.8 − 1:1.2	1:1.2 −
Bank	29	66	63
Cut	27·	22	8
Natural	10	34	18

Since the vertical axis indicates both rain-fall and order of occurrence of disasters, this difference imlies a difference in resistance to rainfall, that is, a bank slope of gradient of 0 to 0.7 is collapsed easily by a little rainfall. However, when the gradient is steeper than 1.3, the re-sistance to disasters varies a little, but the slopes collapse easily.

In case of the cut slope shown in Fig. 10, the points are plotted on the left of cen-ter at any gradient which indicates the tendency forwards surface slip in granite and shale. From the above figures, it can be seen that the number of disasters on cut slope decrease as the gradient becomes more gentle, but this is not related to resistance to rainfall, and collapses occur at any gradient. The graph of the natural

slope is omitted since it is similar to that for the cut slope.

4   CONCLUSION

Rainfall condition and occurrence condi-tions for disasters on road slopes were investigated, and the collapses were ana-lyzed statistically using the density of disasters and quantification theory Type III. The results obtained from this study can be summarized as follows:

(1) The rainfall condition which causes the collapse of igneous rock slope is dif-ferent from the condition which causes collapses of sedimentary rock slope. The collapse of igneous rock slope occurs in both severe local rainfall and preceding

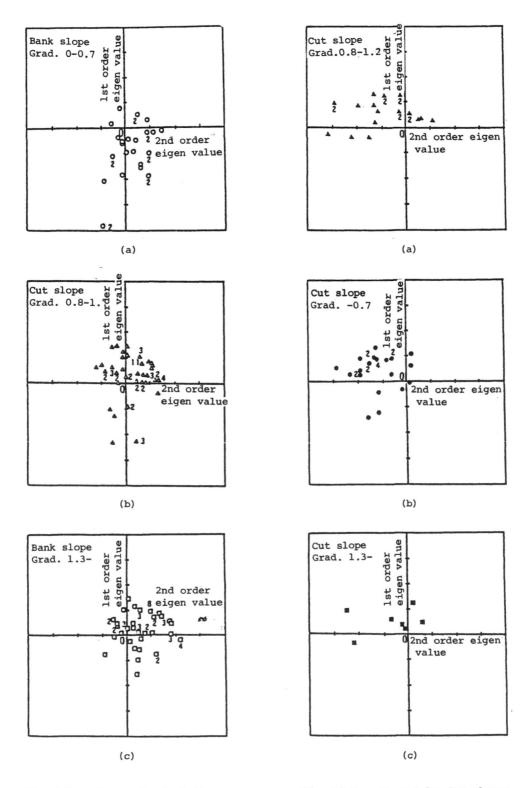

Fig. 9 Case Scores for Bank Slopes

Fig. 10 Case Scores for Cut Slopes.

is presented. However, the relationship
between the rainfall condition and the
collapses of sedimentary rock slope was
not obtained from the study.
  (2) The form of collapse is strongly related to the type of soil. That is river
erosion corresponds to gravelly soil, landslide to clay, flow to clay and cohesive
soil, fall collapse to decomposed granite
soil and surface slip to granite and shale.
  (3) The forms of collapse can be classified into three types. Surface slip is related to cut and natural slopes, flow and
landslide to bank slope, and fall collapse
to both slopes.
  (4) In the case of bank slope, disaster
occurs more easily when the gradient is
steep. Disasters do not occur easily when
the gradient is near 1:0.9, but disasters
cannot be prevented even if the gradient
gets more gentle.

ACKNOWLEDGEMENT

Grateful appreciation is expressed to
Emeritus Professor T. Yamanouchi, Kyushu
University, and the authors are especially
indebted to Mr. K. Machi for his assistance
in the analytical work.

REFERENCES

Kobayashi, R. 1981. An introduction to
  quantification theory (in Japanese).
  Nikkagiren Publisher Co.
Miyake, I., et al. Statistical package
  for social sciences (in Japanese).
  Touyoh Kenzai Shinpoh-sha, (Part 1) 1976
  and (Part 2) 1977.

Environmental Geotechnics and Problematic Soils and Rocks, Balasubramaniam et al. (eds)
© *1987 Balkema, Rotterdam. ISBN 90 6191 785 9*

Soil-pollutant interaction effects on the soil behavior and the stability of foundation structures

Hsai-Yang Fang
Lehigh University, Bethlehem, Pa., USA

ABSTRACT: This paper presents the pollution processes in ground soil and their effects on the geotechnical properties of soil and foundation stability. Laboratory test results are included to support these findings.

1 INTRODUCTION

In recent years, due to population growth, a progressive living standard and industrial progress, much of the air, water and land have become polluted. Open dumps, and chemical and industrial wastes cause these problems, as well as others listed in Table 1.

All types of pollution have direct or indirect effects on ground soil properties. For example, rain falling on a garbage dump will pollute both surface and groundwater systems. The polluted water will attack foundation structures such as footings, caissons, piles and sheet piles. If the polluted water is used for mixing concrete, it will affect the workability and durability of the concrete. In embankment construction, the moisture-unit weight relationship of soil will also be affected.

Numerous examples in the literature have reported on air and water pollution but little effort has been made to determine how the ground soil responds to these hazardous or toxic substances. At present, most geotechnical project designs and construction are based on the test results following ASTM and AASHTO standards. These standards are based on control conditions at room temperature with distilled water as the pore fluid. Since field conditions and the standard control condition are significantly different, many premature or progressive failures frequently occur. To understand soil behavior under in situ conditions, it is necessary to examine soil behavior as close to the actual condition as possible. To accomplish this goal, we must understand the environmental conditions as they exist on the ground and their interactions over a long-term period.

The purposes of this paper are to: (1) discuss the soil pollution process, (2) examine the mechanics of soil-pollutant mechanisms, and (3) anticipate implementation of these concepts to practical problems such as landslides, subsidences, and foundation structure failures, in the hope that the findings lead to improvement of present design and construction aspects of foundation engineering.

2 POLLUTION PROCESSES AND SOIL-POLLUTANT INTERACTION

Due to the complex nature of soil pollution the term "pollution" or "pollutant" is used loosely in this paper. These terms will be reflected in the context of the pH solution in the pore fluid, the temperature of ground soil or pore water, or the ion characteristics of the solution. The major sources of ground soil pollution are listed in Table 1. Regardless of the sources, there are three basic mechanisms by which the ground soil can be contaminated.

(1) Contamination may occur from rainfall such as acid rain for rain falling into a sanitary landfill, or oil or chemical wastes spilled on to the ground.

(2) It occurs when pollutants are introduced as leakage from well disposal or construction of waste disposal facilities such as landfills, septic tanks, laterals and lagoons as described by Fang, et al.(1984b) and Jafek (1985).

(3) It results from hydraulic or chemical alterations which allow polluting substances to move within or between soil layers. In this category, the phenomena cover chemical, physico-chemical and microbiological aspects.

Categories 1 and 2 are direct processes.

TABLE 1   SOME CAUSES OF GROUND SOIL POLLUTION IN U.S.A.

Southwest	South Central	Northeast	Northwest
Acid Rain	Acid Rain	Acid Rain	Abandoned Oil Wells
Animal Wastes	Animal Wastes	Acid Mine Drainage	Acid Drainage & Mine Tailing
Disposal of Oil Field Brines	Disposal Well	Buried Pipelines & Storage Tanks	Acid Rain
Evapotranspiration from Vegetation	Evapotranspiration from Vegetation	Hazardous Chemical Wastes	Brine Injection
Hazardous Chemical Wastes	Hazardous Chemical Wastes	Highway Deicing Salts	Disposal Wells
Injection Wells for Waste Disposal	Irrigation Return Flow	Landfills	Dry Land Farming
Irrigation Return Flow	Landfills	Mine Fire	Highway Deicing Salts
Leaching	Nuclear Wastes	Nuclear Wastes	Hazardous Chemical Wastes
Nuclear Wastes	Oil Field Brines	Petroleum Exploration & Development	Irrigation Return Flow
Saltwater Intrusion	Solid Wastes	Radon (gas)	Landfills
Solid Wastes	Waste Lagoons	River Infiltration	Mine Fire
Spills Hazardous Materials		Saltwater Intrusion	Nuclear Wastes
Water from Fault Zones & Volcanic Origin		Septic Tanks	Septic Tanks
		Surface Impoundments	Sewage Treatment Plant Discharges
			Surface Impoundments

Among these, acid rain (Gunnerson and Willard, 1979) is most common and covers a larger area than any other; therefore, more discussion is presented here. The pH of rainwater is around 5.7, which is slightly acidic because of dissolved carbon dioxide (CO_2) in the atmosphere. Studies of rain in European countries have shown an increasing trend toward higher acidity with pH values varying between 3 to 5. This increased acidity is a result of absorption by rain of sulfuric (H_2SO_4) and nitric acids (HNO_3) formed in the atmosphere from sulfuroxide (SO_2) and nitrogen oxide (NO). Reports from China in 1983, indicate in the southwestern region, the pH of rainwater has dropped to 2.7. As reported by the American Chemical Society in 1969, more than five milligrams per liter of sulfur (S) can be found in rain in the eastern industrial areas in the United States.

Acid rain is related to many interactions between atmosphere, biosphere, hydrosphere and lithosphere. A simple generalized diagram of a model describing the interrelationships of rainfall, evapotranspiration and leachate with the soil solution processes is shown in Fig. 1 (Reuss, 1978).

2.1 Physico-chemical decomposition process

Physico-chemical decomposition, weathering or microbiological factors are all contributing in some degree to the soil pollution process. The major factors are heat, bacteria, percolating rainwater, plant roots and organic matter. The chemical processes of clay mineral decay include: hydrolysis, hydration, carbonation, oxidation and solution (Lyon and Buckman, 1939; Paton, 1978). The soil-forming processes which also contribute to decay processes include: podzolization, lateritization, calcification (carbonation), alkalization, and siallitization (kaolinization). These processes are principally dissolving processes or deteriorating processes of clay minerals which take place in acid mediums in humid regions.

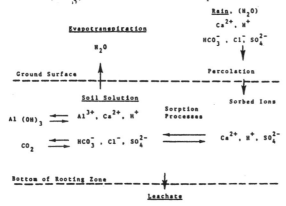

Figure 1   Interaction of Rainfall, Percolation, Evapotranspiration and Leachate with Soil Solution Processes (Based on Reuss)

2.2 Chemical alterations and bacteria attack

Chemical alteration or bacteria attack or both can change a soil property signifi-cantly. For example, decomposition in landfill causes the ground temperature to increase. When temperatures are high, the detention times are long and sulfate con-centrations are appreciable. The diffi-culty is always associated with reduction of sulfates to hydrogen sulfide (H_2S). Hydrogen sulfide is often blamed for the corrosion of various geostructural members. Actually, hydrogen sulfide is considered a weaker acid than carbonic acid and has little effect on concrete or steel members. However, in sanitary landfill areas, biolo-gical changes caused by decomposition con-stantly occur. These changes require oxy-gen (O_2), and if sufficient amounts are not supplied through natural reaeration from air in polluted water, reduction of sulfates occurs and the sulfide ion (S^{2-}) is formed. At the usual pH level of domestic waste waters, most of the sulfide is converted to hydrogen sulfide and some of it escapes into the atmosphere above the water, as shown in Fig. 2.

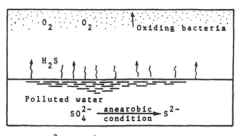

$$S^{2-} + 2H^+ \longrightarrow H_2S$$

Figure 2 Formation of Hydrogen Sulfide in Polluted Soil/Water Area Resulting from Oxidation of Hydrogen Sulfide to Sulfuric Acid

Here it does no damage if the geostructural member is dry. In a poorly ventilated con-dition, however, moisture collects on the surface of walls or structural members and hydrogen sulfide dissolves in this water in accordance with its partial pressure in the sewer atmosphere.

Bacterial attack in the polluted ground soil occurs frequently and is capable of oxidizing hydrogen sulfide into sulfuric acid. It is natural that some of these or-ganisms should infect the structural members at times of high flow. Because of the aerobic conditions normally prevailing in landfill or in polluted soil above the groundwater level, these bacteria oxidize the sulfide to sulfuric acid (Sawyer and McCarty, 1978) as:

$$H_2S + 2\ O_2 \xrightarrow{\text{bacteria}} H_2SO_4$$

and the latter, being a strong acid (H_2SO_4) attacks the underground structures. This effect is particularly serious in the top portion of the structure, as indicated by the arrow in Fig. 2.

2.3 Leaching and ion exchange reaction

Leaching and ion exchange reaction affect soil properties significantly. Soil pro-perty changes are caused by rainwater or drainage processes due to some clay miner-als removed by natural process. During the leaching process, common mineral elements such as calcium (Cl), magnesium (Mg), potas-sium (K), nitrogen (N), and phosphorus (P) are removed. As a majority of calcium is lost, nitrogen is leached from soils in the form of nitrates (NO_3). The quantity of phosphorous lost by drainage is small and comparatively large quantities of sulfur may be found in drainage waters.

Effects of leaching by water and seawater have been studies (Bjerrum, 1954; Rosen-qvist, 1959; Yong and Warkentine, 1966; Woo and Moh, 1977). However, little data are available on how polluted water caused by the leaching process affects soil behavior.

The phenomenon of ion exchange was dis-covered in 1850 and showed that clay miner-als have the property of absorbing certain cations and anions and retaining these ions in an exchangeable state. A significant contribution to the understanding of this important phenomenon was made by the agri-cultural sciences and includes soil acidity and alkalinity, friability, fixation of po-tassium and ammonium, soil fertility, buf-fer action (Weir, 1936; Millar and Turk, 1943) and soil stabilization and ground im-provement (Winterkorn, 1937). The ion ex-change reactions give us two important phenomena during these processes. The first is a reaction that can cause changes in the soil/water structure from dispersive to flocculative structure or vice versa. The second can change water behavior, for ex-ample, hard water into soft water by the re-moval of calcium (Ca^{2+}) and magnesium ions (Mg^{2+}). A great deal of information is available regarding cation exchange; how-ever, very few data were found on anion

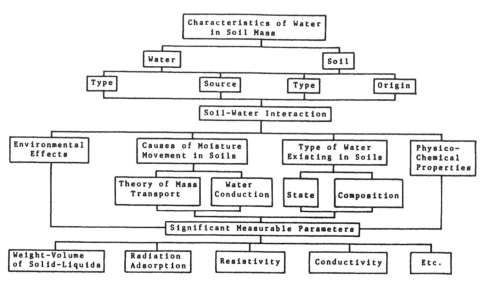

Figure 3   Interaction of Soil/Water System and the Measurable Parameters

exchange due to the complex reactions during the exchange. A comprehensive review of ion exchange in soils is given by Kelley (1948), Grim (1953), Yu (1964), Rich (1968) and Yu, et al. (1984). The ion exchange phenomena in the contaminated soil/water systems have been evaluated by Fang (1985) who reveals that the ion exchanges are an important controling factor in the soil-water-pollutant systems. It is also shown that exchange can take place in partially saturated or even dry soil condition. Both pH and temperature in the soil/water system will have significant effects on the level of ion exchange capacity. Also soil-root systems have shown that ion exchanges can take place from soil to root or vice versa. Other indications are that bacteria can carry some ions from place to place in the ion exchange activity.

Methods for determination of cation exchange capacity have been discussed by Winterkorn (1937), Kelley (1948), Grim (1953) and Yu, et al. (1984). ASTM has proposed standards for cation exchange and related subjects (ASTM, 1981b). More recently, a nondestructive method of studying how metal ions such as lithium (Li^+), sodium (Na^+), calcium (Ca^{2+}) move through the walls of animal or plant cells was developed, using nuclear magnetic resonance technology (Spring, et al. 1985). No direct reliable measuring technique for anion exchange reaction is available at the present time.

2.4 Mass transport phenomena in moist soil layers

Soils are normally composed of solids, li-

quid and gaseous phases. The solid phases are composed of inorganic or organic matter. The liquid phase is usually an aqueous electrolyte solution. The gaseous phases, while normally in contact and exchanged with the atmosphere and biosphere, may have a different composition from the latter, depending on soil depth and biologic activity within the soil mass.

The phenomena of moisture transportation in the soil mass are complex. Figure 3 shows the interaction of the soil/water system and the measurable parameters. If polluted soil/water is introduced, the complexity of each parameter is multiplied by that of the other. The moisture transport from one place to another in the soil mass is due to:

1. Mechanical Causes
   (a) Hydrostatic Potential
   (b) Capillary Action
2. Thermo-Electric Causes
   (a) Hydration Energy of Ions
   (b) Osmotic Energy of Ions
   (c) Kinetic Dispersive Force
   (d) Electro-Motive Force
3. Ion Exchange Reactions
4. Microbiological-Chemical Effects
5. Others
   (a) Vapor Pressure (phase changes)
   (b) Thermo-gradient caused by solar energy, mine fire, thermopipes, cables.

In all cases of moisture transport in hazardous and toxic areas, the thermo-electric (Winterkorn, 1955) and the microbiological-chemical effects are most important.

3 EFFECT OF PORE FLUIDS ON SOIL BEHAVIOR

An understanding of the influence of pore fluids on clay behavior is essential to the design of many components of foundations structures, especially for hazardous and toxic waste containment systems. Without a good understanding of the clay-pore fluid interaction, there is no sound basis on which to project the long-term behavior of these systems. A review of the fundamentals of clay mineralogy and double-layer theory as a basis for interaction is given by Evans, et al. (1981), in which a definite relationship between soil structure and changes in the geotechnical properties of clay has been found and is discussed. Some recent developments (at Lehigh University) in soil-pore fluid interaction are presented in Figs. 4 to 10.

One important phenomenon relating to the effects of pore fluid on soil behavior is the soil crack pattern. Soil cracks are frequently observed in natural and man-made earthen structures. These cracks are the result of an internal energy imbalance in the soil mass caused by nonuniform stress distribution of compaction energy during construction. These small and unnoticed cracks can trigger progressive erosion or landslides in dams, highway embankments, failure of slurry walls, as well as clay liners. The mechanism of soil cracking has been examined by Fang, et al. (1983) and has been classified into four types, namely: shrinkage, thermal, tensile and fracture cracks. The interrelationships among these types of cracks have also been evaluated. However, in all cases the pore fluid between cracks and the soil mass plays an important role in the behavior of cracks, including their pattern, length, width and depth. Further studies on how polluted water affects the cracking patterns and its relation to the leaking of clay liners or slurry walls have been conducted (Fang, et al. 1984; Alther, et al. 1985). Figure 4 shows some of the cracking behavior as it reflects on volume change. In Fig. 4, bentonite shows larger volume changes than muscovite and illite; and the air dried samples have greater variations than oven dried samples. However, in all cases, the polluted pore fluid (pH = 2 or pH = 11) produces larger volume changes. Some acids such as aniline ($C_6H_5NH_2$), acetic acid ($C_2H_4O_2$), and carbon tetrachloride (CCl_4), show greater volume changes than other acids.

Figure 5 shows the typical gradation curves indicating the effects of pH value on soil-water structures and illustrates that the higher the pH value, the smaller the

effective grain size (D_{10}), and the greater the uniformity coefficient (D_{60}/D_{10}).

Other soil parameter changes caused by the pH value such as liquid limit, plastic limit, shrinkage limit, specific gravity have been reported by Waidelich (1958), Andrews, et al. (1967), Torrance (1975), and Evans, et al. (1985). Effects of pore fluid on hydraulic conductivity is most interesting to geotechnical engineers concerned with the hazardous and toxic control system construction. Several technical symposia have been organized in recent years by ASTM (1981a, 1982, 1985) and international organizations. A comprehensive review on pore fluids and hydraulic conductivity is given by Evans, et al. (1981) and Young (1984). Figures 6 and 7 show how both organic and inorganic pore fluids affect the coefficient of permeability. Results shown in Fig. 7 have been obtained by a long-term triaxial cell permeameter developed at Lehigh University (Evans, 1984; Evans and Fang, 1985). The schematic diagram of the permeameter is shown in Fig.8. This apparatus has several unique features. It has the ability to measure both inflow and outflow volumes for pH values varying from 1.0 to 12.6 without damage to the testing apparatus because parts of the components contacting hazardous pore fluid are made of Teflon®. Furthermore, fluids can be changed and inflow and outflow riser tubes can be filled or emptied without changing the state of stress on the sample.

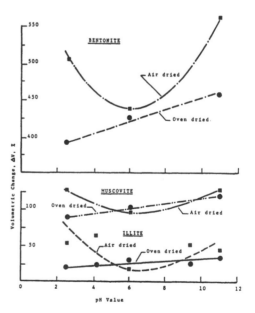

Figure 4 Effect of pH Values on Volumetric Changes of Some Clay Minerals

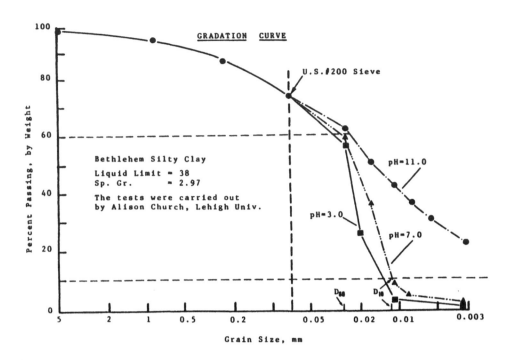

Figure 5 Effect of Pore Fluids on Gradation Curves

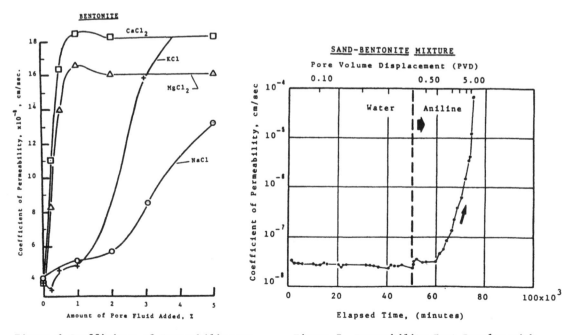

Figure 6 Coefficient of Permeability vs.
Amount of Pore Fluid Added (after
Alther, et al. 1985)

Figure 7 Permeability Test Results with
Water and Aniline (Concentrated)

The system's design accommodates a wide range of stresses up to a maximum of 130 psi (897 kPa). This system has been further developed with a temperature control during testing and another multi-purpose device for the determination of shear stress (Niak, 1984). Figure 9 shows typical curves of the pore fluid and temperature effects on shear strength. The changes of dynamic shear modulus caused by various pH values for a bentonite-sand mixture are given by Du, Mikroudis and Fang (1986) (see Fig. 10).

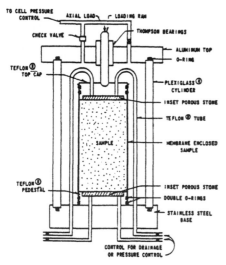

Figure 8 Lehigh Long-Term Triaxial Cell Permeameter for Permeability Testing with Hazardous/Toxic Permeants

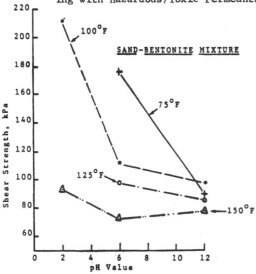

Figure 9 Shear Strength of Sand-Bentonite Mixture vs. pH Value at Various Temperatures

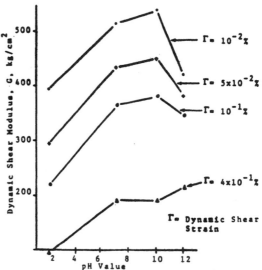

Figure 10 Relationship Between Dynamic Shear Modulus and pH Values (Cell Pressure at 40 psi) (After Du, Mikroudis and Fang, 1986)

4 EFFECT OF SOIL POLLUTION ON FOUNDATION STRUCTURES

As indicated in the previous section, most foundation design and construction is based on ASTM or AASHTO standards. These standards are based on room temperature and distilled water as a standard control pore fluid. Since many underground soil/water systems are polluted to some degree, some difference in soil properties between laboratory and in situ conditions exists. Therefore, certain analysis and design procedures for foundation structures must be revised or modified.

Most foundation structures require friction or adhesive force between structure and soil for earth pressure, pile capacity and bearing capacity computation. Unfortunately, this force changes frequently if the groundwater table fluctuates. This is especially true in the polluted soil/water areas. This fact can be seen in the cracking pattern effects and volume changes during the shrinking process, as shown in Fig. 4. For example, an aniline ($C_6H_5NH_2$) solution can cause clay-water slurry to shrink dramatically. In comparison with distilled water, the aniline solution reduced more than 100% of its volume. Therefore, if groundwater drops and cracking develops, the adhesion between soil and structures can be reduced to minimum or zero.

In examining Fig. 5, the effective size,

D_{10} is influenced by the pH value or other polluted water, thereby, affecting the present filter design concept, which is solely based on grain size distribution. All surface and subsurface drainage and filter design procedures must be modified or revised in the areas of polluted storm water.

Progressive failure or landslides are caused by environmental factors such as acid rain and chemical or waste dumps. This problem has been raised by many investigators (Seifert, et al. 1935; Winterkorn, 1942; Matsu, 1957). Subsidence or settlement analysis should also consider the chemical, physico-chemical and biological effects. Present procedures based on Terzaghi consolidation theory are not sufficient because the theory considers only loading. Settlement caused by decomposition of organic matter and chemical alterations of clay elements is not involved in the calculation.

Since all constants commonly used in geotechnical engineering, that is porewater pressure (u), cohesion (c), friction angle (ϕ), and coefficient of earth pressure at rest (K_o) vary due to local environmental conditions, conventional soil mechanics concepts used in difficult environmental areas must be carefully examined for a proper analysis and design of foundation structures.

5 CONCLUSIONS

An understanding of the interaction between hazardous and toxic wastes and the engineering behavior of ground soil involves three requirements: the measurement of the relevant soil properties; the test equipment, which must be durable and resistant to hazardous waste pore fluid, especially organic acids, as well as usable for long-term performance study; and a knowledge of clay mineral elements, with an understanding of the chemical, physico-chemical and microbiological behavior of soil.

From published data and field observations, it is evident that significant work has been done to provide an understanding of the interaction between soil-pollutant and soil/water systems. To adequately apply and understand these phenomena, a characterization of hazardous and toxic waste is necessary. But the geotechnical engineer must also study site characteristics and understand the general properties of the given waste and how these properties influence the soil behavior from the environmental point of view.

The phenomena investigated here are extremely complex and not all possible influences can be presented in this paper. Further studies must be urged in for further improvement of soil mechanics and foundation engineering. Present soil mechanics concepts are based on short-term performance and consider only loading conditions. Revision or modification of these concepts must include test equipment, testing procedures and constitutive laws of soils, to provide proper design and construction of foundation structures, especially in polluted soil/water systems.

6 ACKNOWLEDGEMENTS

The financial support for these studies was provided by the U.S. Environmental Protection Agency under Grant No. R810992 to Lehigh University and partially supported by Envirotronics Corporation, International. Thanks are extended to Mr. G. K. Mikroudis for his review of the manuscript. The opinions, findings and conclusions expressed in this paper are those of the author and are not necessarily those of the project sponsors.

7 REFERENCES

Alther, G.R. et al. 1985. Influence of inorganic permeant upon the permeability of bentonite. In ASTM STP 876, p.64-74.

Andrews, R.E., Gawarkiewicz, J.J. and Winterkorn, H.F. 1967. Comparison of three clay minerals with water, dimethyl sulfoxide, and dimethyl formalimide. In Highway Research Record No. 209, p.66-78.

ASTM. 1981a. Permeability and groundwater contaminant transport. In Zimmie and Riggs (ed.) ASTM STP 746, 245p.

ASTM. 1981b. Water. ASTM Standards Part 31.

ASTM. 1982. Hazardous Solid waste testing. Conway and Malloy (ed.). ASTM STP 760.

ASTM. 1985. Hydraulic barriers in soil and rock. Johnson, et al. (ed.). ASTM STP 874, 332p.

Bjerrum, L. 1954. Geotechnical properties of Norwegian marine clays. Geotechnique. 5.2:49.

Du, D.L., G.K. Mikroudis and H.Y. Fang. 1986. Effect of pore fluid pH on the dynamic shear modulus of clay. ASTM STP 933.

Evans, J.C., R.C. Chaney and H.Y. Fang. 1981. Influence of pore fluid on clay behavior. Fritz Eng'g Lab report 384.14, Lehigh University. 18p.

Evans, J.C. 1984. Permeant influence on the geotechnical properties of soils. PhD dissertation, Lehigh University. 241p.

Evans, J.C. and H.Y. Fang. 1986. Equipment for permeability testing with hazardous and toxic permeants. ASTM Geotechnical Testing Journal, V.9,n.3, p.126-132.

Evans, J.C.; I.J. Kugelman & H.Y.Fang. 1985. Organic fluid effect on the strength, compressibility and permeability of soil-bentonite slurry walls. In I.J. Kugelman (ed.) Toxic and hazardous wastes, p. 275-291. Lancaster:Technomic.

Fang, H.Y., et al. 1983. Mechanism of soil cracking. Proceedings, 20th Annual Meeting of Society of Eng'g Sciences, 1.

Fang, H.Y., J.C. Evans & I.J. Kugelman. 1984a. Effect of pore fluid on soil cracking mechanisms. Proceedings, ASCE Eng'g Mechanics Specialty Conf. v.2, p.1292-1295

Fang, H.Y., J.C. Evans & I.J. Kugelman. 1984b. Solid and liquid waste control techniques. Majumdar & Miller (ed.) Solid and Liquid Wastes, p.104-118, Pennsylvania Academy of Science.

Fang, H.Y. 1985. Ion-Exchange reactions in contaminated soil-water systems. Envirotronics Corp., Int'l.

Grim, R.E. 1953. Clay mineralogy. McGraw-Hill.

Gunnerson, C.G. & B.E. Willard (eds.) 1979. Acid Rain. New York: ASCE.

Jafek, B. 1985. Trouble in our own backyard. ASCE Civil Eng'g, February, 40-43.

Kelley, W.P. 1948. Cation exchange in soils. New York:Reinhold Pub. Co.

Lyon, T.L. & H.O. Buckman. 1939. The nature and properties of soils. Macmillan Co.

Matsuo, S. 1957. Study of the effect of cation exchange on the stability of slopes. Proceedings, 4th ICSMFE, v.2:330-333.

Millar, C.E. & L.M. Turk. 1943. Fundamentals of soil science. Wiley & Sons.

Naik, D.D. 1984. Combined influence of hazardous pore fluids and heat on the geotechnical properties of clay soils. M.S. thesis, Lehigh University.

Paton, T.R. 1978. The formation of soil material. London:George Allen & Unwin.

Reuss, J.O. 1978. Simulation of nutrient loss from soils due to rainfall acidity. EPA report No. 600/3-78-053, May.

Rich, C.I. 1968. Applications of soil mineralogy in soil chemistry and fertility investigations. Soil Science Society of America SPS No. 3, p. 61-90.

Rosenqvist, I.T. 1959. Physico-chemical properties of soils: soil-water systems. ASCE Proceedings v.85,SM2:31.

Sawyer, C.N. & P.L. McCarty. 1978. Chemistry for environmental engineering. McGraw-Hill.

Seifert, R. et al. 1953. Relation between landslide slope and the chemistry of clay soils. Mitteilung, Preuss. Versuchanstalt für Wasserbau u. Schiffbau, 20.

Spring, C.S. et al. 1985. U.S. Patent 4,532,217.

Torrance, J.K. 1975. On the role of chemistry in the development of and behavior of the sensitive marine clays of Canada and Scandinavia. Canadian Geotechnical Journal 12,3:326-335.

Waidelich, W.C. 1958. Influence of liquid and clay mineral type on consolidation of clay-liquid systems. HRB Special Report 40, p.24-42.

Weir, W.W. 1936. Soil science. Chicago: J.B. Lippincott Co.

Winterkorn, H.F. 1937. The application of base exchange and soil physics to problems of highway construction. Soil Science, 1:93-99.

Winterkorn, H.F. 1942. Mechanism of water attack on dry cohesive soil systems. Soil Science, 7:259-273.

Winterkorn, H.F. 1955. Water movement through porous hydrophilic systems under capillary, electric and thermal potentials. ASTM STP No. 163, p.27-35.

Winterkorn, H.F. 1958. Mass transport phenomena in moist porous systems as viewed from the thermodynamics of irreversible processes. HRB Special Report 40, p. 324-338.

Winterkorn, H.F. & H.Y. Fang. 1975. Soil technology and engineering properties of soils. In Winterkorn & Fang (eds.) Foundation Engineering Handbook, p.67-120, New York:Van Nostrand Reinhold.

Woo, S.M. & Z.C. Moh. 1977. Effect of leaching on undrained shear strength behavior of a sedimented clay. Proceedings 9th ICSMFE Specialty Session on Geotechnical Eng'g & Environmental Control, p. 451-464, v.1.

Yong, R.N. and B.P. Warkentin. 1966. Introduction to soil behavior. Macmillan.

Young, S.C. 1984. Review of leachate induced permeability changes. TVA Water System Development Branch.

Yu, T.Y. 1964. Electro-chemical properties of soils (in Chinese)

Yu, T.Y. et al. 1984. (same as above).

Environmental Geotechnics and Problematic Soils and Rocks, Balasubramaniam et al. (eds)
© *1987 Balkema, Rotterdam. ISBN 90 6191 785 9*

Engineering behaviour of soils contaminated with different pollutants

A. Sridharan & P. V. Sivapullaiah
Department of Civil Engineering, Indian Institute of Science, Bangalore

ABSTRACT : The basic concept of soil, water and air relationships within a soil environment must be considered to allow a more complete understanding of the behaviour of soils. This soil behaviour is further modified in the presence of pollutants, which find their way into soils through the disposal of large quantities of wastes into the land. The pollutants may alter the exchangeable ions held by clays, the nature of pore fluid, the organic matter content of soils etc., affecting the soil properties. The effects of these alterations on the volume change, strength and permeability characteristics of clays are discussed in this paper. The variations in the properties are micro-mechanistically explained.

INTRODUCTION

Large quantities of waste are being disposed of on land, land being an effective medium for the disposal of wastes. The widespread use of waste with natural soil in fills, embankments, road bases etc. necessitates the study of behaviour of different types of soils in different environmental conditions. The properties of soils are modified by the presence of pollutants. The variations in physical and physico-chemical properties together with the mechanisms involved are described by Sivapullaiah and Sridharan (1985). These data are useful for development of a better understanding of engineering properties and establishment qualitative guidelines for soil behaviour in different environments. Major emphasis has been given to the study of clay minerals because clay particles dominate the behaviour of most fine-grained soils.

Though the effects of pollutants on soil are complex, the mechanisms involved may be better understood if the various factors are isolated and considered independently. The primary factors to be considered are (1) ion exchange (2) the nature of pore fluid and (3) organic matter. The engineering behaviour of clays in different environments and their correlations with physico-chemical mechanisms form the basis of the present paper.

The important engineering properties of soils for geotechnical engineers are:

1) volume change characteristics
2) shear strength and
3) hydraulic conductivity (permeability).

The action of various pollutants on the soils and their effects on the above properties will be discussed.

FACTORS INFLUENCING ENGINEERING BEHAVIOUR

Effect of ion Exchange:

Cation Exchange:

In general, the influence of cations on soil properties increases with increasing activity of the clay. The most important char-

acteristics of the cations are
their valence and size. In soils
containing expansive clay minerals,
the type of exchangeable cation
exerts a controlling influence over
the amount of expansion that takes
place in the presence of water.
For exmple, sodium and lithium
montmorllonite can exhibit almost
unrestricted inter-layer swelling
provided water is available, the
conflicting pressure is small, and
the excess electrolyte concentration is low. Di-and trivalent
forms of montmorillonite do not
expand beyond a basal spacing of
about 18A°, regardless of the
environment.

In soils composed mainly of non-
expansive clay mineral, the type of
adsorbed cation is of greatest
importance in influencing the behaviour of the material in suspension
and the nature of the fabric in
sediments formed by them. Monovalent cations, particularly sodium
and lithium, promote defloccula-
tion, whereas clay suspensions,
ordinarly flocculate in the
presence of di-and trivalent
cations.

Volume Change Behaviour :

Volume changes in soils are
important because of their consequences in terms of settlement due
to compression. In addition,
changes in volume lead to changes
in strength and deformation properties, which in turn influence
stability. Compressibility of pure
clays can be accounted for quantitatively by the cosideration of
double-layer repulsive forces.
These forces between paricles are
due to the presence of exchangeable
ions (Bolt 1965). It has been
astablished that electrical double
layer theory of Guoy-Chapman can be
effectively used to describe the
compressibility behaviour.

The consolidation characteristics
of montmorillonite depend upon the
size of the cation present in the
clay-water system. Variations in
pore water electrolyte concentration have little effect on the
void ratio-effective stress relationships for the Ca-montmorillonite in water, apparently

because double-layer effects are
smaller than predicted by classical
theory and because of the formation
of permanent domains. Even in the
case of Na-montmorillonite there is
evidence of domain formation
during swelling at low values of
effective stress though their consolidation curves are in qualitative confirmation of double-layer
theory. This is further explained
by Sridharan and Rao (1973).

Salas and Serratosa (1953) do not
find appreciable differences in the
nature of cations in case of kaolinite because of its low exchange
capacity.

The compressibility order found for
bentonite is

$Li^+> Na^+> K^+> Ca^{++}> Ba^{++}$

in accordance with the characteristics of these ions (Fig.1.).
However, special notice should be
taken of the particular position
occupied by the K-bentonite. By
reason of the characteristics of

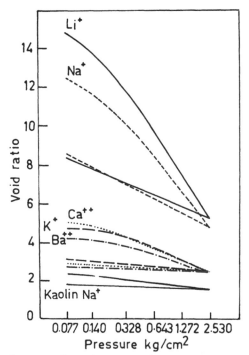

Figure 1. Compression curves of
Bentonites and Na^+ Kaolinite (after
Salas and Serratosa, 1953).

the cation, the K-bentonite should more closely approach in its properties those of Li- and Na-bentonites than those of Ca- and Ba-bentonites. This different behaviour is due to partial fixation of potassium in a non-exchangeable manner (on drying the sample) and becoming colloidally inactive. Winterkorn and Baver (1934) found for a sample of beidellite the following order of swelling

$Li^+ > Na^+ > Ca^{++} > Ba^{++} > H^+ > K^+$.

Strength Behaviour :

Sridharan and Rao (1979) conducted triaxial tests on Ca- and Na-kaolinites and montmorillonites. Friction angels ( ) of 28.4° and 24.4° for Ca- and Na-kaolinites and 21.7° and 11.2° for Ca- and Na-montmorillonites were reported. Zero cohesion intercepts were obtained for all the cases. Thus it si seen that both kaolinite and montmorillonite give higher friction angles when saturated with calcium. Similar results obtained by Mesri and Olson (1970) have been explained on the basis of domain formation.

Permeability :

Soil permeability is one of the most important soil characteristics in the study of subsurface pollution problems, since migration of the pollutant itself is likely to be dependent on the permeability of soil. In the design of a confinement area for waste disposal, permeability characteristics of the linear and strata in the disposal area should be determined in order to prevent pollution of ground water by chemical contaminants.

Salas and Serratosa (1953) presented variations of the coefficients of permeability with void ratios for homoionized bentonite and kaolinites. Linear relationships were observed between void ratio and coefficient of permeability. The slopes of these straight lines have the following order

$Li^+ < Na^+ < K^+ < Ca^{++} < Ba^{++}$.

The variations in the order of permeability between Li^+ and Ba^{++} is more than a thousand fold.

Piping in Earth Dams of Dispersive Clays:

Failure of dams because of the dispersive nature of clays have been reported (Sherard et al 1972). It has been established that clays with an exchangeable sodium percentage (ESP) of 7 to 10 are moderately dispersive and are associated with piping failure in clay dams when the reservoir water is relatively pure. Clays with an ESP of 15 or more have serious piping potential. This can be atributed to the increase in double-layer repulsion resulting in a decrease in effective stress and, hence, strength.

Anion Adsorption :

Anion adsorption changes the physical-chemical properties of clays (Sivapullaiah and Sridharan 1985). Consequently the engineering behaviour of clays is influenced by anion adsorption.

Volume change behaviour :

The compressibility characteristics of various phosphate and acetate adsorbed clays were studied by Sreepada Rao (1982). It has been shown that phoshate adsorption significantly influences the volume change behaviour of clays.

Volume Change - Pressure relationships :

Phosphate adsorbed Na- and kaolnites have shown increased volume on saturation at seating pressure and reduced compressions upon subsequent loading. The quantum of swelling at the seating pressure increases with duration (or degree) of phosphate adsorption, and resistance to compression increase on further loading. The increase in volume at seating pressure has been attributed to the change in the fabric towards a higher flocculation (or random fabric) and increase in volume of individual particles because of phosphate adsorption. An increase in resistance offered by the treated clays

167

has also been attributed to the above changes in fabric and to the reduced plasticity characteristics of the treated clays.

Phosphate treated Ca-kaolinite swells more than treated Na-kaolinite at seating pressure, although the difference is small. The compression for subsequent loading is less for treated Ca-kaolinite than for treated Na-kaolinite. The effect of the associated cation is small for treated kaolinite clays.

Treatment of Na-montmorillonite with phosphoric acid results in reduced swelling at seating pressure and resistance to external loading increases for Ca-montmorillonite. For Ca-montmorillonite treatment causes the clay to swell but the compressibility reduced.

The difference between behaviour of Na- and Ca-montmorillonites is due to the cation effect. While the replacement of Na^+ by H^+ causes less swelling, that of Ca^{++} by H^+ causes increased swelling because of changes in the thickness of the diffuse double layer. This is further confirmed by the essencially similar behaviour of 1000h treated Na- and Ca-montmorillonite, where both the clays become in essence H-montmorillonites.

Treatment causes aggregation of particles and, possibly a change in the fabric towards a greater flocculence. The combinaed effect is for the soil skeleton to resist the external loading with reduced compressions.

The results on H- clays (both treated and untreated) confirm the conclusions drawn for treated Na- and Ca-clays.

For all the different homoionic kaolinite and montmorillonite clays phosphate adsorption causes reduction in compression index values. The reduction is marked for montmorillonite clays. Further, compression index increases with load increment, and the changes are insignificant.

For both treated and untreated homoionic clays the volume increase in the seating pressure upon saturation rises as the initial dry density increases, but the compression - pressure curves on subsequent loading are generally parallel.

Initial, primary and secondary compressions :

For Na-, Ca- and H- kaolinites, treatment with phosphoric acid (phosohate adsorption) causes reduction in initial, primary and secondary compressions and the reduction in primary compression is significant for any load increment. As the duration of treatment increases (phosphate adsorption increase), the total compression for any load increment decreases.

For Na-, Ca- and H- montmorillonite also, the treatment with phosphoric acid causes a significant reduction in primary compression for any load increament. While the primary compression is more than the immediate and/or the secondary compressions for untreated clays, it is the least for treated clays for which the initial compression is maximum. Also the secondary compression decreases with increase in anion adsorption.

The effect of acetate ion on the compressibility characteristics of kaolinite and montmorillonite clays is much less compared with the phosphate ion because of the difference in their configurations.

Shear Strength Characteristics :

Lutz and Haque (1975) report that phosphorus adsorption reduced modulus of rupture of montmorillonite and montmorillonite-kaolinite mixtures. Shroff (1970) finds that the physico-chemical effect of phosphates improves the strength and durability characteristics of clays. Sreepada Rao (1982) confirm, by studying various homoionized saturated clays treated with phosphates, that anion adsorption improves the strength. The following are the typical results :

	C kg/cm^2	degrees
Na-kaolinite		
untreated	0.02	30.5
treated	0.27	38
Ca-kaolinite		
untreated	0	22
treated	0	41
Na-montmorillonite		
untreated	0.07	18.5
treated	0.07	39.5
Ca-montmorillonite		
untreated	0.03	18.5
treated	0.16	44

The phenomenal increase in strength, in the case of kaolinite in spite of reduction in dry density at failure, has been attributed to phosphate adsorption. In case of montmorillonite, the differences in dry densities at failure between treated and untreated soils is small and the increase in strength is due only to phosphate adsorption.

Permeability :

Anion adsorption decreases the permeability of Ca-montmorillonite and increases that of Na-Montmorillonite (Sreepada Rao 1982). Permeability decreases in the case of kaolinites. The increase in the permeability of Na-montmorillonite is due to the replacement of Na+ by reduced cation size of H+ during the initial treatment of phosphate in acid medium. But as the adsorption of phosphate increases, the bulkier phosphate ion reduces permeability. The decrease of permeability of Ca-montmorillonite on phosphate treatment in acid medium is due to the change of divalent cation with monovalent cation by H$^+$ leading to a more effective void ratio. As more and more phosphate gets adsorbed on Ca-montmorillonite, permeability further decreases because of bulky phosphate ion. Thus both cation and anion decrease permeability in the case of Ca-montmorillonite and it increase in the case of Na-montmorillonite. Since adsorption of phosphate in kaolinites is by exchange of OH, the permeability and void ratios decrease and cation exchan-

ges are negligible in the case of kaolinites.

Anion adsorption and differential heaving :

Sridharan et al (1981) report the significant influence of phosphate adsorption on the volume change behaviour of soils leading to extensive damage to superstructure. They establish an uncontaminated virgin soil in a fertilizer plant which does not show any significant swelling on addition of water, exhibits a large amount of swelling phosphate adsorption. The phosphate adsorption took place when the original soil had, over long periods, been exposed to effluent waters contaminated with phosphoric acid through unprotected open drains. Since the ground water level much lower, the phosphate contaminated water drained into the foundation soil. Because the structure build was essentially lightly loaded, with uneven loading in certain cases, the resultant swelling effect caused differential movement leading to distress to floors, beams and so on.

Nature of Pore Fluid

Volume of Change Behaviour :

Effect of dielectric constant : The volume change behaviour of clays can be significantly affected by the nature of the pore fluid. It has been found by several investigators that both swelling and compression can take place with changes in the nature of pore fluid but without any change in the external load. The two important aspects of the nature of pore fluid which have been studied in detail and reported in the literature are the dielectric constant and the ionic concentration.

Sridharan and Rao (1973), using different organic pore fluids, have shown that the effect of dielectric constant is diametrically opposite for kaolinite and montmorillonite clays in their compression behaviour. While montmorillonite clay shows significant swelling with increase in dielectric constant (at constant external load), kaolinite clay showed a decrease in volume (Fig. 2). This behaviour has been explained by two different mechanisms : mechanism 1, where the volume change is governed by the shearing resistance at interparticle level and mechanism 2, where the volume change is governed primarily by the diffuse double-layer repulsive forces. Although these two mechanisms operate simultaneously, the results reveal that mechanism 1 primarily controls the volume change behaviour in non-expanding type clays like kaolinite, whereas mechanism 2 operates in the case of expanding lattice type clays like montmorillonite. In their study Sridharan and Rao also evaluate the results in the light of the modified effective stress concept (equation 1), which takes into consideration the electrical forces in the particulate system.

$$\bar{C} = \bar{\bar{\sigma}} \, a_m = \sigma - \bar{u}_w - \bar{u}_a - R + A \quad (1)$$

where $\bar{C}$ = effective contact stress defined as the effective stress that controls the shearing resistance.

$\bar{\bar{\sigma}}$ = mineral to mineral contact stress

a_m = mineral to mineral unit contact area

σ = externally applied pressure on a unit area

$\bar{u}_w$ = effective pore water pressure

$\bar{u}_a$ = effective pore air pressure

R and A = effective interparticle repulsive and attractive pressures respectively.

Equation (1) may be rewritten as

$$\bar{C} = \sigma' + \sigma''$$

where σ' = conventional effective stress,

and σ'' = intrinsic effective stress
= A - R
= net electrical attractive pressure.

Similar results on the effect of dielectric constant on the volume change behaviour are also reported by Mesri & Olson (1970) and Mesri & Olson (1971). Using the electrical double-layer theory of Gouy and Chapman, the effect of dielectric

cosntant on the pressure - void ratio relationship can be predicted for idealised clays with parallel orientation (Sridharan and Jayadeva 1982). This assumption corresponds to mechanism 1. The theoretical prediction corroborates the experimental findings.

Secondary compression, which can form a significant portion of total compression, is also greatly influenced by the type of pore fluid using several organic fluids. Sridharan and Sreepada Rao, (1982) show that the secondary compression per unit of log (time) to the final thickness of the sample for any pressure increment is directly related to the strength of the soil skeleton at particle level. The secondary compression coefficient decreases with increase in the strength of soil skeleton at particle level, which is governed by the dielectric constant of the pore fluid (Fig.3).

Their analysis shows that the strength of the soil skeleton increases with increase in modified effective stress (eqn.1), which in turn rises with increase in attractive forces and decreases with increase in repulsive forces. It is further shown that as the dielectric constant increases, the net repulsive pressure increases and the attractive pressure decreases.

Effect of Ion Concentration :

As has been stated earlier, the compression behaviour of a clayey soil is significantly influenced by the concentration of ions in the pore fluid. The effect of the concentration of ions in the pore fluid is different for different types of ions. Bolt (1956) was one of the first to emphasise the significant influence of ion concentration of the pore fluid on the e - log p behaviour in the one-dimensional consolidation test. According to double-layer theory (Sridharan and Jayadeva 1982), as ion concentration increases the equilibrium void ratio decreases because the double layer is compressed. The effect of ion concentration is strong at low conslidation pressures when the concentration is

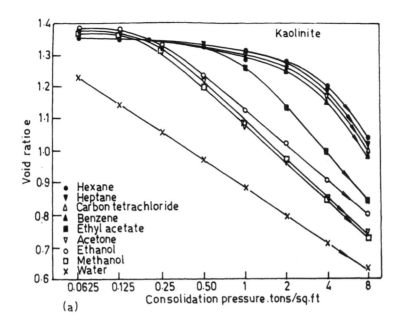

(a)

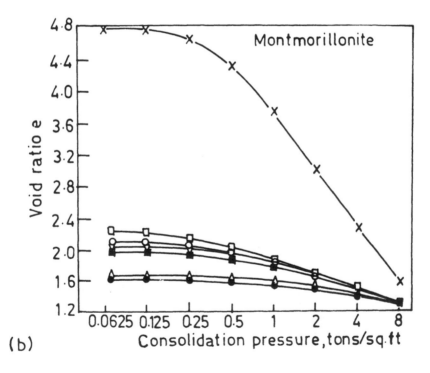

(b)

Figure 2. One-dimensional consolidation curves for (a) Kaolinite, (b) Montmorillonite with different pore fluids

171

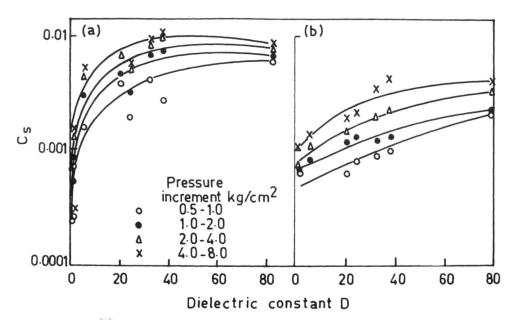

Fig 3. Variation of secondary compression coefficient C_s with dielectric constant D for (a) black cotton soil (b) sodium kaolinite

greater than 10^{-4}M. When the ion concentration is less than 10^{-4}M, its effect on the equilibrium void ratio is negligible at all pressures. The pronounced effect of ion concentration on the consolidation behaviour of Na-montmorillonite (liquid limit = 1116% at 0.0001N) is underlined out by Mesri & Olson (1971) and is shown in Fig.4, where both compression and swelling curves are shown. Comparison of these experimental data with double-layer theory shows reasonably good agreement.

Strength Behaviour :

Along with other factors, strength behaviour is also significantly influenced by the nature of pore fluid. The shear strength of soil which is controlled by modified effective stress (equation 1) is affected by changes in electrical attractive and repulsive pressures. Many factors are responsible for the net attractive and repulsive forces between clay particles. A number of researchers have attempted to improve the understanding

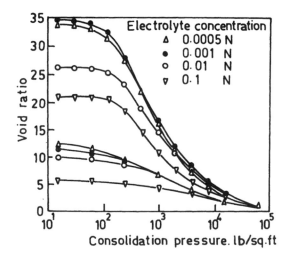

Fig.4 One-dimensional consolidation curves for sodium montmorillonite (pH=7), (after Mesri & Olson, 1971)

of the nature of these forces from the view point of mechanical behaviour (Lambe 1958, 1960, Bolt 1956, Rosenqvist 1971, 1973 and 1979 to name a few). From these investigations it can be concluded that the primary forces responsible for repulsion between two clay particles are due to the interaction of diffuse dobule layers, which is directly proportional to the dielectric constant. A number of phenomena are responsible for the existence of electrical attractive forces among clay particles and these forces are inversely proportional to the dielectric constant of the pore fluid. Sridharan and Rao (1979), using different organic fluids, demonstrate that the shear strength increases with decrease in dielectric constant for both kaolinite and montmorillonite clays (Fig 5 and Table 1) and is in accordance with the modified effective stress concept (equation 1) Corrobrative results are reported by Sridharan, Sreepada Rao and Makan (1983), using black cotton soil.

The effect of ion concentration is not significant on the shear strength of saturated sodium and calcium montmorillonite (Mesri and Olson 1970). However, this aspect needs further investigation with other clays and clayey soils.

Permeability :

For a particular soil at any void ratio, true permeability should be the same irrespective of the type of pore fluid. But because of variations in the effective thickness of diffuse double layer brought about by changes in dielectric constant/ion concentration, the effective pore space for the flow of the fluid varies. It is now well known that the thickness of the diffuse double layer is depressed with decrease in dielectric constant or increase in ion concentration (Sridharan and Jayadeva 1982). Further, the effect of higher valency of ion is to reduce the thickness of the diffuse double layer. Hence, other conditions, including similarity of fabric, the decrease in dielectric constant, increase in ion concentration or increase in the valency of the ion will result in the increase of permeability. Figure 6 shows the void ratio - permeability relationship of montmorillonite

with fluids of different dielectric constant, ion concentration and valency, as reported by Mesri and Olson. The dielectric cosntant of water, ethyl alcohol, carbon tetra chloride and benzene are 80.4, 24.3, 2.28, and 2.28 respectively.

Organic Matter

Various research has reported the effects of organic matter on different engineering properties.

Volume change behaviour :

Orlando and Al-Khafagi (1980) study the compression behaviour of several ;model organic soils prepared from kaolinite and pulp fibre. A plot of organic fraction versus void ratio for selected pressure levels show linear relationships. The two variables, organic fraction and pressure, are uncoupled by plotting the intercepts and slopes against consolidation pressures, giving two void ratio parameters, $C(p)$ and ..(p). Use of these parameters with organic fraction X_f permits prediction of an equilibrium void ratio e :

$$e\ (p,\ X_f) = C(p) + M\ (p)\ X_f$$

For a given consolidation pressure, the equilibrium void ratio can be determined for all organic contents (Fig. 7). This permits settlement prediction in terms of initial and final field over burden pressures and the existing organic fraction, using conventional settlement theory. The void ratio parameters are based entirely on the experimental data for the model organic soils. Excellent agreement was observed between field and predicted equilibrium void ratios for several normally consolidated paper mill sludge block samples.

Strength Behaviour :

Though soil engineers have long recognized the deleterious effects of organic matter on soil behaviour, little is known about the organic matter in soils or about the mechanisms whereby organic matter alters soil behaviour. Arman (1969) reported that slight organic content sometimes acts to

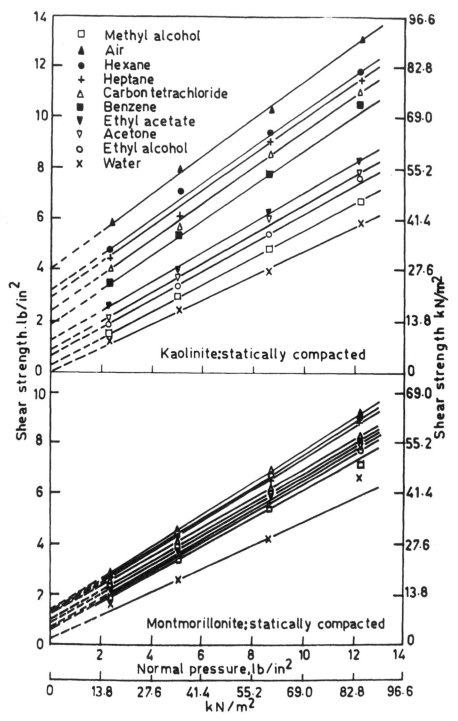

Figure 5. Strength lines for (a) Kaolinite (b) Montmorillonite, with different pore fluids

174

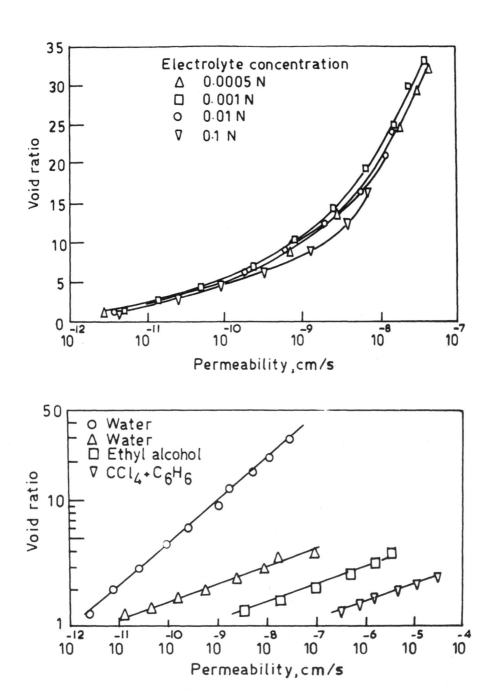

Figure 6. Permeability — void ratio curves for (a) Sodium montmorillonite (pH = 7),(b) Montmorillonite in various fluids (after Mesri & Olson, 1971)

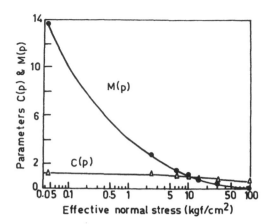

Fig.7 Experimentally derived void ratio parameters for model organic soils (After Orlando and Al-Khafaji, 1980)

increase the strength of soil. It is also possible that it may be beneficial in its effect on other engineering properties. Organic matter and certain materials which aggregate or cement particles can have considerable effects on the soil properties, even though these components are present in relatively small concentrations.

Franklin et. al (1973) have studied the strength characteristics of mixtures of peat and inorganic soils as well as natural organic soils. The effect of organic matter on the unconfined compressive strength was observed to reduce the maximum strength attainable. However, the strength versus moisture curves of the mixutres resembled those of inorganic parent samples rather than those of other organic soils of similar organic content, and the strength versus organic content relationship showed a high degree of scatter. It was inferred that the effect of organic matter below about 20%, was less important than those of minor mineralogical structural differences. Franklin et al also report that as the organic content increases, optimum density reduces and optimum water increases.

CONCLUSIONS

The engineering behaviour of clayey soils are significantly affected by pollutants. For the purpose of this study, the effects of pollutants, are considered in terms of ion exchange, nature of pore fluid (Organic fluids and inorganic salt concentration), and organic matter content.

Volume change behaviour :

Both valency and size of the associated cation significantly influences the compressibility characteristics of montmorillonite clay. Lower valency and larger size of the cation show higher swelling and compression. These effects are marginal for kaolinitic clay. An increase in electrolyte concentration decreases the equilibrium void ratio because of depressed double-layer thickness.

Phosphate adsorption increases the swelling characteristics of kaolinite clays, whereas it reduces swelling in montmorillonite clay. Both types of clays show higher resistance against loads, resulting in reduced compression.

The effect of dielectric constant is opposite for kaolinite and montmorillonite clays. As the dielectric constant of pore fluid increases the swelling of montmorillonite clays increases while in kaolinite clay it decreases.

A good linear relationship exists between organic fraction and void ratio at the same pressure level. More data are needed to fully establish this relationship for various types of organic soils.

Strength Behaviour :

An increase in the valency of cation and/or electrolyte concentration and/or adsorption of anion results in an increase of shear strength. On the other hand, an increase in the ionic size and/or organic content and/or dielectric constant of pore fluid brings down the shear strength. No anamolous behaviour has been reported between kaolinite and montmorillonite clays. However, the effect of these factors is more pronounced for the montmorillonite type of clays.

Permeability :

Permeability of clays depend primarily upon the thickness of the diffuse double layer as established by assoicated valency, ionic concentration and nature of pore fluid. The double layer is depressed with increase in ion concentration and/or valency, and a decrease in ion size and/or dielectric constant of the pore fluid with a resultant increase in the coefficient of permeability.

Two case histories underlining the importance of exchangeable sodium percentage on the dispersibility of clays in a dam leading to piping failure and the effect of phosphate adsorption on soils leading to distress in buildings are cited.

REFERENCES

Arman, A. (1969). 'A Definition of Organic Soils'. Engineering Reseach Bulletin No. 10.1 Lousiana State University, Division of Engineering Research, for Lousiana Department of Highways.

Bolt, G.H. (1956). 'Physico-chemical analysis of the compressibility of clays'. Geotechnique, Vol.6, p.86.

Franklin, A.G., Orozeo, L.F. and Semrau, R. (1973). 'Compaction and Strength of slightly organic soils'. Journ. of Soil Mech. and Found. Division, ASC, Vol. 99, No. SM7, p.541.

Lambe, T.W. (1958). 'The Structure of compacted Clay'. Journal of Soil Mech. and Foundn. Division, ASCE, Vol. 84, SM2, P.1.

Lambe, T.W. (1960). 'A mechanistic picture of shear strength in clay'. Proceed. Research Conference on Shear Strength of Cohesive Soils, Boulders, Colarado, 1960.

Lutz, J.F. and Haque, I. (1975). 'Effects of Phosphorus on some physical and chemical properties of clays'. Soil Science Soc. of Amer. Proc. Vol.39, p.33.

Mesri, G. and Olson, R.E. (1970). 'Shear strength of montmorillonite'. Geotechnique, Vol.20, p.261.

Mesri, G and Olson, R.E. (1971). 'Consolidation characteristics of montmorillonite'. Geotechnique, Vol. 21, p.341.

Orlando, B.A. and Al-Khafaji, A.W.N. (1980). 'Organic material and soil compressibiltiy'. Journ. of the Geotech. Engg. Division. Proc. ASCE, Vol.7, p.749.

Quirk, J.B. (1960). 'The role of interparticle forces in soil structure'. Seminar Int. Particle Forces in Clay-Water Electrolyte Systems'. CSIRO, Melbourne, 2-1 to 2-8.

Rosenqvist, I.Th. (1955). 'Investigation in the clay-electrolyte system'. Norwegian Geotechnical Institute, Publication No.9, p.125.

Salas, J.A.J. and Serratosa, J.M (1953). 'Compressiblity of Clay'. Third International Conference on Soil Mech. and Foundn. Engineering, Vol.1, p.192.

Sherard, J.L., Decker, R.S. and Ryker, N.L. (1972). 'Piping in earth dams of dispersive clay'. Conference on Performance of Earth and Earth supported strucutre, ASCE, Purdue University, p.653.

Shroff, A. (1970). 'Physico-chemical reaction of phosphate with clay'. Journ. of Indian Natl. Soc. of Soil Mech. and Foundn. Engr. Vol. 9, p.341.

Sivapullaiah, P.V. and Sridharan, A. (1985). 'Effect of polluted water on the physico-chemical properties of clayey soils'. 9th Bangkok Geotechnical Symposium and Speciality Session, December 1985.

Sridharan, A. and Jayadeva, M.S. (1982). 'Double layer theory and compressibility of clays'. Geotechnique, Vol.32, p.133.

Sridharan, A., Nagaraj, T.S. and Sivapullaiah, P.V. (1981). 'Heaving of soil due to acid contamination'. 10th International Conference on Soil Mech. and Foundn. Engineering, Stockholm. Ch.6. p.383.

Sridharan, A. and Rao, G.V. (1971). 'Effective stress theory of Shrinkage phenomena'. Canad. Geotech. Journ. Vol.8, p.503.

Sridharan, A. and Rao, G.V. (1973). 'Mechanisms controlling volume change of saturated clays and the role of the effective stress concept'. Geotechnique, Vol.23, p.359.

Sridharan, A. and Rao, G.V. (1979). 'Shear strength behaviour of saturated clays and the role of effective stress concept'. Geotechnique, Vol.29, p.177.

Sridharan, A. and Sreepada Rao, A. (1982). 'Mechanisms controlling the secondary compression of clays'. Geotechnique, Vol.32, p.249.

Sridharan, A., Sreepada Rao, A. and Makan, S. (1983). 'Shear strength behaviour of expansive clays'. Seventh Asian Regional Conference on Soil Mechanics and Foundation Engineering, Technion, Israel. Institute of Technology, Haifa, Israel, Aug. 1983.

Sreepada Rao, A. (1982). 'Physico-chemical properties and engineering behaviour of anion adsorbed clay', Ph. D Thesis of Indian Institute of Science, Bangalore, India.

Winterkorn, H. and Bavor, L.D. (1934). 'Sorption of liquids by soil colloids'. Soil Science, Vol.38, p.29.

Environmental Geotechnics and Problematic Soils and Rocks, Balasubramaniam et al. (eds)
© *1987 Balkema, Rotterdam. ISBN 90 6191 785 9*

Effect of polluted water on the physico-chemical properties of clayey soils

P.V.Sivapullaiah & A.Sridharan
Department of Civil Engineering, Indian Institute of Science, Bangalore

ABSTRACT : The increasing use of clayey soils in civil engineering practice in different environments necessicitates a better rational understanding of the physico-chemical mechanisms controlling their behaviour. Since both solid and liquid wastes are increasingly mixed with soils, soil being an effective medium of disposal for industrial wastes, studies on the modification of properties of soils become important. In this article, the effects of type and amount of clay minerals, the nature of pore fluid, cation and anion exchanges and organic matter are investigated. It is shown that clay properties are modified to different extents and by different mechanisms depending on the pollutant parameters.

INTRODUCTION

The increasing use of clays in soil engineering practice necessitates a better understanding of their behaviour and their changes in various environments. Clays in which surface forces or interparticle electrical repulsive and attractive forces are predominant exhibit a wide range of mechanical behaviour, depending upon changes in physical and physico-chemical properties. However, the physico-chemical properties and processes in soils have attracted only marginal attention from soil engineers. But with land becoming more and more scarce, soils which have hitherto been found unsuitable are also being considered for use for civil engineering purposes. Hence, modification of the properties of these soils becomes inevitable. Secondly, with growing industrialisation, industrial wastes are being disposed of on land. This will result in the modification of soil properties. It is necessary to ascertain the soil behaviour in different environments and the mechanisms involved.

FACTORS INFLUENCING THE SOIL BEHAVIOUR

The behaviour of soil-water systems is primarily controlled by (i) the type and amount of clay mineral, (ii) the nature of pore fluid, (iii) associated cations and anions, (iv) organic matter.

Type of Clay Mineral

The effect of type and amount of clay on soil properties is well researched. In general, the larger the quantity of clay mineral in a soil, the higher the plasticity, the greater the potential shrinkage and swell, the lower the permeability, the higher the compressiblity, the higher the cohesion and the lower the angle of internal friction. But different groups of clay minerals exhibit a wide range of properties. In any particular soil, the different consituents may not influence the properties in direct or even predictable proportion to the quantity present. This is because certain clay minerals which may be present in very small quantities may exert a powerful

soil. A small amount of montmoril- lonite in a clay is likely to provide a material very different to another clay with the same com- position in all ways except for the absence of montmorillonite. Also, both physical and physico-chemical interactions are possible among clays.

The physico-chemical interactions between clay minerals have been demonstrated by Seed et al (1964). Mixtures of bentonite and kaoli- nite, and bentonite and illite were prepared and their liquid limits determined. Although the bentonite- kaolinite mixtures gave values close to their theoretical propor- tions, the liquid limit values for bentonite-illite mixtures were much less than predicted.

Recent work by Sivapullaiah and Sridharan (1985) on the liquid limits of mixtures further elabo- rates this view. Liquid limit values were determined for mixtures of bentonite and angular grained coarse sand; bentonite and round- grained coarse sand; bentonite and fine sand; bentonite and kaolinite; bentonite and different silts. It is concluded that the liquid limit

of mixtures of soils is not a phy- sical but a physico-chemical prope- rty. There are deviations from the theoretical expectations even in clay-sand mixtures; these devia- tions naturally being more for clay-clay mixtures (Fig. 1). It has also been shown that the size has more influence than the shape of the sand particle.

Lambe and Martin (1956), in their studies on the grain size analysis and plasticities of clay mineral mixtures, show the effect of elec- trostatic attraction between diffe- rent clay mineral particles. The clay content and plasticities of mixtures of clay minerals are far lower than one would expect from their compositional data, indicating that clay minerals can intergrow and/or aggregate to form larger and less plastic particles (Fig.2).

Thus changes in soil properties can be effected by the suitable mixing of clays. This emphasises the need for care in evaluating the proper- ties of soil in any particular location where the intermixing of clays can occur.

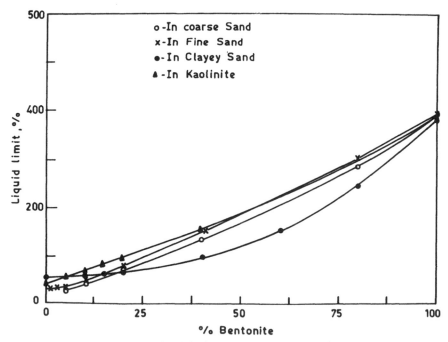

Fig. 1. Liquid limit of Bentonite Mixtures

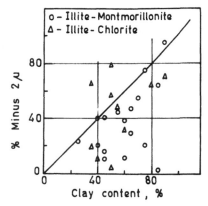

Fig.2a. Clay content vs Clay size

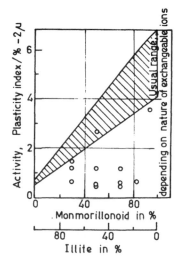

Fig.2b. Plasticity vs Composition (After Lambe 1956)

The Nature of Pore Fluid.

Because of the large quantities of organic solvents being disposed of on land, the properties of soils are significantly changed. Various researchers have reported the effect of dielectric constant of pore fluid on the properties of different types of clays. Such studies also contribute to an understanding of the mechanisms controlling the various physical properties.

The behaviour of clays in different organic solvents was studied by Green et al (1979), mainly to assess the impact of organic solvents on the integrity of clay liners for the disposal of industrial wastes. A number of polar organic molecules were observed to form interlayer complexes with clays and to cause expansion of the clay lattice. Non-polar saturated organic molecules such as n-hexane and n-dodecane are also reported to form interlamellar complexes with air-dried montmorillonite. Bradley (1954) shows that benzene, too, forms a complex with montmorillonite. The extent and ease with which intercalation occurs with non-polar organics may depend on the state of hydration of the clay itself. Ca-montmorillonite heated to 250°C prior to immersion in several organic liquids does not cause interlayer expansion. On the other hand, airdried clays are observed to form interlayer complexes. This is because the presence of a sufficient number of water molecules causes a slight separation of silica layers. The extensive data on clay-organic interaction have been reviewed by Theng (1974). The nature of clay-organic complexes, including the effect of organic molecules on interlayer spacing, is discussed for alcohols, polyhydroalcohols, ketones, aldehydes, ethers, nitrites, amines, aliphatic and aromatic hydrocarbons and organic pesticides. It is indicated out that for some non-polar molecules the interaction of partially dehydrated clays may be limited by the inability of these molecules to replace the interlayer water or to bind with exchangeable cations through aquo bridge.

The characteristic spacing between the layers depends on the nature of the intercalated molecules. The d(001) spacing in montmorillonite can vary from 9.5A° when unexpanded to 23.5A° in the presence of pyridine. The intercalated molecules may orient themselves either parallel or perpendicular to the silicate sheet, the preferred orientation depending upon the external concentration (Greene-Kelly 1975).

The interlayer spacing of montmorillonite in polar organic solvents will be very high, particularly with solvents of high dipole moments and high dielectric constant such as formamide (Olejnik, 1974). The degree of swelling is not in keeping with expectations based on the Norrish swelling index (Norrish, 1954). According to Norrish, swelling should be related to UE/V_2, where U is the solvation energy, E is the dielectric constant of the interlayer liquid and V is the valency of the interlayer cation. This is because the measured bulk dielectric constant of the liquid and dielectric constant of the liquid in the interlayer region of a clay are radically different. A more satisfactory relationship between properties and the octanal/water partition coefficient is reported by Green et al (1979).

The effect of pore fluid on the liquid limit of clays is studied by Sridharan and Rao (1975). Using a cone penetrometer number of liquid limits were measured for kaolinite and montmorillonite in various organic solvents. Since the unit weights of the fluids used differ from one another, the liquid limits were calculated on volume basis to facilitate comparison. The results were plotted against dielectric constant of fluids (Fig 3). Kaolinite and montmorillonite behaved in a strikingly opposite manner with respect to change in porefluid. Whereas a decrease in liquid limit is observed for kaolinite with increase in dielectric constant of pore fluids, an increase in liquid limit is recorded for montmorillonite, for which the liquid limit decreased significantly from a value of 866% for water (dielectric constant = 80.4) to 149.2% for hexane (dielectric constant = 1.89), whereas for kaolinite the liquid limit increased from 127.0% for water to 230.0% for hexane. Based on these observations the mechanisms controlling the liquid limit behaviour of clays are explained.

Though it is widely accepted that the liquid limit is a measure of shearing resistance, with the use

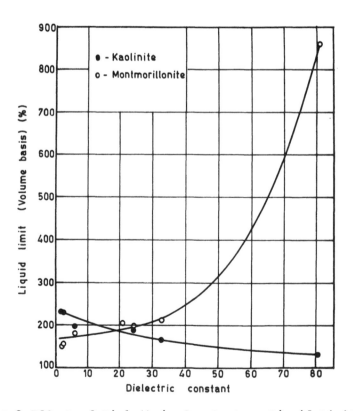

Fig.3 Effect of Dielectric Constant on Liquid Limit

of organic pore fluids and water of different dielectric constants, it was shown that the liquid limit of clays is primarily, controlled by (i) the shearing resistance at particulate level and (ii) the thickness of the diffuse double layer.

An increase in dielectric constant decreases interparticle shearing resistance and increases double layer thickness. A decrease in shearing resistance results in lower liquid limit whereas an increase in double-layer thickness shows higher liquid limit. These effects obviously oppose each other and the liquid limit of a particular clay will depend on which of the two predominates.

For kaolinite, a non-expanding lattice type of clay, the contribution due to diffuse double layer is significant and the liquid limit is primarily governed by the shearing resistance at particulate level. Hence, an increase in dielectric constant results in lower liquid limit.

Although for montmorillonite also, the liquid limit should be governed by the shearing resistance, because it is an expanding lattice type of clay, the contribution to the diffuse double layer overrides and governs the liquid limit. Hence, an increase in dielectric constant results in higher liquid limit.

Given the effect of the various organic solvents on the liquid limit of clays, the effect of these solvents on other mechanical properties can be envisaged since generalisation of soil behaviour through normalisation with liquid limits is possible (Nagaraj and Jayadeva, 1981).

The effect of pore fluid on the shrinkage phenomena of kaolinite and montmorillonite is studied by Sridharan and Rao (1971). It is shown that good correlations exist between the dielectric constant of pore fluid and shrinkage void ratio. For both these clays the srinkage void ratio decreases as the dielectric constant of pore fluid increases. These studies also throw light on the physical mechanisms involved in the shrinkage phenomena, with the aid of modified effective stress concept.

Effect of Electrolyte Concentration

Change in sensitivity from an increase or decrease in salt content of a clay is today recognized by all scientist working with soils, though details of this influence may not be fully understood. Rosenqvist (1955) studies the influence of salt concentration on the liquid limit of some clays.

The addition of sodium chloride has a significant influence on the liquid limit (fig. 4) and the plastic limit. Generally the liquid limit increases steadily until salt content is about 15 g/l when there is a slight drop in the values of the liquid limit. This is due to variations in the electrostatic charge of the clay particle. The effect of the added electrolyte upon the zeta potential or the electric charge of the clay minerals is an important feature in the case of the liquid limit of clays. The liquid limits of two clays are determined after the addition of low concentrations of sodium pyro phosphate and sodium carbonate. At low concentrations of sodium carbonate, some sodium ions are absorbed upon the clay minerals. This way the net negative charge of the clay mineral decreases. At higher concentrations some of the carbonate/phosphate ions combine with the clay mineral by chemical binds. This leads to an increase in the net negative charge. Such chemical binding can continue only until all possible bonds between the anion and the clay minerals are established. The further addition of sodium carbonate will lead to an adsorption of the positively charged sodium ions and hence, to a decrease in the net negative charge of the clay particle. These effects influence the liquid limit and hence the observed behaviour.

The Guoy theory shows that the thickness of the double layer decreases inversely to the valence and the concentration. Increase in electrolyte concentration reduces

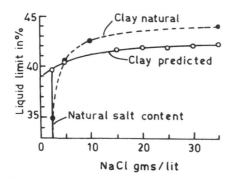

Fig.4 Effect of NaCl on Liquid Limit on Quick Clay (After Rosenqvist - 1955)

the surface potential for the condition of constant surface charge, but the decay of potential with distance is also much more rapid.

The influence with different electrolytes on the flock volumes of clays is studied by Rosenqvist (1955). The significant influence of different ions in the solution is established.

Ion Exchange

It is known that the edges and corners of the clay particles can carry a charge opposite to that of the faces. The surface of the soil particle has a negative charge and the edges and corners have a positive charge. Thus the clay particles have an amphoteric character, i.e., the power to bind both cations and anions. The capacity of soils to exchange cations and to absorb or exchange anions varies greatly with the clay and silt friction, the mineralogical composition, organic matter, the pH value and concentration of ions in solution and colloidal matter. The cations and anions thus associated will play a very important role in the behaviour of the clays.

Cation Exchange

Ions of one type can be replaced by ions of another type, for example, Ca_2+ by $Na+$, $Na+$ by Ca_2+, Fe_3+ by Mg_2+ and so on. The ease with which an ion of one type can replace an ion of another type depends mainly on the valence, relative abundance of different ion types and ion size. Other things being equal, trivalent cations are held more tightly than divalent and divalent more tightly than monvalent. Normally, small cations tend to replace larger cations. A typical replaceability series is

$$Na+ < LI+ < K+ < Rb+ < Cs+ < Mg_2+ < Ca_2+ < Ba_2+ < Cu_2+ < Al_3+ < Fe_3+ < Tn_4+$$

increasingly strong absorption

It is possible, however, to displace a cation of high replacing power, such as Al_3+, by one of lower replacing power, such as $Na+$, by mass action in which the concentration of sodium ion in a solution becomes very high in relation to that of aluminium.

Double layer theory suggests that the type of cations on a clay particle has a major influence on the properties of clays. Changes in these ions may lead to changes in the physical and physico-chemical properties.

The effect of the exchangeable ions on the Attergerg limits of various clays is seen in Table (1).

The proportion of sodium in the adsorbed layer has an important bearing on the structural status of a soil and is often described in terms of the exchangeable sodium percentage (ESP), defined by

$$ESP = \frac{Na+}{total\ exchange\ capacity} \times 100$$

Soils with ESP greater than 2 percent are susceptible to spontaneous dispersion in water.

Anion Exchange

The interaction between anions and the clay is somewhat involved and the anion exchange is as yet far less studies than cation exchange. However, the phenomenon of anion exchange itself is well established, especially for ions like phosphate, arsenates and borates, which have about the same size and geometry as the silica tetrahedra.

Table 1.   Atterberg Limits of Clay Minerals (Lambe and Whitman 1969)

Mineral	Exchange- able ion	Liquid limit (%)	Plasltic limit (%)	Plasticity index (%)	Shrinkage limit (%)
Montmori- llonite	Na	710	54	656	9.9
	K	660	98	562	9.3
	Ca	510	81	429	10.5
	Mg	410	60	350	14.7
	Fe	290	75	215	10.3
	Fe$_a$	140	73	67	-
Illite	Na	120	52	67	15.4
	K	120	60	60	17.5
	Ca	100	45	55	16.8
	Mg	95	46	49	14.7
	Fe	110	49	61	14.7
	Fe$_a$	79	46	33	-
Kaolinite	Na	53	32	21	26.8
	K	49	29	20	-
	Ca	38	27	11	24.5
	Mg	54	31	23	28.7
	Fe	59	37	22	29.2
	Fe$_a$	56	35	21	-
Attapulgite	H	270	150	120	7.6

a   after five cycles of wetting and drying.

Many models have been used to describe anion and cation adsorption in soil systems (Barrow 1980). The Freundlieh and Langumuir equations have been very popular and now the double Langmuir (Mulfadi et al. 1966, Holfred et al 1974, Rajan 1975, Ryden 1975) is coming into vogue. The adsorption of phosphate, citrate and selenite on a clay mineral surface is described by Bowden et al (1977).

The investigations of anion exchange in clay minerals have been to a considerable extent associated with studies of adsorption, fixation and desorption or removal of phosphates by soils and considerably more information has accumulated on phosphate adsorption than that on the adsorption of any other anion.

Effect of Anion Adsorption

The effect of anion adsorption on the physical and physico-chemical properties of clays is not well researched, though some data have become available in recent years. Edmund (1947) explains that anions like phosphates, which fit on the edges of silicate units, may serve as a bond to hold the clay mineral blocks together. The geometric configuration of phosphate ions may also be closely similar to certain water nets of the adsorbed layers and this similarity would permit the anion to fit into the water net and perhaps at the same time also slightly disrupt the water structure. Yeoh and Oades (1981) shows that treatment with phosphoric acid, soils decrease water dispersible clay and improves aggregation by interstitial cement of aluminium phosphate. Thein (1976) reports that phosphoric acid increases stable aggregation leading to higher porosities and water-holding capacities but lowers bulk densities. The porous aggregates are mechanically weaker than those from untreated soils without anion adsorption. Shroff (1970) has studied the behaviour of montmorillonite type clays in the presence of phosphate anions. The adsorption increases the charges on the particles thereby tending to cause dispersion of particles. Side by side with dispersion, the dihydroxy aluminium dihydrogen phosphate formed by the reaction, is known to be a cementitious material. Together with these processes, the compaction brings the particles closer. Shroff reports that the physico-chemical effect of phosphates on clay improves the strength and durability characteristics of clay. Further, there is an optimum amount of phosphate which produces the highest immediate strength and immersed strength in clay. In montmorillonites, phosphate adsorption reduces swelling and shrinkage. However, little work has been carried out on the effect of anion adsorption, though some work has been done on the engineering behaviour of these soils (Shroff 1970).

Recently Sreepada Rao (1982) studied the physico-chemical properties of the anion-treated clays and the mechanisms involved therein. Anion adsorption was caused by treating the clays with phosphate and acetates at low pH. The treatment appreciably changed the physico-chemical properties (Table 2) of both kaolinite and montmorillonite clays, though by different mechanism. The following general conclusions were drawn.

1. The dominant mechanisms of anion adsorption in montmorillonite and kaolinites are surface adsorption and exchange of OH respectively.
2. The effect of cations is negligible in kaolinite clays, while they influence significantly the properties of montmorillonite clays.
3. Specific gravity decreases on phosphate adsorption.
4. Kaolinite clay particles become coarser on phosphate adsorption (Fig. 5).
5. Phosphate adsorption increases the liquid limit, the surface area and the free swell volumes of kaolinite significantly because of flocculation of clay particles. Because of aggregation the liquid limit and free swell volumes of Na-montmorillonite decrease on phosphate adsorption. In Ca-montmorillonite, initially these values decrease with treatment, but subsequently increase because of the exchange of divalent calcium by monovalent hydrogen.
6. All clays become less plastic on adsorption of anion.
7. The changes in soil properties are explained as the formation of flocculant fabric in the case of kaolinite and as aggregation and cation effects in the case of montmorillonite.

Table 2. Physico-Chemical Properties of Phosphoric Acid Treated Clays

Soil No.	Treatment duration	Base Exchange capacities (meg/100g)	Specific gravity G	Surface area m^2/g	Free Swell volume cc/g	L.L. %	P.L. %
Sodium Kaolinites							
1	0	3.5	2.71	39	1.4	52	39.3
2	1000	2.7	2.45	132	2.1	90	N.P
Calcium Kaolinites							
3	0	3.9	2.66	39	1.55	48	35
4	1000	2.9	2.59	129	1.8	93	N.P
Sodium Montmorillonites							
5	0	42.0	2.88	306	3.3	354	52
6	1000	39.5	2.75	383	2.0	73	N.P
Calcium Montmorillonites							
7	0	49.0	2.82	417	2.8	114	49
8	1000	45.2	2.57	380	2.8	106	N.P

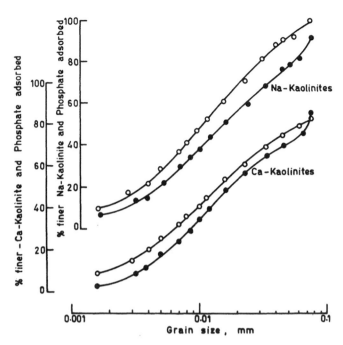

Fig.5 Effect of Phospate adsorption on grain size distribution of Kaolinite particles.

8. The changes in properties are less pronounced on acetate adsorption than on phosphate adsorption because of geometrical configurations of the anions.

The effects of other anions is also known. Tannates are widely used to improve the stability of drilling fluids. The tannate anions are adsorbed at particle edges by complexing with the exposed octahedral aluminium ions. This leads to the change in edge charge from positive to negative and prevents edge to face flocculation.

Sodium polyphosphates are effective soil dispersants which have had application in soil engineering for the construction of high density, low permeability clay blankets for use as storage reservoir liners. A major factor controlling the effectiveness of these materials is the adsorption of the phosphate anion on particle edges.

Anions, particularly bicarbonate, appear to have a role in erosion in soils of low to medium sodium content.

Ion exchange reactions are important in the properties and stability control of drilling muds and slurries. Ground water quality, underground waste disposal and pollutant transport through soils are all influenced by the ion exchange properties of soils. As the composition of the exchange complex relative to that of the pore water is influenced by both temperature and composition of the free water and as the mechanical properties are influenced by adsorbed ion types, suitable controls are necessary if the properties are to be evaluated correctly.

Organic Matter

It is generally accepted that the presence of organic matter in soils is detrimetnal to their engineering properties. For example, organic matter may cause high plasticity, high shrinkage, high compressibility, low permeability and low strength. Arman (1969) reports that organic content sometimes acts as to increase the strength of soils. It seems that in many cases soils with some organic matter content may be beneficial in its effect on the properties. Adequate criteria are not available for distinguishing soils unsuitable for engineering uses in terms of amount and type of organic matter present and conventional practice is over-conservative. It now seems that in many cases soil with some organic matter could be used at great savings.

Unfortunately no detailed studies are available on the influence of various types of organic matter on the engineering properties of soils and present knowledge permits only a few observations. Organic matter in soils is complex, both chemically and physically. Organic particles may range down to Ø.1 m in size. In general organic matter occurs in soils in two ways. It may be present as discrete particles of wood, leaf matter, spores etc., or it may be present as organic molecules adsorbed on the surface of the clay mineral particle. The humic fraction is gel-like in properties and negatively charged.

Organic matter may be adsorbed on the mineral surfaces and edges leading to interparticle bonding, modifying the properties of clay minerals.

Studies on the properties of organic soils are so far largely confined to investigations of highly organic soils such as peats and muck. But it is known that the kinds of organic matter present will significantly differ and affect the properties differently. Thus the study of organic matter in clay is a problem worthy of intensive research.

At high moisture content a decomposed organic matter may behave like a reversible swelling system. At some critical stage during drying this reversibility ceases. This is often manifested by a large decrease in Atterberg limits as a result of drying. This has been attributed to changes in organic colloids under ovenheat so that clay minerals bond together in larger aggregates thereby reducing the plasticity (Casagrande, 1948). Odell (1980) shows that increasing organic content by only about 1-2 percent raises the Atterberg limits by as much as 10-20% in montmorillonite. There is a need for further concentrated work on this topic.

CONCLUSIONS

Soil properties are modified when they are mixed with pollutants. The extent of modification of properties depends upon the type of clay itself and the nature of pollutant. The pollutants may be other clay minerals, organic solvents, inorganic salts or organic matter.

If the pollutant contains other clay minerals, then physico-chemical interactions are possible between original clay and the pollutant and the properties are not as predicted by their theoretical proportions.

Organic solvents, when they come in contact with the clay-water system, alter the behaviour of clay by affecting the dielectric constant of pore fluid. The effect of the variation of dielectric constant of pore fluid is opposite to kaolinite and montmorillonite clays. If the organic molecules are complex with clay minerals, the clay properties are further modified. Studies of different clays in various solvents have contributed to the understanding of the mechanisms controlling the liquid limit and shrinkage behaviour of clays. Changes in electrolyte concentrations also affect the behaviour of clay because of the changes in electrical double layer.

Both cation and anion exchanges can take place for clays in the presence of pollutants modifying the clay properties drastically. The mechanisms and effects are given in details.

Both positive and negative effects of organic pollutants on the engineering properties of clays are possible. The problem of clay organic-molecule interaction needs further concentrated research to improve understanding of the mechanisms involved.

REFERENCES

Arman, A. 1969. A definition of organic soils. Engineering Research Bulletin No.101 Lousiana State University, Division of Engineering Research for Lousiana Department of Highways.

Barrow, N.J., Bowden, J.W., Posner, A.M. and Quirk, J.P. 1980. An objective method for fitting models of ion adsorption on variable charge surface. Aust. Journ. of Soil Research, 18:37.

Bowden, J.W., Posner, A.M., and Quirk J.P. 1977. Ionic adsorption on variable-charge mineral surfaces-Theoretical-charge development and titration curves. Aust. Journ. of Soil Research 15:121.

Bradley,W.F. 1954. Molecular association between montmorillonite and some poly-functional organic liquids. Journ. of Amer. Chemical Society 67.975.

Casagrande, A. 1948. Classification and identification of soils. Trans. ASCE. 43:901.

Edmund, F.P. 1947. Geotechnics and Geo-technical research. H.R.B. Proc. 384.

Green, W.J., Lee, G.F. and Jones, R.A. 1979. Report on 'Impact of organic solvents on the integrity of clay liners for industrial waste disposal pits: Implications for ground water contamination! US Environmental Protection Agency, Ada, Oklahoma.

Greene-Kelly, R. 1975. Sorption of aramatic compounds by montmorillonite Part I - orientation studies. Trans. Farad. Soc. 151:412.

Halford, I.C.R., Wedderburn, R.W.M. and Smattingly, G.E.G. 1974. A Langmuir two-surface equation as a model for phosphate adsorption by soils. Journ. of Soil Science. 25:242.

Lambe, T.W. and Martin, R.T. 1956. Composition and engineering properties of soil. Highway Research Board Annual Meeting, 1956.

Lambe, T.W. and Whitman, R.V. 1969. Soil Mechanics. John Wiley and Sons, Inc. New York.

Muljadi, D., Posner, A.M. and Quirk, J.P. 1966. The mechanism of phosphate adsorption by kaolinite, gibbsete and pseudo-bochmite. Journ. of Soil Science. 17:212.

Nagaraj, T.S. and Jayadeva, M.S. 1981. Re-examination of one-point method of liquid limit determination. Geotechnique. 31:413

Norrish, K. 1954. The swelling of montmorillonite. Trans. Farad. Soc. 18:120

Odell, R.T., Thornburn, T.H. and Mc-Kenzei, L.J. 1960. Relationship of Atterberg limits to some other properties of Illinois soils. Proceed. of the Soil Science, Soc. of America. 24:297.

Olejnik, S., Posner, A.M. and Quirk, J.P. 1975. Swelling of montmorillonite in polar organic liquids. Clays and Clay Minerals. 22:361.

Rajan, S.S.S. 1975. Phosphate adsorption and displacement of structural silicon in an allophene clay. Journ. of Soil Science. 26:250.

Rosenqvist, I.Th. 1955. Investigation in the clay-electrolyte system. Norwegian Geotechnical Institute, Publication No. 9. 125.

Ryden, J.C. and Syers, J.K. 1975. Rationalization of ionic strength and cation effects on phsphate adsorption by soils. Journ. of Soil Science, 26:395

Seed H.B., Woodward, Jr., R.J. and Lundgren, R. 1964. Clay mineralogical aspects of the Atterberg limits. Journ. of the Soil Mech. and Foundation Divison. Proceed. ASCE. 90:107

Shroff, A. 1970. Physico-chemical reaction of phosphate with clay. Journ. of Indian Natl. Soc. of Soil Mech. and Foundation Engineer. 9:341.

Sivapullaiah, P.V. and Sridharan,
A. 1985. Liquid limit of soil
mixtures. ASTM, Geotechnical Testing Journal, 8:111.

Sridharan, A. and Rao, G.V. 1971.
Effective stress theory of shrinkage phenomena. Canadian Geotechnical Journal. 8:503.

Sreepada Rao, A. 1982. Physico-
chemical properties and engineering
behaviour of anion adsorbed clays.
Ph.D Thesis of Indian Institute of
Science, Bangalore, India.

Thein, S.J. 1976. Stabilizing soil
aggregate with phospheric acid.
Proceed. of Soil Science Soc. of
America. 42:432.

Theng. B.K.G 1971. The Chemistry
of clay organic reaction. John
Wiley and Sons, Inc. New York.

Yeoh, N.S. and Oades, J.M. 1981.
Properties of clays and soils after
acid treatment I clay minerals.
Austr. Journ. of Soil Research.
19:147

Environmental Geotechnics and Problematic Soils and Rocks, Balasubramaniam et al. (eds)
© 1987 Balkema, Rotterdam. ISBN 90 6191 785 9

Measurement of traffic-induced vibrations on Lopburi historical monument, Thailand

Ikuo Towhata
Asian Institute of Technology, Bangkok, Thailand

ABSTRACT: Measurement was made of a traffic-induced vibrations at a historical monument site in Thailand. By studying the vertical and horizontal components of the ground motion, it was found that P wave forms the predominant component of the vibration. Although this ground motion looks harmlessly small, it probably accelerates the collapse of the monument, when it continues for a long time.

1. INTRODUCTION

Unfavorable effects of traffic-induced ground vibrations are going to be discussed in this text. Old buildings which were constructed before modern civilization, using stone blocks or bricks, are very often less resistant against vibration than modern ones. Being subjected to continuous influence of traffic vibrations for many years, those buildings suffer cracks and/or separation of blocks nowadays, although they may have been intact for hundreds of years at the sites.

Phra Prang Sam Yod ("Sacred Three Spires") is an ancient temple of Khmer style architecture (Photo 1) which is located in the city of Lopburi 155 km to the north of Bangkok. It was built probably in the twelfth century, made of laterite blocks and bricks. A plan view of the temple is illustrated in Fig. 1. The temple is very close to two main streets of the city as well as to a railway line. These streets are crowded with heavy trucks; particularly so at the time of sugarcane harvest. Thus, this ancient building is continuously exposed to traffic vibrations which are generated by both trains and road traffic.

The temple building is already inclined slightly toward the west, and large cracks are found among laterite blocks in the western side of the building. It is possible that both leaning of the tower and opening of cracks are accelerated by traffic vibrations. Since the cracks are suspected to be growing up inspite of the repair work made by the Fine Arts Department of Thailand in 1926, a need for a complete repair has been pointed out.

The Fine Arts Department carried out a preliminary study on the traffic vibration at the site (Bergado et al., 1984). The subsoil conditions obtained by Dutch Cone Tests are shown in Fig.2. Although there is no information about soil at 0 to 2m deep, it may be found in Fig.2 that the soil beneath the building is quite stiff. Hence, it is reasonable to say that the normal ground settlement has already ceased 800 years after the construction of the temple.

Ground vibration due to traffic were measured around the temple in 1984 (Bergado et al., 1984). Fig.3 shows the relationship between the vertical velocities of the ground surface and the distance from vibration sources. The intensity of vibrations in Fig.3 do not look severe enough to damage buildings when compared with some design code (e.g. Richart et al., 1970).

In Fig.3 it was suggested that the intensity of vibration decayed with distance with the power of -3/2 both for road-traffic and train vibrations. However, the power of -2 may be more appropriate for the case of road traffic as seen in Fig.3a.

Although various studies were conducted at Phra Prang Sam Yod in 1984, the cause of building damages was not clearly understood. In this text a new information is provided which was obtained through field investigation conducted in August, 1985.

2. VIBRATIONS OF ELASTIC HALFSPACE UNDERGOING CYCLIC LOADING AT ITS SURFACE

Analytical solution is available for the dynamic response of ground undergoing traffic loadings, in which the ground is assumed to be an elastic halfspace.

2.1 Ground vibrations caused by line loading

A two-dimensional problem in which a sinusoidal line loading is applied in the vertical direction at the surface is studied. A line loading may be an idealized railway loading. Obviously a plane-strain formulation is reasonable (Lamb's problem as seen in Bath, 1968). The analytical solutions for the surface displacements, u_0 and v_0, in the horizontal "x" direction and the vertical "y" direction, respectively, are given by:

$$U_0 = - \frac{Q}{\mu} H \exp(i(pt - kx)) \quad \text{(I)}$$

$$+ \frac{Q}{\mu} (\frac{2}{\pi})^{\frac{1}{2}} (1 - \frac{h^2}{k^2})^{\frac{1}{2}} \frac{\exp(i(pt - kx - \pi/4)}{(kx)^{3/2}} \quad \text{(II)}$$

$$- \frac{Q}{\mu} (\frac{2}{\pi})^{\frac{1}{2}} \frac{h^3 k^2 \sqrt{(k^2-h^2)}}{(k^2-2h^2)^3} \quad \text{(III)}$$

$$\frac{i.\exp(e(pt - hx - \pi/4))}{(hx)^{3/2}} \quad \text{(Eq.1)}$$

$$V_0 = - \frac{iQ}{\mu} k \exp(i(pt - kx)) \quad \text{(I)}$$

$$+ \frac{2Q}{\mu} (\frac{2}{\pi})^{\frac{1}{2}} (1 - \frac{h^2}{k^2})$$

$$\frac{i.\exp(i(pt - kx - \pi/4))}{(kx)^{3/2}} \quad \text{(II)}$$

$$+ \frac{Q}{2\mu} (\frac{2}{\pi})^{\frac{1}{2}} \frac{h^2 k^2}{(k^2 - 2h^2)^2}$$

$$\frac{i.\exp(i(pt - hx - \pi/4))}{(hx)^{3/2}} \quad \text{(III)}$$

in which Q is the intensity of line loading per unit length, p the circular frequency, μ the shear modulus, Vs the S-wave velocity, and Vp the P-wave velocity. Also

$$h = p / Vp$$
$$k = p / Vs \quad \text{(Eq.2)}$$

and

$$H = - \frac{\kappa(2\kappa^2 - k^2 - 2\alpha_1\beta_1)}{F'(\kappa)}$$

$$K = - \frac{k^2\alpha_1}{F'(\kappa)} \quad \text{(Eq.3)}$$

in which

$$\alpha_1 = \sqrt{\kappa^2 - h^2}$$
$$\beta_1 = \sqrt{\kappa^2 - k^2}$$
$$F(\zeta) = (2\zeta^2 - k^2)^2 - 4\zeta^2 \alpha\beta \quad \text{(Eq.4)}$$
$$\alpha = \sqrt{\zeta^2 - p^2/v_p^2}$$
$$\beta = \sqrt{\zeta^2 - p^2/v_s^2}$$

and

$$F(\zeta) = 0 \quad \text{at} \quad \zeta = \pm\kappa \quad \text{(Eq.5)}$$

The surface displacements in Eq.1 have three components for each. The first component (I) is the Rayleigh wave which propagates along the ground surface, while the second one (II) is the S-wave, and the third one (III) is the P-wave. The Rayleigh terms have a phase difference of 90 degrees between vertical and horizontal components as indicated by the imaginary number, i, in the second equation. Hence, the Rayleigh wave generates an elliptic trajectory of motion. The S-wave terms also have a 90-degree phase difference, resulting in an elliptic trajectory as well. However, the P-wave component shows a 180-degree phase difference, making its trajectory straight (Fig.4).

It should be noted in Eq.1 that the amplitude of the Rayleigh-wave components do not vary with the distance, X. On the contrary, the S- and P-wave components decay with X in proportion to its power of -3/2, which may be called the radiational damping (Table 1).

2.2 Ground vibration caused by point loading

Analytical solutions are also available for the ground vibration caused by a point loading applied vertically. This type of loading is an idealization of road-traffic loading, occurring in an axisymmetric manner. Vibration takes place only in the vertical and radial directions, accompanied by no shaking in the circumferential direction. Both vertical and radial motions consist of Rayleigh-, P-, and S-wave components, having similar trajectories as shown in Fig.4.

Båth (1968) studied the rate of decay with the radial distance, X. In case of point loading, Rayleigh wave decays inversely proportional to square root of X, while amplitudes of P- and S-waves are inversely proportional to square of x.

3. VIBRATION OF FREE FIELD MEASURED AT SURFACE

Traffic vibrations were measured at Station A in Fig.1. Since Station A is located away from the building, soil-structure inter-action does not influence the measured records too much, and the dynamic response of level ground, or free field, may be studied. Also the Station A is much further from the Street B than from the Street A, influence from Street-B traffic being less significant.

Fig.5 illustrates three axes of reference which are vertical, longitudinally horizontal, and transversely horizontal. For each measurement, velocity transducers were fixed at the ground surface, output signals were visualized in an oscilloscope screen, and an photograph was taken of it.

Fig.6 shows the results together with Table 2. A road-traffic loading may be idealized as a point loading for an analytical formulation, and, as discussed before, this loading induces no motion in the circumferential or transverse direction. The vibrations shown in Fig.6(a) and (b) were recorded when a heavy truck was running in Street A. Apparently the transverse component is much smaller than the vertical one. Therefore, it is needless to study the transverse component any more.

Fig.6(c) to (h) indicate a phase difference of 180 degrees between vertical and longitudinal motions. In other words, soil particles moves back and forth in such a straight direction as shown in Fig.7 for a road-traffic vibration, and this direction agrees with the trajectory of P-wave component which was predicted by the analytical solution in Eq.1 (Fig.4). In Fig.3(a) the decay of vibration with distance, X, is more reasonably approximated by X's power of -2. This rate of decay was also predicted analytically for the P-wave component which was generated by a point loading (Table 1). Therefore, it may be said that P wave is dominant in the free field excitations.

Although discussions similar to the above are also possible due to train induced vibrations, there exists one big difference between these two types of loadings. The road traffic is appropriately modelled by a point loading. Conversely, the train loading is a sequence of many point loadings. Although looking like a line loading, those point loadings are not in phase. Each of them generates a wave of different phase and different magnitude, and both the free field and the building are excited by their superposed motion.

Train vibrations measured at the Station A on the free field is presented in Fig.8 and also tabulated in Table 3. Comparison of Tables 2 and 3 reveals that train vibrations are more intense than road-traffic vibrations. Conversely, the number of repetition of loading is much larger for road traffic than trains, because the street is always crowded with vehicles, while trains pass by the site a few times an hour.

Figs. 8(a) and (b) show that the transverse motion is much weaker than the vertical motion, agreeing with the analytical solution in which the transverse motion does not exist. The phase difference between vertical and longitudinal components are basically equal to zero degree, although it is not always clearly seen because of the superposed nature of motion. Therefore, it may be reasonably said that the trajectory of railway-induced vibration is straight and inclined toward the wave source. This nature is same as that of the road-traffic vibration (Fig.7).

Fig.3(b) indicated that the decay of train vibrations with distance is governed by power of $-3/2$; same as the decay of P-wave generated by a line loading (Table 1). When combined with the fact that the train loading is basically a sequence of point loading which generates mostly P-waves as shown before, all the observations derived here suggest that the train vibrations consist of a superposition of P-waves.

4. VIBRATION MEASUREMENT AT FOOT OF BUILDING

The Lopburi temple building is composed of three towers, of which the western side of the central tower seems to be damaged most. In order to study the dynamic behaviour of the building, velocity transducers were placed at the foot of the central tower. Because of the limited number of photofilms which could record the oscilloscope screen, only train-induced vibrations are available for further studies (Fig.9).

High frequency motion which was possibly caused by the soil-structure interaction is overlapped on the significant signals in Fig.9. A phase difference of zero degree between the vertical and longitudinal motions may be observed, although not very clear because of the superposed nature of the train vibration. Hence, the direction of train vibration at the foot of the building may be similar to that shown in Fig.7, suggesting that road-traffic vibration is also similar to what was illustrated in Fig.7. Sight inspection of the oscilloscope screen supported this suggestion when a heavy truck was running in the Street A (Fig.10).

5. VIBRATION MEASURED AT MIDDLE OF BUILDING

As mentioned in section 4, the central tower among three of them is the most severely damaged by cracks and block separations, particularly at its middle height on the western side. Hence, a vibration measurement was taken at this place. Hereinafter, only the road traffic in Street A is studied, and that in Street B is excluded from further discussion. This is because the building damages are seen on the western side of the central tower which faces the Street A, while the south tower near the street B is not damaged so badly as the central one.

5.1 Vibration caused by road traffic

As was indicated before, the road traffic causes a vibration at the foot of the building in the direction as shown in Fig.10. The direction of motion at the middle is going to be studied in this section.

Since the speed control was supposed to be a good measure to reduce the traffic vibrations, controlled traffic tests were conducted employing a heavy truck of 37 ton in weight running along the Street A at specified speeds. The measured vibration is shown in Fig.11. Vibration records in this figure suggest 180-degree phase differences between two components, indicating a motion as is seen in Fig.12, though the record in Fig.11(a) may exceptionally show a zero-degree phase difference.

The vibration intensity was plotted against truck speed in Fig.13. The maximum vibration was observed unexpectedly not at the maximum speed but at the speed of 22 km/h. This is probably because the vibration intensity depends not only on the traffic speed but also on local pavement conditions and vehicle's acceleration. However, the vibration should be small enough, if the speed is very low, whatever pavement and acceleration might be. Hence, a speed limit should be proposed at a very small value so that exceptionally intense vibrations may not occur.

5.2 Vibrations caused by railway trains

Fig.14 shows the vibrations measured at the middle of the tower when a train was passing by. The two records in Fig.14(e) have a parallel configuration with each other, indicating a direction of motion as shown in Fig.15. It should be noted that this direction of motion reasonably agrees with that which was induced at the foot of the building by trains (Fig.10). On the contrary, other records in Fig.15 show a phase difference of 180 degrees. Although the reason for this discrepancy is not yet known, it may be said that, in most cases, the vibration due to trains agree with road-traffic vibrations in their directions (Fig.12). The relationship between train speeds and vibration intensities is illustrated in Fig.16. Basically the vibration increases with train speeds. This trend is especially evident in the longitudinal vibration. Hence, appropriate speed limit may be effective for mitigating the train-induced vibration of

the building. It is noteworthy that little difference is seen between passenger trains and cargo trains.

6. DISCUSSION ON BUILDING DAMAGES

The damage to the Phra Prang Sam Yod building is characterized by the tilting and the separation of stone blocks. Tilting is probably caused by the consolidation of the foundation soils. Since 800 years have passed after the building was completed, the consolidation in its conventional sense is unlikely to continue today. Also the foundation soil consists of stiff clay which hardly undergoes long-term consolidation (Fig.2). On the contrary, a minute vibration due to traffic may help maintain the rate of settlement for many years, although its intensity is far less than damaging a building directly (Table 1).

Crockett (1966) studied the behaviour of over 40 ancient buildings in England, and reported "these buildings invariably . lean over more towards the road carrying traffic than away from it." Probably the same phenomenon is going on in Lopburi. The westward, or streetward, leaning in Lopburi suggests that the road-traffic vibration is more influential on the foundation than the train vibration.

The small vibration which looks harmless to both building and foundation seems to damage them after very long duration. The blocks in the inclined tower are already trying to separate and fall down due to gravity force. When this heap of blocks are subjected to vibration, those tendencies are quite likely to be accelerated. To prove this a model test is needed to be done in future.

7. CONCLUDING REMARKS

A field investigation was conducted at Prang Phra Sam Yod in Lopburi city in order to propose measures which can prevent further damages to the temple building. The conclusions drawn from the investigation are as follows:

1. The traffic vibration in free field is composed of P-waves.
2. Train vibration is more intense than road-traffic vibration. However, the road-traffic vibration involves much more number of repetitions.
3. The damage of the building was probably caused by the road-traffic vibration with a long duration time.

8. ACKNOWLEDGEMENT

This projected was financially supported by the Japan International Cooperation Agency. The heavy truck employed for tests was supplied by Thai Highway Police. The whole field measurements and their interpretations were carried out by field crews of Asian Institute of Technology (AIT). Mr. Thep at the Lopburi Office of the Fine Arts Department of Thailand kindly looked after the field activities. Dr. Bergado and Mr. Sataporn of AIT were enormously helpful in conducting the whole project. It is the author's greatest pleasure to express his sincere gratitude to those people and agency mentioned above.

9. REFERENCES

Båth,M. 1968. Mathematical aspects of seismology. Elsevier

Bergado,D.T., Yamada,Y., Chandra, S., Sataporn Kuvijitajara, & Gauchan,J. 1984. The measurement and monitoring of the level of vibration at three historical sites in Lopburi Province due to railway and highway traffic. Report submitted to the Fine Arts Department of Thailand, Asian Institute of Technology

Crockett,J.H.A. 1966. Some practical aspects of vibration in civil engineering. Symposium on vibration in civil engineering. Imperial College, London, p.253-271

Richart,F.E.Jr., Hall,J.R.,Jr., & Woods,R.D. 1970. Vibrations of soils and foundations. Prentice-Hall. p.308-322

Photo 1. Phra Prang Sam Yod in Lopburi

Table 1. Amplitude-distance relations in Lamb's problem (After Båth, 1968)

Source	Rayleigh	P	S
Line	X^0	$1/X^{3/2}$	$1/X^{3/2}$
Point	$1/X^{1/2}$	$1/X^2$	$1/X^2$

X: Distance from the source

Table 2. Peak-to-peak velocities at surface of free field induced by road traffic

ID in Fig.6	Maximum p-p velocities, mm/sec			Source of vibration
	Vertical	Longitudinal	Transverse	
A	0.062	-----	0.008	Heavy truck
B	0.020	-----	0.005	do.
C	0.033	0.030	-----	do.
D	0.075	0.042	-----	do.
E	0.081	0.081	-----	Medium truck
F	0.042	0.020	-----	Bus & Heavy truck
G	0.028	0.050	-----	Heavy truck
H	0.006	0.008	-----	Medium truck
I	0.011	0.015	-----	Heavy truck

Table 3. Peak-to-peak velocities at surface of free field induced by trains

ID in Fig.8	Maximum p-p velocities, mm/sec			Speed km/h
	Vertical	Longitudinal	Transverse	
A	0.115	-----	0.011	----
B	0.102	-----	0.010	----
C	0.079	0.180	-----	27.5
D	0.063	0.037	-----	27.5
E	0.066	0.091	-----	34.0

Source of vibration : Passenger train

197

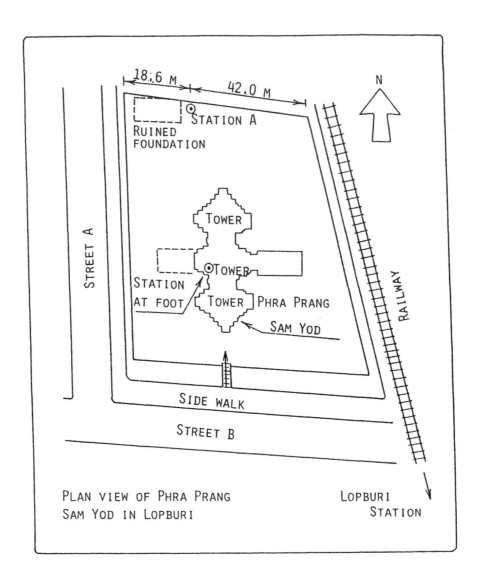

Fig.1 Plan view of Phra Prang Sam Yod in Lopburi

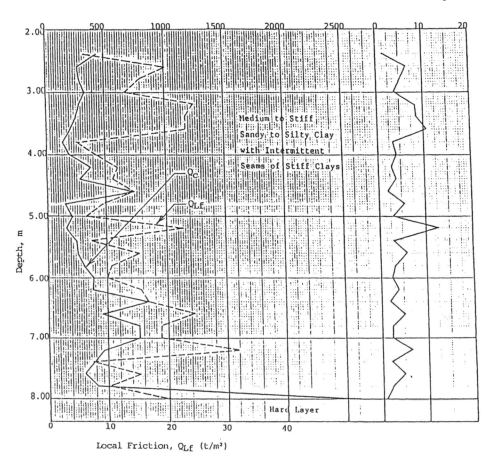

Fig.2 Dutch Cone test profile at Phra Prang Sam Yod (after Bergado et al., 1984)

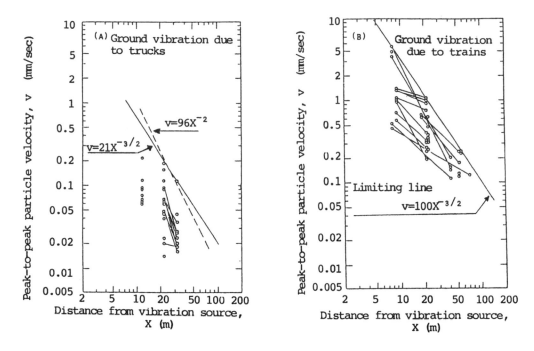

Fig.3 Ground vibrations in vertical direction caused by traffic at Phra Prang Sam Yod
(after Bergado et.al., 1984)

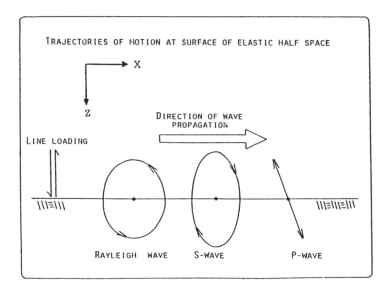

Fig.4 Trajectories of motion at surface of elastic halfspace

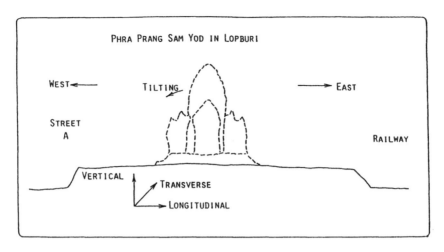

Fig.5 East-west section view through Phra Prang Sam Yod in Lopburi

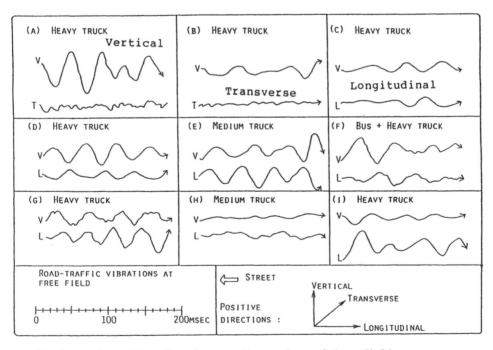

Fig.6 Road-traffic vibrations at the surface of free field

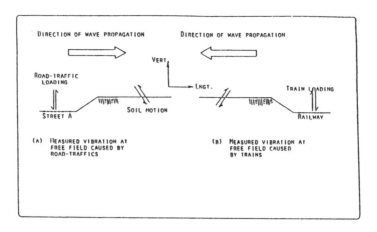

Fig.7 Direction of soil motion at surface of free field

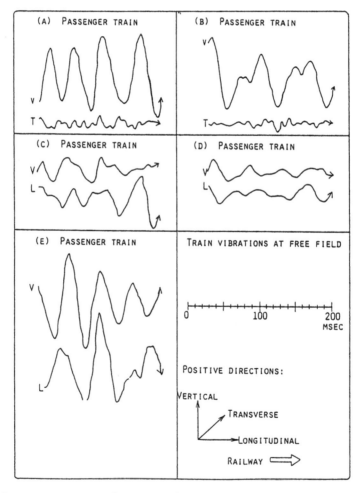

Fig.8 Train vibration at surface of free field

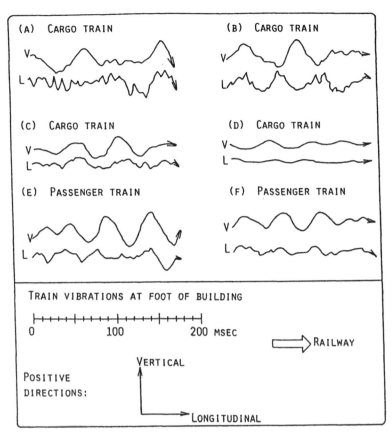

Fig.9 Train vibrations measured at the foot of central tower

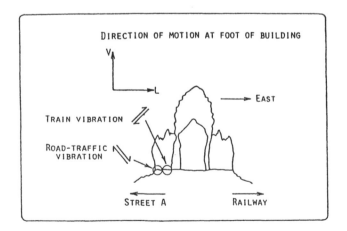

Fig.10 Direction of motion at foot of building

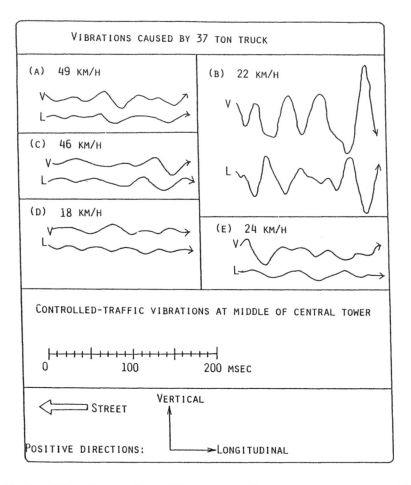

Fig.11 Vibration at the middle of central tower caused by controlled road traffic

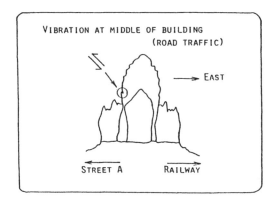

Fig.12 Direction of road-traffic vibration at the middle of central tower

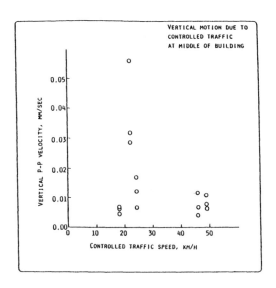

Fig.13 Velocity vs. traffic speed relationship
at the middle of central tower

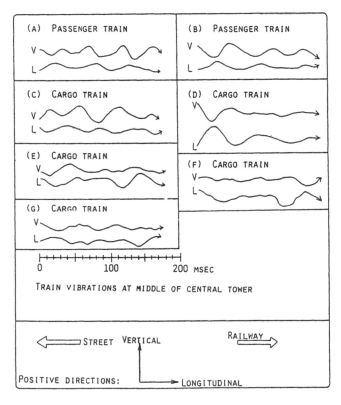

Fig.14 Vibrations at the middle of central tower caused by
trains

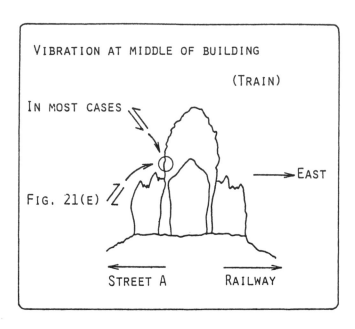

Fig.15 Direction of motion at the middle of central tower
induced by trains

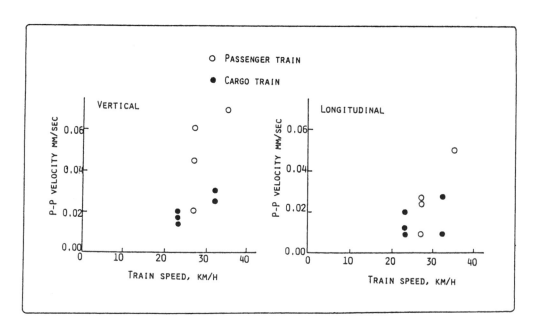

Fig.16 Relationship between train speed and vibration at the middle of
central tower

Environmental Geotechnics and Problematic Soils and Rocks, Balasubramaniam et al. (eds)
© *1987 Balkema, Rotterdam. ISBN 90 6191 785 9*

Relation between physical properties of soils and effects of chemical stabilization

Haruyuki Nakayama & Yuji Miura
College of Science & Technology, Nihon University, Japan

ABSTRACT: When judding or predicting the effect of stabilization on soils, it is necessary, in mauy cases, to test each soil for different mix proportions to judge and confirm the effect on the soil. But this method is highly wasteful of time and money. If a method to estimate the strength of soils after stabilization from their physical properties is established, the effect of stabilization will be able to be estimated with certain level of accuracy from the results of easy-to-make physical tests, etc., though detailed judgement is left to mix proportion test. In the present paper, we try to collect data on chemical stabilization from all over the country ahd find relation between physical properties of soils and the effect of chemical stabilization.

1 INTRODUCTION

In Japan, soils such as volcanic cehesive soils and high organic soils which become a great problem in engineering are widely distributed throughout the country. For example, in the case of volcanic cohesive soils, if they are disturbed by construction machinery or the like, the soil structure will be broken and water confined therein will be liberated, weakening the structure extremely, therefore, utmost care must be exercised at the time of excavation, transportation, consolidation, etc. When utilizing such soils as materials for railway construction, road construction, housing site preparation, etc., it is necessary to perform soil stabilization to make the soil more stable.

Stabilizing weak grounds enhances their mechanical (physical) values, so soil stabilization is of great significance to Japan limited in land space. Various methods are available for soil stabilization. Of these, chemical stabilization has a long history and can be performed by simple and reliable methods. Further, the recent development of new chemical stabilizers, stabilizing methods and equipment has led to wider application of chemical stabilization.

2 PHYSICAL PROPERTIES OF SOILS AND EFFECT OF CHEMICAL STABILIZATION

Factors that govern the mechanical properties of soils are many. For example, external factors include deposition environment related to soil formation, deposition history, strata formation, stress history, weathering history, alteration history, etc., and internal factors include mineral components, contents of organic matters, specific gravity, shape of soil particle, particle size, structure, density, void ratio, water content, degree of saturation, consistency, etc. Generally, it is

assumed that these factors intricately interact to determine the mechanical properties of soil as shown in Fig. 1.

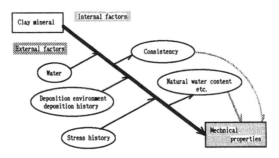

Fig. 1 Mechanical Properties of Soils and Their Factors

On the other hand, the effect of chemical stabilization is said to be dependent upon the difference between the physical and chemical properties of soil before and after reaction with chemical stabilizer. For example, M.R.Thompson statistically examined principal factors that influence the increment in unconfined compression strength by stabilization with lime and assumed these to be (1) Organic carbon content of soil, (2) pH value of soil, and (3) clay mineral having montmorillonite as its dominant component (Thompson 1966). Similarly, it was confirmed by M. Asakawa that these factors are closely related to the effect of lime stabilization (Asakawa 1979).

However, determining these factors except pH requires complicated test procedures, so they are not easy to obtain. In order to estimate the effect of stabilization, it is practical to use only information that is easy to obtain or can be obtained by a simple method. In the present study, therefore, we decided to investigate the relation between easy-to-obtain information and the effect of stabilization. However, usually easy-to-obtain information is limited, so it is very difficult to collect many data. Therefore, we took, as potential indiced that can express the mechanical properties of soil and the effect of chemical stabilization on soil or be related thereto, the following physical properties which are utilizable and whose information is relatively easy to obtain:

(1) Natural water content   Wn
(2) Liquid limit   Wl
(3) Plastic limit   Wp
(4) Plasticity index   Ip
(5) Liquidity index   Il
(6) Consistency index   Ic
(7) Plastic ratio   Pr
(8) Percent passing 74 micrometer sieve
(9) Combination of above indices

Except (1) and (8), these are all related to consistency. Consistency reflects the physical and chemical properties of soil, and its relation with many mechanical properties has so far been feported (Saito & Miki 1975). In addition, the information value of natural water content is considered to be also very high. Namely, the water content of a soil existing stably in the natural state may be affected more or less by the environment there around, the soil nevertheless remains stable at that level of water content. This shows the water-retaining ability the soil possesses, and this ability is considered to include many properties such as physical and chemical properties.

3 COLLECTION OF MIX PROPORTION TEST DATA

Data on mix proportion test for chemical stabilization

Table 1 Mix Proportion Test Data (Cured for seven days)

Test	Complete data	Incomplate data	Total
CBR test	992	65	1,057
Unconfined compression test	172	27	203

were collected from all over the country with the cooperation of Japan Highway Public Corporation, Okuta Ohutama Kogyo Co.,Ltd., Ohbayashi-Gumi, Ltd., Nihon Cement Co., Ltd., Toa Doro Co., Ltd., etc. The total number of collected data was more than about 4,000 sets. However, some of them were incomplete or obscure, some were from a mixture of several soils, and some were different in curing time or curing conditions. Therefore, the total number of data selected for investigation is about 1,200 sets, including CBR and unconfined compression test data as shown in Table 1. Of these, complete data includes all of the proposed indices listed in Section 2, and incomplete data means data that does not include the indices relating to consistency. All the data shown here are from samples cured for seven days.

Fig. 2 shows a plot of the physical test data of sample soils tested in CBR test on a plasticity chart. Many data are distributed on the A line. Of the soils in Japan, particularly alluvial exhibit such data distribution. Distributed below the A line are the data of volcanic cohesive soils. These are very problematic soils that change their properties when disturbed. In Japan, therefore, special methods for preparing physical test samples, for performing compaction test, etc., are used for these soils (JSSMFE 1980).

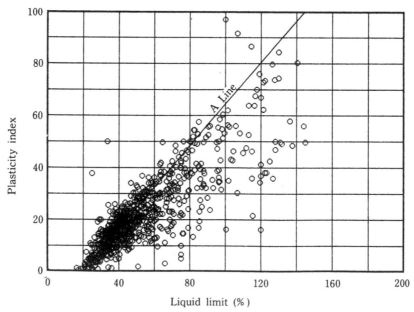

Fig. 2 Locations of Soils on Plasticity Chart (CBR)

4 WHAT TO SELECT AS A PARAMETER OF THE EFFECT OF STABILIZATION

When relating the indices to the effect of stabilization, what to select as a parameter of the effect of stabilization become a problem because this effect is dependent on the kind nf chemical stabilizer, addition rate, curing time etc., as shown in Fig. 3. If these factors are combined into a single parameter, the relation of the effect of stabilization to the indices will be find out.

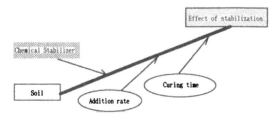

Fig. 3 Soil and Effect of Soil Stabilization

How the strength of soil samples subtected to soil stabilization varies with the addition rate of stabilizer is schematically represented in Fig. 4. Fixing the curing time and examining the strength variations with addition rate, the data can be divided into four types as shown in the figure. That is, A is the type having a peak, B, an approximately straight type, C, a type exhibiting a great increase in strength, and D, a type exhibiting little variation in strength (considered to be included in type B).

As a result of investigating these data, most of them seem to correspond to type B. Even with some data classified as type C and D, the correlation coefficient is found by linear regression analysis to be more than 0.85 and, for type B, it is more than 0.98. Therefore, assuming the strength varies linear as in the case of

type B, we made linear regression analyses of these data by curing time [Equations (1) and (2)]. We defind the slope α thus obtained as the "rate of strength development" and decided to adopt this as a parameter of improvement effect.

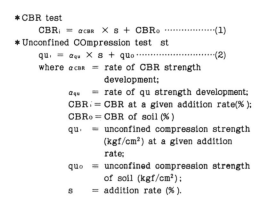

* CBR test

$$CBR_i = \alpha_{CBR} \times s + CBR_0 \quad \cdots\cdots\cdots\cdots(1)$$

* Unconfined COmpression test st

$$qu_i = \alpha_{qu} \times s + qu_0 \quad \cdots\cdots\cdots\cdots\cdots(2)$$

where α_{CBR} = rate of CBR strength development;

α_{qu} = rate of qu strength development;

CBR_i = CBR at a given addition rate(%);

CBR_0 = CBR of soil (%)

qu_i = unconfined compression strength (kgf/cm^2) at a given addition rate;

qu_0 = unconfined compression strength of soil (kgf/cm^2);

s = addition rate (%).

5 ARRANGEMENT OF COLLECTED DATA

5.1 Relations between CBR of soils before stabilization and indices

Investigating the CBR test data that were numerous and complete, we found the relations between the CBR of soils before stabilization and the indices. Fig. 5 shows plot of part of these findings (relation between natural water content and CBR) on a log-log paper. What was most closely related to the CBR of the proposed indices was natural water content, and its correlation coefficient was 0.4. Of the soils having a low water content blow 20% and a CBR over 20 as shown in the figure, many were sandy soils. So, we investigated whether this is or is not related to particle size, taken

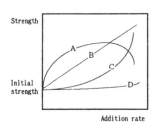

Fig. 4 Strength valuation with addition rate

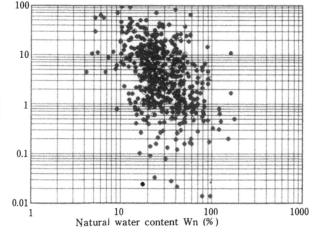

Fig. 5 Relation Between CBR of soil Before Stabilization and Natural Water Content

the passing 74 micromeater fine particles as an index of particle size, but no such relation existed. Further, we investigated the relations between CBR and other indices, but no noticeable correlations were observed, the correlation coefficients of these indices being less than 0.25. This shows that the CBR value is difficult to estimate from the physical indices of soil.

5.2 Rate of the strength development α

There are 1,057 CBR test data. Using these data, we determined the rate of strength development under the 7-day curing conditions, as a result of which it was found that its correction factor shows a fairly high value with a frequency as shown in Fig. 6. This seems to prove the validity of assuming the relation between addition rate and the rate of strength development as pattern B of Fig. 4. Investigating those of the data which show a low correlation coefficient of less than 0.8, we found that most of them except ones are from the samples showing pattern A, many belong to soil categories (SM), (CH) and (MH) and that all are from the samples stabilized with unslaked lime. Therefore, when stabilizing cohesive soils relatively high in natural water content with unslaked lime, it would be necessary to pay attention to the accuracy of the rate of strength development α.

On the other hand, as a result of investigating with what soils or chemical stabilizers a high rate of strength development α is obtained, it was found tht many of such soils are (CL), (ML) and (SM) and that many of such stabilizers are unslaked lime and cement-base stabilizers. While in the case of low values, many of such soils are (OH), (Pt), (VH), etc., that are high in

natural water content, and many of such stabilizers seem to be ones added with gypsum to cope with such types of soils. Nearly the same could be said for the unconfined compression test data.

5.3 Relations between indices and rate of strangth development α

We tried investigate these relations by regression analysis by several functions, as a result of which it was found that the fractional function and exponential function each has a relativery high degree of correlation. With the former, however, the rate of strength development α was considered to become negative depending on the values of the indices. Thus use of this function did not match the actual condttion, so we adopted the exponential function.

Figs. 7 and 8 show the results of a regression analysis by exponential function of CBR and unconfined compression strength, respectively, made with natural water content as an index and using regression equations (3) and (4), respectively, .

$$\alpha_{CBR} = a \cdot W_n{}^v \quad\cdots\cdots\cdots\cdots\cdots(3)$$
$$\alpha_{qu} = a \cdot W_n{}^b \quad\cdots\cdots\cdots\cdots\cdots(4)$$

where α_{CBR} = rate of CBR strength development;
α_{qu} = rate of qu strength development;
a, b = coefficients of regression equations;
W_n = natural water content (%).

Since this coefficient b of the exponential function is derived from Equations (3) and (4):

$$d\alpha/\alpha = \frac{b}{dW_n/W_n} \quad\cdots\cdots\cdots\cdots(5)$$

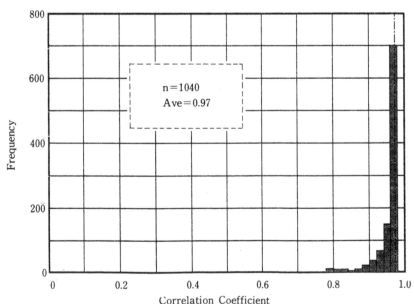

Fig. 6 Correlation Coefficient of Rate of Strength Development α

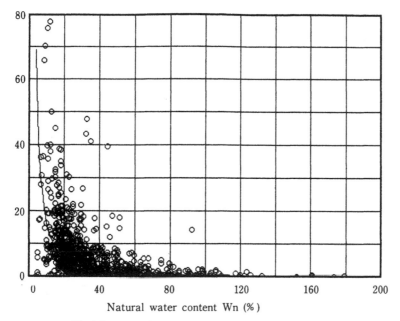

Fig. 7 Regression Analysis by Exponential Function (CBR)

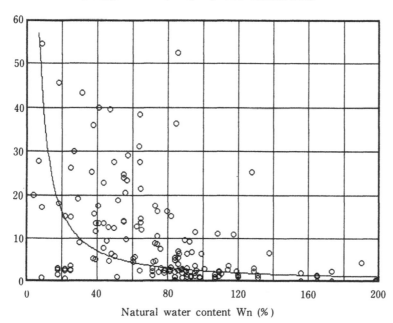

Fig. 8 Regression Analysis by Exponential Function (qu)

it can be seen that coefficient b relates the rate of variation of natural water content to the rate of strength development. For example, when natural water content varies 10% from the present level, the rate of strength a varies 0.1b. Variation of this a is greater with larger absolute value of b. Coefficient a shows the magnitude of the rate of strength development when natural water content is 1.

Table 2 shows the results of regression analysis by Equation (3). Of the physical properties selected as indices, natnral water content Wn showed the highest e degree of correlation and Plasticity index a low degree of correlation. From the relation with the CBR of soils before stabilization given in Fig. 5, it is also seen that natural water content has a higher degree of correlation than other indices. Therefore, the influence of water

Table 2 Results of Regression Analysis by
Exponential Function(CBR), 7 Days Curing

Index	a	b	Correlation coefficient
Wn	704	-1.54	0.62
Wl	6440	-1.97	0.55
Wp	487	-1.52	0.43
Ip	45.3	-0.93	0.45
Pr	2.4	-0.44	0.20
Wn, Pr	35	-0.79	0.51

Table 3 Results of Regression Analysis by Exponential
Function (qu), 7 Days Curing

Index	a	b	Correlation coefficient
Wn	98.7	-1.35	0.55
Wl	6970	-2.30	0.50
Wp	5392	-2.60	0.50
Ip	7.8	-0.92	0.33
Pr	0.3	-0.22	0.07
Wn, Pr	4.9	-0.69	0.39

Table 4 Differences by kind of chemical stabilizers
(CBR), 7 Days Curing

Stabilizer	a	b	Correlation coefficient	Num. of data
Unslaked lime	669	-1.52	0.57	491
Slaked lime	182	-1.39	0.58	109
Cemment	922	-1.58	0.63	60
Special cem.	10458	-2.07	0.66	151
Lime+gypsum	24.6	-0.89	0.34	167

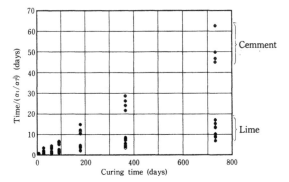

Fig. 9 Relation Between the Rate of qu Strength
Development and Curing Time

content on the mechanical properties of soil is considered
to he so great that it cannot be ignored.

The reason relating to the difference in the degree
of correlation of natural water content between the soils
before and after stabilization mignt be explained as
follows. That is, for the soil before stabilization, it is
considered that there are many factors, other than
natural water content, which influence its mechanical
properties, so natural water content is only one of the
factors. while for the soil after stabilization, it is assum
assumed that the influence of natural water content as
initial water content is increased by the fact adding a
dry chemical stabilizer decreases the water content of
the soil and that the reaction of the soil with the
stabilizer decreases the water content still more.

As for unconfined compression strength, the number
of date is incomparably smaller than that of CBR, about
1/5, but almost the same results as CBR have been
obtained as shown in Table 3, from which it seems that
the degree of its correlation with natural water content
is high.

5.4 Kinds of chemical stabilizers and rate of strength
development α

The CBR data include five kinds of chemical atabilizers:
unslaked lime, slaked lime, cement, special, and lime-
gypsum mixture (including all of various mixing methods).
The results of regression analysis are given by kind of
stabilizers in Table 4. Coefficient b in Equation (3)
shows the degree of influence of natural water content
on the rate of strength development α as is explained

before. And coefficient a shows the magnitude of the
rate of strength development α when natural water
content is 1. Therefore, the effect of soil stabilization
is greater with larger absolute value of b, and the
effect on soils of low water content is greater with
larger value of a.

According to the results of Table 4, it is the special
stabilizer that acts most effectively, and this stabilizer
seems to be particularly effective in stabilizing soils of
low water content. Unslaked lime and cement exhtbit
almost the same effect, while slaked lime acts a little
less effectively. All this substantiates what is generally
said. The lime-gypsum mixture shows extremely low
values compared with others, however, this seems to be
ascribable to the fact that, of the soils stabilized with
this stabilizer, many were problematic soils.

The curing time for the date of Table 4 is 7 days.
We compared in terms of curing time the cement-base
stabilizers cement and special stabilizer and the lime-
base stabilizers such as unslaked lime, as a result of
which an interesting fact was obtained. Fig. 9 shows the
results of investigating the relation between the rate of
qu strength development and curing time. Plotted on the
ordinate are the values of curing time divided by the
ratio of rates of strength development (α_t / α_7) - based
on a curing time of 7 days which show that times the
rate of strength development after a given length of
curing time becomes larger than that after 7 days
curing.

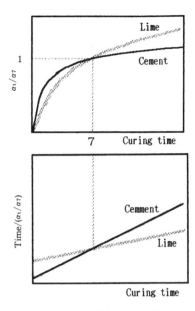

Fig. 10 Schematic Diagram of Curing

(4) The initial effect of soil stabilization is high with special stabilizer, followed by unslaked lime, while the long-time effect is higher with the latter than the former.

(5) As an explanatory variable to be used in judging the effect of chemical stabilization, natural water content is more variable than consistency.

(6) By utilizing natural water content, it becomes possible to judge the effect of chemical stabilization with a certain accuracy.

REFERENCES

Thompson, M.R. 1966. Lime reactivity of Illinois soils, ASCE, SM5.
Asakawa, M. 1979. Simple judgment for effect of lime stabilization, Proc. the 14th Research Meeting of JSSMFE.
Saito, T. & Miki, G. 1975. Swelling and residual strength characteristics of soil based on a newly proposed plastic ratio chart. Soils and Foundations, Vol.15.
JSSMFE. 1980. Testing Procedure of soils, p.642-667. Tokyo:JSSMFE.

According to these data, there seems to be some difference in the rate of strength development over a long period of time between the cement-base and lime-base stabilizers. It can be seen that with the cement-base stabilizer strength develops relatively early, while with the lime-base stabilizers strength develops over a long period of time.

As shown in the schematic diagram of Fig. 10, there is some difference in the slope of straight line between the cement-base and lime-base stabilizers. We think it possible to use this difference to express of curing time.

6. CONCLUSIONS

In conclusion, the results of the present study can be summarized as follows:

(1) The relation between the addition rate of chemical stabilizer and the rate of strength development can be assumed to be linear, i.e in the form of a straight line.

(2) The slope of this straight line is defined as the rate of strength development a. Then, through a regression analysis by exponential function, a high degree of correlation can be observed between this rate of strength development a and natural water content Wn.

(3) Most of the type A data in Fig.4 are data taken from the soils of high water content stabilized with unslaked lime.

Section B
Environmental geotechnical aspects of major infra-structure projects

Environmental Geotechnics and Problematic Soils and Rocks, Balasubramaniam et al. (eds)
© 1987 Balkema, Rotterdam. ISBN 90 6191 785 9

Environmental and geotechnical considerations for the use of bored tunnels in urban areas

Ian McFeat-Smith
Charles Haswell & Partners (Far East), Hong Kong

ABSTRACT: Many advantages are to be gained from the use of bored tunnels, as opposed to shallow excavations, for the development of infrastructure in urban areas. This option is however rarely adopted in developing countries due to a lack of knowledge of the specialist skills and geotechnical processes required to ensure timely and economical construction.

This paper discusses the environmental advantages favouring the use of tunnels, particularly for major contruction projects. The need to establish geotechnical conditions accurately prior to excavation is illustrated in conjunction with appropriate site investigation techniques.

The range of modern tunnelling methods is presented. Emphasis is given to the use of fully mechanised techniques for the excavation of tunnels in both rock and soils.

Finally, the requirement to minimise the costs for tunnelling work is considered, and specific recommendations are made to illustrate how this can be achieved.

1 ENVIRONMENTAL CONSIDERATIONS

1.1 Assessing the cost to the environment

The decision to provide major infrastructure facilities is a major step in the development of any city. At the outset the planner must take responsibility for minimising the environmental impact of the construction of such projects. The advantages to be gained from the completed project will be evident; but will the public, commercial and administrative sectors be willing to accept the disruption induced by heavy construction work? At what level will this become unacceptable? Is this disruption strictly necessary or can it be avoided by favouring certain methods of construction? Clearly these fundamental questions must be asked when the project is conceived for the benefit of all concerned.

When examining the cost of alternative schemes, a number of intangible factors must also be considered. These include assessing the effect of heavy construction work in the city centre, land requirements for such work, the numerous utility diversions that will inevitably be required and the influence of such work on transportation systems. These effects can be considerable and may become key issues in the selection of alternative schemes which are not accurately reflected in the direct financial costing alone.

One of the most significant and cost effective methods of construction available, and one that avoids much of the impact of construction, is the use of bored, or driven tunnels. Technical solutions are now available which allow us to excavate tunnels in almost any type of ground. Where adverse geological conditions exist we can be certain that the construction cost will rise, however this cost is usually a small price to pay to ensure that the construction of the infrastructure will have the minimum effect on the city environment.

1.2 Advantages favouring the use of bored tunnels

Construction of large overhead and cut and

cover works can create chaos in city centres. Traffic congestion, noise, dust, and pedestrian inconvenience all inevitably result as illustrated in Fig. 1. Restrictions have to be imposed on night work, and on work such as sheet piling to make noise levels bearable, and local buildings and traffic have to be protected from excavation hazards such as flyrock from blasting.

Surface excavations are generally extensive, requiring large areas of prime city sites that could be used for other purposes, and may necessitate the demolition of important buildings. These construction sites also act as impassible barriers to surface transport such as busy roads and railways. Widespread traffic congestion inevitably results from the narrowing of roads, lengthy traffic diversions, and site transport obtaining access to excavation sites.

Fig. 1. General impact of heavy construction work for infrastructure in city centre

The use of bored tunnels completely avoids, or at least minimises the effect of all of these constraints. In addition, construction traffic will be considerably lighter as a much smaller volume of excavated material has to be moved, and the extensive use of heavy plant at road level will be reduced substantially.

Construction sites for tunnelling can be small, often requiring as little as 40 m x 40 m for the construction of twin access shafts. Work in these small and compact sites has a negligible effect on the city environment and transportation systems.

Numerous utility services such as gas, water, and sewage pipes, storm drains, and telephone and power cables underlie the streets of every major city. The extensive use of cut and cover works require that many of these utilities have to be located and diverted with the approval of utility companies and government departments. Again, these problems are eliminated by the use of bored tunnels.

1.3 Advantages of tunnels in service

In hilly terrian, bored tunnels can be used to avoid lengthy and expensive detours by providing the shortest route between two points. This reduces the amount of construction work required, may allow more appropriate gradients to be used, and will reduce the running costs of transportation systems.

In comparison with other methods of construction, tunnels provide increased security and reduce the incidence and effect of hazards. In addition to military considerations, accidental death and contact with powerlines, collisions, and derailment of trains by vandalism are all reduced. Similarly, the system is protected from natural hazards such as landslides and severe weather conditions all of which can halt surface transport and require extensive remedial work. For example, in 1981 the Hong Kong Mass Transit Railway (MTR) remained operative during several typhoons, whilst all road and surface rail traffic was forced to a standstill. A similar case can be made for the use of tunnels for harbour crossings to reduce shipping collisions and leave river traffic unhampered.

2 PRELIMINARY STUDIES

There are few disadvantages associated with tunnelling. However, there are several steps that must be taken to ensure the successful construction of a major underground project. Unlike other methods of construction that rely largely upon man-made building materials, the driving of tunnels depends upon the natural and sometimes unpredictable media through which they are driven. It is the expertise of geotechnical and tunnelling engineers in assessing the geological conditions and in selecting the most appropriate methods that will dictate the success or otherwise of a particular project.

During the preliminary studies, emphasis must therefore be placed upon establishing the precise nature of the local ground conditions. This will provide fundamental data for each successive stage of the project, including design work, costing and planning, contract evaluation, and construction of the project.

Basic site investigations will have been carried out during the feasibility stage and an approximate alignment established from this and the engineering requirements. From here, the cost of the main site investigation must be related as much to the tunnel geology as to the envisaged contract cost. A flexible budget is therefore required for the first and most important step, the site investigation, which should be phased and tailored to suit the particular geological conditions revealed, as outlined by McFeat-Smith (1982a). One such approach is outlined as follows:

Feasibility study :
a) survey of existing data
b) aerial photography and surface mapping studies
c) initial boreholes
d) preliminary alignments and schemes

Site investigation stage:
a) final surface mapping in key areas
b) main borehole investigation
c) in-situ testing
d) laboratory testing

Preliminary design:
a) additional site investigation in key areas
b) establishing preferred alignment
c) evaluating engineering constraints
d) preliminary costing and programming

It is of particular importance that a comprehensive evaluation of the site investigation data should be carried out and a full interpretation prepared. This should include sections on the geological and hydrogeological engineering conditions expected, together with an appraisal of the anticipated tunnelling conditions.

These reports can be issued as part of the tender documents, thus providing a sound basis for the tunnel contract, as outlined by McFeat-Smith, Turner and Bracegirdle (1985).

This type of work, together with the main design, and the drawing up of contract documents and contract supervision, is most economically carried out by a consultant experienced in the construction of major underground projects. He will have the appropriate expertise and organisation required to ensure accurate costing and planning of the project and to ensure that the supervision of the works is carried out to the appropriate standards required for tunnelling.

2.1 Investigations at portals and areas of low cover

Drilling at portal areas is carried out primarily to define the geology of the area to be excavated and to measure the properties of each geological unit by in-situ and laboratory testing. Emphasis is given to the need to establish the position of rockhead accurately , and to examine the condition of the rock mass beyond this. Similarly it is important to determine water levels and the overall hydrogeology of the portal area for stability analysis due to the need to cut slopes (to minimise land usage).

At areas of low cover, where penetrative weathering, faulting and other adverse conditions may be encountered, site investigations are carried out for the following:
. design changes
. estimation of quantities
. determination of construction difficulties
. estimation of water inflows
. selection of preferred methods of construction

It is recommended that investigations at these areas include the use of orientated boreholes and seismic surveys. Inclined boreholes are normally drilled in the direction of the tunnels at an angle of $45°$. Their value is considerable, particularly for proving fault zones.

In order to locate rockhead at tunnel portals and to obtain the specific data required on rock quality in the area beyond, the author favours the drilling of long horizontal boreholes. The value of horizontal drilling as opposed to vertical drilling is substantial, as each metre drilled relates directly to conditions within the tunnel. The overall progress for six horizontal boreholes drilled to lengths of 150-220 m for tunnel projects in Hong Kong has been recently described (McFeat-Smith, 1985).

The major difficulty with drilling such boreholes is that they deviate from their intended path. Typical displacements can be calculated from tropari orientation tests, normally taken at 20 m intervals. The average displacements of the holes were 6, 10 and 15 m at drilled lengths of

100, 150 and 200 m. This was either up or down but in all cases to the right. In this case the displacements incurred were not considered to be significant relative to variations in the geology of the local volcanic rocks, and a high geological return close to the tunnel alignment was obtained from each borehole. This included measurement of water outflows, which provides an estimate of the quantities of water expected in the tunnels.

2.2 Investigations in Urban Areas

Tunnels driven under developed areas are commonly required for infrastructure facilities. These are often characterised by the need for easy access, and hence are constructed as close to the surface as possible.

Tunnelling close below major buildings and roads, even in competent rock, induces particular problems and requires detailed investigation. This is often impeded by limited access and the need to minimise disruption in the city centre. In these situations investigations are carried out to obtain data on the following features:
- areas of low rockhead or penetrative weathering
- rock quality particularly where high loadings on the tunnel lining are envisaged
- areas where settlements may occur

Even in rock tunnels where no loss of ground is expected, water inflows into the tunnels may drain the overlying soils and induce high settlements.

Techniques used in these areas generally include the drilling of many short boreholes. Preliminary studies for the Hong Kong MTR Island Line included about 1500 vertical boreholes spaced at distances of 15-20 m in key problem areas. For the Singapore MRT this distance was about 30-40 m.

Access to such tunnel systems is generally gained by shaft. Where portals and caverns are to be constructed or cover is substantial, orientated boreholes can be drilled. Seismic surveying techniques can be employed gainfully in developed areas. Seismic refraction surveying can provide detailed information on variations in the depth of sub-strata and can be used to locate faults (McFeat-Smith et al., 1986).

The principal limitation to the accuracy of seismic surveying is the signal to noise ratio. In developed areas, it is often found that the use of explosives is either not permitted or is governed by very strict controls. Hammer impacts on a metal plate imbedded in the ground can be used as an alternative sound source for short spread lengths of up to 50 m. A hammer survey has the advantage of being quick and cheap, and causes no damage to the environment.

2.3 Selection of methods of ground control

In order to ensure minimal surface disturbance from the settlement of soils due to water ingress into tunnels it is generally desirable that the engineer should specify the techniques to be used. Apart from tunnelling machines fitted with bulkheads, there are four major methods of preventing water ingress. These may be summarised as: compressed air, which remains the best alternative for short drives (Haswell & Campbell, 1983); injection of grouts for grannular soils;

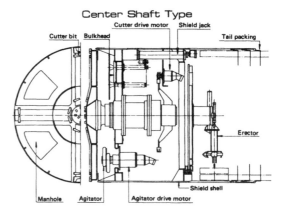

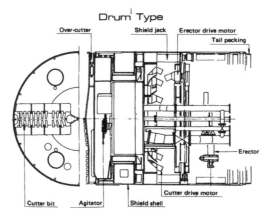

Fig. 2. Mitsubishi Slurry Shield

dewatering in permeable soils where settlement would not be a problem; and ground freezing for unusual conditions.

There is now a wide range of grouting methods available to tunnel engineers. These vary from ground fracturing methods (using cement-based grouts) to chemical gels which are used to fully permeate and consolidate the ground under controlled conditions. For non-grannular soils, such as marine clays, partial replacement of the soil by jet grouting improves tunnelling conditions, although steps must be taken to avoid heave of the ground surface.

3 METHODS OF EXCAVATION

Tunnels in any material have traditionally been excavated by hand. Today this tradition continues, although the pick and shovel of former times have given way to pneumatic tools and to fully mechanised methods of excavation. The particular advantage of these methods over conven-

tional methods for tunnelling in urban areas is that substantially lower ground disturbance can be achieved at high rates of advance.

Methods adopted for excavating tunnels are dependent upon ground conditions, and can conveniently be divided into those appropriate for rock and those suitable for soft ground. Some of the methods described are applicable to both types of ground. The aim of continuing development, especially with mechanical excavation methods, is to make them applicable to as wide a range of soil and rock types as possible.

The method of excavation selected for any tunnel requires consideration based on a number of factors including:
a. ground conditions, especially where fully mechanised systems are contemplated;
b. length of drive;
c. methods of preventing the ingress of water in permeable materials situated below the water table;
d. cost and availability of labour.

Table 1. Soft ground shield types

Excavation method	Face type	Shield type	Face support	Cross section	
Manual	Open	Shield with backhoe digger or boom head	Breasting system/timbering	Rectangular	Horse shoe
		Manual	Breasting system/timbering		
	Closed	Extrusion Shield	Bulkhead, soil pressure		
Mechanical	Closed	Slurry shield	Slurry, water pressure, cutter head	Circular	
		Confined soil shield	Soil pressure cutter head		
		Confined plenum process shield	Compressed air cutter head		
	Open	Fully mechanical shield	Cutter head		
		Boom head type	Breasting system/timbering		
		Pipe jacking shield with backhoe digger	Breasting system/timbering		

221

3.1 Soft Ground

In soft ground, maximum protection is afforded to face workers, and surface settlement is minimised by the use of tunnelling shields (Figure 2) or shielded machines. A rapid rate of advance results in very little ground relaxation at the face, and hence minor settlement, when the permanent lining is erected immediately and fully grouted.

Settlement associated with driving tunnels in soft ground can be negligible or may be several hundred millimetres. Total settlement is dependent on the construction method, the depth of the tunnel, the rate of advance, and the ground conditions. It is therefore essential to identify all structures that could be affected, and to what extent, and select tunnelling methods accordingly. Ground treatment may be considered as a primary means of overcoming such movement due to the flexibility with which it can be implemented.

For short drives of less than 0.3 km where construction can take place in free air (for example through stiff clays), the traditional method of excavation by hand and the erection of a segmental lining is normally best. However this depends on the diameter. For large diameters, or where ground conditions are poor, a shield will be required regardless of the length of drive. As the length of drive increases the level of mechanisation may be increased. For drives from 0.3 to 2.0 km, a shield, an erector and a spoil removal conveyor can be used to advantage. Finally for drives in excess of 2 km a

Table 2. Comparison of different methods of soft ground tunnelling

Item		Manual excavation	Shield-manual or backhoe or boom excavation	Full face machine	Full-face slurry/ bentonite machine
Costs	Initial cost Running cost	Low High	Medium Medium	High Medium	Very high High
Rate of Advance m/100hr wk	Favourable ground	30	100	400	75
	Mixed ground	10-15	10-15		
Installation	New equipment delivery time	Available	6-9 months	12 months	12 months
	Installation time	Nil	4 weeks	8-10 weeks	15-20 weeks
	Space required	No restriction	Shaft or chamber 1.2 x tunnel dia.	Shaft or chamber 1.2 x tunnel dia.	Shaft or chamber 1.2 x tunnel dia.
Application		Traditional method for short drives	Mixed ground conditions	Homogeneous ground conditions	Water bearing homogeneous ground
General access for operations and maintenance		Good	Good	Fair	Fair

fullface machine may be the most favourable solution. Table 1 shows the range of mechanised methods of excavation available for soft ground, and Table 2 gives a general comparison between various methods of excavation.

An extensive range of shields is currently available for different geological conditions, and Fig. 2 shows a slurry shield which can offer a solution to the problem of supporting a water-logged or unstable face without the use of compressed air whilst continuing to advance the drive.

3.2 Excavation in rock

For rock tunnelling, drill and blast remains the most widely used method due to the versatility of the technique in coping economically with different sizes, shapes and gradients of tunnel, with extremely hard rock, and with varied ground conditions.

Figure 3 shows an example of a 200 m long 22 m span rock cavern which was designed to house an MRT station concourse, platform and running line in Hong Kong. This is the largest cavern construction in Asia to-date. The figure shows the various stages of construction, which involved excavation in granite by drill and blast, and illustrates diagrammatically typical displacements monitored in the rock during excavation from anchors installed above the cavern crown.

Vibration limits may be imposed to blasting underground to minimise inconvenience and damage to surrounding structures and installations. For example, a particularly demanding limit of 1/20 mm jump was imposed for excavation under Oslo's Parliament Building in 1975. The answer to minimising vibration effects is smooth blasting, which is effectively controlled blasting of the tunnel perimeter. This can however result in a drop in the advance rates.

For this reason preference is commonly given to the use of fully mechanised excavation systems depending upon the local rock types and hence the overall costs. Table 3 gives a comparative analysis of the various options available.

One of the most economic types of machines for the excavation of tunnels in urban areas is the track-mounted, selective cutting, roadheader tunnelling machine, as shown on Fig. 4. For excavation of the Liverpool Loop underground tunnels in England for example, three roadheader machines were employed for tunnelling in soft sandstones. Their particular advantage was that use could be made of their ability to excavate both the running tunnels at 5 m diameter and the station tunnels at 7 m diameter. A similar heavy-weight machine was also used in the excavation of harder sedimentary rocks for the Tyneside Rapid Transit System in England. Other variations of the roadheader type include shielded machines used for the installation of segmental lining. This type of machine has essentially bridged the gap between soft ground and rock tunnelling, allowing excavation of very weak soils to moderately strong rock in adverse support conditions (McFeat-Smith and Fowell, 1979).

The second type of machine commonly used in rock tunnelling is the TBM, or full-face cutting machine (see Fig. 5). These have been used extensively for the construction of underground railway systems such as the Bay Area Rapid Transit System in San Francisco, for 12 km of tunnels for the Buffalo Subway and many other subway tunnels in New York and throughout the world. This type of machine is capable of high speed tunnelling. In rock, rates of 300-600 m/month can be achieved. Excavation by TBM has allowed the development of projects that would not have been undertaken otherwise. Rapid improvement in the design of TBMs has been made during the last decade in both Europe and the United States. As a result, a range of machine designs are now available for different tunnel diameters and varying geological conditions (McFeat-Smith 1982).

4 CONSTRUCTION PROGRAMMES

Once the decision has been made to construct, the result is often demanded in the minimum possible time. However, before drawing up a detailed construction programme based on current costs and tight construction schedules, the following items should be considered:
1. Corrections or modifications to the design and cost control can be difficult to implement.
2. The demand on local construction companies will be substantial, and short programmes allow them little scope to develop their facilities in conjunction with overseas contractors.
3. Tight programmes result in high tender prices and inefficient engineering.
4. Delays to interdependent contracts

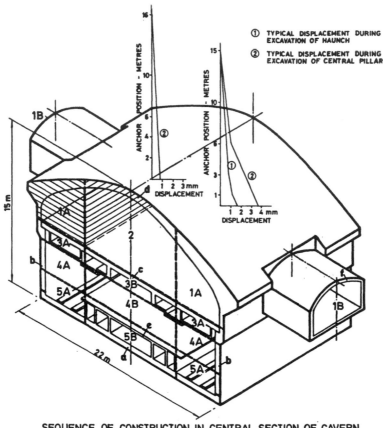

① TYPICAL DISPLACEMENT DURING
   EXCAVATION OF HAUNCH
② TYPICAL DISPLACEMENT DURING
   EXCAVATION OF CENTRAL PILLAR

SEQUENCE OF CONSTRUCTION IN CENTRAL SECTION OF CAVERN

EXCAVATION

1A	Top heading side tunnels (haunch)
1B	Passenger adits
2	Central pillar
3A-5A	Lowering bench at walls, first to third stages
3B-5B	Cavern widening, first to third stages

CONCRETING

a. Laying track or invert slab
b. Lining side walls
c. Concourse slab
d. Lining cavern roof
e. Laying platform slab
f. Lining passenger adits

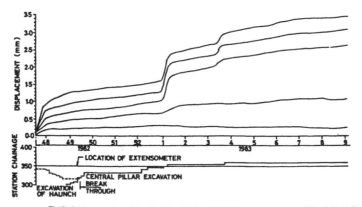

TYPICAL DISPLACEMENTS FROM CENTRAL ARCH EXTENSOMETERS

Figure 3. Construction of Tai Koo cavern, Hong Kong Mass Transit Railway

Table 3. Comparison of different methods of rock tunnelling after McFeat-Smith and Fowell

Item		Drill & Blast	Roadheader	Full face machine
Costs	Initial cost Running cost	Low/medium High	Low/medium Low	High Medium
Typical progress m/100 working hr.wk. (Gt. Britain)	Favourable ground	50	70	130
	Badly faulted ground	30	35	Variable - depends upon design
Installation	New equipment delivery time	1 month	2 months	12 months
	Installation time	2 weeks *	2 weeks *	12 weeks
	Space required	No restriction	Most can be installed within tunnel diameter	Requires erection chamber 1.5 - 3 times greater than tunnel diameter
	Tunnel shape	Any	Most	Circular
Application		Most rock types	Dependent upon rock hardness and abrasivity	Most rock types
Overbreak		High	Medium	Low
General access for tunnelling operations and maintenance		Good	Good	Fair to poor
Temporary support	Position installed	Face	Face	3 - 12 metres behind face
	Type	Any	Any	Most machines restrict use of shotcrete, some prevent installation of arches
Faulted ground		No major restrictions	No major restrictions	Some machines have restrictions to mucking and support
Ground treatment and advance		No major restrictions	No major restrictions	Requires suitable provision
Diameter range (metres)		Any	2 - 8	2.5 - 12

* depends on access

give rise to claims.

Thus whilst a good case exists for rapidly developing a project, particularly one earning revenue, a properly phased work schedule will provide a better quality, lower cost system, with more benefits to local contractors and all concerned.

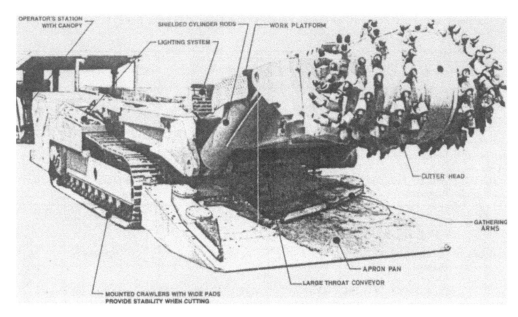

Figure 4. Roadheader rock tunnelling machine

Figure 5. Hard rock tunnel boring machine

5 MINIMISING CONSTRUCTION COSTS FOR UNDERGROUND WORK

The cost of tunnelling is directly related to the time spent underground, and hence to the geology of the tunnel. Where geological conditions are relatively consistent, whether in soils or rock, the behaviour of the ground can be accurately predicted from site investigations, and tunnelling costs will fall. If substantial lengths of tunnels are to be excavated in consistently similar conditions, mechanised excavation is often used and offers the prospect of rapid excavation at low costs.

The final cost of tunnelling will not be accurately known until the ground conditions expected from the site investigation are compared to the conditions actually encountered by the tunnels. This difference in cost can be escalated by the assumption that contractors should bear all of this risk. This leads to higher tender prices and claims, but can be avoided by sound engineering practice and by adopting a more practical and fairer approach. In order to avoid such generating claims and minimise the cost of constructing tunnels, the following steps are necessary:

1. Carry out comprehensive site investigations at an early stage.
2. Avoid placing unlimited liability on consultants as this leads to conservative design.
3. Recognise that underground work requires experienced consultants and construction staff who should be employed at the preliminary stages.
4. Delegate sufficient authority to settle disputes to construction staff.
5. Accept risk arising from unexpected geological conditions, and structure contract documents accordingly.
6. Package the extent of works in relation to geological conditions as well as the nature of the works.
7. Ensure that construction programmes are not unduly tight.

A well organised programme for the construction of underground works and the use of experienced staff will ensure that such works are implemented quickly, safely and at minimum cost.

6 CONCLUSIONS

This paper has outlined the various steps involved in using tunnels for developing infrastructure for major metropolitan cities. The construction cost of such sub-surface works is higher than that of overhead or cut and cover systems, yet this is recognised as being only one of a number of costs to the city. The major issue involved is that of allowing such work to proceed for a period of 3-4 years whilst ensuring that order and normality continues and business life is unaffected. This can be most expediently achieved by utilizing sections of bored tunnels, particularly in central business zones.

Perhaps the most significant and advantageous feature favouring the use of tunnels however is that the ultimate change to the city environment is nil, and city centres, free from the constrictions of long overhead and surface restrictions, can enjoy the full benefits of these facilities and be developed as truly urban environments.

REFERENCES

Haswell C.K. and Campbell J.G., 1983, Compressed air tunnelling with special reference to the Hong Kong Mass Transit Railway. Eurotunnel '83 conf. Switzerland, Pp 173-183.

McFeat-Smith I. 1982 (a), Geotechnical feasibility study and site investigation for the Western Aqueduct Tunnels, Hong Kong. Proc. Seventh S.E. Asian Geotech. Conf. Hong Kong, Pp 171-187.

McFeat-Smith I. 1982 (b), Survey of rock tunnelling machines available for mining projects. Trans. Instn. Min. Metall. (Sect. A: Min. industry), 91, A23-31.

McFeat-Smith I., 1985, The drilling of long horizontal boreholes for site investigation purposes'. Meeting on Geol. Aspects of Site Investigation, Bulletin No. 2, Geol. Soc. of Hong Kong.

McFeat-Smith I. and Fowell R.J., 1979. The selection and application of roadheaders for rock tunnelling. RETC Proceedings, AIME, Atlanta, USA, Pp 261-279.

McFeat-Smith I., Nieuwenhuijs G.N. and Lai W.C., 1986, The application of seismic surveying, orientated drilling and rock classification for site investigation of rock tunnels, Conf. on Rock. Engng., Instn. Min. Metall., Hong Kong.

McFeat-Smith I., Turner V.D. and Bracegirdle D.R., 1985, Tunnelling conditions in Hong Kong. Hong Kong Engineer, Vol. 13 No. 6, Pp 13-30. June.

Environmental Geotechnics and Problematic Soils and Rocks, Balasubramaniam et al. (eds)

Geotechnical modelling – An engineering approach to problems of mechanical interaction of construction work and natural geological environment

Ján P. Jesenák
Slovak Technical University, Bratislava, Czechoslovakia

ABSTRACT: On the basis of theoretical considerations general rules are given for transforming the engineering-geological model of the geological environment into the geotechnical model of an engineering problem within this environment. Categories of geotechnical models are defined and illustrated on examples. The engineering character of geotechnical work is emphasized.

1 INTRODUCTION

Protection of the natural environment has become one of the most serious challenges of Nature to Man: success or failure in solving this problem may decide about the survival of future human generations in large areas of the Planet. Inconsiderate exploitation of natural resources has already caused much incorrigible demage in different parts of the Earth: to restrict further impacts and, wherever possible, to restore and to improve original conditions should be a cathegoric imperative of any human activity.

However, the development of the human society cannot be stopped nor reversed, and so the manifold interaction between natural environment and human life will continue in future, too. Much work has been done in environmental controle. Serious problems are discussed and solved on governemental and even on international level. These trends must continue and increase, and environmental reasoning must become an integral part of general human consideration.

Construction work is one of the branches of Man's activities with most serious im - pacts on the natural environment. The mechanical interaction of construction work with the geological environment is a matter of geotechnics: environmental aspects have been for ever an integral part of well-designed geotechnical work. The last decades present old problems in new dimensions: accelerating development of the human society all over the world have caused an immens demand of construction work, re-

quiring more effectiv methods of site assessment, planning, design and construction.

There are two scientific disciplines concerned with problems of interaction of construction work and the natural geological environment: engineering geology and geotechnical engineering. Engineering geology is a branch of the geological science: it presents models of the geoenvironment defined in different manners, on different taxonomical levels. These model serve as a set of basic informations for regional planning, construction, environmental controle and partly for mining, too. Geotechnics is an engineering science with interdisciplinary character: it solves concretely defined engineering problems in concretely defined geological conditions, modelled by engineering geology. Mutual understanding between these two sciences is often far from an ideal one: different way of deliberation, different terminology and, first of all, different level of accuracy represent problems arising when tranforming the engineering-geological model of the geological environment into the geotechnical model of an engineering task within this environment.

Improvement of the social efficiency of the whole complex of geological-geotechnical works for construction purposes requires fully realistic modelling both in the engineering-geological and in the geotechnical phase of the work. Considering geological and geotechnical work to be one complex problem, the objective of further development will be both technological and economical optimization /starting with geological assessment and ending with completed construction/.

This paper is a short generalization of experience gained during close cooperation of engineers with engineering geologists in the course of the last 25 years in Czechoslovakia.

2 A GENERAL SCHEME FOR GEOTECHNICAL SOLUTIONS

A well-known but often not fully respected peculiarity of geotechnical engineering is that it works not only with manmade materials like steel or concrete and uses not only structures with more or less clear statical effects, but is in the major part concerned with soils and rocks, the result of natural geological processes. The geological structure of the site as well as the considerably varying properties of individual subsoil strata, the scatter of their characteristics must be stated in each individual case by expensive and time-consuming assessment, field tests and laboratory exploration and testing. All these render informations for individual points of the subsoil only /samples, field tests/, or in vertical, eventually inclined or horizontal straight lines /sounding, boring/: to define the engineering-geological model /EG model/ of the site extrapolation on the level of a professional estimate is necessary. Data on the geological environment therefore cannot reach the exactness of informations about the construction work: they represent a limiting factor for any soil-structure interaction problem.

There are two principal decisive factors for any geotechnical solution:

1. correct estimation of the physical substance of the interaction, and

2. a realistic EG model of the site, i.e. an engineering-oriented abstraction of the natural geoenvironment.

On the basis of these considerations a general scheme for the solution of geotechnical problems had been compiled /Fig.1/. The scheme is universal. In individual cases some marginal steps may not be of use. The interdisciplinary side of the solution is emphasized, supposing geotechnicians to be familiar with the engineering parts.

Simultaneous evaluation of the physical substance of interaction, the claims of the designed construction work and the EG model of the site allows to define the geotechnical model of the problem. The term "geotechnical model" /GT model/ was

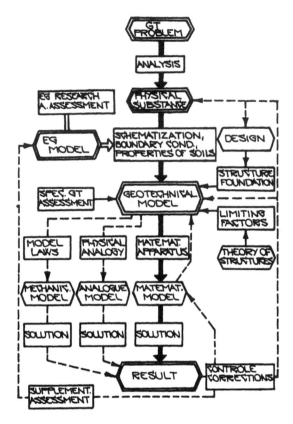

Fig.1 A general scheme for the solution of geotechnical problems

introduced for a basic model for the solution of a given geotechnical problem, defined merely in engineering terms. The GT model represents a further step of abstraction of the natural geoenvironment, evaluated with respect to a certain, often very narrowly defined engineering problem.

The main way of solution is the mathematical one: an adequate mathematical apparatus enables to define the mathematical model /M model/ of the solution. Note that between the EG model and the GT model there is only a one-way dependence, whereas a backfeed exists between the GT and the M model: the GT model must be defined already with respect to a certain appropriate method of mathematical solution /alternativly, in special cases experimental model-solutions may prove advantageous/.

As an important conclusion we can state, that realistic results of geotechnical prognoses depend primarily on a correct estimate of the physical substance of the interaction and on realistic input data. The level of the mathematical solution itself plays a secondary role: the lack of informations cannot be compensated by sophisticated mathematics. The benefits of more perfect mathematical solutions may be used only if adequate input data are available.

3 GEOTECHNICAL MODELLING

3.1 Theoretical considerations

Both EG and GT models interpret the natural geological conditions for purposes of planning and construction. This interpretation is not easy: it must overcome the interdisciplinary gap between the classical geological disciplines working prevailingly with the methods of natural sciences, and civil engineering, which solves its problems with technical /and often technological/ methods and procedures. Engineering geology belongs to the group of geological sciences: EG models are, in principle, geological models, which interprete the natural conditions in terms understandable for technicians, if possible, in a numerical manner. Geotechnics belongs unambigously to the group of engineering sciences. GT models must therefore be defined merely in engineering terms, in spite of the fact, that they reflect not only a technical problem, but also the natural geoenvironment. There are some peculiarities concerning GT models: they are always defined from the standpoint of an individual, often very narrowly specified geotechnical problem, and they must be defined in a manner fully corresponding to a certain supposed mathematical /or experimental/ method of solution.

So the EG model is the basis of the solution, the M model is a medium for it. The GT model, however, is an integral part of the engineering solution, with deciding influence on realistic results.

Both EG and GT models are abstracted from the same natural geological environment: they represent two steps of the assessment and adjustment of this environment for engineering purposes. Primary informations about the natural conditions renders the EG model. The GT model /or - more correctly-the GT models for different purposes and solution methods/ must strictly respect them. In this sense the GT model plays a secondary role. Sometimes it is possible to define the GT model without a previously compiled EG model, e.g. if solving a detailed problem on the basis of a special geotechnical assessment. Of course, this does not alter the secondary, derived character of the GT model as a special purpose engineering interpretation of the natural geoenvironment. It is to be emphasized that interdisciplinary difficulties on the boundaries of geological and engineering approaches may be economically mastered only by strict conservation of the correct sequence of EG and GT modelling in the course of the solution.

3.2 Schematization of the real conditions

A model is an abstraction of the reality which conserves its mean properties, decisive from a given point of view, and neglects the secondary ones in order to make the model open for mathematical /or experimental/ treatment. The mean way of abstraction is schematization. Simplifications refer to
1. geological structure of the zone of interaction,
2. hydrogeological conditions of the site,
3. physical properties and mechanical behaviour of soils and rocks,
4. boundary conditions of the problem,
5. interaction between construction work and geoenvironment.
The way and the size of schematizations depend on several conditions:
1. the relative spatial density of EG and GT informations within the zone of interest,
2. the quality of these informations,
3. the expected accuracy of the solution,
4. the optimum scale of the model in given conditions,
5. the kind of problem,
6. the level of the available mathematical or experimental apparatus.
A detailed analysis of these circumstances, though very important in practice, will be omitted here.

3.3 Transformation of the EG model into the GT model

The EG model of the zone of interest is from interdisciplinary standpoint the most important basis for the GT model. In accordance with the nature of the problem and the requirements of the chosen way of solution /mathematical, experimental and combined methods are possible/ it is necessary to transform the EG model, i.e. to adapt it for a particular engineering problem and to define it quantitatively in engineering terms. This transformation of the EG model of the geological environment into the GT model of a special engineering problem within this environment is the most important part of the solution, at least from geotechnical point of view: it determines whether the GT model will be sufficiently representative, able to fulfil the requirements of realistic and accurate results /on the required level of precision/.

In individual cases the process of transformation will be different. Only technical principles may be generalized. The following steps should be kept to:

1. specification of the physical substance of the interaction,
2. definition of the zone of interaction,
3. estimation of the field of interaction intensity,
4. schematization of the geological structure within the zone of interest,
5. determination of the properties of soils and rocks,
6. definition of the boundary conditions. Some explanations may prove useful:

1. Realistic estimate of the physical substance of interaction and its adequate schematization play a decisive role and require therefore great attention especially in cases exceeding the level of conventional solutions. E.g. the checking of the stability of the slope of a tailings impoundment as sliding along slip surfaces may be from the mathematical side correct, but may lead to confusion, if breakdown by liquefaction or destruction by seepage and suffosion may occure. Is, however, the physical substance of the interaction once correctly defined, even severe simplifications and simple mathematical solutions may lead to realistic results.

2. The EG model figurs, as a rule, a substantially larger zone than that directly participating in the mechanical interaction. The zone of interest of the GT model includes the area affected directly by interaction and its surroundings /not too large/ stated as a professional estimate on the basis of engineering experience, with respect to the requirements of the chosen method of solution.

3. The intensity of mechanical interaction may or may not be uniform within the zone of interaction. The field of interaction intensity has great influence on the admissible measure of schematization: areas with low intensity of interaction tolerate much more schematization without substantial influence on the precision of the results.

4. The model of the geological structure must be geometrically defined with simplifications according to the interaction intensity and with respect to the facilities

of the method of solution.

5. At the time of an EG assessment the method of solution is often still unknown, and sometimes even the GT problems are still not clearly defined. In such cases the properties of soils and rocks are to be determined later by a special geotechnical assessment of the site.

6. Simple calculations are less sensible to correct boundary conditions than complicated mathematical methods. Great attention is to be paid to an adequate choice of boundary conditions when using the method of finite elements /FEM/. Lacking experience should be compensated by checking the influence of the boundaries with parametric studies.

All factors mentioned above are closely connected with the schematization of both natural conditions and interaction and may have considerable influence on realistic results. The fundamental basis for a successful transformation is, however, a well defined realistic EG model of the natural geological environment.

The EG and GT assessment of sites is, as a rule, carried out in several stages. In the course of the assessment and the design the EG, the GT as well as the M model become gradually more and more realistic. The requirements and the conditions of schematizations remain valid, successively always on higher qualitative level.

3.4 Categories of GT models

Geotechnical problems occure at different stages of planning, design and construction, under difficult conditions even in the post-construction period. If planning and design have not been supported by representative geotechnical informations, the consequences appear later, during the construction period, or after the job has been finished. Additional changes in the distribution of structures in housing areas affect the ideas of the designer and the environment of future inhabitants, they may cause trouble in the planned flow of technology in industrial plants, unforeseen changes of foundation methods and additional site stabilization lead always to a considerable increase of costs and loss of time.

It is therefore very important to get geotechnical informations already for the lowest stages of country and urban planning and design, when deciding about the selection of the site, about the distribution of individual objects in new-built housing districts or industrial zones, or about the alignment of roads, railways, channels, levees a.s.o. The EG basis for this stage are EG maps of typological zoning. Lower taxonomical units of EG typological zoning-subrayons and districts - are already directly connected with construction works and they represent on an appropriate level EG models, which are qualitatively homogeneous /Matula 1976/. Transforming them, we get "GT type models". Are the latter represented graphically in plane, maps of geotechnical typological zoning will arise. This kind of GT modelling will be advantageous first of all on large sites with relatively simple geological structure, when a. the detailed distribution of buildings is still unknown, but the types of objects, the loads and their transfer into the subsoil are given, or b. the types, loads and the distribution of objects are known, and the kind and intensity of the interaction and its technological and economical consequences are to be estimated.

The second fundamental case arises at the interaction of a single object with the subsoil. The zone of interaction is relatively small and spatially determined, the distribution of interaction intensity is a. known - when a conventional GT problem in concretely defined geological conditions should be solved, or b. still unknown, when a new unusual case of interaction occures. For such cases the term "special GT model" have been introduced.

The third case originates when the relative influence of variable input data on the result of the solution should be stated. This case leads to "parametric GT models", which solve concretely defined geotechnical problems in altering idealized conditions by iteration. The results of parametric GT models supply valuable informations inevitably necessary when simplifications have to be introduced.

Each of the three categories, GT type models, GT special models and GT parametric models occure in practice in different modifications and at different scales. Each group has, however, its specific properties, which distinguish it clearly from the others. In the following attention will be paid to these categories, illustrating them by several examples.

4 GEOTECHNICAL TYPE MODELS

4.1 The principle hypothesis

The basis of GT type modelling is the hypothesis, that locally variable parameters of the subgrade /i.e. kind, thickness and sequence of soil strata, their properties, the ground water level a.s.o./ may be replaced for construction purposes on limited areas by unified type values of these parameters, in order to obtain on the required level of accuracy a model unit of the subsoil, which may be considered from the standpoint of a certain geotechnical problem /e.g.the foundation of structures/ to be qualitatively homogeneous. The practical consequence of homogeneity is the possibility of solving a certain geotechnical problem throughout the whole area in the same way /e.g. equal type and dimensions of foundations for equal structures/.

The validity of this hypothesis was proved in practice. The advantages of the method both for designers and contractors are evident.

4.2 Geotechnical typology

GT typological systems may be compiled for different purposes and for different scales. They must fulfil the following requirements:

1. universality: they must include all types of soils and subgrade structures, which may occure in practice /at least on a given site/,

2. typification of Quaternary and pre-Quaternary soils and rocks, typification of the geological structure of the subgrade within the zone of interest,

3. unification of the geotechnical properties of each soil type, unification of the subgrade structure within each type, unification of conditions for a certain geotechnical task,

4. compatibility with obligatory standards and regulations, compatibility with an appropriate system of engineering geological typology and zoning,

5. adaptability for different scales, for changes of classification and typification criteria, for various numbers of soil types a.s.o.,

6. system access: logical - when stating suitable types of soils and subgrade structures, and formal - when marking types with brief and instructive code numbers /to be used in maps and sections/.

The compiling of the typological system is one of the most responsible parts of GT typological modelling, with decisive influence on the practical serviceability of the resulting GT type models and the GT typological zoning of the site.

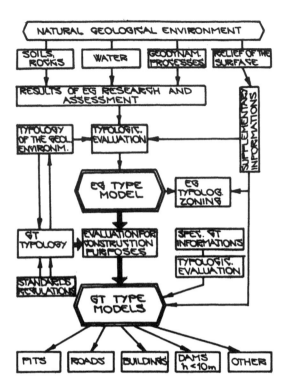

Fig.2 Small scale EG and GT type models

Table 1 Quaternary soil types

No	Type	Code
1	Gravel, sandy gravel	g
2	Sand	p
3	Cohesive soils, gravel with cohesive fill, soft	m
4	Ditto, stiff or hard	t
5	Alternating thin cohesive and loose strata	k
6	Loess, loess loam	s
7	Boulders, cobbles /sandy fill/	b
8	Soils not suitable for foundations /organic, very soft layers a.s./	o

Thickness	Code	Type thickness
1 to 2 m	1	1,75 m
2 to 5 m	2	3,50 m
5 m	−	7,0 m /max/[*]
		14,0 m /max/[*]

[*]Note: max. 7 m if bedrock lies within 5-10 m, max. 14 m if it lies within 10-20 m below surface

Table 2 Bedrock types

No	Type	Code
1	Hard rock, compact	S'
2	Ditto, fissured, wheathered	S
3	Weak rock	B
4	Resudual soils	Z
5	Gravel, sandy gravel	G
6	Sand	P
7	Cohesive soils, hard	T
8	Alternating cohesive and loose strata	B

Depth b.s.	Code /index/
<5 m	1
5 to 10 m	2
10 to 20 m	3

4.3 Small scale GT type models

In Czechoslovakia a number of GT typological systems was developed, for different scales and for different purposes.

Logical connections between small-scale EG type models and the appropriate GT type models are shown in Fig.2.

For systematical engineering geological mapping for the purpose of regional and landuse planning and environmental controle the scale 1:25 000 is used. It proved very useful to complete these by basic geotechnical informations. The typological system used here, a result of cooperation of engineering geologists and geotechnicians, had been developed and used first before 1970, and later improved and entirely coordinated with the system of EG typological zoning. Only main principles may be given here:

8 types of Quaternary soils and 8 types of pre-Quaternary bedrock are distinguished, the first marked with small letters, the latter with capitals /Table 1 and 2/. For Quaternary soils, the thickness is distinguished in three steps, marked with arabic numbers, written behind the symbol of the soil type. The depth of the bedrock below the surface is similarly given in three steps and marked with arabic numeral indices. For preliminary calculations /e.g. of the settlement of appartment houses in new living areas/ the type thickness of Quaternary soil strata is

234

stated in a manner giving results on the safe side. Type values of the basic properties of soils are derived from Czechoslovak State Standards for soil classification and foundation of structures /not given here/.

The introduction of type codes proved advantageous: they are easy to memorize and are instructive. E.g. a section with 1,9 m stiff silty loam and 6,6 m sandy gravel resting upon Tertiary claystone gets a type code

$$t1gB_2$$

This code number marks simultaneously all sections with the same soil types, with the upper stratum being 1 - 2 m thick, the second>5 m thick, and with the bedrock surface lying between 5 and 10 m below the surface of the terrain. A comparison of possible variations of the real section with the unified type section shows Fig.3: at scale 1:25 000 the differences are really not substantial.

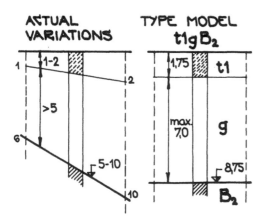

Fig.3 Comparison of possible variations of the actual section with the type section $t1gB_2$

If qualitatively homogeneous type models are presented in plane, maps of typological zoning arise. At scale 1:25 000 the maps of EG and GT typological zoning are identical. Geotechnical data are here given for each distinguished type model of the subgrade /i.e. for EG subrayons and districts/ only as a complement. On the basis of calculations using the type values of soil parameters semi-quantitative scales were compiled in order to allow a preliminary estimate of conditions for different geotechnical works. E.g. the claims of the subgrade on column foots of skeleton

structures were on the basis of the allowable settlement defined as follows:

Degree	Definition
Low	individual footings in minimum depth, contact area < 5 sq.m,
Medium	contact area > 5 sq.m, or foundation in greater depth wanted,
High	pile foundation necessary,
Very high	foundation on large diameter piles or raft foundation are necessary.

Similarly, scales were compiled for structures with load-bearing walls /strip footings/, for pits and trenches /recommended slope inclinations and expected water inflow are given/, for roads /the suitability of the subgrade as a base for pavement constructions, as a material for the construction of embankments, as subgrade for embankments and the workability of soils were estimated/ and for low dams and levees /the suitability of soils for constructing the impervious, resp. the pervious parts of dams and the properties of the subgrade were estimated/.

On occasions, also principally different problems are to be solved. During a long-lasting flood on the Danube river underpiping caused two breaks of the left-bank levee. About 1000 sq.km of most fertile agricultural land, some 40 villages and two towns were inundated for several months, the losses were very high. Afterwards, a geotechnical basis for reconstruction and strengthening of the round 150 km long czechoslovak section of the left-bank alignment of defense was required.

The unusual task had been solved on the basis of archives data /about the geological structure of the subsoil, hydrology, history of dam construction from its beginnings in the 13. century, observations of seepage and piping effects during floods/ and investigation of the terrain. Six factors were considered to be important:

1. cross section of the dam /single or double dam, without or with short or long diaphragm walls against seepage/,
2. height of the dam /up to 7 m/,
3. material and properties of the dam body /old parts only slightly compacted, locally pervious sandy layers/,
4. permeability of the upper soil stratum /sandy or silty/,
5. seepage and piping phenomena registered,
6. situation of dam alignment related to the river/ broad or narrow forefield, material pits, convex or concave side/.

Each of these factors was expressed with several degrees, marked by increasing numeral codes. Sections with constant degrees of all six factors were considered to be qualitatively homogeneous, i.e. guaranteeing equal safety against demage by long-lasting high water level of the river. The codes of the six factors gave the code number of the section: it is instructive and easy to memorize. For mutual comparison of different sections a complex characteristic called the "group index" was introduced: this is a number obtained as the simple sum of six individual codes forming the code number of the section. The group index value served as a criterion for declaring the section to be during long-lasting floods perfectly safe, to be not fully satisfactory, or to be in a critical state. Reconstruction was started on sections falling into the latter category.

The results of the research were graphically presented in a longitudinal section 1:50 000/500. The method of this GT typological zoning was already described in detail /Jesenák 1974/.

Similar cases may be sucessfully solved only if a sufficient amount of informations is available, and if their distribution all over the zone of interest is approximately uniform. Thorough knowledge of the natural geological environment plays a primary role also here.

4.4 Large scale GT type models

The advantages of large scale GT type models /1:5 000 up to 1:500 and more/ are manifest mostly on large sites with relatively simple geological structure, with repeated types of buildings /e.g. appartment houses/, but they are important also if different objects are designed. Zoning of the alignment of roads, railways and levees is also beneficial.

The connections of large scale GT type modelling with the results of EG research and assessment are shown in Fig.4. Presentation of GT type models in plane leads to GT typological zoning, which at this scales may or may not be directly connected with an appropriate EG typological system.

Each GT type model contains an engineering abstraction of real geological conditions. For a certain geotechnical work, however, different geological structure of the subsoil may represent equivalent conditions. E.g. for flat foundations or even for pile foundations of buildings it is often indifferent, whereas the firm subsoil below the compressible upper strata is gravel, hard clay or rock. If, as a further abstraction, in such cases homogeneous

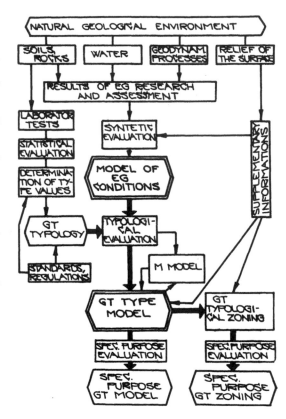

Fig.4 Large scale models of EG conditions and GT type models

units are defined on the basis of a merely geotechnical criterion /e.g. a certain settlement, or the necessary length of piles/, singlepurpose special GT typological modelling and zoning of the site will result. That sort of geotechnical criteria represent from a special geotechnical point of view "complex characteristics" of the subsoil. Only a complex characteristic allows a qualitative comparison of various GT type models /compare with the "group index" in the Danube-levee example/.

Fig.5 demonstrates schematically the procedure of prognosing the probable amount of different types of foundations on the large site of a new housing area. The upper scheme shows the relative series of appartment houses according to the project, the bottom scheme the series of available types of foundations. Typological evaluation of the geological structure of the site renders a series of GT type models /A to N, middle right/, which give after evaluation through a suitable complex characteristic a series of six types of conditions for

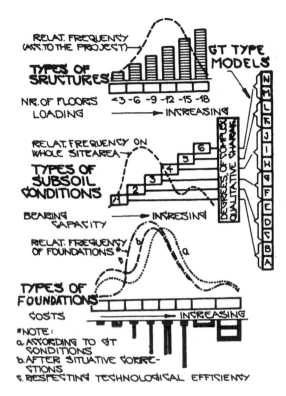

RELAT. FREQUENCY
(ACC.TO THE PROJECT)

TYPES OF
STRUCTURES

GT TYPE
MODELS

NR. OF FLOORS
LOADING

<3 -6 -9 -12 -15 -18

INCREASING

RELAT. FREQUENCY ON
WHOLE SITE AREA

TYPES OF
SUBSOIL
CONDITIONS

BEARING
CAPACITY

INCREASING

DEGREES OF COMPLEX QUALITATIVE CHARACT.

RELAT. FREQUENCY
OF FOUNDATIONS

TYPES OF
FOUNDATIONS

COSTS

INCREASING

*NOTE:
a. ACCORDING TO GT
CONDITIONS
b. AFTER SITUATIVE CORRE-
CTIONS
c. RESPECTING TECHNOLOGICAL EFFICIENCY

Fig.5 Prognosing the amount of different
foundation types on the basis of GT typological zoning

foundations, with a certain relative frequency according to real conditions on the
site /middle scheme/. According to the originally planned distribution of buildings
on the site, for each category of loading
there is an optimum type of foundation:
this gives line "a" for the frequency of
foundations on the site /bottom scheme/.
Should a greater number of heavy structures been constructed in poor subsoil conditions, situation corrections will be
unavoidable /b/. Technological and economical efficiency demands some restrictions
of the number of different foundation methods and specifies so the final distribution of various types of foundations on
the site.

Another example, the typological evaluation of the subsoil of a new housing district, is shown in Fig.6 to 9 /original-
ly at scale 1:2 000/. Fig. 6 presents the
results of the GT typological zoning, in
principle according to Table 1 and 2. Only
the code t̄ was introduced for gravel with
cohesive fill in order to demonstrate,
that the fill material is the same like

the overlying clay /both of Quaternary
age/. The Tertiary bedrock, claystone and
clay-shales, is to depth about 3 m uniformly wheathered /not marked in maps/.
The marginal parts of the site are locally
threatened by stabilized and active slope
sliding. 15 different types of the subsoil
occur on the site, see Fig.6: the GT typological zoning shows a mosaic of irregular
in shape, but qualitatively homogeneous type models of the subsoil. Fig.7, 8 and 9
are parallel maps of single purpose GT zoning: for 2-floor skeleton structures
/schools, store houses, services a.s.o./,
for 6-floor and for 12-floor appartment
houses. As a "complex characteristic" served for flat foundations /block or strip
footings/ different values of settlement,
up to the maximum allowable value for similar structures according to Czechoslovak
State Standards specifications /6 cm/, for
pile foundations the necessary length of
38 cm diameter VUIS-type vibro-bored cast-
in-place concrete piles.

Fig. 6 to 9 show clearly, that the map
of GT typological zoning is more complicated, like the maps of special purpose GT
zoning. At the same time the first has the
advantage of reflecting the real geological structure of the subsoil, and may therefore serve for different GT purposes.
The latter are very simple and very advantageous for designers and contractors, but
they can serve but the only purpose they
were compiled for.

4.5 Diagrams for rapid computations

The evaluation of GT typological zoning
maps requires a lot of geotechnical calculations. E.g. for constructing the special
purpose maps in Fig.7 to 9 the settlement
of all types of designed structures in all
types of the subsoil had to be computed.
Where the resulting settlement exceeded
the maximum allowable value, it was calculated again for pile foundations, for different length of piles.

In order to make such calculations more
rapid and to simplify them so that they
may be done by technicians with lower professional qualification, a series of diagrams had been compiled. To facilitate settlement calculations, first the results for
the elastic halfspace and according to
onedimensional methods were compared for
average conditions. The differences being
smaller then errors caused by oscillations
of soil parameters, for the simple and more adjustable one-dimensional method diagrams for a settlement influence factor "K"

237

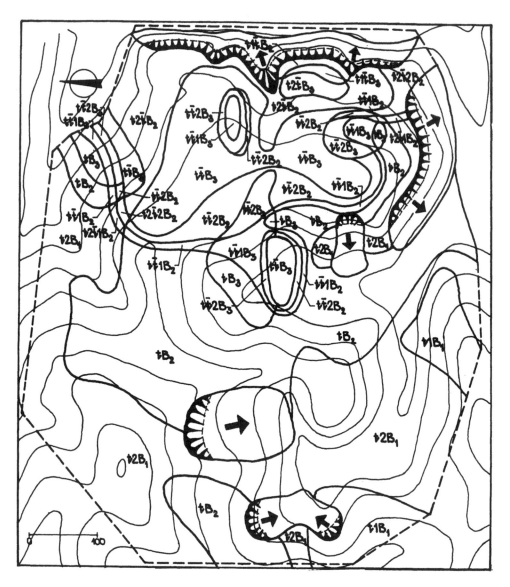

Fig.6 Geotechnical typological zoning map of a large site /originally at scale 1:2000/

were worked out. The following cases were considered: corner of an uniformly loaded rectangle, the characteristic point /Fig.10a/, uniform and triangular strip loads, uniform load on circular area. A lot of similar diagrams may be found in literature, mostly compiled for homogeneous elastic subsoil and therefore hardly to apply for layered geologic structures. The influence of neighbouring loads was also computed and presented in diagrams for different axial distances x/b of interfering footings /Fig.10b/. Similarly, diag-

rams were compiled for the settlement of road and highway embankments with cross sections according to Czechoslovak State Standards specifications /Jesenák 1980/, for arbitrary dam sections /by superposition/, for consolidation, for water inflow in pits and trenches, for the stability of their slopes a.s.o.

When using diagrams at suitable scales, the results are fully equivalent to the results of computations, the probability of errors is much lower, the saving of time is for simple cases about 90 %, for more

238

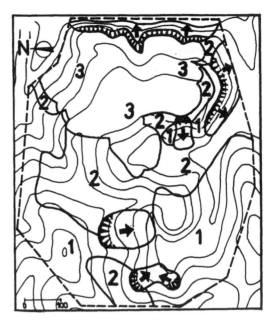

Fig.7 GT zoning for 2-floor skeleton structures: 1-settlement<3,5 cm, 2-settlement
3,5-5,0 cm, 3-settlement >5 cm

Fig.9 GT zoning for 12-floor appartment
house: pile ∅ 38 cm, 1-1=8 m, 2-1=10 m,
3-1=12 m

complex ones about 75-80 % /Jesenák 1978,
1979/. The graphical presentation of generalized results of computations proved as
a powerful tool for mastering large amounts of calculations, connected with the
geotechnical evaluation of subsoil type
models.

5 GEOTECHNICAL SPECIAL MODELS

5.1 Definition and specification

The conventional engineering approach to
soil-structure interaction problems is by
means of GT special models.

The measure of schematizations of the
geological structure, hydrogeological conditions, material properties, boundary conditions and of the kind of interaction is
here restricted to an extent unavoidable
from the standpoint of the solution method.

The GT special model is then an engineering abstraction of an interaction problem,
defined quantitatively with respect to the
physical substance of the interaction and
to the actual structure and properties of
the zone of interaction, in accordance with
requirements of the chosen method of solution and of the expected level of accuracy. The main attributes of GT special models distinguishing them clearly from other

Fig.8 GT zoning for 6-floor appartment
house: 1-strip, s<3,5 cm, 2-strip,
s=3,5-5,0 cm, 3-strip, s >5 cm, 4-piles
∅ 38 cm, 1=6 m

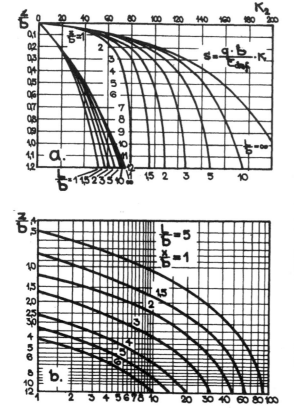

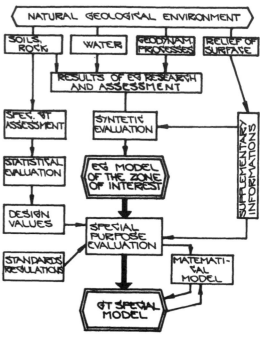

Fig.11 Connections between EG results and GT special model

Fig. 10 Settlement influence factor K:
a – for the characteristic point,
b – for the influence of neighbouring foundations

categories of GT models are
1. true description of real conditions throughout the whole model,
2. unique character, eliminating repeated application.

GT special models are always defined at large scales. They are inevitable when detailed solutions of interaction problems of large and pretentious structures with the subsoil are required /bridges, tall buildings, silos, nuclear, thermal or hydroelectric power plants, heavy and big structures/, or when the behaviour of large dams, tailings dumps or impoundments, open pit mines, a.s.o., is to be prognosed. GT special models are used also for geotechnical solutions at smaller dimensions: retaining structures, sloped, braced and sealed pits, cuts, embankments, detailed research on flat or deep foundations, anchoring, a.s.o. Also problems of

seepage and drainage under natural or artficial conditions are, as a rule, solved on the basis of GT special models.

From the considerations above it follows, that the GT model will be a "special" one in two cases:
1. if there is a unique geotechnical problem to be solved, where typification is not possible or absurd,
2. if on the required level of accuracy, the detailed knowledge of the structure, the behaviour of the geological environment and of their interaction are necessary /the problem itself may be a conventional one/.

Both cases need a detailed model of EG conditions /Fig.11/. Often also special geotechnical assessment and research are necessary, and in individual cases new geomechanical solutions are required: their results are to be evaluated simultaneously with the results of engineering geology.

5.2 Solution methods

The method of solution may be simple and only approximate, when a conventional problem under special geological conditions is to be solved, or when a new simple solution for an unusual case of interaction has been found. Correct mathematical solutions can be derived, as a rule, only for a simple structure of the zone of interaction, and for simple boundary conditions. In spite of this, correct solutions of geomechanics have a fundamental importance: they may serve as reference methods for testing the results of approximate or numerical procedures. Nevertheless, in practice the possibility of their application is limited. The greatest progress in the qualitative and quantitative analysis of geotechnical problems during the last decades have brought numerical methods adapted for computers, first of all the method of finite elements. It applies well not only in the case of complicated geological structure and boundary conditions, but also allows the use highly sophisticated constitutive laws. As a consequence of its universality, the FEM seems at present overestimated: it is often used in cases, where no equivalent input data are available - the results remain inadequate to the pretentious mathematical solution. It must be emphasized, however, that from the standpoint of geotechnical engineering it is indifferent, whether a mathematical solution is approximate or correct, whether it was obtained with an analytical or numerical procedure: decisive is only a better or worser agreement of the results with the actual interaction under natural conditions. Mathematics itself cannot be "good" or "bad", it may be only adequate or not. Besides a physically correct estimate of the interaction the reliablility of results is given by the representativeness of informations. Lower level of input data may be partly compensated only if decisions are made on the basis of results obtained by parametric studies and engineering experience.

It is clearly to be seen, that the amount and the quality of informations concerning the natural geoenvironment represents a limiting factor also in the case of GT special modelling.

5.3 Examples of GT special models

GT special models solving conventional problems under concrete geological conditions contribute to geotechnical knowledge and experience only quantitatively. Qualitative progress in geotechnics is put forward only by new solutions. Because there are still neither exact nor approximate

criteria for deciding whether in any concrete case further more detailed site assessment or the covering of uncertainties by reasonable overdimensioning of the foundations and/or the structure is more efficient, and there are still no proved methods of optimization for the whole complex of geological-geotechnical works for construction purposes, simple approximate but physically realistic solutions will conserve their importance for practical geotechnical engineering even in future. Simple solutions allow rapid checking of the influence of various factors in order to select the best solution, or, if necessary, to decide which alternative should be solved in detail in a more complex way. Some examples - only the results may be given here - may illustrate the foregoing considerations.

1. For the foundation of HT and EHT electric transmission towers on two parallel prismatic walls the ultimate oblique pulling load of a stiff, vertical body embedded in the soil had to be determined.

The problem was solved on the basis of small-scale laboratory model results, field tests and mathematical modelling by FEM for a non linear, elastic medium with joints between foundation and soil, for angles of inclination of the pulling force within the entire range from the vertical to the horizontal: a GT special model solved by pretentious combined methods. On the basis of the results a simple but well suiting approximate method was developed, now incorporated into the appropriate Czechoslovak State Standards. The method was already published and will therefore not be analysed here /Jesenák e.a. 1981, 1986/.

Generalizing we can conclude, that after sufficient experience has been gained, the GT special model may turn into a GT type model, ready for application under different conditions.

2. In another case, the ultimate bearing capacity of saturated clays below the slope of a high dump was to be checked in order to prescribe the allowable slope inclination. The task was solved in assuming deep, vertical cracks through the slope and failure of the base in undrained conditions. For constant strength of the subsoil and triangular strip load the ultimate bearing capacity is

$$q_u = 7,36 \ c_u$$

and the ultimate slope inclination β_u

$$\beta_u = \text{arc tan } 3,23 \ c_u/\gamma d$$

where γ is the unit weight of dump material and d the thickness of the clay stratum.

The investigation was extended also for undrained strength increasing linearly with depth, and for trapezoidal loads, too /Jesenák 1984, 1985/. For rapid calculation, tables, diagrams and approximate formulae were given. Some results are shown in Fig.12.

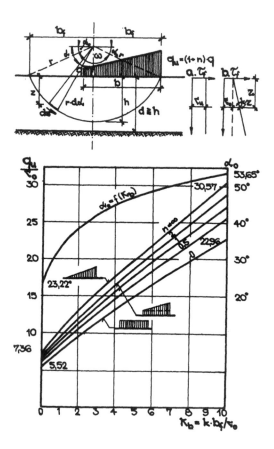

Fig.12 Ultimate bearing capacity q_u and angle α_o in dependence on the parameter $K_b = k \cdot b_f / c_o$

The simple method derived on the basis of laborious calculations suits well for checking the possibility of base failure below rapidly increased dumps and embankments, or below excentrically loaded strip footings. After failure occured, the residual strength of the subsoil can be determined.

The problem - originally a special GT model - was here again tranformed into an engineering type solution.

3. The final settlement of the power station of a new-built hydroelectric plant had to be determined. The dimensions of the power-house are 242,0x78,5 m, the foundation depth 32,5 m below the surface. Since the subsoil is built by an about 300 m thick, very pervious complex of sandy gravels with a single horizontal clay stratum 89 m below the terrain, the power-house is built in a sloped pit sealed with vertical diaphragm walls and a pan-like grouted bottom /at present the greatest sealed pit of the world/. On the basis of loading tests in boreholes, carried out to a maximum depth of 40 m, the deformation modulus of the gravel complex was supposed to follow a parabolic law

$$E_{def} = 27,47 \cdot z^{0,4} \quad [MPa]$$

Oedometric tests gave for the clay stratum an average value of the deformation modulus about 20 MPa. Young's moduli were in general 2,5-times higher.

Settlements were calculated for different stages of unloading and reloading, parallelly with the conventional one-dimensional method /using diagrams/ and with FEM /for a continuously non-homogeneous, linearly elastic medium - sandy gravel - with a single different layer - clay/. The supposition of linear elasticity was justified by the fact, that unloading by excavation exceeds reloading by the structure. The differences between the results of both methods are - surprisingly - quite negligible /see Fig.13 - contact pressure q and vertical displacement s for normal working conditions of the plant, one-dimensional settlement given for completly stiff and for elastic structure. Dipping is caused by the weight of water in the sealed upper reservoir/.

Of course, these results cannot be generalized. But they call attention to the fact, that there are still no criteria for deciding where the relatively cumbersome and expensive FEM is really efficient, and where simple approximate methods render also quite satisfactory results /even when using "not scientific" settlement charts/.

4. The investigation of seepage through the slope of a 90 m high, valley-type tailings impoundment showed a hyperbolic decreas of the average value of permeability toward the pond.

Using Dupuit's simplifications, formulae were derived for two-dimensional seepage through continuously non-homogeneous media, with permeabilities changing according to different laws. For general cases, with permeabilities changing both in horizontal and in vertical direction, an approximate engineering solution respecting the shape of cross-sections of the valley, was

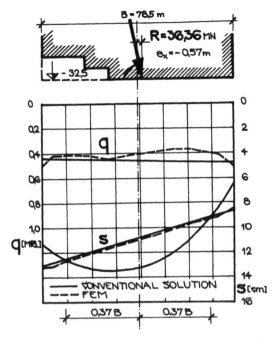

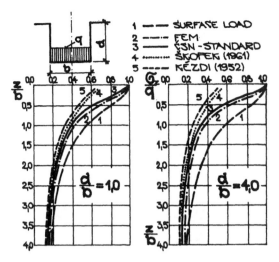

Fig.14 Influence of foundation depth on vertical normal stress σ_z

Fig.13 Final settlement of a hydroelectric power plant: comparison of conventional solution and FEM result

derived /Jesenák 1985/.

This example confirms, that in individual cases even difficult problems may be solved with simple approximate methods.

6 GEOTECHNICAL PARAMETRIC MODELS

GT parametric models do not serve for the direct solution of geotechnical problems in geological conditions connected with a real site. They investigate problems in idealized conditions, which could theoretically occur anywhere. They are a tool for realizing the influence of different variable parameters on the result of an individual method of solution. Another application is the comparing of results when using different solutions for a concretely specified geotechnical problem. Such sort of mathematical experiments may investigate e.g. the influence of varying boundary conditions, geological structure, properties of individual strata, extreme combinations of acting factors, or they may test the reliability of approximate solutions, the boundaries of their application a.s.o.

GT parametric models are time- and work-consuming, but they render directly genera-

lized results, contributing so to theoretical and applied knowledge in geotechnical engineering. For illustration, only two simple examples should be mentioned here.

1. The influence of the foundation depth on the distribution of stresses below flat foundations is often neglected, or respected according to theoretical solutions supposing undisturbed halfspace above the loaded area /Kézdi 1952, Škopek 1961/. The competent Czechoslovak State Standards recommend an approximate method derived originally by Jelinek /1951/. A parametric study was run for homogeneous subsoil according to the mentioned methods, controled by FEM /Fig.14/. The results show clearly, that only the approximate method renders realistic results: the physical substance of the "theoretically correct" solutions is not realistic.

2. The influence of a thin, much stiffer or much weaker layer on the settlement of the subsoil was checked by FEM, for a linearly elastic medium. The width of the uniformly loaded strip was b, the thickness of the stiffer /weaker/ layer b/4, its relative depth below surface varied from $z/b = 0$ to $z/b = 2$, the thickness of the compressible stratum was 4b.

In Fig.15 the integral curves of vertical displacements are given, expressed as a percentage of the entire settlement of a homogeneous layer /dashed line/. A detailed analysis of the results showed, that a stiff layer has only little influence on settlements, but it alters substantially the distribution of stresses. A much weaker layer has an opposite effect.

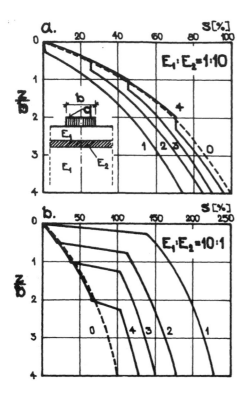

Fig.15 Influence of a stiff /a/ or weak /b/ layer on vertical displacements /related to total settlement of homogeneous medium/

7 CONCLUSIONS

Environmental aspects of geotechnical engineering include also the problems of mechanical interaction between construction work and natural geological environment. The interdisciplinary character of these problems is the reason why there is still often a gap in mutual understanding between geologists and engineers. To overcome interdisciplinary difficulties, engineering geologists should define more accurate and more realistic models of the geological environment, at scales adequate to the given geotechnical problem. The efforts of geotechnical engineering should tend to better understanding of peculiarities of the geoenvironment. Realistic estimates of the physical substance of soil-structure interaction, well-suiting transformation of enginnering geological models of the geological environment into geotechnical models of concretely difined engineering problems within this environment, representative parameters of the mechanical behaviour of geomaterials, as well as

fully adequate mathematical methods are fundamental requirements for successful solutions.

Geotechnical modelling as developed in Czechoslovakia during the last decades, is an engineering approach to problems of mechanical /and hydromechanical/ interaction. GT type modelling at small and at large scales, as well as GT special modelling of detailed problems are closely connected with appropriate engineering geological models of the geoenvironment. GT parametric models serve first of all for better understanding of interaction problems, for conventional as well as for particular solutions. With more engineering geology, more geotechnical experience and with adequate mathematics serious problems were solved: geotechnical informations are included in EG maps already at scale 1:25000, EG and GT typological zoning of large sites is gradually introduced in practice, problems of construction on unstable slopes were repeatedly solved with success, sliding areas stabilized, seepage in granular tailings impoundments controled, geotechnical problems of high dams, bridges, roads, channels, nuclear, thermal and hydroelectric power plants were mastered in difficult conditions in a quite satisfactory manner.

Close cooperation of geologists and engineers leads to better mutual understanding, good understanding is the basis of successful cooperation. The strategic object, optimization of the whole complex of geological-geotechnical work for construction purposes, however, is still waiting for solution.

ACKNOWLEDGEMENT

The author is indebted to pay distinguished thanks to M. Masarovičova M.SC. Ph.D. for energical support in preparing the paper.

REFERENCES

Hulman,R., Jesenák J. 1971, Katalogisierung der Baugrundverhältnisse und statistische Voraussagen im Grundbau, Proc.4th CSMFE Budapest, 629-638

Jelinek,R. 1951, Der Einfluss der Gründungstiefe und begrenzter Schichtmächtigkeit auf die Druckausbreitung im Baugrund, Bautechnik, 125

Jesenák,J. 1974, Geotechnische Rayonisierung von Hochwasserdämmen, Proc.4th Danube-European CSMFE, Bled, 23-27

Jesenák,J., Hulman,R. 1979, Geotechnical
models of the subsoil and their evaluation, Proc. "Engineering geological investigation of rock environment and geo-
dynamical phenomena", ed. M.Matula, VEDA
Bratislava, 279-291 /in Slovak, English
summary/

Jesenák,J. 1980, Setzungen des Baugrundes
unter Strassendämmen, Proc.6th Danube-
European CSMFE, Varna, 2/14, 139-148

Jesenák,J., Masarovičová,M. 1980, Spannun-
gen und Setzungen bei inhomogänem Baugrund, Proc.6th Danube-European CSMFE,
Varna, 2/15, 149-158

Jesenák,J., Kuzma, J., Masarovičová,M.
1981, Prismatic foundation subjected to
oblique pull, Proc.10th ICSMFE, Stockholm, 5/29, 145-150

Jesenák,J. 1983, Contribution to the hydraulics of a continuously nonhomogeneous
medium, Vodohosp. čas. 31, No.3, VEDA
Bratislava, 626-642, /in Slovak, English
summary/

Jesenák,J. 1984, Grundbruch unter einem
Böschungsfuss /φ_u=0-Analyse/, Proc.6th
CSMFE, Budapest, 99-104

Jesenák,J. 1985, Load bearing capacity of
ideal cohesive subsoil below a strip
load, Staveb. čas. 33, No.4, VEDA Bratislava, 275-291 /in Slovak, English
summary/

Jesenák,J. 1985, Contribution to the hydraulics of tailings impoundments, Proc.
11th ICSMFE, San Francisco, 1189-1192

Jesenák,J., Masarovičová,M., Bojsa,M.
/1986/ Foundation of HT and EHT towers
on prismatic walls and piers, CIGRE 1986,
Group 22-Overhead lines, Prefer. subject
22

Kézdi,Á. 1952, Einige Probleme der Span-
nungsverteilung im Boden, Acta technica
Hung., Tom. II, Fasc. 2-4

Matula,M. 1976, Principles and types of
engineering geological zoning, Mem. Soc.
Geol. It., 14, 327-336

Matula,M., Hrašna,M. 1979, Engineering
geological mapping and typological zoning, Proc. "Engineering geological investigation of rock environment and geo-
dynamical phenomena", ed. M.Matula, VEDA
Bratislava, 261-277 /in Slovak, English
summary/

Škopek,J. 1961, The influence of foundation depth on stress distribution, Proc.
5th ICSMFE, Paris, Vol.I., 3A/42,
815-818

Environmental Geotechnics and Problematic Soils and Rocks, Balasubramaniam et al. (eds)
© *1987 Balkema, Rotterdam. ISBN 90 6191 785 9*

Environmental aspects of underground space development in Kansas City, USA

Syed E.Hasan
Department of Geology, Kuwait University (on leave from University of Missouri, Kansas City, USA)

ABSTRACT: Many countries in the world have been utilizing the· underground space for a variety of purposes. While most of such space, in other countries, is primarily used for storage purpose, the underground spaces in the Kansas City metropolitan area, in U.S.A., have been developed for human use and occupancy. With over 5.2 million sq m of developed space and over 2 000 personnel working at least 40 hours per week, Kansas City has the distinction of being the number· one city in the world in terms of human use and occupancy of the subsurface space.

Underground mining of limestone, which began at the end of the last century, is still going on. The conversion of abandoned mines into usable secondary space began in early 1950s. As a result of current mining activity, about 0.25 million sq m of new space is available each year.

Uncontrolled mining techniques of the past coupled with weathering and erosion processes, have caused surface collapse following failures of mine roof. Other environmental problems relate to heaving of mine floors and shaking of surface structures and possible acid discharge from the underground sites. The details of these environmental problems are discussed in the paper. A brief discussion of history of underground mining and the geology of the area is also presented.

1 INTRODUCTION

Global trend for urbanization has been placing heavy demand on the limited land resource. Not only the western nations are facing a myriad of problems related to urbanization, but it is predicted that the developing nations will also face urban population explosion by the year 2025 (Fox, 1984). Figure 1 shows the past and projected world wide increase in the number of urban centers each with a population of five million or more. Figure 2 shows the urbanization trend in the developing countries; urban area being defined as one having a population of more than one million.

In order to deal with the multiplying need of more land, many large cities, throughout the world, have been utilizing the vast resources of the space below the ground surface for a variety of purposes. With the physical limitations of horizontal and vertical - above ground-expansion, it is only logical that the cities would consider exploiting the other dimension of locating facilities in the underground.

A survey of underground space utilization, on a world wide basis, reveals that most of large cities have been using the subsurface space for purposes which either do not require human beings spending any time (such as storage facilities), or a very minimal amount of time, (such as in underground parking lots, etc.). In places where people do spend some time in the underground space, such as, underground power plants, aircraft repair and maintenance facilities, and other military installations, their number, at any location, is very small. However, in the Kansas City metropolitan area, in the United States of America, more than 2 000 people work full

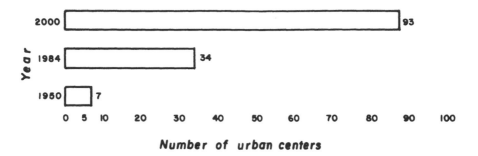

Figure 1. Growth of urban centers on a global basis.

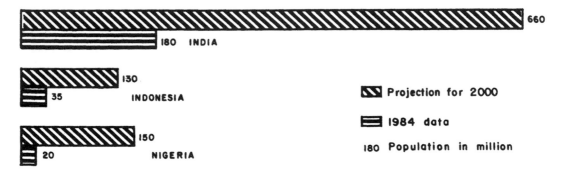

Figure 2. Urbanization trend in developing countries.

time (40 hours or more per week) on
a variety of job located underground
(Ward, 1983). Nowhere else so many
people spend as much time in the
underground space. In this respect
Kansas City is world leader in human
occupancy of the underground space.

1.1 Location

Kansas City is located in the midwestern United States. It straddles
the boundary between the states of
Missouri and Kansas (Figure 3).
Kansas City is one of the 36 metropolitan areas in the United States
with a population of more than one
million people. Metropolitan Kansas
City covers more than 5 120 sq km
area, encompassing seven counties.
The City serves as the transportation hub, agribusiness related industries, and service center for the
Great Plains and the Midwestern
United States.

2 GEOLOGIC SETTING

Sequence of sedimentary rocks, comprising repetitive layers of clayshale and limestone with thin carbonaceous beds, with a total thickness
of 274 m, occur in the Kansas City
area. The exposed part of this
sequence, totalling 91 m, outcrops
along valley slopes. These rocks
belong to the Missourian Series of
the Pennsylvannian System (Figure 4).
   Almost all underground space development has occurred in a limestone
bed, known as the Bethany Falls
Member of the Swope Formation, of
the Pennsylvannian System (about 300
million years old).
   Part of Kansas City was glaciated
during the Pleistocene time. Loess
deposit mantles the bedrock in the
northern part of the City, averaging
9 to 12 m in thickness; decreasing
in the southern part.
   The topography of Kansas City is
characterized by gently rolling hills
and valleys with limestone escarpments. Relief in the area is about
154 m. Bluffs formed due to erosion

248

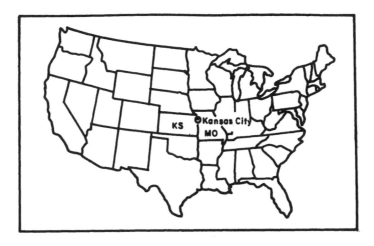

Figure 3.   Location map of Kansas City, U.S.A.

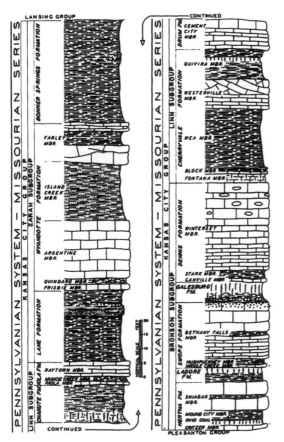

Figure 4.   Geological column of rocks in Kansas City area (after Howe, 1961).

of limestone provide ready access to the underground mines and facilitate extension of highways and railways into the underground facilities.

The rocks in the area have not undergone large scale deformation, except for some normal high-angle faults occurring in the southern part of the City. The beds are nearly horizontal with well developed joint sets. Solutioning is very minimal or missing altogether.

3 HISTORY OF UNDERGROUND MINING IN KANSAS CITY

The history of limestone extraction by surface quarrying dates back to the 1800s when the rock was used mainly for construction purposes. Later, between the late 1800s and early 1900s, underground mining of limestone commenced. Up until early 1950 underground mining was done solely for the purpose of extracting the limestone which, being of high quality and suitable for a variety of uses, found ready market. The large scale mining operations, extending over several decades, created a large volume of void spaces in the sub-surface which remained as a wasteland for long time. Nonetheless, just before the Depression of 1928, some enterprising businessmen recognized the potential of these spaces and envisioned the great possibility of converting the abandoned mines into usable space (Dean,

1975). This dream, however, could not become a reality due to economic and war reasons; and it was not until 1944 that the first underground facility for storage purpose was developed at Atchison, Kansas, about 80 km west of Kansas City. This development brought about a radical change in the purpose of limestone mining: from extraction of resource alone to a combination of mining and creation of usable secondary space. This approach initiated controlled mining techniques aimed at providing structurally stable and well laid-out space in the underground.

Currently there are many companies in the Kansas City area that are engaged in limestone mining and development of underground space for secondary use. As of December 1983, a total of 5.17 million sq m of developed space was available, to which another 0.25 million sq m of mined-out space was being added every year (Ward, op.cit.).

4 MINING METHOD

Room-and-pillar method is employed in the development of underground spaces. Conventional techniques of drilling and blasting are used to communite the limestone; ammonium nitrate is the most commonly used explosive.

During the early period of mining, because of uncontrolled excavation methods, random pattern of room and pillars were created. However, since the 1950s when secondary use of the mined-out spaces became one of the main considerations, controlled drilling and blasting techniques are being used to develop underground space in a grid-like pattern. The pillars, ranging in size from 6.1 to 9.1 m are located approximately 15.2 m apart. This configuration allows for recovery of 85 percent of the limestone, leaving a 9.1 m wide corridor having an area of 195 sq m. Rock bolts, with or without steel straps, are used whereever needed, to provide stability to the roof. Practice of measurement of convergence between the roof and the floor is increasingly getting popular in all planned excavations. Other than alerting the geologists or the engineer about potential roof instability, it gives

the prospective user of the secondary space a much-needed assurance of safety and integrity of the underground spaces.

5 ENVIRONMENTAL PROBLEMS

Major environmental problems related to the underground space development in the Kansas City area could be grouped under: (a) mine floor heave and (b) roof failures resulting in ground subsidence. Minor problems include: (a) alleged shaking of surface buildings due to blasting operations, (b) dust control in the winter season, and (c) possible acid discharge from the underground space. The problems are discussed below.

5.1 Mine floor heave

The floor in most of the underground space is the top of a black shale called the Hushpuckney Shale. This shale, upon exposure to atmosphere, undergoes expansion resulting in heaving of the floor. Noticeable expansion occurs anytime between two and five years after exposure. Generally the amount of heave keeps increasing with time but attains the maximum value in about 10 years. The total floor heave ranges from 2.5 cm to 20 cm. Maximum heave occurs in the middle of passageways or rooms and gradually decreases away from the center, becoming negligible at the pillars. This produces an inverted V-shaped profile where the floor slopes away from the center towards the pillars.

Heaving of the shale causes warping and cracking of concrete floors, block walls, doors, and windows. Several factors are believed to cause heaving of the shale. These include: (a) conversion of primary sulfides in the Huspuckney Shale into gypsum with attendant stress and volume increase and (b) plastic deformation of the shale in the unconfined middle part of the room. Various remedial methods have been used to control the heave with varying degree of success. For a complete discussion of the causes and control of heave, refer to the paper by Hasan (1982).

5.2 Ground subsidence

Instances of ground subsidence, though not frequent, do occur once in a while. Because the area involved in subsidence has usually been built upon, the environmental hazards become all the more serious. Two examples of recent subsidence are discussed below.

5.2.1 Blue hills sink hole

This subsidence, with the resulting sink hole, occurred on July 19,1983, on the east side of Kansas City. The land on the surface had been developed for a 48 unit mobil home trailer park in the early 1970s. On the morning of July 19, 1983 a sink feature appeared in the back of a mobil home. The subsidence hole measured 5 m by 3 m. Fissuring of ground, comprised of loess, extended about 3 m from the edge of the hole. Part of the concrete pad of one of the homes was swallowed by the hole; damage to utility lines was also reported (Kansas City Times, 1983), and electricity to six homes was disconnected. Immediate evacuation of six homes was ordered by the City officials.

The cause of subsidence is directly related to mining of limestone below the mobil home park. The ground cover, above the roof of the mine, is about 10 m which includes a thin layer of loess on the surface. Mining operations had ceased by 1950; and the weathering and erosion of the rocks by percolating surface water, for about 25 years, weakened the roof rocks. The erosion alongwith uncontrolled mining, caused the failure of the roof rocks which culminated in the subsidence.

An interesting aspect of this subsidence relates to inadequate engineering geologic evaluation of conditions in the abandoned mines before granting permission to develop the mobil home park. It is reported that the City Engineer inspected the site and in his report, dated September 19, 1971, concluded: "....the overall condition of the mine appears relatively sound. There are several areas of roof collapse. This was caused by mining too close to the overlying Galesburg Shale. This plot.... should not have prob-lem supporting floating concrete trailer pads." (Kansas City Times, 1983).

5.2.2 Independence subsidence

On April 21, 1984, subsidence of a large area that included two houses, occurred in Independence-one of the several cities that comprise the metropolitan Kansas City area. The ground surface involved in the subsidence measured about 100 m across with a vertical drop of more than one meter in the center. Two homes that suffered heavy damage were located on the edge of the subsidence trough. Failure of concrete slabs and foundation walls, together with ground fissuring, was widespread; tensional cracks upto 12 cm were common (Fig. 5 and 6).

Like the other subsidence, the Independence collapse is also related to weathering and erosion aided by uncontrolled mining that had resulted in creation of odd shaped pillars with random spacing. The limestone was mined during the 1960s and the mine was abandoned some 15 to 20 years ago. The ground cover above the roof of the mine is about 15 m with a number of roof failures in the mine. A large area of the mine was under 1.2 m of water at the time of the collapse. Continued water percolation with attendent erosion and weathering of the shale and other roof rocks, resulted in the loss of support to the ground up above, which consequently collapsed.

5.3 Ground vibration

Extraction of the limestone in the subsurface is done by drilling and blasting. Depending upon the amount and type of explosive used, thickness of the ground cover, and the nature of the geologic material; shaking and vibration of the surface structures may occur. Vibrations generally result in rattling of dishes and minor nuisances; some residents, however, allege that shock waves have caused cracks in the ceilings and basements of their homes. These damages may be due to design, construction or environmental factors rather than blasting

Figure 5. Failure of concrete foundation; center of subsidence is toward the right.

Figure 6. Shearing of concrete slab.

(Kansas City Times, 1985). Further studies are needed to ascertain the cause.

5.4 Dust control

This is a seasonal problem that bothers the owners of the developed spaces. During the winter months trucks and cars entering the underground sites, pick up snow which has fine dust in it. After coming inside and with warmer temperatures, dirty snow melts from the underside of the vehicles and releases the dust. This is minor problem that is easily solved by periodic removal of accumulated dust from the roadways.

5.5 Acid discharge

No study has been done to evaluate the geochemical characteristics of the water that drains out of the mines. It is known that the Hushpuckney and other black shales, that underlie and overlie the Bethany Falls Limestone, contain 5-6 volume percent of sulfide minerals, such as pyrite, sphalerite, chalcopyrite, etc. (Coveney and Parizek, 1977). It is very likely that leaching of these sulfides would result in increased acidity of water. The environmental effects of such acid water, that ultimately gets into the creeks and streams, need to be investigated.

6 CONCLUSION AND RECOMMENDATIONS

Mining of limestone and subsequent conversion of mined-out areas into usable secondary spaces, while adds to the economic advantages of an urban area, also creates special environmental problems. Ground subsidence and floor heave are serious problems that result in loss and damage to property and entail large maintenance expenditure. Ground vibration and dust control are minor problems; acid water discharge may cause serious problems and needs to be further studied.

These environmental problems can be eliminated by adopting appropriate measures, such as, (a) proper mine design with optimum size and spacing of the pillars, (b) roof reinforcement by rock bolts, wherever necessary, (c) leaving about 2 m of Betahany Falls Limestone in the roof, (d) survey and preparation of mine plans to delineate the area that were mined in the past, (e) drawing up suitable land use plans so that buildings and other critical structures are not located above abandoned mines which have questionable stability, and (f) periodic monitoring of the geotechnical condition in the abandoned mines, above which construction has already been done.

REFERENCES

Dean, L. 1975. How Underground Space Started in Kansas City area. In Stauffer and Vineyard (eds.), Proceedings of the Symposium on the Development and Utilization of Underground Space, p. 25-28. Kansas City: University of Missouri.

Coveney, R.M. and Parizek, E.J. 1977. Deformation of Mine Floors by Sulfide Alteration. Bulletin of the Association of Engineering Geologists. 14(3): 131-156.

Fox, R.W. 1984. The World's Urban Explosion. National Geographic 166: 179-185.

Hasan, S.E. 1982. Engineering Geology of Underground Space Development in Kansas City, U.S.A. In Proceedings, 4th International Congress, International Association of Engineering Geology. (5) 247-257. New Delhi: Oxford & IBH.

Howe, W.E. 1961. The Stratigraphic Succession of Missouri. Missouri Geological Survey and Water Resources. 401 (2nd series): 99-101.

Kansas City Times, July 22, 23 and 30, 1983. Blue Hills Mobile Home Park Subsidence. Kansas City: Gannett
_____, April 24, 25 and 27, 1984. (Independence Subsidence.
_____, August 9, 1985 (Blasting Problems).

Kansas City Star, April 24, 1985 (Independence Subsidence). Kansas City: Gannett.

Ward, D.M. 1983. Underground Space: Inventory and Prospect in Greater Kansas City, 19 p. Kansas City: Center for Underground Space Studies, University of Missouri.

Environmental Geotechnics and Problematic Soils and Rocks, Balasubramaniam et al. (eds)
© *1987 Balkema, Rotterdam. ISBN 90 6191 785 9*

Study on the utilisation of steel slag as ballast material

S.D.Ramaswamy, S.C.Kheok & Y.Tanaboriboon
Department of Civil Engineering, National University of Singapore, Kent Ridge

ABSTRACT: Steel slag has always posed an environmental problem for smelters and it is usually disposed off as a landfill material. Investigations on possible economic usage of steel slag has been conducted by researchers including its usage as a railroad ballast material.

Steel slag which is obtained as a by-product from a steel mill in Singapore is studied with the objective of utilisation as railway ballast for the mass rapid transit system in Singapore.

The laboratory test results indicated that the Singapore steel slag has all the desirable qualities for use as railway ballast material and its potential use for elevated tracks and tracks on ground is very encouraging.

INTRODUCTION

Until only a few decades ago, metallurgical slag which is a by-product of the smelting process, was recognized only as a waste-product and a nuisance material of little intrinsic value. Smelters were concerned only with the desired metallic product and their interest in slag was restricted to disposal by the most convenient and cheapest way, such as landfilling.

However, as available landfill sites became scarcer and disposal costs sored higher slag waste posed, an environmental problem.

The slag at the steel mill in Singapore is a waste-product generated from the electric arc furnance. Due to the large quantity of the slag produced, an environmental solid waste disposal problem, has to be tackled preferably putting the slag to constructive uses.

In the case of Singapore, the depleting natural aggregate (crushed stone, sand and gravel) resources also is of increasing concern to the local construction industry. The general concensus in Singapore is that granitic materials should be conserved, as far as possible, and other materials should be sought for replacement. One such material that could be considered as most logically suitable is steel slag. The steel slag being a

waste product could be more economical as compared with natural granitic aggregate only if it satisfies the required specifications at least to the level of natural aggregates. Its use could be well justified if it could be put to good use. It not only helps preserve precious natural aggregates to some extent but also helps in disposing off a "solid waste" in an economical and constructive way.

Steel slag has been used as ballast in USA and Europe for many years. In a recent communication with Mr H.K. Eggleston, President of the National Slag ssociation of America, the authors were informed that steel slag is being used extensively in USA in the past quite satisfactorily as railway ballast and presently the demand for steel slag for this purpose surpassess supply.

This paper describes a feasibility study on the utilization of Singapore steel slag as railway ballast for possible use in the Singapore Mass Rapid Transit Project.

TYPICAL TRACKS ON BALLAST

For a long time rail track has been supported by ballast. Only recently, ballast-free track has been introduced where tracks are placed on concrete slab. Though ballast-free track has been acclaimed for its advantages, but it has

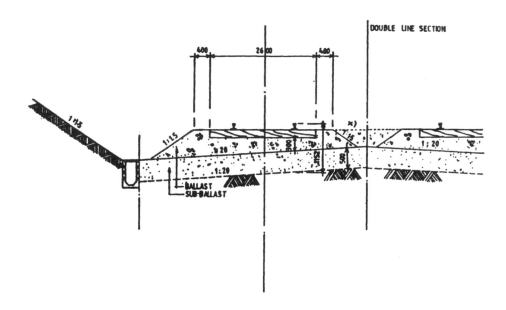

Fig.1. Typical cross-sectional view of ballast track

never been able to replace ballast track completely.

Typical cross-sectional view of ballast track is shown in Fig. 1. The rails are tied onto wooden or steel or precast concrete or prestressed concrete sleepers and then supported by ballast. The ballast must be able to withstand high dynamic loadings and must be able to drain fast. Various national rail authorities have drawn up specifications for ballast materials to fulfil the above requirements and these will be discussed later.

Recent developments involve the use of geotextiles for railroads. It has been claimed that non-woven fabrics provides an inherently stronger track system that last longer and require less maintenance. Fig. 2 illustrates the use of non-woven geofabric. Installed between the ballast and subgrade the fabric effectively reinforces the railbed and drains water quickly so that the ballast remains dry and functional.

The material for ballast has commonly been derived from igneous rocks. However, steel furnace slag has been used extensively and successfully on main railroads in the United States for more than 25 years. According to A.W. Johnston (personal communications, 1984) of the Association of American Railroads, steel slag is being used on many railways quite popularly in the United States.

The physical characteristics and ballast

Fig. 2. Soil stabilized with continuous filament needle-punched, nonwoven fabric.

specification of steel furnace slag as a railway ballast are presented in Table 1. Annually, these slags are used as ballast at a rate of about one million tonnes (USBM, 1976) much of it on main line with 90 metric tonne car loadings.

Steel slag when used as railway ballast has been found to have better operational maintenance because of its following qualities:

1. Steel slag is more resilient compared with granite and this provides better riding comfort.

2. Steel slag because of its pronounced micro-texture and macrotexture has better keying effect in a spread mass and is less likely to be dislodged under dynamic loading to the extent a granite ballast would.

3. Again because of its surface texture as well as its irregular shape, steel slag provides a free drainage path to keep the track above water-level which may otherwise result during heavy downpour.

Most slag can meet all the specification limits of the American Railway Engineering Association (AREA). Actual use on high density railroads has confirmed the long range economics of their use. However steel slag from any new source has to be thoroughly investigated for its suitability for such applications.

Perhaps the best evidence of the acceptability and value of steel slag in general as railway ballast is the fact that it has been used successfully for a long period of time on all classes of track of major railroads in the United States. This general acceptance and use has developed primarily from the good record of steel slag ballast in actual use.

Reservations on the use of steel slag as railway ballast has been raised regarding its iron and other metallic content which may cause short out track circuitry.The American Railway Association specification requires conductivity test on steel slag ballast and that the resistance should be more than only 2 ohms per 305 m of track length. With regard to steel slag from Singapore, tests on resistivity were thoroughly conducted and tests results to be presented later indicated that such problem of short circuitry should not be a matter of concern.

BALLAST-FREE TRACKS

Instead of the rail track being supported by ballast, an alternate system is to have it supported by a concrete slab.

A continuous pad of about 10mm thick is laid beneath the rail to provide vertical resilience and the material must have electrical resistance to provide an adequate rail-to-ground insulation for track circuiting purposes.

One major disadvantages of concrete track is that it cannot be installed in the typical time interval available for track renewal which is normally 6 hours at night when the trains are not running. This limitation is a consequence of the need for the in-situ concrete to acquire sufficient strength prior to trafficking and of the time required after paving the concrete slab to check and correct the correct levels and install the rail fastenings pads and rails, although the time taken for these post-paving operations is largely dependent on the manpower available (Cope, 1982).

The main advantages of concrete slab track compared with conventional ballast track may be summarized as follows:

(a) It retains the initial line and level of the track without the need for periodic maintenance.

(b) In the event of a derailment, damage to the track strucure is likely to be much less than with ballasted track.

(c) Construction depths in tunnels can be lower than with ballasted track and this can produce substantial savings in tunnel construction costs if new tunnels are designed from the outset with slab tracks.

(d) In existing tracks where increased clearances are required, slab track can eliminate the need for costly rebuilding of bridge and tunnel structures.

(e) The installation of slab track can provide as part of the construction a well controlled and accessible drainage scheme.

The main disadvantages are as follows:

(a) All slab track system are more expensive in cost than ballasted track.

(b) Paved concrete track requires long possession times to install and cannot be trafficked shortly after laying.

(c) Other forms of slag designed for early trafficking cannot be installed at the same rate as ballasted track.

(d) Once installed slab tracks cannot be realigned or recanted for higher speeds.

Based on the discussions on ballast and ballast-free tracks, it can be concluded that ballast-free tracks will find a role in replacing ballast tracks in particularly difficult maintenance situations such as tunnels. According to Cope (1982) the widespread replacement of ballast tracks by slab (ballast-free) tracks in the open is extremely unlikely because of economic considerations.

With reference to the mass rapid transit systems such as the one being built in Singapore, the underground routes should preferably adopt concrete slab as trackbeds. After considering the advantages and disadvantages discussed earlier, those in the open on the ground and elevated tracks should use ballast for tracks.

SPECIFICATIONS FOR RAILWAY BALLAST

Grading

Type and gradation of the material to be used for ballast are important with respect to the cost of maintaining line and surface. In general, ballast are specified according to their nominal top

Table 1   The Gradations Specified by the
American Railway Association (AREA)

Screen size, m.m. (in.)	40 ($1\frac{1}{2}$)	25 (1)	20 (3/4)	12.5 (1/2)	10 3/8	4.8 No. 4
% by weight passing	100	90-100	40-75	15-35	0-15	0-5

Note: 1 in. = 2.54 m.m.

Table 2   Gradations Specified by British Railways Board

Screen size, m.m.	63	50	28	14	1.18
% by weight passing	100%	97-100	0-20	0-2	0-0.3

and bottom particle size. Individual railroads have different preferences for size of ballast. As tables 1 to 3 show the gradations specified by the American Railway Association, The British Railways Board, and the Taiwan Railway Administration, respectively.

Although there are some slight differences in grading required by different organizations, the Singapore slag can be crushed and screened to any particular required grading size.

Quality Requirements

The prime function of railway ballast is to spread the wheel loadings from the base of the cross tie to the subgrade at pressures that will not cause subgrade failure. Thus, ballast must be capable of withstanding many forces. Without exception every ballast specifications attempt to assess the quality of ballast particles under loading. Ideally this quality measure reflects both the hardness and toughness of the ballast particles. Typical tests are the Los Angeles Abrasion test, the Impact test, and the Crushing Value test.

In addition, some types of slag ballast material may contain metallics which can cause the failure of the track circuitry. The American Railway Engineering has set the resistance requirement for ballast as 2 ohms per 305 m (1000 ft) of track length. Table 4 summarized the physical properties requirement for railway ballast by the American Railway Association.

PROPERTIES OF SINGAPORE STEEL SLAG

As mentioned earlier, the steel slag must be capable of withstanding many forces. The primary requirement is to test the hardness and toughness of steel slag as railway ballast. Tests were carried out on all three typical methods (Los Angeles Abrasion, Impact and Crushing Value) of testing the toughness and hardness. However, it is only necessary to use one of these tests as required by the AREA, the Los Angeles Abrasion test.

Test results indicated the Singapore slag has the Los Angeles value of 34% which is well within the specification of 40%. As discussed in the previous section, slag's metallic content may short out track circuitry. To check on this possibility, the study was carried out on the conductivity test on Singapore slag. The test results showed that the resistance of the slag is well above the minimum requirement of 2.0 ohms per 305 m. length as specified by the AREA.

Table 5 summarises all the results of tests conducted on the slag along-side with the AREA specifications.

As can be seen in Table 5, all the requirements for Singapore slag are well within the specifications set by the AREA. It is therefore obvious that there is a great potential for the slag to be used as railway ballast.

STEEL SLAG AS SUB-BALLAST

For track-section on embankments or cutting along the railway route, steel

258

Table 3   Gradations Specified by Taiwan Railway
Administration

Screen size, m.m.	60	40	25	15
% by weight passing	100	25-60	3-10	0

Table 4   Physical Properties of Railway Ballast

Property	AREA Specifications
Weight	1602 kg per cm^3
Los Angeles Abrasion Loss	40% max.
Sodium Sulphate Soundness Loss	10% max. (5 cycles)
Electrical Resistance	2.0 ohms min per 305 linear m. of track

Table 5   Summary of Test Results and the AREA Specifications

Property	Singapore Slag	AREA Specifications
Weight	1986 kg/cm^3	1602 kg/cm^3 min.
Los Angeles Abrasion Loss	34%	40% max.
Electrical Resistance	> 1000 ohms	2 ohms/305 m.
Compressive Strength	110 N/mm^2	NA

slag could be advantageously used as sub-
ballast.   The slag can be crushed and
supplied to meet the grading and
permeability requirements.   It can be
placed and compacted according to required
specifications. Since the steel slag sub-
ballast is purely granular in nature, the
effect of water content on compaction will
not be a concern and therefore compaction
during wet weather or rain should not
cause any problem. The steel slag will be
far better material for use as sub-ballast
as compared with sandy gravel, sand or
silty sand materials which are also likely
to be contaminated with considerable
percentage of fines and such fines are
prone to be washed away during rains along
with the draining water.   Better keying
effect of steel slag sub-ballast would add
to the stability to key the cross section
in shape unlike the other sandy materials
which are likely to be lost by erosion
etc.   The steel slag sub-ballast is
expected to be relatively more economical
as well.

CONCLUSIONS

1)   Steel slag from Singapore has all the
desirable qualities to be considered
seriously for use as railway ballast
material.
2)   For railway tracks within tunnels, it
is preferable to use ballast free track
while for tracks outside tunnels on ground
and for elevated tracks use of ballast is
still preferred.
3)   Ballast on ground is preferably to be
laid over a layer of non-woven geotextile
in order to provide better stability and
reduced maintenance.
4)   Use of steel slag as ballast in place
of granite would have the economic
advantage of using a much cheaper waste
material as well as deriving the technical
advantage of a material which meets all
the required specifications for such a
purpose.   Another indirect advantage is
that a waste material would be put to a
productive use and thus assist conserve a
useful depleting aggregate resource as
well as help solve disposal problem.
5)   Considering   the   limited   supply

available, steel slag can be recommended for use primarily at the yards and for replenishment purposes. Because of its high inter-locking and drainage properties the steel slag can be most advantageously used as sub-ballast or as ballast mat.

6) Use of steel slag as railway ballast is recommended from the point of view of national interest while not relaxing in anyway the standards of quality or specifications in its intended usage.

7) Steel slag being a waste product, its unit cost of crushing to the required gradation and supply to the site is expected to be much lower than that of granite ballast. Although steel slag because of limited availability may be used only as sub-ballast/ballast mat for certain section, considerable economic advantage may be derived.

REFERENCES

British Railways Board, Track Ballast Specification (Draft) 1983.

Cope, D.L., "Concrete Support For Railway Track: Precast And In Situ Slabs", Proc. Instn. Civil Engrs., Part I, 1982.

Eggleston, H.R., "Use Of Blast Furnace And Steel Slags As A Rail Ballast Material In USA", Utilisation of Steelplant Slags Symposium, Illawara, Australia, Feb., 1979.

Johnston A. W. Personal Communication, 1984.

Taiwan Railway Administration, The Specification Of Crushed Stone Ballast. 1983.

U.S. Bureau of Mines (USBM), 1952-1976. Minerals Yearboks, Slag-Iron and Steel, 1976.

ACKNOWLEDGEMENT

The authors express their sincere thanks to M/S Jurong Industries, Singapore a subsidiary of the National Iron and Steel Mills, Singapore for having supported the study and to have granted permission to publish the material contained in this paper.

Environmental Geotechnics and Problematic Soils and Rocks, Balasubramaniam et al. (eds)
© 1987 Balkema, Rotterdam. ISBN 90 6191 785 9

Geotechnical considerations in storage sewer operations

D.J.Hagerty
University of Louisville, Ky., USA

ABSTRACT: Many sewers in older cities are combined sewers which carry storm water and sanitary wastewaters. Consideration is being given to storing storm discharge in these pipes, with subsequent low-flow release to avoid overloading wastewater treatment facilities. Such storage has very serious geotechnical implications, explained in this paper, which can lead to failure of the pipe and overlying facilities.

1 INTRODUCTION

In many older cities of the world, existing sewer lines carry both stormwater discharges and sanitary or industrial wastewater flows. Attempts to add or improve wastewater treatment facilities thus are handicapped by the need to process the high flows which occur after storms, whereas the long-duration flow of wastewaters is of much lower magnitude. One alternative method of meeting the need to handle the high storm flow economically is to store the storm runoff temporarily, and discharge it later at a rate consistent with the processing capability of the wastewater treatment plant in the system. Utilizing existing sewer pipes as temporary storage reservoirs has been suggested as a low-cost means of obtaining temporary storage of runoff in combined sewer systems. This concept may appear to have low costs in a short-run analysis, but in the long run this technique may lead to failure of the sewer pipes and facilities over the pipes, with consequent astronomical costs for repair and replacement.

One of the most important considerations in this situation is the mismatch between the conditions for which older combined sewer pipes were designed and the conditions which will prevail in a temporary storage operation. The sewers were designed as long-duration low-flow, short-duration high-flow, low pressure systems. The sewers were designed to carry low flows of wastewater in most cases, with the wastewater flowing in the bottom of the pipe or in a special low-flow channel formed in the bottom or lower side quadrant of the pipe. The pipes were sized to carry the combined wastewater-runoff flow under full-pipe gravity flow, or under pressurized flow for only very short periods of time (after storms). The pipes were never intended to remain full at high pressures for prolonged periods of time.

Another factor of importance in the response of the combined sewer systems to the change in operational conditions is the general lack of appreciation of soil-structure interaction which existed at the time of the design and construction of the older sewer systems. In many cases, these systems were built at a time when the importance of good bedding and proper backfill was not recognized. Also, the crucial need for a suitable filter material around the pipe was not identified until after most of these systems were completed.

Finally, in most of the older combined sewer systems the joints between the lengths of pipe were not tightly sealed. Bell-and-spigot joints packed with mastic or a similar material were the rule, not the exception, for such systems. Such joints were not capable of resisting high pressures when placed, and the filler materials deteriorated seriously with the passage of time. This deterioration was not always of great significance because of the low pressures which occurred during dry-weather flow and because of the short durations of higher flows, but occasional failures of such systems as a result of infiltration and soil loss are a common feature of combined sewer systems wherever they have been installed. Modifications of the flow system in the combined sewers can have very detrimental effects on the system, and problems which were unimportant formerly may assume great severity.

1.1 Operational modifications

Addition of a treatment plant to a combined sewer system or modification of an existing plant to furnish a higher degree of treatment presents a design dilemma. The wastewaters for which the plant is required, constitute only a small fraction of the combined flow of wastewater and storm runoff which the sewers carry for relatively short periods of time after significant precipitation events. Providing the plant capacity to treat the high flows after storms is extremely expensive compared to the cost of providing only enough treatment capacity to accomodate the wastewater flows. One means of solving this dilemma is to provide some form of temporary storage for the large volumes of runoff, and to discharge the runoff and wastewaters through the treatment facility at a low rate over a long period of time. This low flow rate may be only slightly greater than the sustained flow rate for the wastewaters, and, consequently, the added cost of the needed capacity at the treatment plant (in excess of the capacity needed for the wastewaters alone) may be much less than the extra cost of facilities sufficient to process the combined flows without temporary storage.

The relatively conservative design of many of the existing combined sewer systems in older cities presents a sore temptation for present-day system managers. The sewer pipes were sized to flow full only under rather extreme circumstances. Thus, the sewer pipes themselves can function as temporary reservoirs for storage of storm runoff if appropriate valves and gates are installed in the system. The costs of such appurtenances obviously is much less than the cost of additional storage facilities, and the use of the existing pipes for temporary storage appears to be very advantageous and economical. However, temporary storage of storm waters in combined sewers can create conditions never considered in the design of those sewers.

The storage of combined flows after storms creates much higher pressures in older combined sewers than the design

water pressures. If water is stored in vertical risers such as man-holes in addition to the sewer pipes themselves, the resultant water pressures may be three or four times the pressure for which the pipe system was designed. Moreover, and perhaps more importantly, the duration of storage will be much longer than the previous duration of combined flow. This longer duration of pressurized conditions is extremely important with respect to leakage out of the pipe system. Because of the storage, gradients for leakage will be much higher than for the gravity flow system, and because of the long duration of storage compared to previous combined flows, the potential quantity of leakage will be much greater than for the gravity flow system.

Another change in system parameters compared to the original combined sewer operations is the increase in loading on the bedding beneath the sewer pipe because of the greater depth of water in the pipe over longer periods of time. Compared to the changes in pressures and gradients for leakage, this change is of much less significance. Nevertheless, the leakage may create changes of much greater importance in the loading on the pipe. The significance of the leakage lies in the fact that when sewers are modified to serve as temporary storage reservoirs, quite often the pipes are sealed to prevent only leakage <u>out</u> of the pipes.

1.2 Implications

The use of older combined sewers as temporary storage reservoirs is in essence a use of a facility for which the facility was not designed and in a way which may lead to failure in individual pipes or in the entire system if conditions are especially adverse.

The central difficulty in the use of combined sewers in this way arises from a lack of understanding on the part of system modifiers of the interaction between sewer pipe and the surrounding support media, namely the bedding below the pipe and the backfill around the remainder of the pipe. Relatively minor deflections of the pipe can be caused by small changes in loading on the pipe or by alterations in the bedding material beneath the pipe. The effects of so-called minor deflections may not be minor at all. Such deflections can create cracks in the joint sealant if that sealant is a brittle material and the cracks can lead to serious alterations in the flow system and the support system for the pipe.

2 GEOTECHNICAL CONSIDERATIONS

As mentioned above, the leakage of water out of the pipe system is much more likely in a sewer used for temporary storage than in a combined sewer used only in a gravity flow mode with no storage of storm waters over prolonged periods of time. This leakage can be very important to the continued functioning of the pipe, since it may cause changes in the stiffness and regularity of the support for the pipe itself. Changes in support can lead to greater deflections and cracking at joints, with subsequent increased leakage. The increased leakage can produce soil loss around the pipe and may lead to the ultimate collapse of the pipe.

2.1 Stiffness of support

The stiffness of the pipe support is a crucial element in the load-deflection behavior of a sewer pipe. In many older sewer installations this stiffness is quite variable because of the irregular quality of the bedding on which the pipes were placed. Even if proper bedding materials were placed at relatively uniform density under the pipes, the underlying subsoil beneath many sewer lines can vary significantly in stiffness as well as strength and permeability along the length of the pipe. In any case, even if support conditions were virtually uniform under the older pipes, operation of the system has led to changes in support because voids and soft spots have been created near the joints in the lines.

In the older sewer installations, joint sealing materials and techniques were far from perfect. Additionally, deflections in the pipes caused by the inequalities of support mentioned above were reflected in movements at the pipe joints. Loss of joint filling material and opening of joints led to leakage of water out of the pipes. In areas where the ground-water table was above the elevation of the pipe, infiltration of groundwater into the pipes occurred. Leakage of water out of the pipes saturated the soil zones around the joints and, in the case of cohesive soils, often decreased the soil strength and stiffness in those zones. More importantly, leakage of water out of the pipes usually led to later leakage

of water back into the pipes. This reversal of flow could occur because the greatest gradient for exfiltration was created after storms, when the pipes carried combined flows at pipe-full conditions. When combined flows receded or where the groundwater table was high, water flowed into the pipe. Inflow was important to the long-term behavior of the pipe in many instances because the concentrated inflow at joints often brought soil particles into the pipe with the infiltrating water. This internal erosion produced voids around the joints, which eventually could enlarge and undermine overlying surface facilities.

In most older combined sewer systems the effects described in the preceding paragraphs were concentrated at the joints and were not necessarily of great importance because the duration of exfiltration was relatively short. Of course, in systems where the groundwater table was high and the soil around the sewer was highly erodible (fine sand or silt), the effects could be much more dramatic. However, in most older systems, leakage around joints led to limited formation of voids, some weakening of support of joints, increasing deflection and widening of openings at joints, and, in turn, more leakage. The problem was most severe after periods of prolonged intense rainfall or other events leading to heavy runoff. It was seasonal in many localities because of the seasonal nature of runoff in those cities. For the most part, it was a hidden ever-increasing problem recognized only by maintenance personnel. Improvements in joint sealing materials and techniques, and in sewer bedding and backfilling practices have minimized the problem in sewers installed in recent years. In the older sewers, however, changes in operation to include temporary storage of combined flows could increase greatly the danger of system collapse. These systems have not been designed or modified, in most cases, to function properly under the changed conditions.

2.2 Joint sealing for storage

In the cities where older combined sewers are being modified for temporary storage, municipal budgets are already strained. If funds for sewer construction were ample and available, new sewers would be constructed and temporary storage would not be used. In such a situation, it is almost certain that the attention given to modification of existing sewers for temporary storage purposes will be minimal. The most common and the cheapest methods will be used to seal the sewers for use as storage reservoirs. Joint sealing most often will be done by the application of gunite, shotcrete or some sort of mortar to the inside of the pipes at the locations of joints. Minimal effort will be expended in injecting grout into the joints themselves.

The effectiveness of these sealing operations is dubious. The inexpensive sealing materials in common use all tend to be brittle, stiff compounds which crack at relatively low tensile strains. Such materials undoubtedly would be cracked if even small deflections occurred at joints in the sewer pipes.

Placement of joint sealing compounds on the insides of the sewer pipes also is not likely to create a durable long-term seal, since such placement may be effective against pressures exerted from within the pipe but will not be effective against pressures exerted by water in the soil surrounding the outside of the pipe. Many of the common sealing materials, if low cost, do not possess strong adhesive properties. Even in cases where grout is injected into pipe joints, the long-term effectiveness of the seal must be virtually perfect to prevent gradual exfoliation of the sealing materials and increased leakage.

2.3 Effect of Leakage

Leakage of water into soil zones around pipe joints may drastically alter the stiffness and strength of the soil if that soil was unsaturated and susceptible to softening (e.g., cemented soil or preloaded sensitive cohesive soil). Also, transfer of water into the soil around the pipe can create a serious change in loading conditions when the pipe is drained. In combined sewers modified for temporary storage of runoff and wastewaters, the longer duration of pipe-full flow compared to unaltered operations, and the much higher internal pressures created in storage operations lead to the transfer of large amounts of water out into the surrounding soil unless the groundwater table is already high in the surrounding soil. Discharge of the stored water from the sewer pipe takes place more slowly than discharge in unaltered sewers, but much more quickly than seepage of water into the pipe from the surrounding soil. Consequently, a condition is created in which considerable uplift is exerted on the emptied sewer pipe. This uplift condition is very different from the previously prevailing loading conditions for many combined sewers, and the resultant deflections in the pipe will be different from the deflections previously produced. Cycles of filing and emptying of the pipes thus can produce cycles of deflection, cracking of the joint seals, and opening of the joints of much greater severity than in any previous operations.

The locally high gradients around openings at pipe joints also can cause serious internal erosion in the soil (Terzaghi and Peck, 1967: 613-614). Most natural soils are susceptible to internal erosion (Aitchison and Wood, 1965; Ingles, 1968: 42; Sherard et al., 1977). Uniformly sized silts are most vulnerable to internal erosion, although fine sands and dispersive clays also are subject to this form of erosion (Sherard et al., 1972). Only well-graded granular soils are resistant to internal erosion (Sherard et al., 1984). This internal erosion is most widely known as "piping" because of the appearance of the voids and cavities that are created by the internal erosion if the soil possesses sufficient cohesion to support voids. Geomorphologists use the term "spring sapping" to describe the same phenomenon. Jones (1981) gives an excellent summary of the worldwide occurrence of this phenomenon and a comprehensive qualitative description of the mechanism. More quantitative descriptions are availabe in the engineering literature (e.g., Hagerty et al., 1986; Ullrich et al., 1986).

2.4 End result

The long-duration high pressures created during temporary storage of combined flows lead to exfiltration of large quantities of water around pipe joints except in situations where the groundwater table is much higher than the sewer pipe (in that case, infiltration already may be a serious danger to the pipe system). Rapid emptying of the sewer can produce very high inflow gradients near pipe joints, with consequent internal erosion of soil (and possibly bedding material) around the pipe joints. Creation of voids and saturated zones around the pipe joints changes the stiffness of the support system and modifies the soil-pipe interaction. Increasing and cyclic deflections in the pipe lead to cracking of joint seals and greater opening of pipe joints. High pressures in the water around the pipe cause intensified internal erosion as well as exfoliation of the sealant compound on the insides of the pipe. Voids may increase in size, pipes may be damaged, surface facilities may be undermined, and sewers may be partially plugged by infiltrated bedding and/or soil. Plugging, or the accumulation of debris in the bottom of the pipe, may lead to ponding of wastewaters, generation of gases of decomposition, and the creation of a serious explosion hazard. Problems with using combined sewers as storage reservoirs can be illustrated with a brief case history.

3 CASE HISTORY

The author was requested to inspect the interior of a section of sewer system in a medium-sized city in the Midwest United States. The purpose of the inspection was to determine the conditions of the joints between pipe sections, and to investigate the possible causes of any distress seen in the joints. The sewer pipes were 72 inches in inside diameter, and were installed in 1958. An asphaltic mastic compound was used to pack the bell-and-spigot joints of the pipes. In 1980, a monitoring program was initiated, with flows measured at selected locations in the system. Later, a scheme was developed to store combined runoff and wastewater flow in the main sewer lines to avoid overloading a wastewater treatment facility being added to the system. Storage of the combined flows would create as much as 25 feet of head in the section of sewer under inspection. Prior to use of the sewers for temporary storage, the system had been inspected, and deterioration of the pipes at joints had been noted; mastic was missing at many locations and some pipes were cracked near joints where deflections had caused compression in the alignment. The joints appeared to have been leaking significantly, as indicated by the presence of debris including bedding material and soil in the pipes at several locations. Some sections of the pipe had deteriorated so seriously that they were replaced. The soils surrounding the pipes in question were fine sands, sandy silts, and silts of alluvial origin. Flow of these soils into the sewer pipes was of such magnitude that a section of street had been undermined and had collapsed. At that location, sheetpile enclosures were built around the collapsed zone, the damaged pipe

sections were removed, and steel H-piles were driven to support the replacement sections of pipe. A contract was let to repair all leaking joints by the use of gunite applied to the inside surface of the pipe at all deteriorated zones and wherever mastic was missing from joints. The loss in flow cross section caused by application of the gunite was considered insignificant in comparison to the benefits of sealing the leaking joints. The work was completed in 1983. However, minor instances of collapse and deterioration of overlying pavements continued and, in 1984, a contract was let calling for the removal of all debris in the pipes, and repair of any defects at pipe joints. At joints where mastic was missing, the joints were to be sealed with a fast-acting mortar injected into the voids. Loose pieces of gunite were to be removed, and exposed voids were to be filled with the non-shrinking high-strength mortar. All work was done under the supervision of a representative of the municipal sewer agency. At several locations, infiltration was so significant that mortar application was difficult; at those locations, three-quarter inch-diameter pipes were installed in the leakage zones to act as "weep holes" to relieve inflow pressures and facilitate placement of the mortar. These pipes were installed at the suggestion of the inspector, and were not plugged, in all cases, after the mortar had hardened. Subsequent to the completion of this repair work, a controversy arose about the quality of the work because distress in the soil surface, sidewalks and pavements continued during and after the repair work. The author was asked to inspect the insides of the pipes along the lines where surface settlements were continuing, and to try to determine the cause of the continuing distress.

The inspection revealed conditions which were diagnostic of the cause of the distress. In the section of pipe where piles had been installed and new pipes had been placed, the joints were found intact and sealed, with no relative deflection between the three sections of pile-supported pipes. However, large relative deflections were found between the pile-supported pipe sections and adjacent soil-supported sections, and between soil-supported sections at other locations. Several thousand feet of sewer line were inspected, and a number of joints were found in deteriorated condition. At these joints, mastic was found exposed and bare, unsealed areas were found. Discolorations on the inside of the pipes, and the presence of large pieces of dislodged gunite in the bottoms of the pipes nearby, indicated that gunite had been removed from the inside of the pipe joints at those locations. All voids or gaps in the gunite seal coat were filled with mortar at the time of the 1984 repair work, under the supervision of the city inspector, so existing gaps (at the time of the 1985 inspection) indicated either a failure of the mortar off a repaired zone, or a failure of the gunite off the mastic or out of a void. The difference in color between the gunite and the mortar permitted easy evaluation and determination of which material had failed. With only one exception, the exposed areas of mastic or gaps showed no staining from the mortar, but showed characteristic discoloration from the gunite. At a number of locations where gaps were found in the seals, mortar staining around the outsides of the area indicated that mortar had been applied to cover gunite, but the gunite had failed between 1984 and 1985. At several locations where gaps in the joints were found, silty sandy debris was found along the bottom of the sewer pipe. At many other locations, mortar was found in patched zones and over older gunite, essentially intact and adhering to the surfaces to which it had been applied, but broken by cracks oriented perpendicular to the axis of the pipe. In other words, the cracks followed the alignment of the joints. At a number of joints, the mortar which had been applied to adjacent sides of the joint (i.e., to pipe ends on opposite sides of the joint) was found offset, indicating relative deflection of the pipe sections. These pipe sections had been installed in 1958, more than 25 years before the mortar was applied to the joints. It is highly unlikely that continuing long-term consolidation settlement of the soils under the pipes was responsible for the continuing deflections in the pipes. In several instances, new offsets between adjacent sections of pipe were as much as one-quarter inch; if ongoing consolidation settlements of the pipe had been responsible for these offsets, it would be logical to expect settlements of as much as 8 to 12 inches between the same sections of pipe during the previous 25 years. No such gross misalignments were found. Moreover, although the soil under the sewer lines is variable and more compressible in some spots than in other areas, such widespread occurrence of differential settlements of such magnitude (one-quarter inch in one year) is extremely unlikely. A much more probable explanation of the offsets and cracking in the pipes, and of the soil found in the pipes, is the occurrence of internal erosion in the bedding and backfill, as well as the natural soil, around the pipes. Evidence for such erosion was sought.

The surface areas above the sewer pipes in question showed considerable distress in sidewalks, curbs and street pavements. The cracks in the curbs, sidewalks, and streets "appeared" to be of considerable age, with relatively minor exposure of fresh, unweathered surfaces along the cracks. The appearances of the cracks was consistent with long-term occurrence of distortion coupled with much more recent movement. The recent movements, judging by the appearances of the cracks, were of magnitude equal to or greater than the previous long-term movements, in some places, while in other locations, the long-term movements appeared to have been of greater magnitude. This was consistent with the appearance of the sewer lines themselves, in which some of the joints showed offsets of several inches between pipe sections but no offset in the mortar placed over the joint (indicating long-term settlement but no recent deflection), while other joints showed fresh offsets in the mortar placed less than a year before the inspection.

Further inspection of surface areas over the sewer lines disclosed the presence of a number of sinkholes and voids over the pipes. Several of the sinks were excavated and were found to extend down to a joint in the sewer pipe. Near the location of one sink, a smaller tributary sewer pipe near the surface was found clogged with silty soil which had flowed into the pipe from around a broken joint; in this pipe, the flow patterns in the silty soil were still visible and migration of the soil into the pipe could be traced very easily. The sketch shown (next page) ilustrates one of the sinkholes found near the sewer.

The offsets found in the sewer pipes, the cracks at the joints, the inferred removal of sections of gunite, and the presence of soil debris in the pipes all indicated that internal erosion of soil around the pipes had occurred, with an increase in deflections at pipe joints, and further leakage and soil loss to be anticipated. It is significant, perhaps, that many of the cracks in the joint seals were found at the tops of

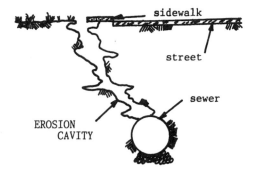

Figure 1. Erosion around sewer pipe.

the joints. Deflection of sewer lines
upward between locations of manholes
would induce joint-opening movements
at the tops of the joints, particularly
near the midpoints of the lines between
manholes. Leakage of water out of the
pipes during temporary storage of runoff
would have created saturated zones in
the surrounding soil and would have produced significant uplift forces on the
pipes when they were emptied. All of
the evidence gathered during the inspection indicated that the distress in these
sewers was not caused by simple failure
of mortar seals at pipe joints, but was
the result of a much more serious and
complex problem involving soil loss.
Continued leakage into the pipe system
was anticipated, with increasing soil
loss and undermining of surface facilities. Continued operation of the old
combined sewer line as a temporary reservoir for storm runoff will lead, in all
likelihood, to aggravated distress and
possible failure in the system.

4 CONCLUSIONS

Use of old combined sewers as temporary
storage reservoirs for runoff as well
as wastewater flow is ill-advised, and
likely to cause problems far more costly
than the expenditure necessary for
properly designed separate sewer systems.
Sewer line failure, or collapse of overlying facilities, would be much more
expensive than the design and construction of wastewater treatment facilities
with capacity sufficient to process
combined flows.

Variable support conditions under
older sewers has led to differential
settlements and deflections between pipe
sections, permitting leakage at joints.

Deterioration of joints exacerbates this
problem. Leakage at joints leads to
subsequent infiltration, with possible
inflow of soil into the pipe.

Temporary storage of combined flows
in old sewers creates water pressures
higher than the sewer joints were designed
to resist, for periods of time much longer
than the duration of pipe-full flow in
the unaltered sewer systems. These high
pressures for long periods of time aggra-
vate the outflow-inflow problem, and
intensify internal erosion of the soil
around the pipes. The erosion, in turn,
causes loss of support for the pipe,
and increased deflection between pipe
sections. Uplift forces caused by saturation of the soil around the pipes produce
further cracking and deterioration of
joint seals. Progressive failure of
undermined surface facilities, as well
as the sewer system itself, can occur.

This problem could reach severe proportions if existing plans for operation
of combined sewers as temporary storage
reservoirs are carried out in many of
the older cities of the world.

REFERENCES

Aitchison, G.D. & Wood, C.C. 1965. Some
interactions of compaction, permeability, and post-construction defloccu-
lation affecting the probability of
piping failure in small earth dams.
Proc. 6th ICSMFE, Montreal, 2: 442-446.
Hagerty, D.J., Spoor, M.F. & Kennedy,
J.F., 1986. Interactive mechanisms
of alluvial-stream bank erosion. Proc.
3rd Int. Symp. River Sedimentation,
Jackson, MS, p. 1160-1168.
Ingles, O.G., 1968. Soil chemistry rele-
vant to the engineering behavior of
soils. In C.K. Lee (ed.), Soil
mechanics-selected topics, p. 40-43.
London: Butterworths.
Jones, J.A.A. 1981. The nature of soil
piping-a review of research. Norwich:
GeoBooks.
Sherard, J.L., Decker, R.S. & Ryker,
N.L. 1972. Piping in earth dams of
dispersive clay. Proc. ASCE Spec.Conf.
Perf.Earth & Earth Supported Struc.,
W. Lafayette, IN, 1:589-626.
Sherard, J.L., Dunnigan, L.P. & Decker,
R.S. 1977. Some engineering problems
with dispersive clays. In J.L. Sherard
& R.S. Decker (eds.), Dispersive clays,
related piping, and erosion in geotech-
nical projects, ASTM STP 623, p. 3-12,
Philadelphia, ASTM.

Sherard, J.L., Dunnigan, L.P. & Talbot, J.R. 1984. Basic properties of sand and gravel filters, and filters for silts and clays. Jour.Geot.Engrg.Div. ASCE 110(6): 684-718.

Terzaghi, K. & Peck, R.B. 1967. Soil mechanics in engineering practice. New York: Wiley.

Ullrich, C.R., Hagerty, D.J. & Holmberg, R.W. 1986. Surficial failures of alluvial streambanks. Can.Geot.J., 23, 3, p. 304-316.

Environmental Geotechnics and Problematic Soils and Rocks, Balasubramaniam et al. (eds)
© 1987 Balkema, Rotterdam. ISBN 90 6191 785 9

Environmental aspects related to Sabarmati river bank structures

R.C.Sonpal
Applied Mechanics Department, L.D. College of Engineering, Ahmedabad, India

ABSTRACT: Old city walls on the eastern side of the Sabarmati river and the present form high-rise structures and residential colonies on the western bank of the river have proved recently an alarming situation for stability of foundations of structures in Ahmedabad. Washing out of residential colonies and approaches to bridges, imminent danger to stability of bridges and river bank structures were witnessed just before three years.

Based on studies for the river flow characteristics and bank deposits, analysis reveals adopting the reinforced earth structures for protection against loosing the stability of structures founded on river banks. This proposition besides providing an economical solution has extended water-front development, additional civic activity spaces on both banks and offers a well-integrated programme of environmental aspects related to river bank structures.

1 GENERAL

Development of Ahmedabad, the state capital of Gujarat has been quite phenomenal since last two decades. The city having orientation as Latitude 23° 02', Longitude 72° 37' average altitude 55m above M.S.L. is situated on the bank of river Sabarmati at about 80 km. north of the Gulf of Cambay. The river Sabarmati has a catchment area of 12953 sq.km. upstream of Ahmedabad and the average rainfall in the area is of the order of 60 cm. Index and location plans for the city are shown in Fig.1.

The rapid growth of city could be evident from the population figures stated in Table No.1. As per 1971 census, the population was 1.59 million and the same stands today near to 2.00 million. The growth of the city is largely due to the spread of industrialisation in the city and consequently this has led to heavy schedule on construction activity to cater to the needs of the industries, commercial establishments and the housing accomodation.

Table 1. Population of Ahmedabad city

Year	Population (in lacs)
1901	1.95
1911	2.30
1921	2.70
1931	3.83
1941	5.92
1951	7.88
1961	11.50
1971	15.94

2 GEOTECHNICAL DATA

Ahmedabad is an in-land city in the semi-arid zone of India and has the basic soils of alluvial deposits. Bore hole details at the seven bridge sites indicate the distinct layers of soils as river bed sand silt, soft clay, hard clay and sand layers. A typical bore hole details available at Railway bridge site is shown in Fig.2. River Sabarmati generally is dry except during monsoon season. Ground water levels have fallen considerably due to

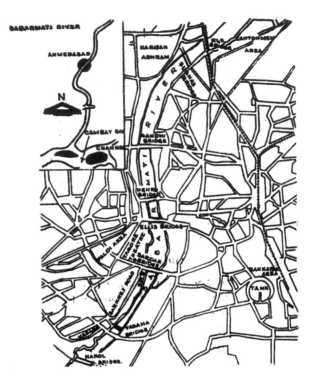

Figure 1. Index and location plans

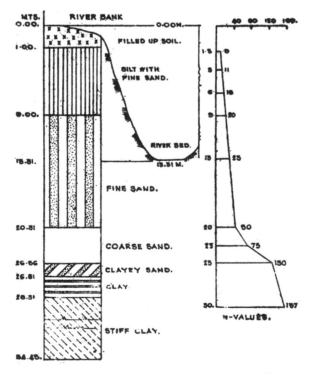

Figure 2. Bore hole details at Railway bridge site on right bank

heavy drawal of water by about 200 public tube wells grouped through tube well stations numbering around 40. The drawal from these bore is of the order of 196.15 MLD which forms nearly 73 percent of the total domestic water supply of the Corporation per day. Per capita supply is about 208 LPD and gets reduced to 163 LPD during summer when it is needed the most. In addition, most of the industries have their own private bores which number to 150 including both the large and small bores and drawing nearly 113.25 MLD.

For the city, far eastern side largely forms the industrial zone, central zone forming the old city wall premises comprises of commercial establishment and housing, high rise commercial structures recently established on both the banks of the river and the large number of housing societies established on the western side of the city. Topographical features of the city reveal low lying positions for the areas lying between Subhash bridge and Gandhi bridge interiors. Nehru bridge and Sardar bridge bank sites and a couple of local spots on the eastern side. On the western side, low lying areas are largely on the south-west of Sardar bridge side and are densely populated.

Bore hole details taken on many of the spots on both the sides of the river banks for foundation analysis of industrial and multistoreyed structures reveal the soils consisting of silt, silty sand, medium sand to coarse sand with kankar. Typical Index properties of these soils as available near Gandhi bridge west end are shown in Table 2 (3, 1976).

High rise structures and residential colonies which have come up since last about a decade and a half have not presented any problem so far as the stress transmission is concerned to the foundation soils. However, the meandering nature of Sabarmati river and the phenomena associated with it during floods have posed problems for the stability of foundations of high rise structures, bridge approaches, under passes and residential colonies situated in low lying areas.

3 1973 FLOODS

River Sabarmati forming the life line for nearly the then 1.8 million population and having sober flow characteristics turned into a devouring menstress in high floods on August 31 and September 1, 1973. Sabarmati in spate inflicated untold miseries on the population inhabiting on the right bank upto its tale on down stream. Flood waters skirted the residential colonies of low lying areas and eroded considerably the left bank from Ellis bridge to Narol bridge. Low lying areas were inundated by the floods waters on the right bank of the river. Buildings of Girnar Society that had inroad in the river bed had been swept away causing unhealing wounds of woes to the people who resided in the society. Vitrag Society under construction had undergone very heavy damages on the buildings, Fig.3.

Underpass provided below the Railway bridge on the western side was washed out, Fig.4. The Railway bridge pier on the eastern side of the river had also shown movement indicating the effect of floods. Narol bridge on the National Highway link had its approach completely missing during the floods, Fig.5. Eastern approach of Sardar bridge had also started giving way to the flood waters, Fig.6. Danger was imminent for stability of Nehru bridge, the widest (23m) balanced cantilever span bridge across the Sabarmati river, Fig.7. Many of the structures having locations comparatively at higher levels on the buildings largely due to the effect of flood water on the foundation stability, Fig.8. At Narol bridge on the National Highway link downstream of the barrage, at about 1.00 hours on September 1, 1973, the flood depth was of 3.11m which gradually rose to 6m within 6 hours and to 7.44m in another 10 hours time. The flood intensity was so high that the same level continued for nearly twelve hours. At Sardar bridge, the National Design Institute (NDI) was inundated completely and the road adjoining to it was under several meters of water. Tagore theatre, Fig.9, opposite to National Design Institute near Sardar bridge and residential colonies on low lying areas near bank, Fig.10; were also inundated

271

Table 2. Gandhi bridge west end auger hole soils index properties

Sr No	Depth of sample (m)	Sample No.	Field density g/cc	Field mositure content %	Grain size analysis				L.L. %	P.I.	Soil classification	C kg/cm²	∅
					Gravel %	Sand %	Silt %	Clay %					
1	1.50	MAHS31	1.61	15.60	0	28	65	7	31.6	6.2	Filled up soil		
2	3.00	MAHS32	1.71	14.29	0	29	68	3	28.3	3.6	ML	0.08	26°
3	6.00	MAHS33	1.50	13.00	2	33	60	5	26.6	4.2	ML	0.09	31°
4	9.00	MAHS34	1.50	05.26	0	95	05	-	-	-	SW	-	35°

Figure 3. Damaged Vitrag Society buildings

Fig.4. Washed out underpass

Figure 5. Missing approach of Narol bridge

Figure 6. Yielding approach of Sardar bridge

Figure 7. Nehru bridge across Sabarmati

Figure 8. Damages to river bank structures

Figure 9. Inundated Tagore Theatre

Figure 10. Residential colonies in low lying areas

and water started spilling the right approach of the Sardar bridge. The spill had crossed the Şarkhej road at Vasna village and traffic was clossed beyond it. The National Highway upto Sarkhej was under submergence. Paldi area near the river bank was completely inundated and flood waters had reached near to main Paldi-Sarkhej road. The New Wadej area was completely isolated. Within the old city walls, low lying areas of Khanpur and Raikhad were also inundated to more than 1.5 to 2 m of water.

4 DEVELOPMENT PLAN

Metropolitan Development Plan for the city is in the active consideration of the civic authorities. Ahmedabad, as it immerses as a major industrial city alongwith the developments, it brings with it many of the problems associated with industrialisation. Slums grow in increasing numbers, environment gets polluted through industrial operations, effluents are discharged in the river, smoke nuisance in the eastern industrial sector enlarges the tentacles and increased traffic further add to spoil the environment. Commercial zones enchroach upon earlier residential zones leading to high rise structures aggravating further the situation. With a view to preserve and improve upon the environment of the city and also to have a beautifying programme, a proposal of developing Sabarmati river front was mooted many years ago by the civic body of

the city. Before about a decade and a half, Town Planning Group, an informal citizens committee interested in welfare development of the city undertook this assignment, Bernard Kohn (1, 1967).

The proposal of Bernard Kohn envisaged the integrated planning and development of the Sabarmati river frontage with the area under consideration from Sardar bridge to Gandhi bridge. Present day development programme needs to extend this region on both the sides, on the upstream as well as downstream of the river from Sabarmati village to Narol bridge site.

5 PROPOSITION AND ANALYSIS

The proposal, at one time, of the civic authority was to have water level raised in Sabarmati river from 1.5 m as of Vasna barrage to around 3.5 m to develop a very good water fron for the city. This proposal had in its fold of analysis, the stability criteria for structures situated on the banks of the river as well as the structures situated on the low lying areas. Hence the protective work in the form of retaining walls either of R.C.C. or coursed rubble wall was considered on both the banks to provide adequate safety to the river bank structures and also the residential colonies of the low lying areas.

The proposition is to have protection work on both the banks now with the area limits from Sabarmati village to Narol bridge on western side of the river and from Cantonment to Pirana farm on eastern bank

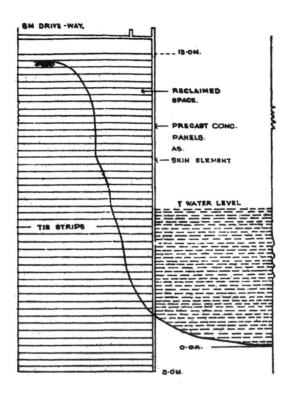

Figure 11. Proposed reinforced earth
retain wall

of the river. This would call for
analysing retain walls having the
running length of 1300 m. on the
eastern side and 1450 m. on the
western side of the banks of river
Sabarmati. The average height from
the bed of the river to the high
level marks on the banks is of the
order of 13 m. Solutions (2, 1973)
are worked out for the reinforced
earth retaining walls for providing
protection to the foundation of the
river bank structures and also to
provide a cover to the low lying
areas on both the banks.

The section of the reinforced
earth retaining wall with details
is shown Fig.11. Details for an
optimal design are worked out keeping in view the soils met with on
the banks for steel plate sizes and
their spacings. The whole project
when conceived with the reinforced
earth retaining walls on both the
banks provides an economy to the
tune of about 50 percent.

The present proposition is in

harmony with the earlier scheme of
water front development of the city
(1, 1967) and is extended to cover
additional areas on both the upstream side upon Sabarmati village
on western bank, cantonment area on
the eastern bank and on the downstream side to Narol bridge on the
western bank to Pirana farm on the
eastern bank of the river. This
incidentally will also provide
additional driveway lengths on the
banks with a very good water front.
Further the present scheme offers
a very sound solution for providing
adequate safety to the multistoreyed
river bank structures and also an
effective and reliable protective
cover to the low lying areas on
both the sides of the river bank.

6 CONCLUSION

Based on the studies for flow
characteristics of the river
Sabarmati and also the geotechnical

data of this city region, environmental aspects as related to the
stability of the multi-storeyed
structures and protective cover to
the low lying areas of the city are
appropriately dealt with in the
proposition of having reinforced
earth retaining walls on both the
banks of the river Sabarmati. The
cost is considerably reduced by this
proposition wherein the original
water-front development plan is
extended to have better drive-way
on the water-fronts, additional
civic activity spaces, a well
integrated programme of environmental
aspects related to river bank
structures.

REFERENCES

Kohn, B. 1967. Proposal for the
    integrated planning and develop-
    ment of the Sabarmati river
    frontage, Report of the Citizens
    Action Committee.
Lee, K., Adams, B.D. & Vagneron, J.
    (1973). Reinforced earth retain-
    ing walls. Proceedings of A.S.C.E.,
    SM & F Division, Vol.90, No.10,
    pp.745-764.
Sonpal, R.C., Thacker, K.C. &
    Prajapati, A.H. 1976. Foundation
    soil investigations for flyover
    bridge sites. Report No.IV,
    forwarded to Ahmedabad Municipal
    Corporation, Ahmedabad.

Environmental Geotechnics and Problematic Soils and Rocks, Balasubramaniam et al. (eds)
© *1987 Balkema, Rotterdam. ISBN 90 6191 785 9*

Cakar Ayam construction system – Pavement and foundation system for structures on soft soil

Rijanto P.Hadmodjo
PT. Cakar Bumi, Jakarta, Indonesia

1 CAKAR AYAM CONSTRUCTION SYSTEM

1.1 Introduction

During the last decades many new foundation system have been developed using varied and sophisticated equipment. All of them have the same goal, i.e. more bearing capacity, quicker and easier construction, increased economy and fewer problems with neighbouring buildings. But very few, if any, of the system are foundation systems which can solve without underlying harder layers.

The system describes here is one that is designed specifically to meet this "soft soil problem". It is a simple method with a good bearing capacity, named by its inventor the Cakar Ayam (literally Hens Claw) foundation.

1.2 Cakar Ayam in brief

The Cakar Ayam construction sistem was invented by the late Prof.Dr.Ir. Sedijatmo in 1961, when he was constructing high-tension electric towers in the swampy Ancol area, to supply electric power to the Asian Games facilities. Time was running short, so he had to find a new way to complete this job in time.

The Cakar Ayam construction system consists of a reinforced concrete slab of 10 to 15 cm thick, depending on the load to be borne and the condition of the soil underneath. Underneath the slab reinforced concrete pipes are fitted monolithically with a distance between the axes of each pipe of 2.00 to 2.50 meters, while the length of these pipes varies between 1.50 to 3.50 meters. The length of these pipes also depends on the load and the soil condition. The pipe diameter is 1.20 or 1.50 meters and its wall-thickness is only 8 cm.

The reinforcement of the pipe is single, for the slab double (see Figure A).

In principle Cakar Ayam construction is applicable on any kind of soil, but experience indicates that it is most economical if applied on soil with a bearing capacity of 1.50 to 3.50 tons/m^2. Where it remains economical, however, we have also applied on harder soil.

1.3 Basic Idea

The basic idea of the Cakar Ayam construction system is to utilize the soil characteristic that is not properly utilized by by other types of construction or foundation, namely passive-soil pressure.

The thin concrete slab floats on the soil and its stiffness is obtain through the pipes underneath; the pipes stay vertical because of the passive-soil pressure in the soil. This combination gives the slab and the pipes in the soil a stiff construction that does not easily bend.

If we compare this system with a slab with stiffening beams the latter gets its stiffness through the concrete beams, while Cakar Ayam construction obtains stiffness through passive-soil pressure. It is clear that this will mean a reduction in volume of concrete and consequently greater economy for the same bearing capacity.

It is important to note that the function of the pipes under the slab is not to support the slab as in the case with a pit-foundation, but to keep the slab stiff and flat; these pipes do not reach the hard layers under the slab for support, but rather hang on the slab. Thus the main difference between the pit-foundation and the Cakar Ayam foundation is that the pipes of the latter one are slab-stiffeners and not slab supporters.

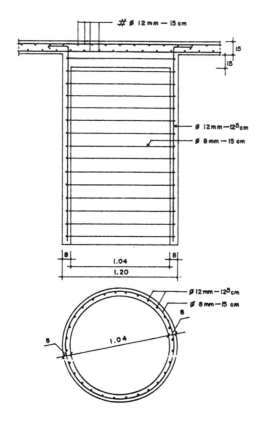

Figure A

1.4 Ease of construction

We have applied three methods in constructing this system, namely:
a. manual
b. semi-mechanical
c. mechanically
In constructing this system, especially in remote areas and on building sites with no accessiblity for equipment (like high-tension electric towers in swampy areas or wet paddy fields), the manual method is the fastest and most economical. The concrete pipes are cast on the spot where they will be embedded. Steel formworks for casting the pipes and portable concrete-mixers are required. The embedding of the pipes is done by digging out the earth inside the pipes so that the pipes go down into the pit by their own weight. The pouring of the slab is also an easy job. Since passive - soil pressure is crucial for the stiffness of the slab, care should be taken that soil outside the pipes remains undisturbed.
Also direct contact between the pipe-wall and the earth outside the pipes should be maintained. This is not difficult because one is mostly dealing with soft soil.

Where the site is accessible for building equipment, a choice between the three methods or a combination may be made, whichever is the most economical as regards time and cost.
The mechanical method was developed recently during the construction of Jakarta International Airport, Soekarno - Hatta.
The pipes are fabricated in a specially build workshop on the site and transported to the location. A specially designed device called a CHADU drills an annular hole in the ground, exactly the size of the pipe. CHADU stands for Cengkareng Hydraulic Auger Drive Unit (Cengkareng is the name of the building site). Another device name a CHUPP grabs the 1.5-ton pipe and presses it into the annular hole, where it fits neatly. CHUPP stands for Cengkareng Hydraulic Unit Press Pile. These two devices together make one set and can embed twelve pipes an hour or around 120 pipes a day, which was the target of the project.
The ideal method is to insert the pipe into the ground without making a hole before-hand so that the pipes make full and firm contact with the soil. This method is feasible if the soil is soft enough.

CHUPP and CHADU in action

1.5 Load and settlement

For equally distributed loads the Cakar A-yam is particularly useful. But as building foundation it has to support concentrated loads through building columns which may reach around 500 or 600 tons.
In such cases, we apply concrete footing under the column to reduce the shear in the slab to permissible values.
As building foundation the Cakar Ayam system has been successfully applied for various constructions (see Appendix I below).
Like every shallow foundation the construction system undergoes settlement within

278

allowable ranges. It will settle less than the conventional full slab and beam foundation because the pipes under the slab improve the bearing capacity of the soil by preventing it from moving sideways. We can compare this system with an open concrete-box which is turned upside down.

Experience with the projects mentioned in Appendix I is that no excessive settlement has occured so far.

Another advantage is that this system considerably cuts construction costs.

For a four-storey building in Jakarta where 20 to 25 meters long concrete piles are used, for example this system reduces the building costs of the foundations by around 30 %.

2 CAKAR AYAM CONSTRUCTION AS PAVEMENT

2.1 Principle

The computation principles are now considered to improve understanding of this system.

Figures 1a and 1b demonstrate clearly the forces and the moments below the slab or pavement created by an excentric concentrated load Q on the slab.

The excentric load Q (fig.1a) is assumed to be replaced by a concentrated one Q_1 at the centre of the slab with a moment $M = Q_2 \times \frac{1}{2}L$ where L represents the width of the slab (fig.1b), while $Q = Q_1 = Q_2$.

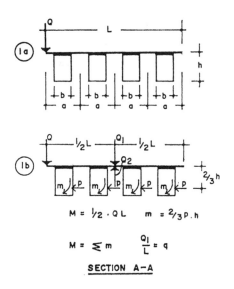

$$M = \frac{1}{2} \cdot QL \qquad m = \frac{2}{3}P \cdot h$$

$$M = \leqslant m \qquad \frac{Q_1}{L} = q$$

SECTION A-A

Because of the concentrated load Q, an uniform ground counter - pressure is created $q = Q_1/L$, while the moment M is kept in balance by moment $m = 2/3 \times P.h$, where P, created by M, represents the passive - soil pressure working at each pipe such that: $M = m$ or $\frac{1}{2} Q.L = 2/3 P.h$, where h is the height of the pipe. Due to these forces and moments the slab remains flat, while below a uniform ground - pressure is created in spite of the excentric concentrated load Q.

With the help of a simple artifice from

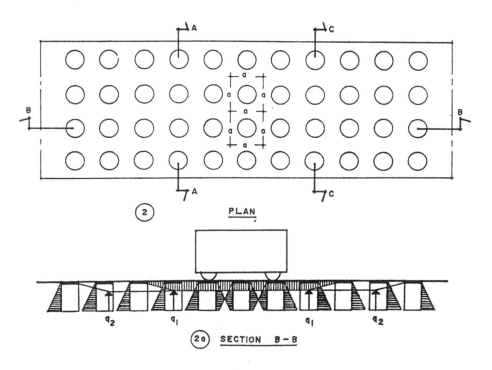

PLAN

SECTION B - B

279

the above moment-equation a formula can be derived which includes the mutual relations between the dimension of the slab, the pipes and their respective distances, as well as the properties of the soil and the ground pressure.

$$b.h^3 = \eta.a^3.y ; \qquad y = f(q, \gamma, \lambda), \text{ where}$$

a = distance between pipes in meters,
b = the outer diameter of the pipes in meters,
h = height of the pipes in meters,
γ = the specific weight of soil,
λ = soil-contact, dependent on the internal angle of friction,
q = the ground counter-pressure in tons/m^2,
η = the safety coefficient, amounting to between $1\frac{1}{2}$ and 2

2.2 The bearing ability of the pavement/ road-surface

Figures 2, 2a and 2b represent the plan, the longitudinal and cross - section of a road according to this system.
They show the shape of the ground counter-pressure, here simplified as a trapezium owing to the weight of two trucks placed side by side on the road, and also the shape of the passive ground-pressure againts the pipes.

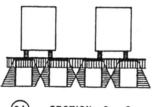

(2b)   SECTION C - C

The hatched trapezium q, represent the uniform ground counter-pressure below the slab created by the weight of two trucks of, say, 20 tons each (bruto weight), total 40 tons while the unhatched trapezium q_2 represents the uniform ground counter - pressure below the slab due to the weight of two trucks of two times heavier bruto weight each, i.e. 40 tons, total 80 tons.

In both cases the uniform ground counter pressure q is (more or less) the same, which means that the trapezium q_2 is twice as large as trapezium q,.

It is evident the theoritically the bruto weight of the trucks can be increased infinitely without the ground counter-pressure becoming greater, since the slab has been calculated based upon a predetermine q, for example 500 kg/m^2.

One of the most important benefits of the road-surface of the system is that increasing traffic weight will not necessitate any reinforcement given that the increasing traffic weight will be distributed over a larger road-surface.

2.3 Settlement of the road-surface

As is well known, the road-surface of highways or expressways is calculated to withstand uniform mobile weights of 400 to 500 kg/m^2, including shocks, which is equivalent to the maximum mobile weight applicable to first-class bridges.
For airport-runways uniform load of 1 ton/ m^2 should be assumed.

If the ground pressure below the shoes of a playing child is assumed to be 0.1 kg/ cm^2 or 1,000 kg/m^2 (which can be easily checked), then the maximum allowable ground pressure below the road-surface of this system will be only half of these figures.
The conclusion can be drawn that there will not be any sag (settlement) under the road-surface caused by traffic. Soil sag below the road-surface is then only possible if there is settlement of the original soil caused by the weight of the embankment of the road, or by the sag of the soil used for the embankment (which is also the case for conventional road-surfaces), but not due to the traffic load.

For the sake of completness it should be mentioned that the slab according in this system is only 8 (eight) cm thick, even if it is calculated to bear a relatively high load of 500 kg/cm^2.

2.4 Absence of expansion joints

A big drawback of conventional concrete roads is that expansion joints have to be provided to avoid cracks caused by shrinkage due to the large differences in temperature between the concrete slab and the base below.

The most important advantage of this construction system is the total absence of expansion joints. This is made possible by three important factors:
a. The presence of the pipes, in this case acting as a very effective anchorage of the slab to the subsoil.
b. Owing to the existance of hollow open pipes, the contact-surface between the concrete part of the construction and the soil becomes larger, about 80%, and can thus be almost eliminated, especially if an asphalt concrete layer of 3 to 5 cm be laid on top of the concrete slab.
c. Due to the reinforcement, possible ten-

sions caused by shrinkage will only create fine hair-cracks over the length of the slab. Even with the application of close double-net reinforcement the consumption of steel remains extremely low, at around 60 kg per cubic-meter concrete.

2.5 Conclusion

The major advantages of the Cakar Ayam system are as follows:
a. The almost unlimited bearing capacity and consequently also the long durability of the road-surface, which will have a great impact on the economics of transport by road, for example reducing considerably the cost per ton per kilometer.
b. The danger of sags and undulations of the road-surface owing to traffic is non-existent since the pipes keep the slab completely level. The soil below the slab, between and inside the pipes becomes more compact under the influence of vibrations caused by the traffic.
c. The expensive and elaborate maintenance of expansion joints required for the existing roads is obviated.
d. The system is applicable to any kind of soil except rock and hard soil with the advantages mentioned above.

3 APPLICATION FOR AIRPORT-PAVEMENT AND ROADS

The applications of the system so far include:
  1. Apron-extension pavement, Juanda Airport, Surabaya, in 1979
  2. Runway-extension pavement, Polonia Airport, Medan, in 1981
  3. All pavement at Jakarta International Airport, Soekarno-Hatta, in 1980-1984
  4. Second apron-extension pavement, Juanda Airport, Surabaya, in 1983
  5. Access-road pavement, from Jakarta to JIA Soekarno-Hatta, in 1983-1984
  6. Apron pavement of Garuda Maintenance Facilities, at JIA Soekarno-Hatta, in 1985

Now in preparation: the taxiways and apron for the second terminal of Jakarta International Airport, Soekarno-Hatta. The total slab area is around 2,150,000 square meters
  Given their special characteristics pavements no. 1, 3 and 5 deserve further brief discussion.

3.1 Juanda airport pavement extension

The construction of this Cakar Ayam pavement was started in April 1979 and was completed in August the same year. It is only 100 m wide and 120 m long. The original soil was inclined and silty sand was used as filling material. The production and the embedding of off pipes were carried out 100% manually while for the pouring of the slab concrete-train and finishers were used.
Unlike the other airport pavements, slabs, where double-layer reinforcement was installed, only single-layer reinforcement was used in the slab here. The intention was to prove that the moment in the slab was very small, the most important role being played by the tensile stress which has to be met by the reinforcement and the shear stress which is taken care of by the thickness of the concrete. The compressive-strength of the concrete was about 300 kg/cm^2, while mild steel U24 was used as reinforcement.
It emereged that using mild steel as reinforcement was not appropriate due to the small diameter of the steel, which consequently very flexible. This is why for all the other pavements, high - tensile - strength (HTS) steel (U50) wire-mesh is installed. Another charactristic is that there are no dummy joints in the slab, so that this 12,000 m^2 slab forms one jointless concrete pavement.
  With regard to the moments in the slab, the following can be noted. After three months fine cracks appeared as an anticipated. The size of most of the cracks was ¼ mm, with some wider, but not exceeding ½ mm. Now after six years of operation some cracks are wider (about 1 mm) but the rest remain the same and require no significant repair. The conclusion is that almost all the cracks are shrinkage cracks. No cracks due to moments in the slab are observed and the whole apron functions well.
  Twenty months after completion, ten loading-tests were carried out at various locations, on the slab with cracks as well as without, between four, between two and right above the pipes. The load consisted of 80 tons on a loading plate of 30" diameter. The test results showed that the bearing capacity of the pavement is still very good, with an average diraction of 4 mm.

3.2 Jakarta International Airport, Soekarno - Hatta

This new airport is located in Cengkareng, some 20 km west of Jakarta. Originally the site was occupied by fish-ponds, paddy fields, low swamp and a small area of dry land. The original design by Aeroport de Paris, advocated a sand-filling of about 3 meters above the existing level to increase teh soil-bearing capacity, which meant a fill-material of about 8 million m^3.

After the government decision to apply the Cakar Ayam construction system to all pavement rather than construction conventional flexible pavement, a new design was made for the final level of whole project. The Cakar Ayam does not need heavy sand-filling.
Cut and fill supplies because the ground-level is still above the flood-water level. Only 1.5 millions m³ fill-sand were needed.

Three French contractors cooperating with an Indonesian contractor won the tender late in 1980 and had to finish the job within four years. Aeroport de Paris did the general supervision and for the Cakar Ayam construction the supervision was entrusted to PT Cakar Bumi.

The technical data of this projec can be found in Appendix 2, while Figure 3 gives the details of the construction.
Previous to the construction of this project a full-scale tests had been conducted.
A special test-slab was made on the design to the actual pavement. A range of the gauges and meters were installed to enable us to monitor the behaviour of any part of the structure. As test-loads hydraulic jacks on four plates representing the four tire-prints of a Boeing B-747 were used to simulate one main landing-gear of this jumbo-jet. Two sets of these jacks were placed at a distance similar to that of real airplane main landing-gear.

The test results can be summarized as follows:
a. The direction of the load-composition in relation to the Cakar Ayam pipe-position does not matter; the Cakar Ayam is thus an almost isotropic system.
b. Static loads with two main landing-gears of 90 tons each, equivalent to a full B-747 load, gave a deflection of not more than 2.54 milimeter.
c. Repetitive loads with two main landing-gears of 90 tons each with 1000 repetitions @ 1 minute gave a deflection of 2.348 milimeters ± 0.64 milimeters.
d. A static load on single wheel load (¼ of one main landing-gear) of 150 tons (equivalent to one main gear load of ± 450 tons) gave a slab deflection of 10 milimeters with no cracking.
e. The test-slab can withstand three times or more than the load of a full loaded Boeing B-747 without any problem.

The construction of the pavement went smoothly and using the mechanical method we were able to finish on time (four years).

3.3 Access road Jakarta to Soekarno-Hatta Airport

This is the first application of the Cakar Ayam system as road pavement. Most of the

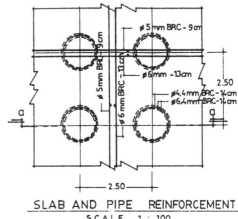

SLAB AND PIPE REINFORCEMENT
SCALE 1 : 100

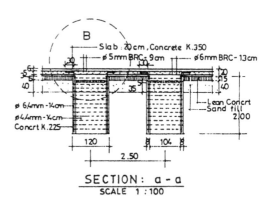

SECTION: a - a
SCALE 1 :100

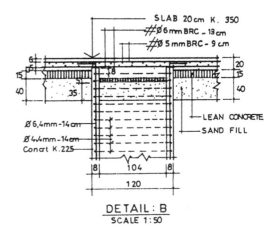

DETAIL: B
SCALE 1:50

area where the road must go are swamp and soft land with fish-ponds here and there.
To obtain a flood-free level an embankment of one to three meters high must be constructed. The access road, 13.50 km long, has a double pavement 10 meters wide with a median between. Most of the fill-material for the embankment is coral-sand, transport-

ed from the island, north of Jakarta.

At some places where the soil is very soft and the fill is high (3 meters or more), we were forced to lay geotextile before starting pouring the snad fill. The purpose of this geotextile is to obtain a firm layer and to minimize the quantity of fill-material. The whole job was done mechanically by an Indonesian contractor with experience in this field. The technical data of this project can be found in Appendix 2.

In all the pavements-slab which form the continuous reinforced concrete slab, fine hair-crakcs have developed, mostly in transverse direction. we can divide these cracks into three catagories:

a. fine cracks, smaller than 0.5 mm wide
b. medium cracks with a width 0.5 mm to 1.5 mm
c. wide cracks, wider than 1.5 mm.

In this road project we have found that of all these cracks 94% belong to the fine, 4% to the medium and only 2% to the wide catagories.

Based on the American Manual of Conrete Practice (1976) and the British, Guide to Concrete Road Construction issued by the Department of Transport and Road Research Laboratory, and our previous experiences, we maintain the slab to as follows:

1. For the fine cracks we do nothing.
2. On the medium cracks we put sealant to prevent water penetration.
3. The wide cracks are repaired by widening the cracks to 6 mm - 10 mm, and then filling them with epoxy material.

So far these methods of maintenance have been reasonably successful.

This paper concerns a relatively new construction system which is relevant to construction problems on soft soil. It is hoped that it will find application in countries outside Indonesia in areas where soft-soil problems are to be met.

Jakarta, September 1985

APPENDIX I
List of constructions using
The Cakar Ayam System

1. Hundreds of electric high-tension towers in Jakarta, Palembang, Medan, Banjarmasin, Central and East Java
2. 40 m^3 and 100 m^3 watertank-towers
3. A 35-meter high Catholic church, Jakarta
4. Four-storey office building for the Bank of Indonesia in Pakanbaru, (floorspace 3,000 m^2)
5. Three-storey office building for the State Electricity Authority in Jakarta with 2,800 m^2 floorspace
6. Four aeroplane hangars with 64 m span each, in Jakarta and Surabaya (total floorspace 4 x 5,120 m^2)
7. Slipway in Tanjung Priok, Jakarta, with 1,200 m^2
8. Gas-turbine electric power station, in Palembang (floorspace 900 m^2)
9. Custom office building in Semarang, with 2,500 m^2 floorspace
10. Two bonded ware-houses in Tanjung Priok, Jakarta (floorspace 2 x 2,800 m^2)
11. Six-storey office building for Department of Justice, Jakarta (2,600 m^2)
12. Pavement for Juanda airport, apron extension with 12,000 m^2 slab
13. Four-storey office building for Cipta Karya, Surabaya with 1,900 m^2 floorspace
14. Pavement for Polonia iarport, runway extension with 30,000 m^2
15. Swimming pool and tribune, Samarinda, with 1,000 m^2 pool and 600 m^2 tribune
16. Three-storey office building for Customs in Surabaya with 3,000 m^2 floorspace
17. Six-storey building for Governor of Samarinda, East Kalimantan (8,400 m^2)
18. Five-storey office building for Jarum Cigarette Factory, Kudus, Central Java with 4,800 m^2 floorspace
19. KONI swimming poll in Surabaya, with 1,000 m^2 pool
20. Four-storey rental office building in Jakarta with 3,000 m^2 floorspace
21. Two runways, taxiways and aprons of Jakarta International Airport, Soekarno-Hatta, Cengkareng (1,200,000 m^2 pavement)
22. Pavement for the second extension in Juanda airport, Surabaya (30,000 m^2)
23. Pavement for access-road from Jakarta to JIA Soekarno-Hatta, 13.5 km long and 2 x 10 m wide
24. Pavement for the aprons of Garuda Maintenance Facilities area, in JIA Soekarno-Hatta (270,000 m^2 pavement)

Technical data of the Cakar Ayam
Airport and Road Pavement

	Jakarta International Airport Soekarno - Hatta	Access-road Prof.Dr.Ir. Sedijatmo
1. SLAB		
1.1. Concrete characteristics		
- Aggregate size	0 - 20 mm	0 - 3/4"
- Water-cement factor	0.40 - 0.45	0.45 - 0.50
- Slump	Max. 5 cm	Max. 5 cm
- Compressive strength (after 28 days)	350 bar	350 bar
- Flextural strength (after 28 days)	45 bar	45 bar
1.2. Slab area		
- Total Slab area	1,175,070 m²	261,585 m²
1.3. Concrete volume		
- Total volume	257,810 m³	30,595.50 m³
1.4. Reinforcement (U-50)		
- Upper wire-mesh	ϕ 5 mm - 9 cm	
- Lower wire-mesh	ϕ 6 mm - 13 cm	
- Single wire-mesh		ϕ 5 mm - 9 cm
- Total steel weight	9,123 tons	941,7 tons
2. CAKAR AYAM PIPE		
2.1. Concrete characteristics		
- Aggregate size	0 - 14 mm	0 - ½"
- Water-cement factor		
• mechanical	0.50 - 0.60	0.35 - 0.45
• manual		0.60 - 0.65
- Slump		
• mechanical	0 - 2.50	
• manual		5 - 10
- Compressive strength (after 28 days)	225 bar	225 bar
2.2. Total Cakar Ayam pipes		
- 2.00 meter long	166,050 pcs	23,535 pcs
- 2.20 meter long	35,550 pcs	
2.3. Reinforcement (U-50)		
- Vertical	ϕ 6.4 mm - 14 cm	ϕ 6.4 mm - 14 cm
- Circular	ϕ 4.4 mm - 14 cm	ϕ 6.4 mm - 14 cm
- Weight per pipe	21.76 kg/2m pipe	21.76 kg/2m pipe
- Total weight	4,470 tons	461.29 tons
2.4. Concrete volume (per pipe)		
- 2.00 meter pipe	0.563 m³	0.563 m³
- 2.20 meter pipe	0.619 m³	
3. LEAN CONCRETE		
3.1. Concrete characteristics		
- Aggregate size	0 - 20 mm	0 - ½"
- Water-cement factor	0.65 - 0.70	0.60 - 0.80
- Slump	7.50 - 12.5 cm	10 cm
- Compressive strength (after 28 days)	500 pci	100 bar
3.2. Concrete volume		
- Total volume	134,851 m³	13,229 m³
4. DUMMY JOINTS		
- Class	Spandex	Fuel resistant sealant
- Depth	50 mm	25 mm
- Width	(13 + 3) mm	(13 + 3) mm
- Distance	15 m	Max. 14 m
- Total length	187,70 m	

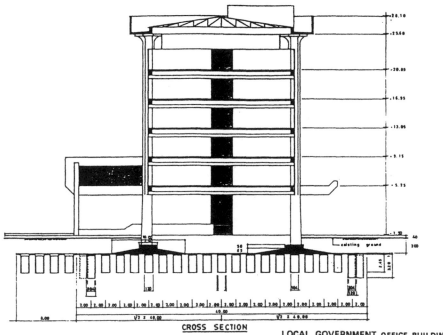

CROSS SECTION

LOCAL GOVERNMENT OFFICE BUILDING SAMARINDA
EAST KALIMANTAN.

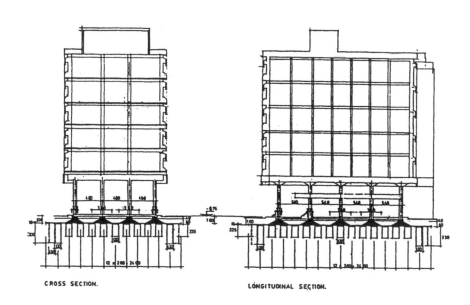

CROSS SECTION.

LONGITUDINAL SECTION.

CAKAR AYAM FOUNDATION
FOR MIN. OF JUSTICE BLDG JAKARTA
Drwg. no. TA. 01001

285

Environmental Geotechnics and Problematic Soils and Rocks, Balasubramaniam et al. (eds)
© 1987 Balkema, Rotterdam. ISBN 90 6191 785 9

Experimental study on concrete block retaining walls

K.Makiuchi
Nihon University, Funabashi, Japan

S.Hayashi & T.Yamanouchi
Kyushu University, Fukuoka, Japan

S.Mino
Sumitomo Construction Co., Tokyo, Japan

ABSTRACT: Investigations were conducted into the behavior of two concrete block retaining walls, one a conventional T-type block wall and the other a newly developed H-type block wall. A deformation and a failure pattern in the retaining walls subjected to both surcharge loads and back-fill water pressure were measured by a prototype field test. It was observed that almost the same earth pressure and displacement were generated on both block retaining walls. The experimental results also indicated that the H-type block retaining wall was able to stand a larger deformation than the T-type and had a higher toughness.

1 INTRODUCTION

Concrete block retaining have been used for sheathings and revetments together with the concrete retaining walls as a substitute for conventional stone masonry from around 1960 in Japan. They have excellent characteristics such as stable quality and supply, easy work, low cost, etc. The concrete blocks were standardized by JIS (Japanese Industrial Standard) in March 1974. However, since JIS specified only the dimensions of block surface and depth, weight, and quality of cement concrete, various attempts to improve the shape of the blocks have been tried. Many types of counterfort (supporting structure in depth) of blocks are employed at present; there are about 750 types of the blocks in Japan. Among these blocks, two types are widely used. One is the T-type block, which has the same shape as the T-type rock (*Kenchi-ishi*, a conventional pyramid or T-shaped stone for masonry in Japan), and the other is the H-type block, which has a rear edge length the same as its front face.

Block retaining walls in many places have been broken by; e.g., the earthquake near Izu-Ohshima in 1978, and it emerged as a problem that the protective costruction works had been done based on experience only, without the basic design and construction method, including earth pressure and strength of concrete block retaining walls being validated. In particular, it was pointed out that the basic factors related to the collapse of the block retaining walls were the bearing capacity of the ground, bonding strength between base footing, block and joints, sound building work in body-filling and back-filling, quality control, etc. However, it is difficult to treat such factors theoretically; thus there are still many problems in the rational design of the block retaining walls. It is necessary, therefore, to obtain reliable quantitative data through various experiments and research.

The purpose of the experiment here was to compare the behavior of the T-type and h-type blocks (including efficiency of placement of base footing, body-filling, back-filling, etc.) through prototype experiment in the field, and to study a deformation and failure mechanism in block retaining walls subjected to both earth pressure and water pressure.

2 TESTING FACILITIES AND METHOD

2.1 Testing ground

The testing ground was a section (about 50 m × 20 m) of the excavation ground for decomposed granite soil in Maebaru, Fukuoka city. The plan of the testing ground, and the front view of the block retaining wall and shown in Fig. 1. The T-type and H-type block retaining walls were on each side of the partition wall. The height of each wall was 2.0 m and the length was 12.0 m (Length of base footing 16.0 m).

For the back of the retaining walls, decomposed granite soil (*Masa-do*; natural soil in excavated ground) was embedded, and

10 m thick of sand was spread on the loading area. To parallel fall of the retaining walls (assuming infinite length), the front counterfort was placed on the heads of the blocks at the partition wall, and two impervious vinyl sheets were laid under the ground at each end to prevent water leakage a erosion and flow of fine soils caused by deformation of the retaining walls.

2.2 Section retaining walls and construction method

A sectional view of the retaining walls is shown in Fig. 2. The relatively hard ground of weathered granite was excavated down to 40 cm, where cobble stones and foundation concrete were placed. Then both T-type and H-type blocks were built up with a gradient of 1:0.2. The block walls were constructed by means of uncoursed masonry, and block installation, body-filling, and back-filling were performed as far as possible according to the actual work procedure.

The back ends of the second and higher layers of the T-type block were supported by chockstones to form the back frame, then the space between the blocks was filled with concrete. After the concrete had been cured to a certain level, the back frame was removed and the back-filling sand banking soil were compacted. Since the blocks had to be piled one layer at a time and it took

time to make and remove the back frame and to cure the concrete, the construction progressed very slowly. The standard block piling speed was 1 to 1.5 layer/day.

On the other hand, the H-type block has the same face edges on the front and rear ends, thus no time for installing chockstones and making the back frame is necessary. In addition, several layers of blocks can be filled with concrete at a time. However, they were piled three layers to compact the back-filling sand and the embanking soil under the same condition as for the T-type block.

For the body-filling concrete, aggregate with a maximum diameter of 40 mm was used, and the slump was 8 cm ($\pm$ 2 cm) and σ_{28} was 16 MPa. No water drain holes were made so the water level in the back-filling sand would rise evenly.

2.3 Measuring instruments

As shown in Fig. 2, the piezometers, earth pressure cells, and displacement devices were installed on or in the block retaining walls. Fig. 3 shows the positions of the piezometers, pressure cells, and reading lines. As the piezometer, a transparent vinyl pipe with an inside diameter of 4 mm was used, and a porous stone was attached to the pipe end. Seven piezometers were installed on the H-type block wall.

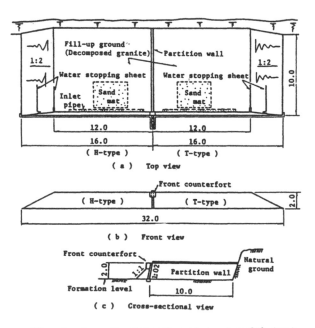

Fig. 1 General view of prototype model test

288

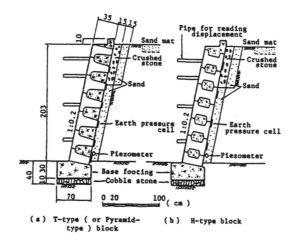

Fig. 2   Cross-sections of retaining walls

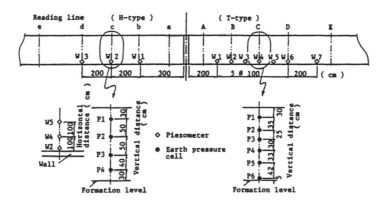

Fig. 3   Arrangement of earth pressure cells and piezometers

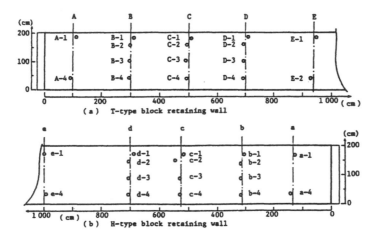

Fig. 4   Arrangement of displacement reading pipes and reading lines

BE*2KE pressure cells of strain gauge type were employed. The measuring capacity was 20 kPa. The pressure cells were stuck to the back side of retaining walls with proper adhesive, the pressure receiving surface covered with fine sand and the soil compacted on them. Six pressure cells were embedded in the T-type block retaining wall, and four cells were embedded in the H-type block retaining wall.

As displacement devices, ruled PCV pipes with an outside diameter of 5 cm were attached to the surface of the retaining walls, and observed through a transit. The reading lines and positions of the displacement devices are shown in Fig. 4. Sixteen displacement devices were installed to each on T-type and H-type retaining walls.

2.4 Testing procedure

After construction the block retaining walls were left for 14 days to cure the concrete. Then the initial earth pressure, water pressure, and displacement were measured. After that, to load the retaining walls, 1,248 sand bags (30 kg each) were placed on the back side of the retaining walls. They were lifted by truck crane in twenties, then piled carefully by hand up to 17 layers (2.6 m).

The sandbag piling work was stopped for about 30 minutes after piling up to 3, 9, 15, and 17 layers for the necessary measurements. The number, height, and surcharge pressure of the sandbags at each layer are shown in Table 1.

After piling all the sands and measuring the necessary items, water was applied to the sand and gravel layer of back-filling through the water supply pipe, and the water level, displacement, and change of earth pressure were measured. As the water level was increased, the displacement increased and cracks made on the retaining walls and enlarged. When the water leakage from the cracks balanced with the water supply, the feeding water was stopped.

2.5 Material for back-filling

The back side of the retaining walls was filled with decomposed granite soil obtained from the tesing ground. After compacting the backfill soil, a thin wall tube of 7.5 mm diameter and cm height was inserted to measure the density in the field and the water content of the backfill soil. Then the backfill material was brought to the laboratory to measure specific gravity and gradation and to perform a triaxial compression test (submerged, $\overline{CU}$ test). The results of the measurement of the specific gravity and the density in the field shown in Table 2, and the grain size distribution curve is given in Fig. 5.

For the triazial compression test, unsaturated specimens were made and set in the triaxial pressure chamber. Then the specified pressure was applied and the water was supplied to the bottom of the specimens until they were saturated, simulating the condition of the back-filling for actual construction. After the supplying of water, the axial compression test was performed and measuring the pore water pressuure was measured in undrained condition. Fig. 6 shows the triaxial test results for backfill soil of the average dry density, 1.38 g/cm³. For the reason mentioned above, Mohr's stress circle was a little uneven; the shearing strength constant ranges were from (c' = 3 kPa, φ'= 26°) to (c' = 4 kPa, φ' = 24°).

3 RESULTS OF EXPERIMENT

3.1 Earth pressure and displacement caused by the sandbag load

Fig. 7 shows the measured earth pressure due to the surcharges (sandbags) at the stages given in Table 1 and the Coulomb's earth pressure calculated assuming a perfect evenly distributed load. Although it seems that the earth pressure generated on the H-type block retaining wall is a little

Table 1 Surcharge load

Layer of sandbag	Number of sandbag	Mass (Mg)	Height (cm)	Stress at center (kPa)	Mean stress (kPa)
3	280	8.4	42	5.1	4.7
9	720	21.6	130	15.3	12.0
15	992	29.8	225	25.5	16.5
17	1248	37.5	260	28.9	20.8

Table 2 Specific gravity, moisture content and
         density of backfill soil

Section of wall retaining	Specific gravity Gs	Moisture content w (%)	Dry density ρ_d (g/cm³)	Wet density ρ_t (g/cm³)
T-type		29.0	1.13	1.46
		31.0	1.21	1.50
	2.63	34.5	1.30	1.75
H-type		29.3	1.10	1.42
		32.8	1.20	1.59
		33.0	1.33	1.79

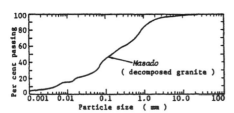

Fig. 5 Gradation curve of back-fill
        soil

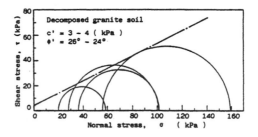

Fig. 6 Result of triaxial compression test
        (submerged, $\overline{CU}$) of back-fill soil

higher than that on the T-type block retaining wall, it can be supposed that the
earth pressure caused by sandbag load and
applied to each wall is almost the same.
The shearing resistance angle is equivalent
to $\phi' = 24°$ to $26°$, with it is compared with
the calculated Coulomb' earth pressure.
   The maximum displacement caused by the
sandbag load of each wall was measured at

C-1 and D-1 (c-1 and d-1), and was 3 mm in
the case of the T-type block retaining wall
and 5 mm in the case of the H-type. However,
displacement of 2 to 3 mm were measured at
the bottom of walls C-4 and C-5 (c-4 and
d-5), indicating that the displacement
caused by the sandbag load was similar to
a parallel movement (slide) which involves
the displacement of the base footing of the

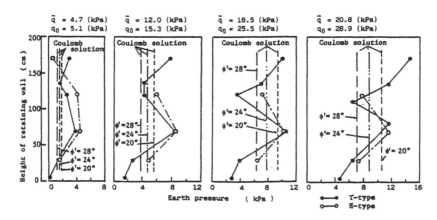

Fig. 7 Calculated and measured earth pressures subjected to surcharge load

retaining wall.

3.2 Displacement caused by rising water table

The water level measured in the backfill of the retaining wall is shown in Fig. 8. Although the water level varied depending on the positions of the water supply pipes and water leak points, it rose relatively evenly. Fig. 9 shows the relationship between the mean water level in the backfill after applying the load and supplying the water and the displacement at each reading line. Figs. 10 and 11 show the relationship between the increase in the mean water level and the displacement at reading

points 1 and 4 along the reading lines.
The displacement of the retaining walls caused by the water in the backfill is similar to a parallel movement of the walls and the base footing when the water level is low, but the displacement of the upper part of the walls enlarges to cracks as the water level at which the first crack occurred was 98.6 cm in the case of the T-type block retaining wall, whereas for the other wall it was 111.2 cm.

3.3 Acting point of resultant force

The acting point of resultant force of the earth pressure and water pressure applied to the retaining wall when it was cracked

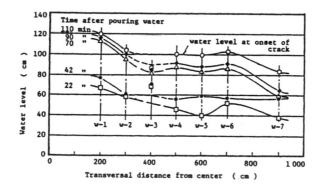

Fig. 8 Water level behind the T-type block retaining wall

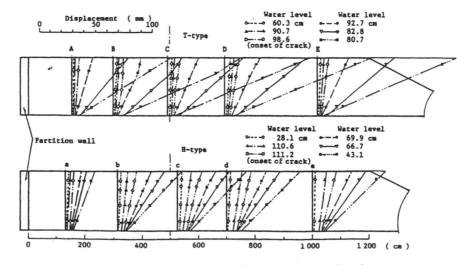

Fig. 9 Deformation of wall due to water level

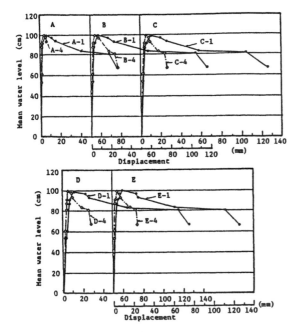

Fig. 10 Relationships between mean water
level and displacement of the T-type
block wall

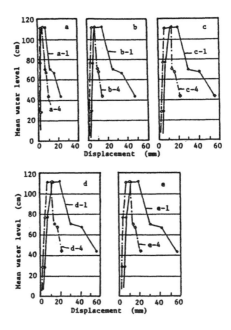

Fig. 11 Relationships between mean
water level and displacement
of the H-type block wall

is explained below. Fig. 12 shows the relationship between the acting point of the
so-called active earth pressure resultant
force of the earth pressure, water pressure,
and the dead weight of the retaining wall
and the load and the water level in the
backfill. If the shearing strength constant
ranges from (c' = 4 kPa, ϕ' = 24°) to (c' =
3 kPa, ϕ' = 26°), the surcharge load q is
20.8 kPa, thus the acting point of the
resultant force is a little outside the
edge of the bottom side of the blocks when
the water level in the backfill is 98.6 cm
and 111.2 cm.

3.4 Failure pattern of retaining walls

Figs. 13 and 14 illustrate the final failed
front view of the T-type and H-type retaining wall, respectively. The failure
pattern of the T-type block retaining wall
is a little more comlicated than the H-type
one. The T-type block retaining wall is
weak as a slab structure, thus it breaks
into small blocks easily. The cracks on the
H-type block retaining wall developed along
the joints, but those on the T-type were in
the blocks itself, as well as along the
joints.

4 DISCUSSION OF THE EXPERIMENTAL RESULTS

The earth pressure on the T-type block retaining wall from the sandbag load is almost
the same as that on the H-type one, and the
displacement caused by the sandbag load is
assumed to be a parallel movement caused by
the base footing. Thus the effects of the
shape of blocks on the earth pressure on
the retaining wall and the deformation
caused by the sandbag load itself neednot
be discussed. However, the effects of the
shape of blocks on the relationship between
the water level in the backfill and the dis-
placement/failure are clear. They are considered below.
The measured displacement includes the
parallel movement caused by the slide of
the base footing. The relative displacement
obtained by subtracting the movement of
base footing at each reading line from the
displacement measured at the top of each
reading line is shown in Fig. 15. this relative displacement curve seems to indicate
characteristics of the block retaining wall
itself. If the curve of the T-type in (a)
or Fig. 15 is compared with that of the H-
type in (b), it can be seen that the cracks
in the T-type wall occur when the maximum
relative displacement is 5 to 6 mm. But the
maximum relative displacement causing the

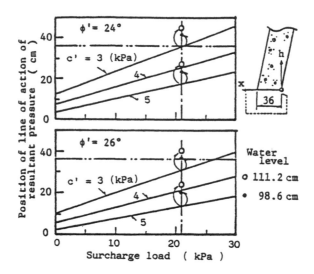

Fig. 12 Diagram illustrating effects of surcharge load and height of water
level on position of line of action of resultant pressure

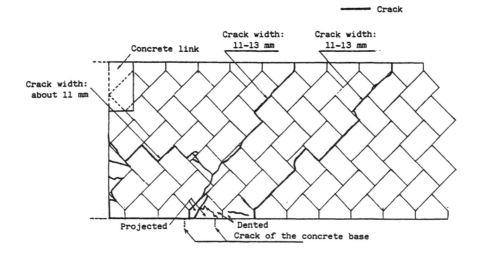

Fig. 13 Sketch of final failure of the T-type block wall

cracks in the H-type wall is 10 to 12 mm,
which is almost twice as long as the former.
From points E-1 and e-1, where the displacement is the maximum shown in Fig. 16,
the so-called quasi-elastic behavior is
revealed, and there is little difference in
the behavior between T-type and H-type
walls. however, there is a remarkable difference between them when the water level
is raised and the deformation becomes high.
That is, a crack occurs in the T-type when
the water level is 98.6 cm and displacement

is 10 mm, but there are no cracks in the
H-type until the water level is increased
to 111.2 mm and the displacement is increased to around 15 mm. This means that
the H-type block retaining wall has a higher toughness than the T-type one. That is,
the body-filling concrete and the blocks
bond to each other tightly and form a single
body, and the mechanical behavior is similar
to that of one concrete slab until the deformation is increased to a relatively high
level.

294

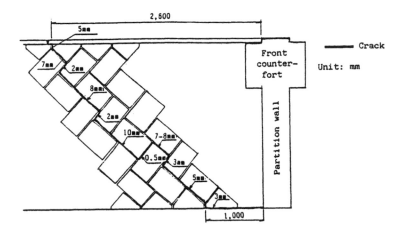

Fig. 14 Sketch of final failure of the H-type block wall

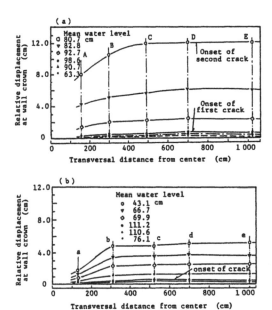

Fig. 15 Relative displacement at wall crown due to water level

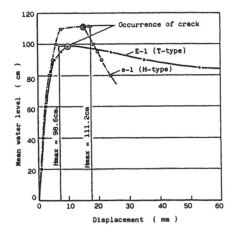

Fig. 16 Comparison of displacements of two walls due to water level

5 CONCLUSION

The results obtained from the experiments on concrete block retaining wall deformation and failure are summarized here.

(1) Almost the same earth pressure was generated by the top load on both T-type and H-type block retaining walls, and it was similar to Coulomb's earth pressure based on $\phi' = 24°$, as in the result of the triaxial compression test. The maximum

displacement was 3 to 5 mm, and the deformation was equivalent to a parallel movement caused by the base footing, etc.

(2) In the relationship between the increase in the water level and displacement of the retaining wall, there was a peak value like a yield point. The displacement of the retaining wall was small and the water level and displacement were related almost linearly to each other until the water level reached that point. After a crack was made at the yield point, the water level decreased, the displacement increased rapidly, and the crack was enlarged gradually. The H-type block retaining wall was able to stand a larger deformation than the T-type one, and had a higher toughness.

(3) The water level in the backfill at
which the retaining wall cracks was 98.6 cm
in the case of T-type, and 111.2 cm in the
case of H-type. This means that H-type block
retaining walls can withstand about 1.13
times higher water level or about 1.27
higher water pressure than the T-type.

(4) When the cracks occurred, the acting
point of the resultant force of the earth
pressure and water pressure was outside the
outer edge of the bottom sides of the
blocks.

ACKNOWLEDGEMENT

Sincere gratitude is expressed to Messrs.
T. Takaghi and H. Onoue for their cooperation in the field experiments, and
appreciation is also due Mr. M. Nakajima,
staff member of Kyushu University.

Environmental Geotechnics and Problematic Soils and Rocks, Balasubramaniam et al. (eds)
© *1987 Balkema, Rotterdam. ISBN 90 6191 785 9*

Geology of a stream terrace deposit in Massachusetts and its utilization as aggregate

Oswald C.Farquhar
Department of Geology and Geography, University of Massachusetts-Amherst, USA

ABSTRACT: The geology of a stream terrace deposit was studied at Pochassic, near Westfield, Massachusetts. This is roughly circular with an area of about five km^2 and a thickness of 30 m. It was formed in the Late Pleistocene as an alluvial fan. The fan extends southward from a south-flowing stream at the edge of the mountains surrounding the former glacial Lake Westfield. The upper surface lies about 40 m above the present river system, which is the second highest level of five stream terraces recognized in the local fluvioglacial sequence. The outer margin of the deposit has been truncated by a post-glacial river, and other parts have been incised by small streams. The main exposures examined are extensive sand and gravel workings. The alluvial fan has three main environments, the distal parts being composed mainly of bedded sand with a pebble content of less than five percent. The mid fan is built of sand layers interbedded with gravel layers, both the thickness and the coarseness of the gravel layers increasing toward the proximal area on the rim of the original lake basin. The materials consist of stream-flood deposits and channel deposits, constituting two more or less distinct facies. Higher up at the head of the stream is the upland source area of glacial till where the ice melted.

The petrography was also studied. The sand and gravel are derived from metamorphic and igneous rocks. The particles are divided into three groups: dark rocks, including certain schists; light rocks, mainly granite and quartzite; and quartz. Pebbles in the gravel layers are mostly smooth and ellipsoidal. Those composed of banded rocks tend to be discoidal in shape, and those of rocks with uniform texture are more nearly spherical. Pebble sizes in the mid fan are mixed and vary from layer to layer, the proportion over 12.5 cm exceeding ten percent toward the proximal area. In the sand layers of the distal part about nine percent is finer than 300 μm, consisting of fine sand and silt. There is no material of clay size in the deposit. About ten percent of the dark rocks are weathered, only about two percent of the light rocks, and none of the quartz.

The processing equipment being used on the site consists of gyratory and cone crushers, dry and wet screens, and washers. The chief products are concrete sand and several sizes of crushed stone aggregate. There is normally enough sand from the matrix in gravel units to meet the demand, and sand units are used to only a small extent. The main workings are in the mid fan, but are heading toward the proximal area. The treatment plant is described, and tables are included showing the sieve fractions, composition, and condition of the processed and washed concrete sand. Largely because of the similar geologic origin of all of the materials and the fact that both flood and channel deposits were transported only by stream flow, uniform aggregates and processed sand can be produced in large quantities by conventional treatment procedures. Environmental concerns are diminished because the deposit is in a remote, undeveloped area.

INTRODUCTION

This report describes the geology and the related use of sand and gravel at Pochassic, near Westfield, Massachusetts (Fig. 1). The material consists of detrital sediment transported southward from the steep slopes between Tekoa Mountain and Ball Mountain by late glacial streams that preceded the modern Moose

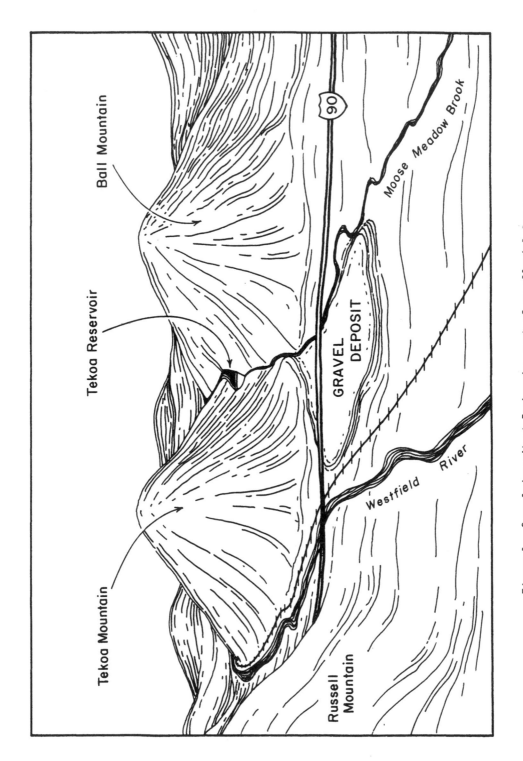

Figure 1. Gravel deposit at Pochassic, part of an alluvial fan near Westfield, Massachusetts. The surface is at the second of five glacial stream terrace levels, and its area is about five km².

298

Meadow Brook. As the streams emerged from the mountains south of the Tekoa Reservoir, the sediment formed an alluvial fan, and this was later terraced by river erosion.

Fluvioglacial beds also occur east of the brook but are seen only in small pits and building foundations. They have not been explored or tested. The working area is west of the brook and utilizes deposits about 30 m thick (Fig. 2).

Samples of both the natural, bank run materials and the processed aggregates and concrete sand are reviewed and analyzed. Procedures and equipment in the treatment plant are also discussed.

1 GEOLOGY

1.1 Shape of the deposit

The sand and gravel deposit widens southward from its steep valley source, the Moose Meadow Brook of glacial time, into a roughly circular alluvial fan. This belongs to the second highest of five stream terrace deposits formed in the Late Pleistocene at different heights above the modern streams. The upper surface of this terrace is about 90 m above sea level and some 40 m above the present Westfield River, into which Moose Meadow Brook drains. The surface is an erosional feature, now randomly pitted. Large areas have been stripped of soil to allow sand and gravel to be removed. The deposit is about 30 m thick.

1.2 Glacial history

The alluvial fan is a glacial meltwater deposit consisting of mixed sand and gravel. Before it was formed, the lower parts of the Westfield area were occupied by glacial Lake Westfield. Three types of sedimentary deposit are recognized along the present rivers: (1) stream terrace or glaciofluvial beds, as at Pochassic; (2) glaciolacustrine beds in the area of the former Lake Westfield; and (3) deltaic beds.

As noted by Stone (1982) in an overall assessment of late glacial materials in Massachusetts, "coarse glaciofluvial valley fills may be included in the glaciolacustrine map units" and "a fluvial gravel facies may lie on the eroded surface of bottomset or foreset beds of an older delta."

The Pochassic deposit consists almost entirely of stream terrace or glaciofluvial beds. Lake beds are not exposed, and deltaic beds are limited to small ponds dammed by bars and other obstructions within the developing alluvial fan. However, in the adjacent Connecticut River Valley, which had a much larger glacial lake, lake beds and delta deposits are extensive.

The classification of late glacial sediments in selected nearby quadrangles varies slightly. In the West Springfield quadrangle (Colton and Hartshorn, 1970), stream terrace deposits are no more than six meters thick, but the authors recognize a series of seven Late Pleistocene stream and lake deposits, also described as stratified drift deposits. In addition, outwash deposits are distinguished.

In the Southwick quadrangle (Schnabel, 1971), four of the same seven glacial age deposits are found, again designated as stratified drift deposits. Stream terraces are described as partly erosional and partly depositional features.

In the Mount Tom quadrangle, Larsen (1972) mapped a numbered sequence of lake and delta deposits as stratified drift. He also described recent unpaired stream terrace deposits ranging from one to three meters in thickness.

On a provisional map of the Woronoco quadrangle, which includes the Pochassic area, Holmes (1968) showed a sequence of five Pleistocene stream terrace deposits at heights of about 50 m, 40 m, 30 m, 20 m, and 10 m above the present streams. These are regarded as distinct from an earlier deposit of ice-contact, stratified drift, which consists mainly of sand and gravel in kames and kame terraces.

The Pochassic deposit, as noted, is part of the second stream terrace with an upper surface 40 m above the present Westfield River. It is primarily a depositional feature with the shape of an alluvial fan. The western and south fronts, which form a single curve, have been truncated by the action of the Westfield River swinging laterally across its valley. This has destroyed the extremities of any protruding lobes. Along its northeastern and eastern margins, the deposit has been incised by the modern Moose Meadow Brook and by an east bank tributary of equal size called the Cooley Brook.

1.3 Mode of deposition

The sand and gravel of Pochassic were transported by meltwater streams along

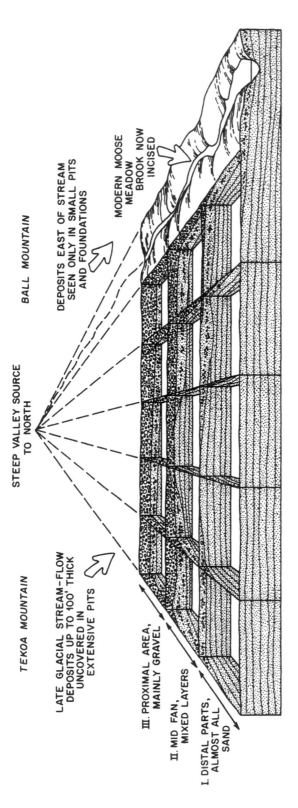

TEKOA MOUNTAIN

STEEP VALLEY SOURCE
TO NORTH

BALL MOUNTAIN

LATE GLACIAL STREAM-FLOW
DEPOSITS UP TO 100' THICK
UNCOVERED IN EXTENSIVE
PITS

DEPOSITS EAST OF STREAM
SEEN ONLY IN SMALL PITS
AND FOUNDATIONS

MODERN MOOSE
MEADOW
BROOK NOW
INCISED

III. PROXIMAL AREA,
MAINLY GRAVEL

II. MID FAN, MIXED LAYERS

I. DISTAL PARTS,
ALMOST ALL SAND

WESTERN AND SOUTH FRONT OF ALLUVIAL FAN

Figure 2. Distal, mid, and proximal fan sediments of the Pochassic deposit, the material having been transported by meltwater streams. The area crossed by sections in this diagram is about three km^2.

the former Moose Meadow Brook and then built into an alluvial fan, the different parts of which are recognized according to the dominant grain size and the form of bedding. The extremities are the fan base and the distal parts of the fan farthest from the source. These consist of sand units with no more than five percent pebbles and were formed as stream-flood deposits. Channels with coarser deposits extend into the distal parts.

The sections now being excavated are in the mid fan and are composed of mixed layers of gravel and sand. The proximal area, lying closer to the source, is made up almost entirely of gravel, with larger particles than in the mid fan. This is due to the inability of the stream to move the heaviest material far from the upland source. Although mostly covered by trees beyond the range of the present operations, the proximal area is cut by streams which expose gravel beds in their banks.

Together, the various parts of the fan form a coarsening-upward morphosequence. In the fan head section, large stream-transported boulders are seen in stream cuts lying directly on bedrock. They, in turn, are overlain by glacial till that has collapsed down the valley walls. The upper part of the stream has followed the same course since the alluvial fan was formed.

In the early stages of fan formation, sand was removed from glacial till and from melting ice at various levels in the watershed, leaving heavier and larger fragments behind. Much of this sand was transported by channel flow to the distal edge, where it formed mainly stream-flood deposits.

When the stream flow was strong enough, gravel could be transported but hardly reached beyond the mid fan area except in channels. Sand was added to this material from glacial till and melting ice high up in the watershed. Thus, the gravel beds have a matrix of sand and also are interlayered with the sand beds. The mid fan is characterized by these mixed layers. Fragments of 12 cm are common, and many in the fan head area exceed 20 cm.

1.4 Facies present

Two principal facies are distributed across the fan. First, the distal parts, including the fan base, are stream-flood deposits transported from the source through the oldest channels.

Second, channel deposits stretch across the fan from the fall line toward its now-truncated margin. These stream flow beds include particles of all sizes from silt to boulders, but there is no material of clay size in the deposit. Debris flows have not been observed, and their absence may be due to the lack of clay and mud to provide a support medium for sediment masses to move coherently.

Lenses of gravel commonly occur in the Pochassic sand beds, and vice versa. These are attributed to bars along stream channels. Both types of lens are formed of particles that have been transported as a fluidized sediment rather than as a coherent mass. The mechanism was one of grain flow and grain interaction. Longitudinal and/or sheet bars alternating from side to side and migrating within the stream channels may also account for large sets of cross bedding. Comparable models have been proposed by McCabe (1977) to explain cross-bedded sets in Namurian rocks of northern England and by Ramos and Sopena (1983) as the origin of six distinct facies in Permian and Triassic rocks of central Spain, most of them gravels.

Another minor facies in which grain size varies on a regular basis consists of lobes of fine material fringed by much coarser material evidently deposited at the same time. Because these overlie coarse, permeable sediment, it seems that the load-carrying stream water may have rapidly drained downward, stranding both types of material. Such sediment pairs resemble the sieve deposits described by Hooke (1967) in alluvial fans in arid regions.

1.5 Other facies

By contrast with the stream-flood and channel deposits of many alluvial fans and in this case Pochassic, some fans also exhibit extensive debris flow deposits. In recent work, Filipov (1986) and Filipov and Hubert (in prep.) describe debris flows on the surface of alluvial fans in the White Mountains of California, with a maximum gradient of about 6 degrees.

Some of these California deposits contain no more than 15 percent silt and clay. A constant supply of material from high up in the canyons is generated by active faults and then travels across the fans from apex to toe without much change in composition, sorting, or size. The debris flows have a recurrence interval

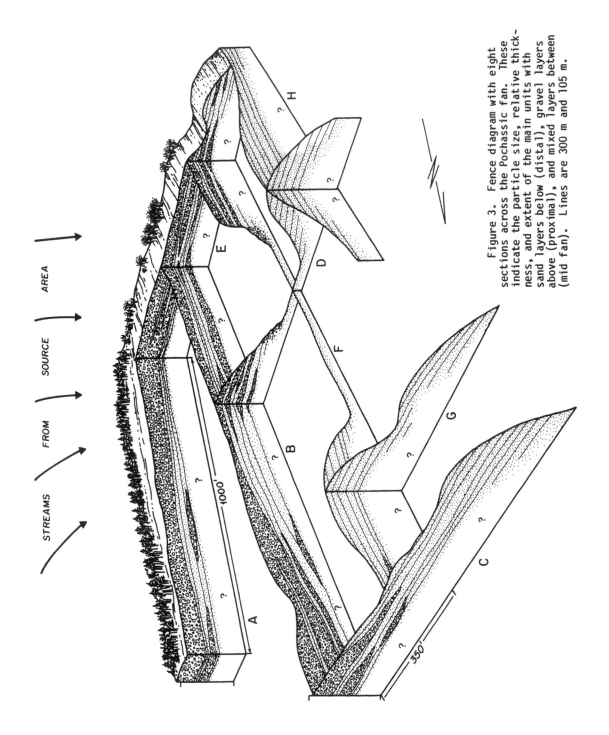

STREAMS FROM SOURCE AREA

Figure 3. Fence diagram with eight sections across the Pochassic fan. These indicate the particle size, relative thickness, and extent of the main units with sand layers below (distal), gravel layers above (proximal), and mixed layers between (mid fan). Lines are 300 m and 105 m.

302

of about 300 years.

Their motion compares with that of ice or lava and contrasts with that of stream-flow deposits involving free motion of separate particles. Some of the largest boulders in most debris flows ride high in the flow because of matrix support, while others act as obstructions around which a debris flow may diverge.

1.6 Cross sections

The Pochassic deposit is worked in one large pit with many more or less straight vertical faces at five principal levels. Old sand pits to the northwest were virtually abandoned by about 1969, and present activities are mainly conducted in the gravels of the mid fan. A fence diagram, prepared from eight sections across the working areas, shows the size, attitude, and extent of the main sedimentary units (Fig. 3).

Each section intersects with one or more of the others. Most are drawn just a few feet behind the face while others cross the faces at right angles. Almost the full thickness of the exposure is displayed, from the thin forest soil down to the floor of the pit. The lines of the cross sections, A to H, are shown on a map of the processing plant and related installations (Fig. 6).

1.7 Bedding

Typical bedding features are seen in a series of drawings made from photographs of the working faces at Pochassic (Figs 4 and 5). Material coarser than sand is well rounded, including granules, pebbles, cobbles, and boulders. Most of the sand, however, consists of subangular particles. The shape and size of the particles, as well as their composition and density, affected their flow behavior and the resulting bedforms.

1.8 Discussion

Except for differences in particle size and variations in bedding that mark the distal, mid, and proximal areas, the materials exposed are notably uniform. The fact that stream flood and channel deposits make up the entire sequence ensures a consistent flow of material through the treatment plant. Clay banks, dune deposits, and biogenic constituents that could alter the composition of the

sediments are virtually absent.

Streams across the fan came from the north and northeast, with the direction varying as channels filled and overflowed. The various bedding directions can be seen as gravel faces are excavated.

Although stream terrace deposits in the area of the former glacial Lake Westfield may rest upon lake beds, eroded deltas, or bedrock, only the alluvial fan itself is exposed on the Pochassic property. The nearest bedrock at the surface is in the upper watershed of Moose Meadow Brook, in a tributary of similar size called Cooley Brook, and on adjacent mountains.

The Pochassic fan, lying astride the modern brook and created by its predecessor and head zone tributaries, is a separate feature not overlapped by other fans. It rests on lower level, stream terrace deposits on the banks of the Westfield River. Because of its small size of only about five km^2, the different portions of the fan can be examined in a single pit. This compares with a borrow pit in Maine, described by Retelle and Konecki (1986), in which all three facies also are recognized although in that case the distal portion is marine.

For Maine and adjacent regions, a comprehensive series of reports on the specific mechanisms of deglaciation that produced such features has recently been published (Borns and others, eds., 1985). As apparent from a table (op. cit., p. 25) on the history of deglaciation in the mountains of northeastern New England, the main distinctions among the various glacial deposits there are related to the different flow phases of ice and water. During the earlier phases in the Late Pleistocene with the ice becoming increasingly stagnant, the volume of ice far exceeded the amount of meltwater. As the ice thickness was reduced to less than the local relief in isolated valleys, the proportion of meltwater increased and outwash sequences of sediment were formed. These included terrace and fan alluvium as well as landslide deposits. The Pochassic case in southern New England is similar except for lower relief and, at least in the mid and distal parts of the fan, an absence of landslide and debris flow deposits.

Modern accounts of work related to alluvial fans include one on the interaction between turbulent flow, sediment transport, and bedform mechanics in channelized flows (Leeder, 1983) and a new

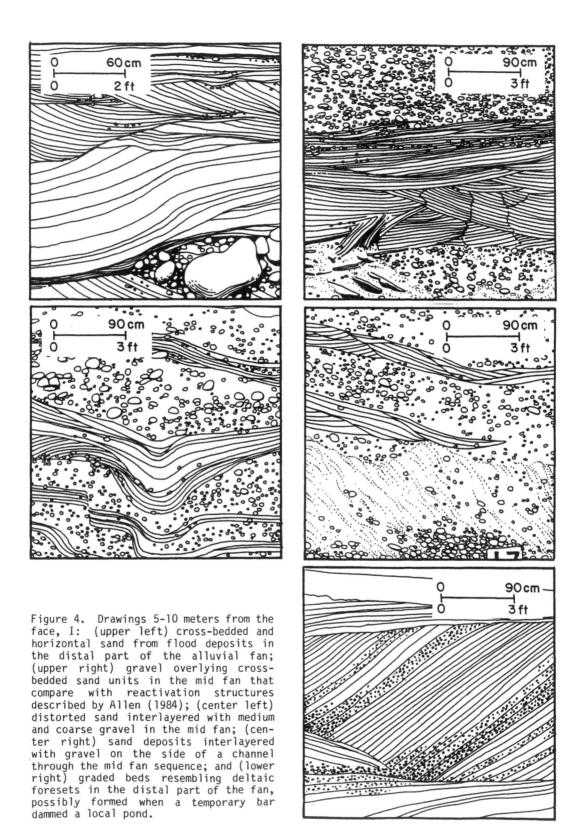

Figure 4. Drawings 5-10 meters from the face, I: (upper left) cross-bedded and horizontal sand from flood deposits in the distal part of the alluvial fan; (upper right) gravel overlying cross-bedded sand units in the mid fan that compare with reactivation structures described by Allen (1984); (center left) distorted sand interlayered with medium and coarse gravel in the mid fan; (center right) sand deposits interlayered with gravel on the side of a channel through the mid fan sequence; and (lower right) graded beds resembling deltaic foresets in the distal part of the fan, possibly formed when a temporary bar dammed a local pond.

Figure 5. Drawings 5-10 meters from the face, II: (upper right) medium and coarse gravel beds with a few sand layers in the mid fan; (center left) coarse and very coarse gravel beds with one distinct sand unit in the proximal area--the largest fragment is over 5" in length (12.5 cm); (center right) medium and coarse gravel beds alternating with cross-bedded sand layers in the mid fan; (lower left) gravel beds and gravel lenses within sand units in the mid fan; and (lower right) channel deposits in the proximal area with many pebbles and boulders upended in an imbricate pattern and cross-laminated sand beds that curve at the channel edge.

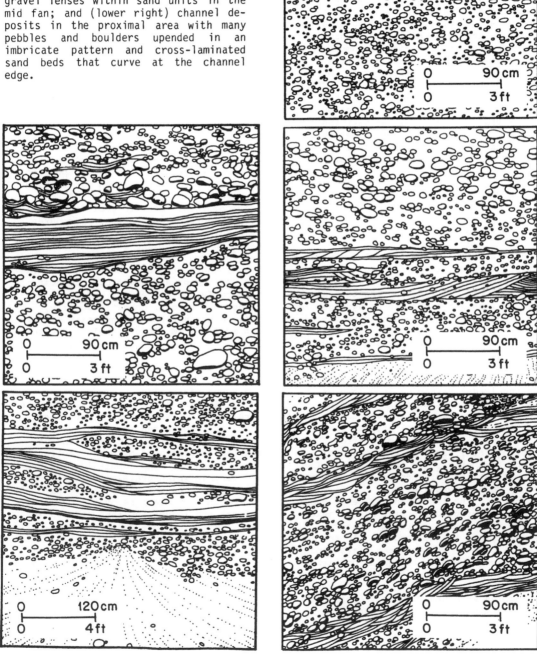

classification for sediment gravity-flow deposits based on flow conditions (Postma, 1986). A collection of papers on the sedimentation of coarse-grained terrigenous clastics (Koster and Steel, eds., 1984) and an exploration guide to alluvial fans and fan deltas (Fraser and Suttner, 1986) have been issued recently. Revised editions of Klein's treatment of sandstone depositional models (1985) and Reading on sedimentary environments and facies (ed., 1985) also are available.

In the development of models for basin analysis and for specific items of alluvial architecture, Miall (1983) has included a number of variables, among them climatic changes. In Massachusetts, climatic changes in the Pleistocene controlled the deluges of meltwater. Structural changes, including faults, may result in different terrace levels, but in the Pochassic area drainage patterns connected with ice dams and fluctuating lake levels provide a simpler explanation for the eroded surface of the alluvial fan.

2 PETROGRAPHY

2.1 Purpose

The materials of the alluvial fan were examined for two purposes. One was to compare samples with each other and with the source rocks. The second was to evaluate the various sizes for use as aggregate and as concrete sand. The aggregate includes crushed rock as well as particles that have gone through appropriate screens with little treatment except for washing.

The sand from the plant passes through the same screening and washing treatment as the larger particles, and some of it is derived from crushing. This processed, concrete sand is regarded as distinct from unwashed, natural sand obtained as bank run material from the working face, although the essential properties of the two may overlap. The specifications for industrial uses commonly require washing.

In fact, the last event for each sand/gravel layer before burial by the next layer was washing in the meltwater streams that transported it. In the interval of over ten thousand years since that time, the only change in condition has been the minor weathering of limited numbers of fragments.

Samples were analyzed on the basis of the size, composition, and condition of constituents.

2.2 Procedure

Because the materials were being evaluated for construction and industrial uses, the analysis was conducted according to American Society of Testing and Materials (ASTM) procedures, particularly those in: C-295, Standard practice for petrographic examination of aggregates for concrete. Comparable guidelines are issued by the American Association of State Highway Officials.

Other applicable ASTM documents are: C 33, Specifications for concrete aggregates; C 117, Test methods for materials finer than the 75 μm (no. 200) sieve in mineral aggregates by washing; C 136, Method for sieve analysis of fine and coarse aggregates; C 294, Descriptive nomenclature of constituents of natural mineral aggregates; C 702, Methods for reducing field samples of aggregates to testing size; C 851, Recommended practice for estimating scratch hardness of coarse aggregate particles; D 75, Practice for sampling aggregates; and E 11, Specifications for wire-cloth sieves for testing purposes.

All of these refer to geologic materials or geotechnical testing procedures, or both. In addition, C 823 and C 856 refer to the sampling and examination of hardened concrete in which sand and gravel aggregates have been used. Various government and commercial agencies may modify the procedures and specifications when special conditions arise, while other nations may advocate different measures to achieve quality control.

2.3 Particle size

Two types of samples were collected. The first was bank run material from the working face. Many samples were examined, the one chosen for detailed analysis being taken from the mid point of the face represented by cross section E (Figs 3 and 6). The sample is from the mid fan, in which gravel is interlayered with sand. The untreated sample was collected in a single pass by the bucket of a front end loader and then reduced in size. The fractions examined were: (1) gravel, coarser than 6.25 mm, and (2) sand, which passed through the 6.25 mm sieve. Tables on these samples are included in this section.

The second type of sample came from stockpiles of concrete sand that had been screened and washed in the treatment plant. The sizes are those re-

tained on the 4.75 mm and smaller sieves.
Tables on these samples are included in
the section on processing.

2.4 Composition: general divisions

The constituents of the various samples
were separated according to size fractions and composition. ASTM C 294, Standard descriptive nomenclature of natural
mineral aggregates, provides guidelines
that cover the entire range of materials
found in rock formations. In a single
deposit like Pochassic only a small number of such materials is likely to be
represented. In this discussion the
range of materials is briefly reviewed
and then those which apply to Pochassic
are selected.

There are four main divisions to consider. The first is minerals: (1) the
light silica minerals--quartz, opal,
chert, and chalcedony--and the feldspars,
and (2) the dark silica minerals, most
of which are ferromagnesian silicates.
Other minerals that might be significant
if present in a concrete sand are micas,
clays, zeolites, carbonates, sulfates,
sulfides, and oxides.

The second division is that of igneous
rocks, again quite readily subdivided
into light and dark with the light including granite, syenite, diorite, and
pegmatite. The chief dark varieties are
gabbro, peridotite, and basalt.

Third, the sedimentary rocks fall into
four main subdivisions: (1) conglomerate,
sandstone, and quartzite, (2) greywacke
and subgreywacke, (3) claystone, shale,
argillite, and siltstone, and (4) carbonate rocks.

The fourth division consists of the
metamorphic rocks, marble, metaquartzite,
slate, phyllite, schist, amphibolite,
hornfels, gneiss, and serpentinite. Marble, metaquartzite, and the more silicic
gneisses are light, and the rest are dark.

2.5 Three groups of material at
Pochassic

Materials found in the Pochassic deposit
can be assigned to three main groups that
cut across the four general divisions
noted above. The groups are based on
compositional types that have been identified in the various sand/silt sieve
fractions. The dark igneous and dark
metamorphic rocks containing mostly dark
minerals form the first group (D) with
essentially one set of physical and chem-

ical features.

The light igneous and light metamorphic
rocks containing mostly light minerals
form a second group (L) with a slightly
different set of physical and chemical
features: they are more durable and more
silicic.

Third, the (Q) group consists of quartz
and feldspar particles from broken up pegmatite, metaquartzite, and other rocks.

The (D) group is the most abundant of
the three groups in fractions retained on
#16 (1.18 mm) and coarser sieves. The (L)
group is less abundant than the (D) group
and also less abundant in fractions retained on #16 (1.18 mm) and finer sieves
than the (Q) group. The (Q) group is the
most abundant of the three groups in fractions retained on #50 (300 μm) and finer
sieves.

Among the light igneous rocks in the
Pochassic gravels, granite, syenite, and
diorite would be distinguishable in thin
sections of sample pebbles. However,
these rock types, while they are geochemical variants of each other, are mechanically so similar that their relative abundance would have no effect on the quality
of the gravel, whether crushed or uncrushed. Therefore, they were grouped together and added to the light metamorphic
rocks, i.e., metaquartzite and the more
silicic gneisses. Marble was not found.

Among the dark igneous rocks, gabbro
and peridotite are present, although they are
not easily distinguished from amphibolite.
Because of basalt in the area, albeit
mainly south of the source area of the
glacial till and melting ice that gave
rise to the alluvial fan, a special watch
was kept for this fine grained material.
None was confirmed, but a small quantity
may have been included with the dark mica
and hornblende schists.

2.6 Separate mineral particles

Minerals as separate particles have been
left to last because, in many cases, they
have resulted from disintegration of rocks
and become most abundant in the finer
fractions. Quartz is the leading member
in this group (Q), some feldspar also occurring. Micas were found only in trace
amounts, but when an occasional flake
shows up it may be of fingernail size.
In fact, one sample of concrete sand had
as much as one percent of mica, mostly in
finely divided form.

Clays have not been found in any of the
sieve fractions nor in the pans and settling ponds, nor has the decomposition

of feldspars and other silicates to clays been recognized. The nearest clay beds now exposed are four miles away on the far side of the mountains. Their position suggests that at the end of the glacial era they were covered by ice, glacial till, or lake water, thus being an unlikely source.

No lignite or other traces of past vegetation were found. The climate was undoubtedly too cold and the conditions of fluvial sedimentation unfavorable even if there had been trees in the upland watershed. Lignite, if present, could have been used for carbon-14 dating. Fortunately, other means have been found to establish a local timetable for deposits formed when the ice sheet was melting, namely the relative levels of several stream terraces.

Zeolites, because of their comparative lack of durability, are unlikely to have survived sedimentary transport by stream flow. Carbonates form a substantial belt to the west and the northwest in the directions from which glacial till originated. Therefore, a comprehensive search was made for calcite grains, both optically and using hydrochloric acid, but none were found. Sulfates would have been dissolved, and sulfides would have been oxidized: none were present.

Traces of magnetite were found, probably as residues from the disintegration of granite by glacial erosion. Other heavy minerals also may be present, but separation yielded nothing that could be recognized. The source rocks do not suggest that any substantial suite of heavy minerals is likely, although some pegmatites are present. Still, there is always a chance that heavy minerals may occur, even though special conditions may be required to preserve them. One example is the trapping of gold in algal mats in the distal portions of fans in the vast Witwatersrand Basin.

The possibility of assigning some of the coarse sand and gravel in the alluvial fan to the category of diamicton was considered. This non-genetic name for non-lithified, essentially non-sorted, terrigenous deposits composed of sand and/or larger particles in a muddy matrix may fit parts of numerous fans, both ancient and modern. Examples have been mapped and described in considerable detail in the McConnellsburg quadrangle of Pennsylvania (Pierce, 1966). However, without the essential mud and clay of the diamicton type, this category does not fit the Pochassic sedimentary unit, which is characterized only by the presence of stream flood and channel deposits.

2.7 Condition

The condition of the particles was examined. The three conditions are: (1) fresh, dense; (2) moderately weathered; and (3) very weathered.

Those in group (D) have enough weakened or weathered particles to be counted, whereas those in group (Q) have none. Materials in group (L) are intermediate in terms of the condition of the particles. Their count of crumbly or weathered particles is low. It was considered possible that weathered material would be appreciably higher in the unprocessed sand than in the product that was screened and washed. However, using samples of unprocessed sand from at least 10 m below the surface, only a minor difference was noted between the two types. Of the small amount of weathered material in the washed concrete sand, most of it was from (D) group rocks, and almost all of it was on #16 and coarser sieves, which account for only one third of the sample by weight. The rest of the weathered material was reduced in size and washed out with the fines.

2.8 Other factors

Coatings on pebbles were rarely seen. A slight discoloration was observed in only a few cases. Any particle with significant surface alteration was included in a weathered category (condition 2 or 3). The entire deposit was formed by water action, and no deleterious substances such as soluble salts that might form coatings survived.

As for particle shape, fragments larger than sand grains are in general ellipsoidal, although some are spherical. The more elongated, flattened, and discoidal particles are chiefly of schistose rock types, the shortest axis being across the plane of foliation. Sizes smaller than granules usually are subangular and are not as smooth as larger particles. Shape is of geologic interest in imbricate and other flow structures. It also affects the way in which particles fit through the various screens, whether squares, hexagons, or slots, and it may influence the rate of weathering and disintegration.

The color of the deposits as a whole

is mainly brown, and darker when moist.
Among the constituent rock types, of
which there are differing proportions in
different sizes, the (D) group is dark,
the (L) group is light, and the (Q) group
is very light or gray in color. Fragments of reddish brown sedimentary rock
rarely occur.

There is no natural cementing material,
either original or secondary. The texture of the deposit is unconsolidated,
with open pore space and loosely packed
grains. Each particle was transported
freely and separately by flowing streams
of water, moving faster within the channels and less so in the stream-flood
areas. No studies of pore geometry were
undertaken in this near-surface deposit,
but they might be useful in the exploration of buried alluvial fans.

Information about significant discontinuities is standard in evaluating geologic materials with industrial uses.
Layering among the gravel beds and inter-
layering with sand strata are both notable, chiefly in terms of the volume of
gravel that can be recovered from a particular working face. The higher the
stone content of a given unit, the
greater its value. The matrix of gravel
beds normally provides enough sand to
meet the company's quota for concrete
sand, while the stone furnishes reserves
of gravel aggregate to keep the stockpiles active. Sand units that are of
little value because of the virtual absence of pebbles are another type of
geologic discontinuity at Pochassic.

2.9 Sampling

The first sample is bank run gravel.
This was separated on five screens or
sieves as shown in Table 1. This untreated and uncrushed gravel consists of
boulders, cobbles, and pebbles. Dark
rocks (D) exceed the other two groups,
(L) and (Q). Seven percent of the whole
sample is in a moderately weathered condition, and two percent is very weathered. Most of the weathered rock is
removed from the processed product by
screening and washing.

The Pochassic products are intermediate
in size when compared with the ASTM standard sieves. Thus, one is 37.5 mm rather
than 50.0 mm or 25.0 mm. The others are
18.75 mm and 9.5 mm. These are offered
as blends in which a certain proportion
of the particles is smaller than stated.
In addition, the 18.75 mm grade is offered straight, with all particles being

about the same size.

The main differences between the untreated gravel and the plant output of
stone aggregate are that, in the latter,
most of the fragments are crushed and
weathered material is removed by successive screenings and washings.

Table 2 shows the sieve fractions of
bank run sand. This particular sample
was from a distal part of the alluvial
fan which is almost all sand, but the
sand in units richer in gravel is comparable. This is basically because the
source materials and the stream flow mode
of transport are the same. The specifications for the percent passing the seven
sieves are as given in ASTM C 33. Both
this untreated sand and the concrete sand
meet these specifications.

2.10 Discussion

This discussion is on the general nature
of weathering as it affects particulate
materials of differing composition. The
condition of the materials has been
briefly noted as a geologic factor that
influences the quality of the aggregate
produced at Pochassic.

Washing of aggregate is essential for
most uses as a means of lessening the
proportion of fine material in the final
product. An excess of fine material may
prevent the aggregate from being freely
drained, form an unwanted coating on the
coarse particles, and require more binder
to be used in the cement.

Most sand and gravel deposits are notable for having been washed into place by
fluvial or glaciofluvial processes. In
the case of Pochassic, intensive stream
washing under natural geologic conditions
was the last event that occurred before
deposition. Between then and now, an
interval of 10,000-20,000 years, what
changes have occurred? Fluvial processes
have continued but, without glacial meltwater of high volume and turbulence, at
a much slower rate. Streams have cut
deep into the fan, and minor slumping
has occurred on the slightly undulating
surface. Thin soil and a forest cover
also have developed.

Another change is in the condition of
the material. The remarks that follow
apply only to stream-flow deposits and
not to debris flow deposits. Only
stream-flow deposits are present at
Pochassic. It may be assumed that all
of the particles larger than sand were
strong and unweathered and had the ability to withstand the rigors of stream

transport. Otherwise, they would have broken down into smaller sizes.

However, a small quantity of the larger fragments, of pebble and even cobble size, are now soft and weathered when removed from a working face. About seven percent of the dark materials are affected in this way, five percent of them being moderately weathered and the other two percent being very weathered. Some of the five percent splits into small particles most of which display no weathering. About four percent goes through the screening and washing equipment as less than 100 mesh (150 μm). This is washed over the lip of the spiral washer into a trough and then a pipe leading to the settling pond.

Questions about why some large fragments are rotten and disintegrate while the majority remain sound are not easily answered. If seven percent of one group of materials is weathered in 10,000-20,000 years under a given set of conditions, how long would weathering take to affect all of the deposit? While a search for solutions to these questions continues to be of geologic interest, the results also could be of geotechnical value in understanding the quality and the treatment of such resources. In fact, alteration of pebbles may not be due solely to atmospheric effects but probably also involves acidic ground water and various other chemical and physical causes of decay.

Flat and elongated fragments are more prone to weathering than spherical fragments. Usually, the incipient weakness is due to layering, particularly from consolidation pressures. The lighter fractions of aggregate material at Pochassic, including granite and quartzite pebbles, show little or no weathering, even when foliated. If a fluvial deposit contains only these rock types, as many of them do, the amount of washing and screening that is needed is much less. However, crushing is still required to produce an appropriate range of sizes.

3 PROCESSING

3.1 Materials

Materials processed in the plant are of two types: gravel with sand as a matrix (GR), and sand with little or no gravel (S). The ratio in (GR) of gravel to sand varies from 70:30 to 5:95. If there is

less than five percent of gravel, a layer would be classed as sand. There are considerable thicknesses of sand alone with no more than five percent of gravel, but complete layers of gravel having no sand do not occur. Minor accumulations of pebbles without much matrix are of no geotechnical significance.

Approximate plant capacity under normal operating conditions is 130 tons per hour.

3.2 Map of operations

The working areas are shown on a map with the lines of the eight cross sections, A to H (Fig. 6). Current working faces are marked as either GR (for gravel [and sand]), or S (for sand), or both. No bedrock is exposed. The plant area, stockpiles, main conveyors, and new and old pit areas are indicated. Fixed features include a railroad line, two buildings, and two ponds. Contours of the site are being modified as gravel (GR) is excavated at a fairly uniform rate and sand (S) is removed at a much slower rate.

If there is not enough sand in the GR beds to meet the plant's requirements for sand products, sand from the S beds can be utilized for it is within the same range of size, composition, and condition as the sand matrix in the GR beds. In practice, loads of pure sand rarely are used.

The small amounts of stone that the S beds contain can be recovered by using a portable screen that is towed to the working face. This is convenient when labor is available during the winter shut-down of the main plant. The screen, a Read Screen-All, is equipped with 6.25 mm slots rather than squares. It has a capacity of about 125 tons per hour. The sand is not washed before sale, but the stone is processed in the spring and summer when the plant reopens.

The Massachusetts coordinate system was used in preparing the map.

3.3 Plant flow diagram

Figure 7 is a plant flow diagram showing the equipment used to process the sand and gravel and to separate it into various sizes. The diagram was prepared in 1986 during a period of painting and annual maintenance. [After this report had been completed, a new computer-controlled plant was installed to improve overall efficiency and increase output.]

3.4 Equipment

The major equipment includes a variety
of hoppers, feeders, silos, gyratory and
cone crushers, bins, chutes, dry screens,
wet screens, and washers. There are four
main units: gravel pits; tower one, primary feed; tower two, cone crushers; and
tower three, final screens and product
handling.

Transfer of material from each unit to
the next is by conveyor. Gravel enters
and leaves the system by truck. Sand
products leave by conveyor to form piles
on the ground. Water is piped to tower
three, and after being used for washing
it overflows into a pipe and carries the
fines in suspension to a settling pond.

3.5 Gravel pits

Bank run material, consisting of gravel
and sand in varying proportions, is
trucked to a field hopper. The pit faces
in 1986 are no more than 300 m from the
hopper, some being much closer. Conveyors extending to one or more new hopper
sites may be installed later as excavation extends to the northeast. Transfer
from excavation to hopper is by front-end
bucket loader. At intervals during the
working day oversize boulders from the
gravel banks are removed from the grizzly
before they can enter the hopper.

The base of the hopper leads into a
Syntron feeder #20 with a 230-460 H.P.
motor running at 1750 R.P.M. and a
Hewitt-Robins reducer with a back stop.
Material is fed to a 212 m Douty conveyor
with a 60 cm belt that runs almost level
about one meter above the ground. At
the end of this conveyor, the material
is delivered into what until 1983 was
the field hopper. This is now arranged
so that the quarried material drops
about four meters on a stepped chute to
the next conveyor which travels upward,
crossing over the road for trucks going
to the gravel stockpiles. Another conveyor completes the primary feed into
the twin silos of tower one.

3.6 Tower one, primary feed

A limited reserve of material to begin
supplying the primary crusher can be
stored in the twin silos when the plant
is shut down at the end of a shift or
during a power failure. Most items such
as belts, the crusher itself, screens,
and washers normally remain empty until

the next start-up. From the silos,
material falls onto a Syntron vibratory
feeder plate, one half of which is a
screen with 75 mm hexagonal openings.
Particles retained on the feeder plate
fall into a 33 cm Superior-McCully gyratory crusher. Smaller sizes pass through
the screen into a drop box. Both of
these pathways lead to a conveyor to
tower two.

3.7 Tower two, cone crushers

Material on the conveyor from tower one
enters a two-deck Allis Chalmers dry
vibrating screen, measuring 120 x 130 mm.
Each deck is divided into halves. The
upper deck can be set with a 37.5 mm
square mesh for one half and a 31 mm
square mesh for the other, depending on
product demand. Material retained on
the upper screen may enter a 6-36 Allis
Chalmers cone crusher.

Alternatively, this material can pass
without being crushed to a chute and conveyor to the next device that can effect
a separation of sizes. This is the wet
screen of tower three. All of the materials from tower two are transferred to
the wet screen via the same belt, whether
they have been through a cone crusher on
tower two or not. Periodic decisions on
stockpiling the larger sizes are based
on anticipated demand. The gates that
follow the screen are designed to be
easily swiveled for material to move
either to a crusher or to the conveyor.

On the lower deck, one half is normally set with a 22 mm square mesh and
the other with a 24 mm hexagonal mesh.
Material that is retained may enter a
4-36 Allis Chalmers cone crusher. The
output from both crushers is returned on
a 45 cm belt to the main conveyor, from
which it passes again on to the two-deck
screen. Crushed materials passing
through the screens will be added on the
conveyor to those that have not been
crushed.

No wetting agents are used in tower
two. The material being processed does
not generate large amounts of dust,
partly because it enters the plant moist
and also because the proportion of fine
material is minimal and clay particles
are virtually absent.

3.8 Tower three, final screens and
     product handling

Whatever sizes of material are permitted

Table 1. Composition and condition of untreated, natural, bank run gravel at Pochassic

Constit-uents	Percentages of the different rock types present							
	In fractions retained on sieves shown below					Condition of whole sample		
	3" (75.0 mm)	2" (50.0 mm)	1" (25.0 mm)	1/2" (12.5 mm)	1/4" (6.25 mm)	1 fresh, dense	2 moderately weathered	3 very weathered
D	54	70	76	79	76	61	5	2
L	26	19	17	11	15	19	2	-
Q	20	11	7	10	9	11	-	-
Totals	100	100	100	100	100	91	7	2

(D) dark rocks	(-) rare or absent	(L) light rocks	(Q) mineral fragments
mica schist & gneiss; phyllite amphibole schist & gneiss amphibolite; gabbro	peridotite (trace) hornfels; chert marble; lime- stone; dolomite other sedimentary rocks (trace)	granitic & related gneisses unfoliated gran- itic rocks pegmatite metaquartzite	quartz feldspars micas amphiboles magnetite; other heavy minerals

Table 2. Sieve fractions of unwashed, natural, bank run sand at Pochassic

Sieve Size		Weight in Grams		Percent Passing	Specifications	Percentage of Sample
		Retained	Passing			
			575			
4.75 mm	(#4)	22	553	96.2	95 - 100	3.83
2.36 mm	(#8)	69	484	84.2	80 - 90	12.00
1.18 mm	(#16)	101	383	66.6	55 - 80	17.57
600 μm	(#30)	186	197	34.3	30 - 60	32.35
300 μm	(#50)	138	59	10.3	10 - 25	24.00
150 μm	(#100)	40	19	3.3	2 - 8	6.96
75 μm	(#200)	12	7	1.2	0 - 2	2.09
PAN		7				1.20
		575		296		100.00

Dried to Constant Weight   Fineness Modulus 2.96

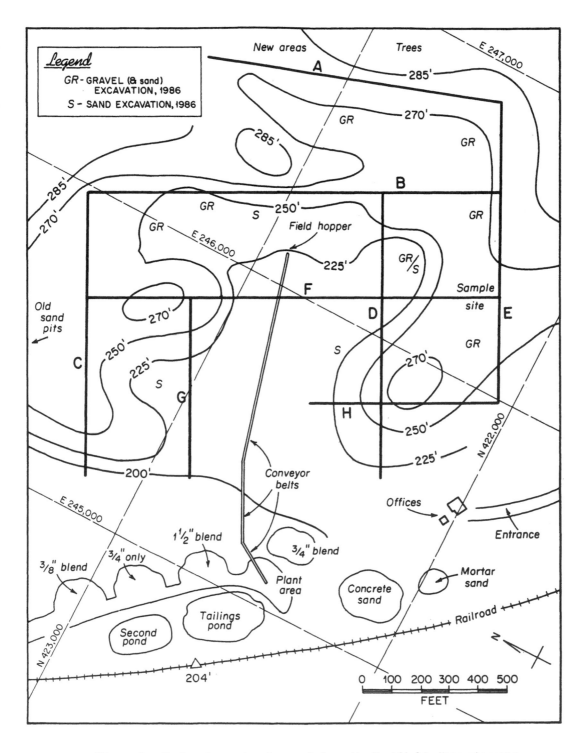

Figure 6. Pochassic sand and gravel deposit, Westfield, Massachusetts. Map of operations showing lines of the cross sections and the areas where gravel (GR) and minor amounts of sand (S) are currently being excavated. Ft (') to m, x by .3048; in (") to cm, x by 2.5.

313

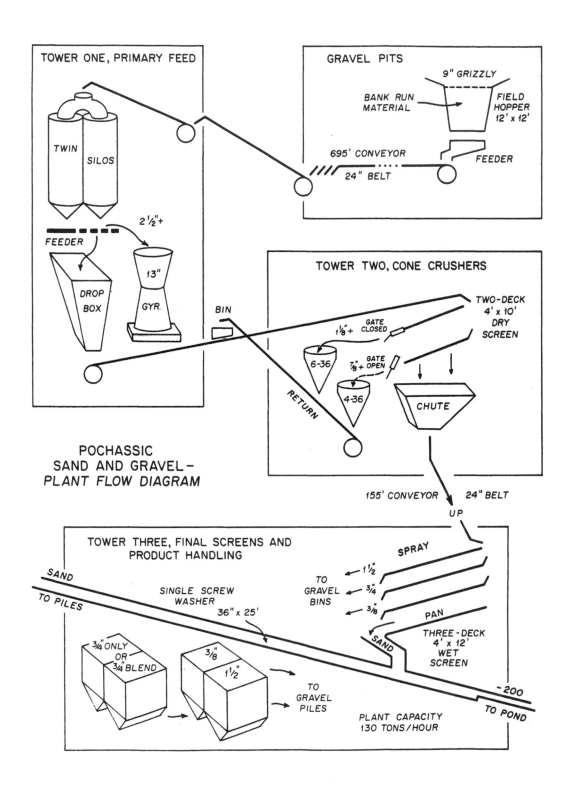

Figure 7. Pochassic sand and gravel: plant flow diagram showing equipment and use. Ft (') to m, x by .3048; in (") to cm, x by 2.5.

to pass through or to bypass the dry screens of tower two are all transferred by conveyor to tower three. Here they reach a three-deck Allis Chalmers wet vibrating screen, measuring 120 x 360 mm. Each deck is divided into thirds. The top deck begins with a wear-resistant, rubber-covered screen with 25 mm round holes, and is followed by two sections with a 22 mm square mesh. The middle deck is usually equipped with screens all having 14 mm round holes. The bottom deck has one third with 6.25 mm round holes and two thirds with 5 mm round holes. Products are blends classed in the 37.5 mm, 19.0 mm, and 9.5 mm ranges, the proportions and amounts depending on market needs. Screens can be readily exchanged. The sizes reaching tower three are controlled by routing material to the crushers of tower two or by diverting material from them.

Sand-size fragments small enough to pass through all of the screens reach a steel pan plate from which they are washed down a pipe into a box that leads to a spiral washer.

Water for the wet screen is pumped from the Westfield River 300 m south of the plant area. This is done with a Worthington pump and a Lincoln motor. Water first washes material from the conveyor onto the head of the screen. More water then is sprayed from eight bars above the screen. No water is returned to the river. At the lower end of the spiral washer, the water carries fine material of #200 size or less by flotation into a pipe that leads to a settling pond. These fines are classed as rewash. A second pond was dug out some years ago to take the overflow. Water from both ponds seeps into the subsurface through surrounding deposits of natural sand.

The spiral washer is a 90 cm Eagle single screw with a gear reducer, driven by a Louis Allis motor. It occupies an eight meter long open box and is mounted above the gravel storage bins. Both sets of equipment are located on the same tower for convenience and economy. The washer is inclined at an angle of 17 degrees. At its upper end, the sand becomes progressively dewatered, and then it falls onto a boom with a conveyor, also called the sand line.

The boom is attached by cables to the top of the tower. The tower end is linked to a curved rail, the weight of which is supported by a central bearing. When its anchoring chain is released the boom can be swung by hand through almost one-quarter of a circle, until its cradle reaches some rails on the tower. This allows an arc of sand piles to be formed on the ground below. A front-end bucket loader is used to rearrange and extend these piles whenever necessary. This type of vehicle also is used for loading highway trucks with gravel, crushed stone, and sand.

Gravel products from the three-deck screen are transferred to separate hoppers for removal by 10-wheel dump trucks to piles on the ground just beyond the plant. When the piles are small, the material can be dumped on the ground, but as the piles increase in height, the trucks run along an embankment to dump from above. Location of these piles along the main access road through the property makes it easy to fill customers' vehicles, many of them 18-wheel trucks, using one of the gravel company's front-end bucket loaders.

When gravel is loaded from the hoppers into a truck it is again washed. The truck driver first turns the water on for a motor-driven Niagara screen. Material passing through this screen is removed with the wash water, becoming a type of rewash, although distinct from the rewash that is separated from sand in the spiral screen washer.

3.9 By-products

During the winter, operations are suspended because of freezing conditions. The plant cannot function without water, and both excavation and transportation would be difficult in the harsh New England climate. In any case, the time is needed for plant maintenance. Among the many tasks involved, fines are removed from the settling pond to locations near the weighing scale for customer pick-up in the summer. This material is marketed as rewash, finding wide use as an inert soil base for lawns for example.

Another by-product of the sand and gravel operation is classed as fill, a fine sand from original sedimentary layers in bank deposits, some of which are conveniently close to the weighing scale. Material of this type is not put through the plant and is virtually regarded as waste. In fact, the excavation and truck routes normally go around such deposits, leaving them as mounds or islands. Their presence may lengthen each truck trip from pit face to hopper, and to some extent they complicate siting, grading, and overall design of surface

Table 3. Sieve fractions of processed, washed, concrete sand at Pochassic

Sieve Size	Weight in Grams Retained	Weight in Grams Passing	Percent Passing	Specifications	Percentage of Sample
		561			
4.75 mm (#4)	18	543	96.8	95 - 100	3.21
2.36 mm (#8)	84	459	80.9	80 - 90	14.97
1.18 mm (#16)	84	375	66.8	55 - 80	14.97
600 μm (#30)	139	236	42.1	30 - 60	24.78
300 μm (#50)	145	91	16.2	10 - 25	25.85
150 μm (#100)	70	21	3.7	2 - 8	12.48
75 μm (#200)	16	5	0.9	0 - 2	2.84
PAN	5				0.90
	561		307		100.00
Dried to Constant Weight			Fineness Modulus 3.07		

Table 4. Composition and condition of processed, washed, concrete sand at Pochassic

Constit- uents	Amount, as number of particles in percent								
	In fractions retained on sieves shown below						Condition of whole sample		
	4.75 mm #4	2.36 mm #8	1.18 mm #16	600 μm #30	300 μm #50	150 μm #100	1 fresh, dense	2 moderately weathered	3 very weathered
D	55	59	62	39	20	10	35	1	trace
L	22	22	12	27	5	-	13	1	-
Q	23	19	26	34	75	90	46	-	-
Fines							4		
Totals	100	100	100	100	100	100	98	2	trace

(D) dark rocks	(-) rare or absent	(L) light rocks	(Q) mineral fragments
mica schist & gneiss; phyllite amphibole schist & gneiss amphibolite; gabbro	peridotite (trace) hornfels; chert marble; lime- stone; dolomite other sedimentary rocks (trace)	granitic & related gneisses unfoliated gran- itic rocks pegmatite metaquartzite	quartz feldspars micas amphiboles magnetite; other heavy minerals

facilities. This problem of good material surrounding relatively undesirable areas is common on many mining properties.

Oversize boulders that are too large to enter the field hopper are utilized as another by-product, either on the property where embankments are needed or for sale to customers.

3.10 Main products

The main products are: (1) gravel aggregates of four size grades (9.5 mm, 19.0 mm, and 37.5 mm blends, and 19.0 mm only), which consist of natural pebbles as well as crushed rock, and (2) sand, mainly concrete sand, but also mortar sand. As noted in section 1.6, old sand pits on the northwest side of the property were virtually abandoned by about 1969, and present activities are mainly conducted in the gravels of the mid section of the alluvial fan.

Table 3 shows the sieve fractions of concrete sand which has been screened and washed in the plant. The stockpile and material shipped from it are sampled daily, or as needed. Samples are dried before weighing and sieving. Records are kept in the scale house. The fineness modulus is calculated according to ASTM C 136, Sieve analysis of fine and coarse aggregates, although at least one other method seems to be in general use. The tolerances for concrete sand are small, normally 0.2. In this sample, no more than about one percent finer than the #200 sieve remains on the pan, and the total finer than the #100 sieve is about four percent, both of which are within allowable limits.

Table 4 shows the composition and condition of the same concrete sand sample in percent by particle count. The worksheets, with counts as high as 800 particles per fraction (300 particles being a commonly accepted minimum number), have been omitted for lack of space. The preponderance of dark rock particles (D) on the three coarser sieves and of quartz particles (Q) on the two finer sieves is notable. These two constituents and the light rock particles (L) share almost equal thirds of the 600 μm fraction.

3.11 Environmental concerns

Environmental concerns at Pochassic are minor, for the deposit is in mountainous terrain and isolated at the end of a side road. Although concrete sand is a principal product, concrete itself is not manufactured at this remote site.

Many years ago, a feeder canal crossed the property, but its route has long since been abandoned. The relationship between the level of the canal and the water table is not known. In the 1950's a major interstate highway, the Massachusetts turnpike, Route 90 on Figure 1, with no access in the Pochassic area, further isolated the property.

Because there is no clay or other impervious material to impede the downward percolation of surface water, there are few perennial springs in the immediate area. The only ponds are two that receive water with fines from the spiral screen washer at the end of the treatment cycle. This water, pumped to the plant from the nearby Westfield River, seeps into the underground aquifer through natural sand that forms the banks of the ponds. When washing and other operations cease with the onset of winter, no more water enters these ponds, and the level subsides. After a few weeks, the fines are exposed on the bottom of the first pond, the water table being below the full pond. The fines are then removed as tailings to be stockpiled and sold as rewash. The same settling pond can be used season after season. The second pond is for overflow.

Finally, in regard to the environment and the use of this property, land reclamation is not planned at this point. Any eventual reclamation such as restoration to forest land, conversion to agriculture, landfill use, or housing development could involve several applications of environmental geotechnology. Under different conditions in Colorado, West Germany, and many other locales, full reclamation of open workings of gravel and coal has been undertaken as part of a long-term plan.

3.12 Discussion

The principal products from the Pochassic plant are aggregates of gravel and crushed stone and aggregates of sand. Among the latter, concrete sand is the most in demand and is the type discussed in this report. Mortar sand, rewash, and fill also are marketed as lower cost items.

Background references to the industry include Schenck and Torries (1975) on crushed stone aggregates and Dunn (1975) on sand and gravel aggregates. These may be regarded as at least semi-official by reason of their publication by the American Institute of Mining, Metallurgical, and Petroleum Engineers.

Sand and gravel have played a major part in industrial development, and production figures are useful as an economic index. In the United States, the annual value of sand and gravel surpassed $1,000,000,000 for the first time in 1968, and has remained at more than twice that sum since 1977. The quantity mined closely approached 1,000,000,000 short tons in 1978, but then declined. These statistics, while clearly incomplete, combine both construction and industrial categories. Comparisons of the figures quoted with those for other mineral commodities may help to put them in perspective. Thus, in 1968 the value of phosphate rock produced was $250,000,000, and in 1978 the quantity of portland cement was 80,000,000 short tons.

Sand and gravel formations on the surface have numerous uses, for instance as aggregate at Pochassic, as large-scale filters in municipal water-supply systems, and as long distance links between islands (Farquhar, 1967). Their value, when covered by impervious strata, may also be considerable, both as source rocks for oil and gas and for such purposes as the underground storage of compressed air (Karalis and others, 1982). In these cases, information on hydraulic activity and specific yield usually is more important than the precise composition of the material.

Concrete with stone aggregate and/or concrete sand and cement also have numerous well-known uses in building, transportation, and other industries. Among novel uses was the reported dousing by sand from helicopters and post-construction containment and sealing by concrete of one of the crippled nuclear reactors at Chernobyl near Kiev in 1986.

As far as Massachusetts is concerned, Upper Wisconsinan meltwater deposits ended the Pleistocene era by covering 37 percent of the area (Stone, 1982). Over 2000 morphosequences are mappable at quadrangle scale, including the Pochassic deposit. The total can be combined into more than 200 map units that have formed the basis for a regional chronostratigraphic framework, which is still being refined. The Pochassic deposit has been noted as part of the second stream terrace in a series of five. Although primarily depositional in nature as an alluvial fan, its upper surface was terraced by erosion. What percentage, if any, of the deposit was lost by this process has

not been assessed.

Because of numerous glacial and fluvial deposits, Massachusetts continues to be a major producer as well as consumer of sand and gravel in its several forms. Although rock in any form is a high bulk, short haul commodity, various aggregates are transported by rail and truck, both in and out of the state. Four regional reports are pertinent: on sand and gravel supplies for the Boston area (Marden, 1967); on geologic methods used to test and explore gravel deposits (Sinnott, 1967); on quality control of aggregates for concrete construction (Van Epps, 1967); and on siting for aggregate production (Barton, 1982).

SUMMARY

In this study, equal emphasis is placed on the geology, petrography, and processing of sand and gravel comprising an alluvial fan in Massachusetts. The geology, with only stream-flood and channel beds represented, controls the types of materials available for use as aggregates. Petrographic analysis is needed to observe and monitor both the geologic materials and the various aggregates which are, respectively, the input and output of the processing plant.

The report considers the separate but related geotechnical and resource aspects of the geologic unit, also briefly focusing on environmental concerns. A suite of rocks in fragmental form is traced from its source in glacial till and melting ice through successive stages of sedimentary transport and layered deposition in the alluvial fan to river terracing, marginal erosion, and limited weathering. Then, the suite is followed through sampling and analysis to large scale pit extraction, processing, further analysis, and finally stockpiling of crushed stone and treated sand.

ACKNOWLEDGMENTS

Sincere thanks are due to reviewers who made suggestions about the manuscript, including Allan Filipov, Donald Houghton, and John Hubert. Thanks are also due to Muriel Boisseau, Ester Cressotti, Marie Litterer, Danny Nason, Fred Nason, and Karen Thatcher. The support and encouragement of Harry Lane are gratefully acknowledged. Lastly, contacts with A.S. Balasubramaniam and Sarvesh Chandra were always a pleasure.

REFERENCES

Allen, J.R.L. 1984. Sedimentary structures, their character and physical basis. Amsterdam: Elsevier.

American Society for Testing and Materials. 1983. Annual book of ASTM standards, 04.02, Concrete and mineral aggregates. Philadelphia: ASTM.

Barton, W.R. 1982. Siting for aggregate gate production in New England. In O.C. Farquhar (ed.), Geotechnology in Massachusetts, 375-377. Amherst: Univ. Massachusetts Graduate School.

Borns, H.W., Jr., and others (eds.). 1985. Late Pleistocene history of northeastern New England and adjacent Quebec. Boulder: Geological Society of America, SP 197.

Collinson, J.D. and D.B. Thompson. 1982. Sedimentary structures. London: Allen & Unwin. Not cited.

Collis, L. and R.A. Fox (eds.). 1985. Aggregates: sand, gravel, and crushed rock aggregates for construction purposes. London: Geological Society, Engineering Geology SP 1. Not cited.

Colton, R.B. and J.H. Hartshorn. 1970. Surficial geologic map of the West Springfield quadrangle, Massachusetts-Connecticut. Washington: U.S. Geological Survey, GQ 892.

Dunn, J.R. 1975. Aggregates--sand and gravel. In Industrial minerals and rocks, 97-108. New York: AIMMPE.

Farquhar, O.C. 1967. Stages in island linking. In H. Barnes (ed.), Oceanogr. Mar. Biol. Ann. Rev. 5, 119-139. London: Allen & Unwin.

Filipov, A.J. 1986. Sedimentology of debris-flow deposits, west flank of the White Mountains, California. M.S. thesis, Univ. Massachusetts, Amherst.

Filipov, A.J. and J.F. Hubert. In preparation.

Fraser, G.S. and L.J. Suttner. 1986. Alluvial fans and fan deltas: an exploration guide. Boston: International Human Resources Development Corporation.

Holmes, G.W. 1968. Preliminary materials map of the Woronoco quadrangle, Massachusetts. Washington: U.S. Geological Survey, OFR 68-136.

Hooke, R. LeB. 1967. Processes on arid-region alluvial fans. Journ. Geol., 75, 438-460.

Hunt, C.B. 1986. Surficial deposits in the United States. New York: Van Nostrand Reinhold. Not cited.

Karalis, A.J., E.J. Sosnowicz, O.C. Farquhar, and R.B. Schainker. 1982. Procedures for evaluating a compressed air energy storage system in a utility system. Amer. Inst. Aeron. Astron., Inter. Conf., San Francisco, 82-1638, 25-31. New York: AIAA.

Klein, G. deV. 1985. Sandstone depositional models for exploration for fossil fuels. Boston: International Human Resources Development Corporation.

Koster, E.H. and R.J. Steel (eds.). 1984. Sedimentology of gravels and conglomerates. Canadian Society of Petroleum Geologists Memoir 10.

Kukal, Z. 1971. Geology of Recent sediments. New York: Academic Press. Not cited.

Larsen, F.D. 1972. Surficial geology of the Mount Tom quadrangle, Massachusetts. Ph.D. thesis, Univ. Massachusetts, Amherst.

Leeder, M.R. 1983. On the interaction between turbulent flow, sediment transport, and bedform mechanics in channelized flows. In J.D. Collinson and J. Lewin (eds.), Modern and ancient fluvial systems, Spec. Publs. Int. Ass. Sediment. 6, 5-18. Oxford: Blackwell.

Lewis, D.W. 1984. Practical sedimentology. Stroudsburg: Hutchinson Ross. Not cited.

Marden, P.R. 1967. Sand and gravel supplies for the Boston area. In O.C. Farquhar (ed.), Economic geology in Massachusetts, 67-70. Amherst: Univ. Massachusetts Graduate School.

McCabe, P.J. 1977. Deep distributary channels and giant bedforms in the Upper Carboniferous of the central Pennines, northern England. Sedimentology 24, 271-290.

Miall, A.D. 1983. Basin analysis of fluvial sediments. In J.D. Collinson and J. Lewin (eds.), Modern and ancient fluvial systems, Spec. Publs. Int. Ass. Sediment. 6, 279-286. Oxford: Blackwell.

Pierce, K.L. 1966. Bedrock and surficial geology of the McConnellsburg quadrangle, Pennsylvania. Atlas 109a. Harrisburg: Pennsylvania Geological Survey.

Postma, G. 1986. Classification for sediment gravity-flow deposits based on flow conditions during sedimentation. Geology, 14, 291-294.

Ramos, A. and A. Sopena. 1983. Gravel bars in low-sinuosity streams (Permian and Triassic, central Spain). In J.D. Collinson and J. Lewin (eds.), Modern and ancient fluvial systems, Spec. Publs. Int. Ass. Sediment. 6, 301-312. Oxford: Blackwell.

Reading, H.G. (ed.). 1985. Sedimentary environments and facies. 2nd ed.

Oxford: Blackwell.

Retelle, M.J. and K. Konecki. 1986.
Proximal-distal glaciomarine facies
relations: sedimentation at a pinning
point in the lower Androscoggin valley,
Maine. Geol. Soc. Amer. Abs. with
Programs 18, 1, 62.

Schenck, G.H.K. and T.F. Torries. 1975.
Aggregates--crushed stone. In Indus-
trial minerals and rocks, 66-84. New
York: AIMMPE.

Schnabel, R.W. 1971. Surficial geologic
map of the Southwick quadrangle, Massachusetts and Connecticut. Washington:
U.S. Geological Survey, GQ 891.

Selley, R.C. 1985. Ancient sedimentary
environments and their sub-surface
diagnosis. 3rd ed. Ithaca: Cornell
University Press. Not cited.

Sinnott, J.A. 1967. Geologic methods
used to test and explore gravel deposits. Same volume as Marden, 137-139.

Stone, B.D. 1982. The Massachusetts
state surficial geologic map. Same
volume as Barton, 11-27.

Van Epps, R.J. 1967. Quality control
of aggregates for concrete construction. Same volume as Marden, 62-66.

Zuffa, G.G. (ed.). 1985. Provenance of
arenites. Dordrecht: Reidel. Not
cited.

Environmental Geotechnics and Problematic Soils and Rocks, Balasubramaniam et al. (eds)
© *1987 Balkema, Rotterdam. ISBN 90 6191 785 9*

Typical landforms and slope stability of the Zagros Mountains along the Shiraz-Boushire road in Southwestern Iran

Asian Institute of Technology, Bangkok, Thailand

G. Rantucci

Asian Institute of Technology, Bangkok, Thailand

ABSTRACT: The instability of landforms is a main conditioning aspect in selecting the alignment of roads in mountainous areas. This paper describes typical structurally-controlled landforms of the Zagros Mountains along the Shiraz-Boushire road in Iran and related slope stability problems.

1 INTRODUCTION

In highway design practice roads are broadly classified as flat, rolling and mountainous, depending on the type of topography crossed by the alignment. Obviously along any road one or more different types of morphology may be encountered with different effects on design and construction.

The influence of the landforms is, in general, quite remarkable at the decision-making stage since the morphology is both a conditioning factor for the selection of the alignment and a determining input for the design of concrete structures, not to mention the cutting of the natural slopes.

In the case of flat morphology in tropical and subtropical zones, one can expect a number of major and minor drainage structures, as well as protection works to prevent erosion problems at the foot of embankments, but only limited or irrelevant alignment constraints. In other words the hydrological aspects may have to be investigated very carefully for an adequate design of concrete structures and a safe embankment height to prevent flooding of the road platform.

In mountainous areas the irregular topography is generally the main constraint on the alignment and the design engineer usually has to compromise between elevation variations affecting the alignment, stability of slopes, and major and minor drainage structures.

The intermediate condition, that is the case of undulating terrain may often result in a physical environment in which the alignment can be easily defined among a number of acceptable alternatives, being constraints of the topography, hydrology and geology relatively relevant.

There are cases, however, as in some areas of Southwestern Iran, where morphology and geology can combine in such a way that landforms are unstable on a regional scale. The largest recorded landslide on earth, occurred in Iran (Saidmarreh), in the Lurestan region of the Southwestern Zagros Belt (Voight 1981). The slide, which affected the northern lank of the elongated Kabir Kuh anticline, intersected the Asmari limestones forming a high hogbak ridge. The volume of the Saidmarreh landslide was estimated at nearly 20 cubic kilometers. On a smaller scale but in the same region a widespread instability of landforms occured along the Shiraz-Boushire Road mainly in the Kazerun-Dalaki section (Fig. 1).

During the design stage of that section several alignment alternatives were studied, each of them characterized by remarkable differences in topography, geology and stability of slopes.

The purpose of this paper is to describe the main landforms along the Kazerun-Dalaki section, both for the interesting problems met with during the design and the construction of the road and because this area is characterized by the most typical morphologies of the Zagros Mountains.

2 IMPACT OF ROAD CONSTRUCTION PROJECTS ON THE NATURAL ENVIRONMENT AND AN INTERDISCIPLINARY APPROACH TO ROAD DESIGN

In the realm of physical modification of the landscape, highways, are by far, one of the most evident signs, both for the extent of the land used and for the widespread landscape reshaping involved.

The impacts of road projects on the physical environment can be summarized as follows:

- intense modification of the ground surface, due to the huge quantities of earthworks often involved in modern road construction work; quarry sites usually remain as deep wounds in the environment even many years after the construction has been completed and the road has adapted to the environment;
- alteration of the natural drainage system; although this is done in such a way that both surrounding landscape and road platform should not suffer any negative effect, neverthless, due to difficulties in hydrological predictions, the induced modifications can result in inadequate design, with negative impact on the environment;
- modifications induced by rock blasting activity; in sound rocks vibrations do not usually induce stability problems, but in weak formations and loose soils a marked destabilizing effect can be induced;
- modification of the existing stability of natural slopes, by steepening or undercutting them; this may induce potential failures in slopes which are close to the limiting equilibrium condition;

- modification of the existing vegetation mantle and alteration of its protective function against surface waters.

Since the physical environment is a complex system resulting from the interactions of factors such as geology, tectonics, morphology and climate, only an interdisciplinary approach can provide a proper evaluation of individual aspects and overall behaviour.

3 THE SHIRAZ-BOUSHIRE ROAD

3.1 General

The design of the Shiraz-Boushire road started in the mid-sixties but the middle zone, between Kazerun and Dalaki, was not studied until the beginning of the seventies, mainly due to difficulties in selecting the alignment (Fig.1). This area is in fact characterized by a very irregular morphology, with complex but typical landforms and widespread instability on the natural slopes. A detailed final design study was performed during 1970-71, based on aerophoto interpretation of geological and geophysical investigations, mainly divided towards defining the stability of the slopes.

3.2 Climate, rainfall and vegetation

The present climate of the Shiraz-Boushire zone is characterized by high temperature variations and low rainfall. In the Kazerun-Dalaki zone, during the field investigation campaign, minimum temperatures close to 5 degrees C in winter and maximum temperatures close to 45 degrees C in summer were recorded, with a relatively high daily excursion range (10 to 20 degrees C). This day-night variation of temperature, usually producing intense mechanical weathering with surficial disintegration of rock outcrops, affects sandstone sequences with a devastating exfoliation process.

The Asmari limestones forming the core of most Zagros anticlines are less sensitive to the thermal effect than other formations. Marine clays and soft rocks, such as the marls, react to temperature variations by drying and cracking, both favourable to surface wash and water penetration.

The mean annual precipitation is between 300 mm in Shiraz and about 200 mm in Boushire, which is typical for the hot-dry climate of the central Zagros. Precipitations are usually of short duration but of high intensity with marked erosional effects. Little vegetation is present except near villages and rivers. The ground surface is thus totally unprotected from the rain action.

3.3 Geology and tectonics

The Tertiary and Mesozoic sediments of the southwestern Iranian basin were folded into a sequence of plunging elongated anticlines and sinclynes during the latest Alpine movements in Plio-Pleistocene times (Fig.1). During that period the Arabian Shield moved towards the Russian Platform and as consequence of this compression the whole area was faulted, folded and thrusted with the build up of the Zagros Mountains (Fig. 2). The geological sequence involved in the folding was basically composed of thick well-bedded limestones (Asmari formation) of Creta-Eocene age, forming the core of the anticlines, covered by alternations of sandstone, clay and marls of younger age and interbedded with evaporites. This stratigraphic sequence, characterized by a very rigid calcareous core underlain by alternances of resistent and weak, locally soluble, rocks was intensely worked and modeled by tectonics and climatic agents.

3.4 Selection of the alignment

During the design stage particular attention was devoted to the Kazerun - Dalaki zone, where the presence of very unstable landforms made the selection of the alignment quite problematic. Due to the unfavourable combination of surface features and weakness of materials, several alignments were considered and two major alternatives among them were studied in detail (Fig. 3).

On these two main routes (Fig.4) geological and slope stability mapping as well as seismic refraction and drilling investigations were performed.

The major difficulty in selecting the alignment was that the whole area has relatively marked altimetric variations and unstable morphologies, which are often difficult to adapt to the design requirements accepted in modern road projects.

3.5 Typical landforms

As result of the interactions between geology, tectonics and surface erosion, a structurally-induced morphology, evolved on sedimentary sequences, was built up in Southwestern Iran. The Zagros Mountain Belt, is in fact characterized by a NE-SW sequence of plunging, intact or breached anticlines and synclines. Together with these typical morphologies, cuestas and homoclines, hogbacks ridges, bad lands and karst landforms are encountered (Figs. 4 and 5).

The Kazerun-Dalaki zone represents an ideal example of folded style morphology imposed over deep-sea sediments, with typical cases of selective erosion. Mechanical weathering, the absence of vegetation and a devastating surface wash all contribute to the intense disintegration of surface rocks and the denudation of the landscape.

A different rate of change currently affects the landforms and the landscape is under a marked geological impact. Various types of slow and rapid mass-movements take place along the natural slopes because of the combined effects of geology, topography and rainfall.

Some cases of typical Zagros Mountains morphologies, related evolutionary processes and slope-stability problems, are described below:
a) Sequence of anticlines and synclines (Alternatives 1 and 2). Undulated and partially breached and plunging anticlines separated by synclinal

walleys characterize the land-scape near the town of Kazerun (Fig.4). The anticline in Fig.6 is gently undulated, the structure is solid, the rock is composed of medium bedded limestones, but the flanks are locally incised by streams with marked erosion phenomena. A dipping angle of 35-40 degrees usually characterize both the natural slope and the sedimen-tary strata along the flanks of the anticline; the slope may be locally unstable since ideal plane failure conditions exist and upper slope tension cracks are frequent on the fold crest because of the past tensile-stress condition. The floor of the Kazerun valley, situated between two parallel anticlines, is entirely covered by thick gravelly-sand water-bearing sediments.

Whenever possible the align-ment was located on synclinal valleys, which are stable and rich in good quality construc-tion materials.

b) Tabular morphology (Alterna-tive 2). Examples of tabular landforms with varietes such as homoclines and cuestas are widely represented along major and minor alternatives.

Quasi-horizontal sequences of sedimentary strata form this type of morphology, which is characterized by the presence of a resistent layer on top and soft or easy-erodible la-yers below (Figs.4 and 7).

Normally the slope retreats for the fall of blocks forming the resistent layer and for the consequent exposition of underneath materials, usually quickly eroded by the surface waters. If the the foot slope is eroded by a stream, as in Fig.7, and the scarp is formed of a melange of blocks and loose soil, rotational and translational types of fai-lure can occur. This type of slope is basically unstable since the hard sandstone layer on top has a number of weak-ness planes, esily turning into tension cracks.

The fall of blocks from the top resistent layer is the fundamental reason for keeping the alignment as far as possi-ble from the foot of the slope.

c) Badlands (Alternative 1, km 8) Bad lands are frequent in quasi-horizontal sedimentary sequences whenever a thick layer of unconsolidated clay, originally packed between so-lid rock strata, become ex-posed by the fast retreat of the upper protecting roof (Figs.4 and 8).

The surface exposed to rain-drop impact and to the rainwash action becomes dissected into a closely-spaced network of major and minor channels. This type of landscape, characteri-ed by a fine-textured gully dissection, caused by the low permeability of the materials is usually dominated by cata-strophic surface erosion.

Depending on local conditions, such as river erosion at the toe of the scarp slope retreat may occur under rota-tional and translational types of failure.

Extensive areas of badlands (Fig 8) present a serious problem for road alignment location since the surface soils are characterized by very low mechanical properties, while the morphology is repre-sented by a closely-dissected type of relief.

d) Karst landforms (Alternative 1, km 5-13). The left bank of the Shapoor River between km 5 and km 13 is mainly eva-porites composed of marine clay strata alternating to gipseous levels (Figs. 4 and 8). The most characteristic feature of this zone is the presence of numerous sinkholes 10 to 20 m in diameter and several tens of m in depth, connected to the underground drainage system controlled by the Shapoor River.

The several well-shaped and equally-spaced sinkholes form an intricate network of cavities imposed over evaporitic sedimentary strata. The collapse of such cavities was found to have occurred along the proposed alignment, and for this reason the area was considered quite unsuitable for the location of the road.

e) Hogback ridges (along the Darreh Shur and Dalaki Rivers on both Alternatives).

Wide zones of hogbacks exist on the two Alternatives (Fig.4). The first zone, along the Darre Shur River on the Alternative 2 at the km 10 - 15, is really the most spectacular and is represented by very steep limestones ridges parallel to the river stream.

The second zone along the Dalaki River (Figs.4 and 9) is similarly made of triangular shaped facetes (flatirons) resulting from the differential erosion of limestones. The evolution of this type of landform is basically slow and the conditions are ideal for the toppling of blocks. Very few cases of footslope erosion with the plane type of failure were observed. A variety of local conditions can combine with hogback ridges. Basically, however, their slope evolution occurs at a slow rate due to the fact that the slope is mostly composed of a very steep resistent layer acting as a retaining wall. Given the marked steepness of this type of slope, hogback ridges are unsuitable for the location of the alignment.

4 CONCLUSIONS

The final design of the Kazerun-Dalaki road section, based on aerophoto interpretation, geology and tectonics on one side, and seismic refraction and drilling investigations on the other side, enabled to select the final alignment. The solution adopted, for Boushejan-Kamarej-Banat-I.P.46-Kunar Takteh-Point K-Dalaki, combined sections of major and minor alternatives in an attempt to minimize problems, but keeping the construction costs at a reasonable level.

Neverthless, after the construction, the instability of some slopes and swelling of soft rocks (marl) on the floor of tunnels close to the Dalaki River were major problems.

REFERENCES

Dov Nir, 1983. Man a geomorphological agent. An Introduction to anthropic geomorphology. D. Reidel Publishing Co.
Spencer E.W, 1977. An introduction to the structure of the Earth. McGraw Hill.
Twidale C.R., 1971. Structural landforms. Australian National University Press, Canberra.
Voight Barry, 1979. Rockslides and avalanches. Elsevier.
Young A., 1972. Slopes. Oliver & Boyd, Edinburgh.

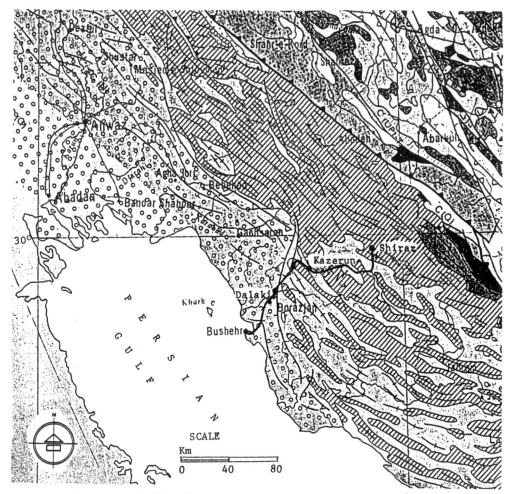

Fig. 1 - Tectonic Map of Southwestern Iran showing the folded
Belt of Zagros, the Thrust Zone and the location of
Shiraz-Boushire Road.

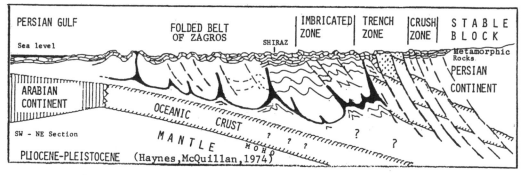

Fig. 2 - Tectonic model for development of the Zagros Mountains
Folded Belt. According to plate-tectonics theory the
movement of the Arabian Continent towards the fixed
Persian-Russian Plate produced first subduction and
deep-water sedimentation and later intense shearing
and imbricate overthrusting (E.W. Spencer, 1977).

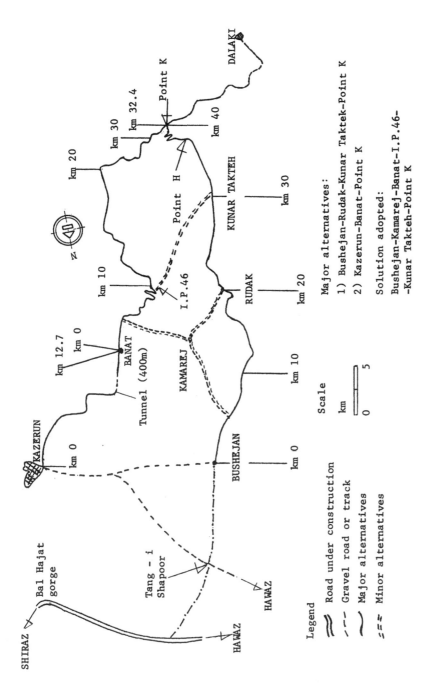

Legend

⟋⟋ Road under construction

⟋⟋ Gravel road or track

⟋⟋ Major alternatives

⟋⟋ Minor alternatives

Scale

km
0   5

Major alternatives:

1) Bushejan-Rudak-Kunar Taktek-Point K

2) Kazerun-Banat-Point K

Solution adopted:

Bushejan-Kamarej-Banat-I.P.46-
-Kunar Takteh-Point K

Fig. 3 - Overview of major and minor alternative alignments studied along the
Shiraz - Boushire Road, in the zone between Kazerun Valley and Dalaki
Village.

327

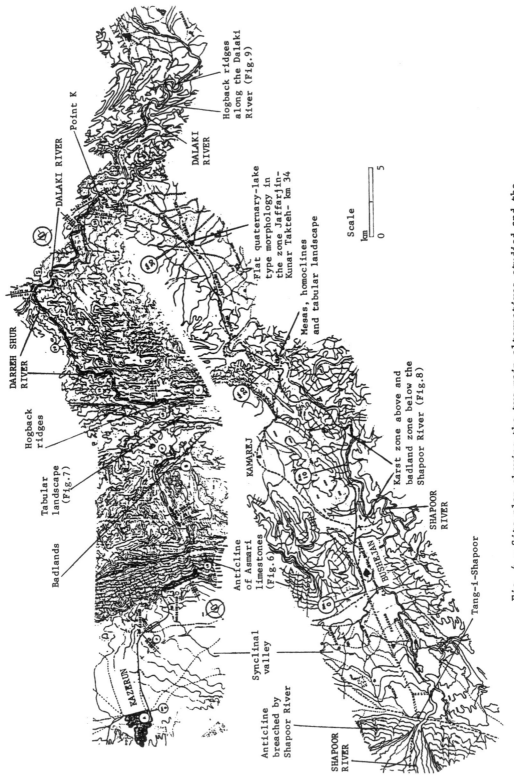

Hogback ridges
along the Dalaki
River (Fig.9)

DALAKI RIVER

Point K

DALAKI
RIVER

Flat quaternary-lake
type morphology in
the zone Jaffarjin-
Kunar Takteh- km 34

DARREH SHUR
RIVER

Mesas, homoclines
and tabular landscape

Hogback
ridges

Scale

km   5
    0

Tabular
landscape
(Fig.7)

KAMAREJ

Karst zone above and
badland zone below the
Shapoor River (Fig.8)

Badlands

SHAPOOR
RIVER

Anticline
of Asmari
limestones
(Fig.6)

BUSHKAN

Synclinal
valley

Tang-i-Shapoor

KAZERUN

Anticline
breached by
Shapoor River

SHAPOOR
RIVER

Fig. 4 – Site plan showing the two main alternatives studied and the
most typical landforms

328

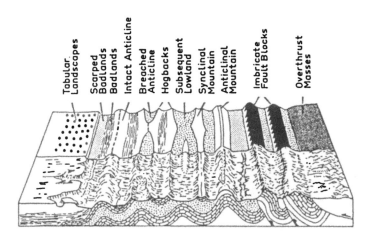

Fig. 5 - Block-diagram of fold patterns and typical landforms
in the Zagros Mountains (Oberlander, 1965).

Fig. 7 - A flat-laying, massive bed of sandstone forming a wide
structural plane and resting on soft marine clays. Debris
of the collapsed resistent layer cover the slope on the
right hand side (Alternative 2, km 8-10).

329

Fig. 6 – Typical features of the elongated plunging anticline between the villages of Kazerun and Banat. The Asmari limestone formation of Oligo-Miocene time forms the resistent core of the structure, which is breached by incised streams locally named "Tangs".

Fig. 8 - Typical landforms at about the km 8 of the Alternative 1. The
upper zone, above the Shapoor River,is characterized by karst
morphology with numerous sinkholes, 10 to 30 m in diameter,
cutting through evaporites interlayered with gray clay. The
lower zone shows the closely-dissected and fine-textured topography of badlands, imposed over the gray marine clay formation

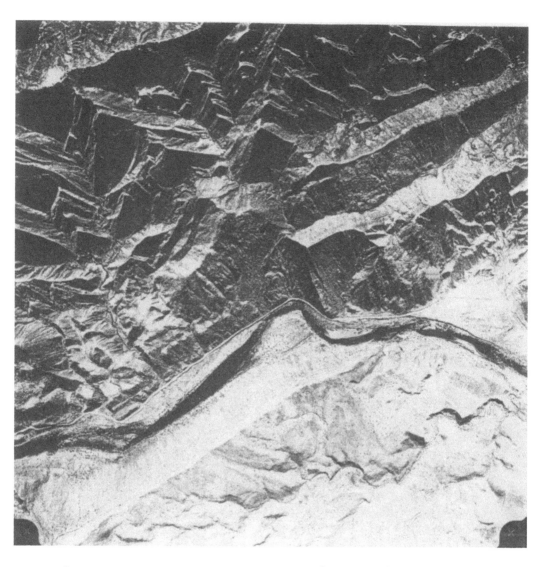

Fig. 9 - Hogback ridges along the Dalaki River (see Fig. 4). The sedimentary
sequence is basically made of steeply-dipping resistent strata alternating to weak layers. Ridges of triangular limestone facets, produced by differential erosion, are isolated by V-shaped incised valleys.

Section C
Environmental geotechnical aspects of natural hazards

Environmental Geotechnics and Problematic Soils and Rocks, Balasubramaniam et al. (eds)
© *1987 Balkema, Rotterdam. ISBN 90 6191 785 9*

Effects of forest roads on watershed function in mountainous areas

Walter F.Megahan
United States Department of Agriculture, Forest Service, Intermountain Forest and Range Experiment Station, Boise, Idaho

ABSTRACT: Numerous examples from mountainous areas in the interior Western United States are used to illustrate the effects of forest road construction on watershed functions. Impacts can occur at onsite and downstream locations. Onsite effects include reduced forest productivity, increased runoff, and accelerated surface and mass erosion. Downstream, roads can change streamflow rates, water quality and channel characteristics. Accelerated erosion and resulting sedimentation are the most common and serious kinds of watershed damages caused by road construction. Basic principles for reducing road erosion and sedimentation impacts are presented.

1 INTRODUCTION

The effects and impacts of forest road construction on watershed function and associated downstream damages are a major concern to forest land managers and other interested parties at many locations. In my attempt to present an overview of how watersheds respond to road construction, I will limit my discussion to road construction activities on forested lands in the mountainous areas of the interior Western United States, which includes the Rocky Mountain chain, the various ranges in the basin and range province, plus interior mountain ranges in the upper Columbia River basin. Results of studies from within this area will be used to document my discussion whenever possible. Otherwise, studies from other areas, mostly in the Western United States, will be used. Although the research cited is regional in nature, trends in watershed responses and general principles for reducing impacts have widespread application.

In this discussion, a road is defined as a vehicle corridor built to specific design standards and requiring movement of excavation material during construction. This definition discriminates between roads requiring cut and fill construction operations and the skid trails and landings required for timber harvest activities. The distinction is made because the hydrologic impacts of the two operations are different enough to warrant treating them separately.

Effects of road construction on watersheds are manifest both onsite and downstream. Onsite effects are measured in terms of changes in site productivity. Downstream effects occur because of changes in the hydrologic function of the watershed and are measured in terms of changes in (1) the volume and timing of streamflow, (2) water quality, including water chemistry and sediment load, and (3) channel morphology.

2 SITE PRODUCTIVITY

Forest roads require preemption of timber producing lands for roadbeds, cuts, and fills. Presumably, construction limits site productivity by exposing low nutrient subsoils, reducing soil water holding capacity, and compacting surface materials. A common method of evaluating the effects of roads is to equate productivity loss to the total area of road disturbance. For an individual road, the amount of area disturbed depends on the road design features and the hillside gradient. For a timber sale, the number of roads required varies with the type of logging system. A summary of 11 studies in the Western United States documenting the percentage of area involved in road construction showed a range from a high of 30 percent for early tractor and jammer logging systems to a low of 1 percent for a helicopter operation (Table 1). Most current timber harvest operations in the interior West usually do not exceed 10 percent of the area in roads.

However, the total area of road disturbance may not be a good measure of the loss of site productivity. One important consideration is the fact that road fills, although composed mainly of lower nutrient subsoils, are deeper and usually have more available water than adjacent undisturbed soils with a vegetation cover. This provides a good growth medium for trees (Megahan, 1974b) and grasses, herbs, and shrubs (Monsen, 1974). Fills comprise at least one-third of the total road width, so that actual productivity losses from road construction might amount to about two-thirds of the total road area.

Evidence shows that productivity losses amount to even less than the remaining two-thirds of the road area because of changes in the moisture regime and the thinning effect when trees are removed along the road right-of-way. The increased light and moisture stimulates the growth of trees adjacent to the road compared to preroad conditions. Pfister (1969) found that increased growth below outsloped jammer roads completely offset the timber volume lost to road construction on 50-year-old white pine (Pinus monticola, Douglas) stands in northern Idaho. The total area of roads accounted for 10 percent of the timber stands in the area. These mitigating effects may vary depending on road width, whether the roadbed is sloped into or away from the hillside, and timber stand properties (stands of mixed ages and species as compared to pure stands and plantations). However, they do suggest that productivity losses from roads can be considerably less than the area disturbed by construction.

3 WATERSHED HYDROLOGIC FUNCTION

Road construction can change the hydrologic function of watersheds leading to possible changes in the rates and quality of streamflow and in channel morphology. Such changes reflect modifications of hydrologic processes within the watershed caused by (1) reduced evapotranspiration within the road right-of-way; (2) changes in snow accumulation and melt rates; (3) removal or reduction of protective soil cover; (4) destruction or impairment of natural soil

structure and fertility; (5) increased slope gradients created by construction of cut and fill slopes; (6) decreased infiltration rates on parts of the road; (7) interception of subsurface flow by the road cut slope; (8) decreased shear strength, increased shear stress, or both, on cut and fill slopes; (9) concentration of runoff and intercepted water on and below the road; (10) detachment and movement of sediment particles by traffic; and (11) direct encroachment in the stream channel by construction activities. Many of these adverse effects tend to be proportional to slope gradients. Thus, compared to other locations, road construction in mountainous areas provides greater potential for hydrologic impacts.

3.1 Streamflow Rates

Reduced evapotranspiration within the road right-of-way plus changes in the runoff patterns within the road prism can affect streamflow rates. Studies throughout the world document that reduced evapotranspiration associated with forest removal tends to increase the total volume of streamflow (Bosch and Hewlett, 1982). However, not one of the 94 studies reported showed detectable water yield increases for removal of less than 15 percent of the total timber stand. Only 7 percent of the studies showed increases for removal of less than 30 percent of the total timber stand. Most current road construction involves less than 15 percent of the total timber stand, so measurable increases in water yields are not expected from road construction simply in response to reduced evapotranspiration.

Other factors such as compaction, removal of permeable surface soils, and increased gradients on cut and fill slopes all tend to reduce infiltration rates on roads. In most forest stands, this causes overland flow runoff where none existed before. In addition, incision of the slope by construction of road cuts may intercept subsurface flow zones at many locations in the mountains (Burroughs et al., 1972; Megahan, 1983). Water intercepted by this means is transformed from slow subsurface flow to rapid overland flow on the road. Such changes in runoff within the road prism have the potential for changing watershed streamflow rates. Studies in the coast range of Oregon (Harr et al., 1975) and California (Ziemer, 1981) reported no increases in streamflow from rainstorms following road construction if less than

Table 1.--Soil disturbance from roads for alternative methods of timber harvest in the Western United States and Canada.

Method of Harvest	Percent of Logged Area Bared			Location	Reference
	Roads	Skid roads & landings	Total		
Tractor - clearcut	30.0	---	30.0	British Columbia	Smith 1979
Tractor - selection	3.0	6.0	9.0	California	Rice & Wallis 1962
Tractor - selection	2.2	6.8	9.0	Idaho	Haupt & Kidd 1965
Tractor - group selection	1.0	6.7	7.7	Idaho	Haupt & Kidd 1965
Tractor - partial clearcut[1]	1.8	---	1.8	Idaho	King & Tennyson 1984
Tractor - partial clearcut	3.0	---	3.0	Idaho	King & Tennyson 1984
Tractor - partial clearcut	4.3	---	4.3	Idaho	King & Tennyson 1984
Tractor - selection	5.0	10.0	15.0	California	Ziemer 1981
Tractor & helicopter - fire salvage	4.5	0.4	4.9	Washington	Klock 1975
Tractor & cable - fire salvage	16.9	---	16.9	Washington	Klock 1975
Jammer - group selection	25-30	---	25-30	Idaho	Megahan & Kidd 1972
Jammer - clearcut	8.0	---	8.0	British Columbia	Smith 1979
High lead - clearcut	14.0	---	14.0	British Columbia	Smith 1979
High lead - clearcut	6.2	3.6	9.8	Oregon	Silen & Gratkowski 1953
High lead - clearcut	3	1	4.0	Oregon	Brown & Krygier 1971
High lead - clearcut	6	1	7.0	Oregon	Brown & Krygier 1971
High lead - clearcut	6	---	6.0	Oregon	Fredriksen 1970
Skyline - clearcut	2.0	---	2.0	Oregon	Binkley 1965
Skyline - clearcut	1.0	---	1.0	British Columbia	Smith 1979
Skyline - partial clearcut[2]	3.9	---	3.9	Idaho	King & Tennyson 1984
Skyline - partial clearcut	2.6	---	2.6	Idaho	King & Tennyson 1984
Skyline - partial clearcut	3.7	---	3.7	Idaho	King & Tennyson 1984
Helicopter - clearcut	1.2	---	1.2	---	[3]

[1] Percent watershed area in roads, approximately 16 to 34 percent of watershed clearcut.

[2] Percent watershed area in roads, approximately 36 to 52 percent of watershed clearcut.

[3] Estimated by Virgil W. Binkley, Pacific Northwest Region, USDA Forest Service, Portland, OR.

337

12 percent of the area disturbed was in roads. Recent research on our study watersheds in northern Idaho suggests more variable results in the snow zone of the interior West. King and Tennyson (1984) found statistically significant changes in spring snowmelt runoff on two of six study watersheds with road areas ranging from 1.8 to 4.3 percent. Flow increases for selected time intervals were detected on one watershed with 3.9 percent road disturbance, whereas decreases for selected time intervals were found on another watershed with 4.3 percent disturbance. These changes were attributed to subsurface flow interception. In one case, the intercepted water augmented the normal timing of snowmelt runoff to cause increased flows. In the other case, intercepted flows desynchronized the normal snowmelt runoff pattern for the watershed, thereby leading to decreases in flows.

3.2 Water Chemistry

Road construction has the potential for influencing water chemistry in two ways: by disrupting the normal forest chemical cycling pattern on the watershed, and by introducing chemicals to the watershed system. No published evidence exists showing that roads alone cause detectable changes in forest chemical cycling that are detectable in streamflows. Some studies show increased outflows of natural chemicals following timber harvest. However, these responses were attributed to the effects of timber removal on the ecosystem. Isolation of the effects of roads comprising only a small percentage of the total disturbance has not been done to date.

Chemicals introduced during forest road construction or maintenance, such as bituminous products and calcium chloride, also have a potential for influencing downstream water chemistry. In addition, opportunities for chemical spills on forest roads are becoming greater as mining and other developments increase on forest lands. Although the potential for damage exists and is growing, I have not been able to find any published material documenting changes in water chemistry from either standard construction and maintenance practices or industrial spills on forest roads.

3.3 Channel Morphology

Direct channel encroachment is probably the most common cause of changes in channel morphology following road construction in mountainous areas. Encroachment can occur either at channel crossings or by lateral constriction caused by construction along the length of the stream. Channel crossings, using bridges, culverts, or fords installed without changing the stream gradient, have minimal effect on channel morphology. Constriction of the flow or relocation of the flow path can cause some localized bank or bed erosion or deposition. Channel crossings that alter the stream gradient can have more widespread influence on channel morphology depending on the nature of the construction, including bed and bank erosion or deposition at both upstream and downstream locations. Another important change in channel morphology occurs when road construction in valley bottoms requires lateral channel encroachment. The reduced channel cross-sectional area within the zone of encroachment increases flow velocities that can lead to accelerated bank erosion, downcutting of the streambed, or both. Morphology changes are not only limited to the area of encroachment but may carry far downstream as the channel adjusts to material eroded from the upstream zone of encroachment.

Roads may also have indirect effects on channel morphology. The morphology of a stream channel is dependent on the streamflow and sediment load supplied from the watershed as well as the channel gradient, soils, vegetation, and geology of the area. Schumm (1971) indicated that as sediment loads increase, channel width, gradient, and wavelength tend to increase, and channel depth and sinuosity tend to decrease. After a large storm in 1964, Lyons and Beschta (1983) measured an increase in landslide frequency of 24 and 27 times on logged areas and roads, respectively, compared to undisturbed forested areas in the Upper Middle Fork of the Willamette River in Oregon. The authors showed statistically significant increases in channel width ranging from 25 to 250 percent, predominantly at the junction of the river with its major tributaries. They attributed the increased width to the deposition of accelerated landslide material in the river.

Channel aggradation caused by accelerated erosion in mountainous watersheds is often accompanied by a reduction in the particle size distribution of the streambed. This is because the particle sizes of the eroded watershed soils tend to be smaller than the particle sizes of channel bottom alluvium. Although a minor change in channel morphology, decreases in the particle size of streambed materials have important biological implications especially for fish survival and growth. Megahan et al. (1980) documented maximum sand percentages of about 40 percent on the surface of salmon spawning areas in the South Fork Salmon River in Idaho following large deposits of sediment during the mid-1960s. Much of the accelerated erosion leading to the aggradation of sands was attributed to road erosion on the watershed. A basin-wide road rehabilitation program coupled with a moritorium on new developments rapidly decreased the inflow of sediment to the river. Reduced sediment inflows provided more energy for the removal of fine sediments stored in the river channel. From 1966 to 1979, sand percentages on the surface of spawning areas dropped from an average of 40 percent to an average of about 10 percent. At the same time, concomittant increases in the coarser gravel and rubble materials were documented as the original channel substrate was uncovered.

3.4 Erosion and Sedimentation

Here, the term erosion implies a net loss of lithic materials from a given site. This is important because a net loss in one location often causes a net gain at another location. The term sedimentation refers to the movement of lithic material past or deposition at a given down drainage reference point. As such, sedimentation is not synonymous with erosion--an important point often ignored when evaluating the effects of roads on erosion and sedimentation.

Two types of erosion are affected by road construction: surface erosion and mass erosion. Surface erosion is simply the movement of individual soil particles by a force--the most common being water (either raindrop impact or overland flow) and wind. Mass erosion is the movement of numerous soil particles en masse primarily under the influence of gravity. The dominant mass soil movements (landslides) acting in the interior Western U. S. take the form of (1) failure of finite masses of soil and forest debris that move rapidly (debris avalanches or debris flows) or slowly (slump-earth flow) along planar or concave surfaces or both, (2) pure rheological flow with minor shifting of mantle materials (creep), and (3) rapid movement of valley bottom materials and water down stream channels (debris torrents).

Under undisturbed forest conditions, surface erosion processes are generally secondary and mass erosion processes dominate. With road construction, however, surface erosion may accelerate greatly along with mass erosion. Sediment at some downstream reference point of interest represents the integrated interactions of all erosion processes and storage responses above that point.

A number of studies document the effects of roads on erosion and sedimentation in the interior Western U. S. These are summarized in Table 2. For the most part, the studies have dealt with on-site erosion or downstream sediment responses. Until recently, studies linking the overall erosion-sedimentation process have been largely ignored.

Surface erosion is a function of three factors: (1) the amount of force available to move soil particles; (2) the inherent erosion hazard for the site in question; and (3) the amount of protection available on the soil surface. Under undisturbed forest conditions, surface erosion is low to nonexistent because considerable material is available on the forest floor to protect the soil surface and because soil permeabilities are normally high. Following road construction, however, surface erosion accelerates in response to disruption of soil structure, removal of protective cover, increased raindrop impact and wind movement, reduced infiltration rates that create overland flow of water, increased slope gradients on cut and fill slopes, interception of subsurface water flow zones, concentration of overland flow of water on the road and in channels, and detachment and movement of road surface material by traffic. The amount of increase varies tremendously depending on road design features and site characteristics.

A study by Megahan and Kidd (1972) illustrates the effects of road construction on steep granitic slopes in southern Idaho. Soils derived from granitic materials are some of the most erodible in the West (André and Anderson, 1961). Surface erosion rates per unit area of road averaged 220 times greater than the undisturbed values for a 6-year study. A number of other investigators

Table 2.—Studies documenting the effects of forest roads on erosion and sedimentation in the Interior West.

Reference	Location	Type of Erosion		Accelerated Sediment			Comments
		Surface Erosion	Mass Erosion	Yes	No	Not Measured	
Trimble and Tripp—1949	Wyo. S. Mont.	X	-	-	-	X	Observed erosion on previously disturbed areas.
Hetherington—1976	S. British Columbia	X	-	X	:	-	Mostly from channel encroachment.
Leaf—1966	Colorado	X	-	X	-	-	Some road erosion did reach stream.
Haupt & Kidd—1965	S.W. Idaho	X	-	X	-	-	Most erosion from road crossings.
Megahan & Kidd—1972	S.W. Idaho	X	X	X	-	-	Used erosion plots and debris basins.
Haupt et al.—1963	Idaho	X	-	-	-	X	Ocular comparisons of erosion on inslope vs. outslope roads.
Megahan—1974a	Idaho	X	X	-	-	X	Developed a model for time trends in road erosion.
Gonsior & Gardner—1971	S.W. Idaho	-	X	-	-	X	Postmortem slope stability study.
Rosa & Tigerman—1951	S.W. Idaho	X	-	X	-	-	Used sediment rating curves to predict sediment yield.
Megahan et al.—1978	Idaho	-	X	X	-	-	Inventory of 1,418 landslides, sediment delivery estimated.
Jensen & Cole—1965	Idaho	-	X	-	-	X	Survey of landslides during water year 1965.
Ohlander—1964	Idaho	X	-	-	-	X	Erosion plots on road fills.
Bethlahmy & Kidd—1966	Idaho	X	-	-	-	X	Erosion plots on road fills.
Haupt—1959	Idaho	X	-	-	-	X	Measured downslope sediment flow below roads.
Hartzog & Gonsior—1973	Idaho	X	-	-	-	X	Ocular evaluation of road design practices.
Gardner et al.—1978	Idaho	X	-	-	-	X	Ocular evaluation of road design practices.
Packer—1967	Idaho, Mont.	X	-	-	-	X	Erosion on road tread and downslope movement.
Burroughs & King—1985	Idaho	X	-	-	-	X	Rainulator studies on road tread and cuts.
Megahan et al.—1983	Idaho	X	-	X	-	-	Long-term erosion on road cuts compared to watershed sediment yields.
Cook & King—1983	Idaho	X	-	-	-	X	Measured trap efficiency of slash on fills.
Heede—1984	Arizona	X	-	X	-	-	Sediment delivery in a small basin up to 0.3 ha size.
Campbell & Stednick—1983	Colorado	X	-	-	-	X	Downslope sediment movement below fills.
Megahan et al.—In press	Idaho	X	-	X	-	-	Developed construction phase sediment budget on three watersheds.
King—1984	Idaho	X	-	X	-	-	Sediment measured on cuts and fills, at road crossings, and at basin outlet.

have documented increased surface road erosion following road construction in the interior West (Table 2), but for the most part the amount of increase was less than that reported by Megahan and Kidd.

In response to vegetation regrowth and seasoning of erosion surfaces, rates of surface erosion tend to decrease over time after disturbance. Megahan (1974a) found that over 80 percent of the accelerated surface erosion on roads in granitic soils occurred within the first year after disturbance. Within 5 years, erosion rates were greatly reduced but still greater than undisturbed conditions primarily because of erosion on road cut slopes. Trimble and Tripp (1949) reported long-term accelerated erosion on 20- to 35-year-old roads constructed in lodgepole pine forests in Wyoming and Montana. Surface erosion rates in excess of undisturbed conditions are probably the rule for roads in the interior West as long as the roads remain in use. The amount of erosion increase tend to be directly proportional to the amount of traffic and inversely proportional to the level of road maintenance. Following proper road closure, however, surface erosion rates should return to predisturbance conditions within some indeterminant period depending on the amount and type of road closure activities and site conditions.

Important factors influencing mass erosion (landslides) include hillslope gradient and gradient of road cut and fill slopes, depth of the saturated soil zone relative to the total soil depth, soil and/or parent material strength properties including cohesion and friction angle, and strength provided by vegetation roots. Given a large enough storm, landslides can occur even on undisturbed forested slopes and may in fact be the major erosion process at many locations. Vegetation removal for road right-of-way clearing tends to increase landslide hazards because of increased soil water contents in response to reduced evapotranspiration and increased total snow accumulation and snowmelt rates. Landslide hazards are also increased because of decreased vegetation root strength caused by root decay. In addition, roads increase slope gradients on cut and fill slopes, cause possible changes in soil strength, and may increase the relative depth of the saturated zone over and above that caused by vegetation removal alone.

Landslides must exceed specific threshold conditions before they occur. Thus, roads don't necessarily accelerate mass erosion even though construction in mountainous areas does tend to create conditions that are closer to the threshold. In fact, accelerated mass erosion on roads is rare in many locations where inherent mass erosion hazards are low. Some indication of this is shown in Table 2 where only 20 percent of the studies reported mass erosion. In high hazard areas, however, dramatic increases in mass erosion can occur, especially during extreme rain or snowmelt. For example, Jensen and Cole (1965) reported on an inventory of landslides occurring in the South Fork Salmon River drainage in central Idaho during a large rain-on-snow storm in April 1965. Only nine out of 89 landslides were not road associated. A similar proportion of road effects was found on the inventory of 1418 landslides by Megahan et al. (1978) on the Boise and Clearwater National Forests; 88 percent of the landslides inventoried were road associated.

Additional data from landslides on the Boise and Clearwater National Forests was used to better define the road mass erosion problem and what portion of the road prism was causing problems. Road-associated slides were classified as cut slope or road fill slope failures. Almost two-thirds of the slides occurred on road cuts, the rest were on fills. Slides on road cuts averaged smaller than those on fills so that only 47 percent of the slide volume occurred from road cuts whereas 52 percent occurred on fills.

Not all forest roads are constructed according to the same standards. Standards vary depending on the intended type, volume of use, and the design speeds. Standards in use on the Boise National Forest at the time of the landslide survey by Megahan et al. (1978) were used to classify roads, thus making it possible to evaluate the effects of differences in design. The five road standards ranked from high to low are arterial, collector, service, terminal, and temporary. Arterial roads were designed for 30 to 35 mi/h (48 to 56 km/h); collector roads for 10 to 20 mi/h (16 to 32 km/h); and service roads for 5 to 10 mi/h (8 to 16 km/h). In mountainous areas, increased design speed means increased excavation to provide a wider running surface and more gentle curves. Although terminal and temporary roads had the same design speed as service roads, they more closely conform to the terrain and so require less excavation. Temporary roads are simply terminal roads that are closed after use.

In terms of potential influence on slope stability, the major effect of road standards is the volume of excavation required for construction. One might expect landslide occurrence to be directly proportional to volume of excavation, and this is the case. The average number of slides per kilometer for arterial, collector, service, and temporary and terminal roads was 2.2, 1.2, 0.6, and 0.2, respectively. Thus, slide density approximately doubled for each step in road standard above the terminal road.

Accelerated sedimentation was reported for only 40 percent of the 24 studies (Table 2) documenting accelerated surface or mass erosion following road construction. We have no way of knowing whether accelerated sedimentation occurred in the other 60 percent because sediment yields were not measured. However, barring direct channel encroachment immediately upstream, we do know that any sediment increase that does occur at downstream points will be less than the amount of material eroded within the road prism. This is because much of the eroded material is trapped on the slopes and in the channel below the road. Data from the Silver Creek study area in southwest Idaho (Megahan et al. in press) illustrate the effectiveness of downslope sediment retention. Large increases in surface erosion occurred during construction on the steep, granitic slopes in the area. However, an average of only 7 percent of the material appeared as sediment at the mouth of the three study watersheds. Most of the eroded material (85 percent) was stored on the slopes below the road; the remaining 8 percent was stored in stream channels above the basin outlet. Thus, although surface erosion rates within the construction area were increased by a factor of about 500 times, the sediment yields at the basin outlet were increased by a factor of only five times. This difference is caused by slope storage and by the fact that only 3 percent of the total watershed area was disturbed by road construction.

Many types of mass erosion on roads involve a sudden release of a large amount of material on steep slopes. Thus, one would expect greater downstream delivery of eroded material from mass erosion than from surface erosion. Data from the landslide inventory by Megahan et al. (1978) support this hypothesis but show that delivery of slide material to streams also varies by road source. Much of the material from road cut slides is commonly deposited on the roadbed, so only about 33 percent of the cut slide material was delivered to streams. Because road fill failures have no such downslope catch points, about 67 percent of the material from fill failures reached streams. These figures contrast with the downslope delivery values of 7 percent reported above for surface erosion.

Considerable variation is reported for studies of the effects of forest road construction on sedimentation at the mouth of the drainages (Table 2). Results are not directly comparable because of differences in study procedures. However, differences in site conditions and the nature of the disturbances are apparent in the results. For example, Hetherington (1976) compared water samples for concentrations of suspended sediment above and below logged areas on two watersheds near Penticton, British Columbia. Although suspended sediment was slightly higher downstream from the disturbance, the increase was not statistically significant. The principal potential source areas for sediment were associated with skidding operations in and across two small tributaries and at one road crossing. The lack of increased sediment was attributed to buffer strips along channels, winter logging on snow, and gentle slopes.

Haupt and Kidd (1965) reported on postlogging sediment yields at the mouth of 16 small drainages on granitic soils in south central Idaho. Accelerated sediment yields induced by logging were detected in only four of the 16 drainages. The bulk of the accelerated sediment occurred within 3 years after disturbance. Log haul roads were a major contributor of sediment to channels. Relatively gentle slopes and the width of buffer strips between log haul roads and streams was the primary factor influencing sediment delivery to channels. Leaf (1966) found sediment yields averaging 200 lb/acre (224 kg/ha) following road construction and logging on a high elevation watershed in Colorado. This represents a 127 percent increase compared to the long-term (14-year) average annual sediment yield of 88 lb/acre (99 kg/ha). Sediment yields were high in the year during and immediately after logging but decreased rapidly in subsequent years toward preharvest levels. Surface erosion on roads was identified as the primary erosion process following disturbance. King (1984) measured average sediment yield increases ranging from 19 to 156 percent for 3 years following road construction on six study basins in northern Idaho. Differences between watersheds were attributed to different road design and road erosion control measures.

Fortunately, the erosion and sedimentation impacts of road construction need not be passively accepted. A variety of practices are available to reduce impacts. Megahan (1977) summarized these practices into four basic principles:

1. Avoid construction in high erosion hazard areas.

2. Minimize the amount of disturbance caused by road construction by controlling the total mileage of roads and reducing the area of disturbance on the roads that are built.

3. Minimize erosion on areas that are disturbed by road construction by a variety of practices designed to reduce erosion.

4. Minimize the off-site impacts of erosion.

All four principles must be weighed to reduce total erosional impacts. This is important because stress on individual factors may not meet this goal. For example, a shorter road may have to be lengthened to avoid high erosion hazards. In this case, total erosional impacts may be minimized although the area disturbed is increased. Erosion control practices are certainly beneficial, and considerable effort has been and should be devoted to their development and implementation. However, prevention (principles 1, 2, and 4), rather than control, is usually by far the most efficient and cost effective means to reduce erosion impacts. Prevention can have an added benefit by avoiding possible irreparable damage or costly repairs that may exceed original construction costs.

The first basic principle for reducing road erosion impacts is another matter of prevention rather than control and consists simply of avoiding high erosion hazard areas. Examples of serious erosion problems caused by road construction in high erosion hazard areas are common, especially where landslide hazards are high. Here, even minor location changes of 50 or 100 ft (10 or 20 m) may eliminate a major erosion problem. Problems of this type arise most often because road locators fail to recognize high erosion hazard conditions. Sometimes adding to the problem is the adoption of and strict adherence to traditionally accepted road standards (e.g., alignment standards for speed purposes) rather than providing flexibility that would allow the road location to be adjusted to the site

properties of a landscape. A number of guidelines have been developed to help identify high landslide hazard situations: Burroughs et al. (1976), Megahan et al. (1978), Foggin and Rice (1979), Swanston (1974, 1981), and Ward et al. (1982).

The second basic principle emphasizes measures designed for erosion prevention rather than control. Minimizing road mileage and areas of disturbance help reduce erosional impacts considerably. This is particularly true on forested lands where the total length of road required is often regulated by the distance capabilities of logging systems and the silvicultural practices prescribed for the timber stands.

Reductions in the area disturbed by road construction can also be made by careful road location and design. For example, use of flexible horizontal and vertical alignment standards during road location to avoid steep slopes can considerably decrease the width of area disturbed. To illustrate, total width of disturbance by a road 12 ft (4 m) wide increases from about 22 ft (7 m) on a 40 percent slope to 51 ft (16 m) on a 60 percent slope; on a 65 percent slope the width increases to 102 ft (32 m). For a given slope, additional reductions in area disturbed can be made by minimizing road and ditch width and by maximizing the gradient of cut and fill slopes (assuming the steeper slopes do not increase other erosion hazards).

The third basic principle is to reduce erosion on the areas that are disturbed by road construction. This is the traditional approach using a multitude of road design practices to help reduce erosion. Some examples of the effectiveness of surface erosion control measures used on forest roads in the interior West are summarized in Table 3.

Successful design of erosion control practices requires considerable knowledge of erosion processes, including the major type of erosion that is occurring and the individual factors that control erosion. To illustrate, little benefit results from attempting to stop mass erosion by mulching or surface erosion by installing subsurface drains. Likewise, mulching a road fill slope may have little value if improper road design, failure of the road drainage system, or both, cause large quantities of water to flow over the fill. In addition, erosion control measures vary greatly with respect to both costs and effectiveness. For example, Haber and Kadoch (1982) report costs of $124 per acre for dry seeding and $5,662 per acre

Table 3.—Examples of the effectiveness of surface erosion control on forest roads in Idaho.

Stabilization Measure	Portion of Road Treated	Percent Change in Erosion[1]	Reference
Tree planting	Fill slope	-50	Megahan 1974b
Hydromulch, straw mulch & dry seeding[2]	Fill slope	-24 to -58	King 1984
Straw mulch	Fill slope	-72	Bethlahmy and Kidd 1966
Wood chip mulch	Fill slope	-61	Ohlander 1964
Exselsior mulch	Fill slope	-92	Burroughs and King 1985
Paper netting	Fill slope	-93	Ohlander 1964
Asphalt-straw mulch	Fill slope	-97	Ohlander 1964
Straw mulch-netting -planted trees	Fill slope	-98	Megahan 1974b
Gravel surface	Road tread	-70	Burroughs and King 1985
Dust oil	Road tread	-85	Burroughs and King 1985
Bituminous surfacing	Road tread	-99	Burroughs and King 1985
Terracing	Cut slope	-86	Unpublished data[3]
Straw mulch	Cut slope	-32 to -47	King 1984

[1] Percent change in erosion compared to similar, untreated sites.

[2] No difference in erosion reduction between these three treatments.

[3] Intermountain Forest and Range Experiment Station, Forestry Sciences Laboratory, Boise, Idaho.

for placement of plastic netting to control erosion on road fills in southern Idaho. A careful evaluation of site erosion hazards, timing, and effectiveness of individual erosion control measures and downstream sedimentation risk is needed in order to decide if dry seeding is more or less desirable than placement of erosion net.

Some guidelines for controlling surface erosion include Gallup (1974), Highway Research Board (1973), Horton (1949), Kochenderfer (1970), Larse (1971), Megahan (1977), Packer and Christensen (1964), and U.S. Environmental Protection Agency (1975). Guides for reducing mass erosion on forest roads include Keener and Bell (1973), Prellwitz (1975), and Burroughs et al. (1973, 1976).

The fourth basic principle is to minimize the off-site impacts of erosion that does occur. Essentially this amounts to reducing sediment delivery to stream channels by (1) keeping disturbed areas as far from channels as possible, (2) providing a maximum of obstructions to catch and retain sediment before it reaches the drainage system, and (3) recognizing that the efficiency of a downslope area to deliver sediment varies considerably depending upon its form and structure. Careful evaluation of the potential and control of downslope sediment delivery is necessary in order to select the most cost effective erosion control measures described in principle 3 above. For example, use of the more costly erosion net could be limited to areas where downslope delivery is high (such as in the vicinity of channel crossings) and dry seeding could be used elsewhere.

Development of guidelines for control of downslope delivery of sediment below roads has been much more limited compared to guides for controlling erosion on the road. However, some material is available: Haupt (1959), Trimble and Sartz (1957), Packer (1967), Cook and King (1983), and Campbell and Stednick (1983).

4 SUMMARY AND CONCLUSIONS

The effects and impacts of forest road construction watersheds can be measured in terms of changes in both site productivity and the hydrologic function of the watershed. It is tempting to equate productivity losses with the total area of road disturbance. However, studies suggest that much of the productivity lost to tree removal required for the road right-of-way is counter balanced by accelerated rate of growth of trees adjacent to the road. This thinning effect can completely mitigate production losses in some cases, and in all cases it will probably reduce productivity losses to levels below that suggested by the area of disturbance.

Road construction often causes major changes in hydrologic processes on mountain slopes that can lead to potential changes in the rate and quality of streamflow from the watershed. Studies to date suggest only minor changes in streamflow rates from road construction involving up to 12 percent of the watershed area. No adverse effects were reported for the streamflow changes that did occur. I could find no published information documenting changes in water chemistry following road construction involving either natural chemicals or introduced chemicals. In contrast, channel encroachment at road crossings and by direct encroachment in valley bottoms

parallel to the stream commonly causes changes in channel morphology including bank and bed erosion, deposition, or both. Most impacts are minor and localized except in some cases of extensive lateral encroachment where effects may carry far downstream.

Without exception, road construction accelerates surface erosion rates compared to undisturbed conditions. Erosion rates are highest during and immediately after construction and decrease rapidly within the first few years after construction. Although greatly reduced, surface erosion rates continue to exceed undisturbed conditions for at least as long as the road remains in use depending on the type and amount of traffic and road maintenance.

Accelerated mass erosion is rare in many locations where inherent mass erosion hazards are low. In high hazard areas, however, dramatic increases in erosion can occur, especially during extreme rain or snowmelt.

Many studies document increased surface and mass erosion following road construction. Many of these studies also reported downstream increases in sedimentation. Thus, accelerated erosion and consequent increased sedimentation is by far the most serious impact of road construction on watershed function. In extreme situations, increases in sediment production can cause changes in channel morphology such as bank erosion, loss of water depth, and decreases in the particle size of streambed materials.

Fortunately, ways to minimize the erosional consequences of road construction can be summarized by four basic principles:
  1. Avoid construction in high erosion hazard areas.
  2. Minimize the amount of disturbance caused by road construction by controlling the total mileage of roads and reducing the area of disturbance on the roads that are built.
  3. Minimize erosion on areas that are disturbed by road construction by a variety of practices designed to reduce erosion.
  4. Minimize the off-site impacts of erosion.
A variety of erosion control practices are available to greatly reduce both surface and mass erosion rates. However, prevention (principles 1 and 2) and minimizing off-site impacts (principle 4) are usually, by far, the most efficient and cost effective means to reduce downstream sedimentation impacts.

REFERENCES

André, J.E. & H.W. Anderson 1961.
Variation of soil erodibility with
geology, geographic zone, elevation and
vegetation type in northern California
wildlands. J. Geophysical Research.
66:3351-3358.

Bethlahmy, N. & W.J. Kidd 1966.
Controlling soil movement from steep
road fills. Research Note INT-45. Ogden,
Utah: U.S. Department of Agriculture,
Forest Service, Intermountain Forest and
Range Experiment Station; 4 p.

Binkley, V.W. 1965. Economics and design
of a radio-controlled skyline yarding
system. Research Paper PNW-25. Portland,
Oregon: U.S. Department of Agriculture,
Forest Service, Pacific Northwest Forest
and Range Experiment Station; 30 p.

Bosch, J.M. & J.D. Hewlett 1982. A review
of catchment experiments to determine
the effect of vegetation changes on
water yield and evapotranspiration. J.
of Hydrology. 55:3-23.

Brown, G.W. & J.T. Krygier 1971. Clear-cut
logging and sediment production in the
Oregon coast range. Water Resources
Research. 7(5):1189-1198.

Burroughs, E.R., M.A. Marsden & H.F. Haupt
1972. Volume of snowmelt interupted by
logging roads. J. of the Irrigation and
Drainage Division, Proceedings of the
American Soc. of Civil Engineers, Vol.
98, No. IR1, March 1972: 1-12.

Burroughs, E.R., G.R. Chalfant & M.A.
Townsend 1973. Guide to reduce road
failures in western Oregon. U.S.
Department of Interior, Bureau of Land
Management; 111 p.

Burroughs, E.R., G.R. Chalfant & M.A.
Townsend 1976. Slope stability in road
construction. Portland, Oregon: U.S.
Department of Interior, Bureau of Land
Management; 102 p.

Burroughs, Jr., E.R. & J.G. King 1985.
Surface erosion control on roads in
granitic soils. Proceedings, American
Society of Civil Engineers, Watershed
Management Symposium, Denver, Colorado;
April: 183-190.

Campbell, D.H. & J.D. Stednick 1983.
Transport of road derived sediment as a
function of slope characteristics and
time. Research report, Water Quality
Laboratory, Department of Earth
Resources, Colorado State University,
Fort Collins, Colorado; 46 p.

Cook, M.J. & J.G. King 1983. Construction
cost and erosion control effectiveness
of filter windrows on fill slopes.
Research Note INT-335. Ogden, Utah: U.S.
Department of Agriculture, Forest

Service, Intermountain Forest and Range
Experiment Station; 5 p.

Foggin, G.T. III, & R.M. Rice 1979.
Predicting slope stability from aerial
photos. J. of Forestry. 77(3):152-155.

Fredriksen, R.L. 1970. Erosion and
sedimentation following road
construction and timber harvest on
unstable soils in three small western
Oregon watersheds. Research Paper
PNW-104. Portland Oregon: U.S.
Department of Agriculture, Forest
Service, Pacific Northwest Forest and
Range Experiment Station; 15 p.

Gallup, R.M. 1974. Roadside slope
revegetation - past and current practice
on the National Forests. Equipment,
Development, and Testing Report 7700-8.
San Dimas, California: U.S. Department
of Agriculture, Forest Service,
Equipment Development Center; 37 p.

Gardner, R.G., W.S. Hartsog & K.B. Dye
1978. Road design guidelines for the
Idaho Batholith based on the China Glen
road study. Research Paper INT-204.
Ogden, Utah: U.S. Department of
Agriculture, Forest Service,
Intermountain Forest and Range
Experiment Station; 20 p.

Gonsior, M.J. & R.B. Gardner 1971.
Investigation of slope failures in the
Idaho Batholith. Research Paper INT-97.
Ogden, UT: U.S. Department of
Agriculture, Forest Service,
Intermountain Forest and Range
Experiment Station; 34 p.

Haber, D.F. & T. Kadoch 1982. Costs of
erosion control construction measures
used on a forest road in the Silver
Creek watershed in Idaho. Research
report, University of Idaho, Dept. of
Civil Eng., Moscow, Idaho; 79 p.

Harr, R.D., W.C. Harper, J.T. Krygier, &
F.S. Hsieh 1975. Changes in storm
hydrographs after road building and
clear-cutting in the Oregon Coast range.
Water Resources Research. 11(3):436-444.

Hartsog, W.S. & M.J. Gonsior 1973.
Analysis of construction and initial
performance of the China Glen Road,
Warren District, Payette National
Forest. General Technical Report INT-5.
Ogden, Utah: U.S. Department of
Agriculture, Forest Service,
Intermountain Forest and Range
Experiment Station; 22 p.

Haupt, H.F. 1959. A method for controlling
sediment from logging roads.
Miscellaneous Publication No. 22. Ogden,
Utah: U.S. Department of Agriculture,
Forest Service, Intermountain Forest and
Range Experiment Station; 22 p.

Haupt, J.F., H.C. Rickard & L.E. Finn

1963. Effect of severe rainstorms on insloped and outsloped roads. Research Note INT-1. Ogden, Utah: U.S. Department of Agriculture, Forest Service, Intermountain Forest and Range Experiment Station; 8 p.

Haupt, H.F. & W.J. Kidd, Jr. 1965. Good logging practices reduce sedimentation in central Idaho. J. of Forestry. 63(9): 664-670.

Heede, B.H. 1984. Overland flow and sediment delivery: an experiment with small drainages in southwest ponderosa pine forests (Colorado, U.S.A.). J. of Hydrology. 22:261-273.

Hetherington, E.D. 1976. Dennis Creek: a look at water quality following logging in the Okanagan Basin. Canadian Forest Service, BC-X-147; 33 p.

Highway Research Board 1973. Erosion control on highway construction. National Cooperative Highway Research Progress. Synthesis of Highway Practice 18. 52 p.

Horton, J.S. 1949. Trees and shrubs for erosion control in southern California Mountains. Berkeley, California: California Division of Forestry in cooperation with U.S. Department of Agriculture, Forest Service, Forest and Range Experiment Station, California; 72 p.

Jensen, F. & F. Cole 1965. South Fork of the Salmon River storm and flood report. U.S. Department of Agriculture, Forest Service, Payette National Forest, McCall, Idaho; 15 p.

Keener, Q.R. & J.R. Bell 1973. An investigation of the feasibility of a cutbank slope design method based on the analysis of natural slopes. In: Proceedings of the 11th annual engineering geology and soils engineering symposium; 1973 April 4-6; Idaho State University, Pocatello, Idaho; 131-155.

King, J.G. & L.C. Tennyson 1984. Alteration of streamflow characteristics following road construction in north central Idaho. Water Resources Research. 20(8):1159-1163.

King, J.G. 1984. Ongoing studies in Horse Creek on water quality and water yield. In: National Council of The Paper Industry for Air and Stream Improvement, Inc. Technical Bulletin No. 435. 28-35.

Klock, G.O. 1975. Impact of five postfire salvage logging systems on soil and vegetation. J. of Soil and Water Conservation. 30:78-81.

Kochenderfer, J.N. 1970. Erosion control on logging roads in the Appalachians. Research Paper NE-158. Upper Darby,

Pennsylvania: U.S. Department of Agriculture, Forest Service, Northeastern Forest Experiment Station; 28 p.

Larse, R.W. 1971. Prevention and control of erosion and stream sedimentation from forest roads. In: Proceedings, Symposium on forest land use and stream environment; 1970 October 19-21; Oregon State University, Corvallis, Oregon; 76-83.

Leaf, C.F. 1966. Sediment yields from central Colorado snow zone. In: J. of Hydraulics, Division of American Society of Civil Engineers, Volume 96, No. HY1. Proceedings Paper 7006; 87-93.

Lyons, J.K. & R.L. Beschta 1983. Land use, floods, and channel changes: Upper Middle Fork Willamette River, Oregon (1936-1980). Water Resources Research. 19(2):463-471.

Megahan, W.F. & W.J. Kidd 1972. Effect of logging roads on sediment production rates in the Idaho batholith. Research Paper INT-123. Ogden, Utah: U.S. Department of Agriculture, Forest Service, Intermountain Forest and Range Experiment Station; 14 p.

Megahan, W.F. 1974a. Erosion over time on severely disturbed granitic soils: a model. Research Paper INT-156. Ogden, Utah: U.S. Department of Agriculture, Forest Service, Intermountain Forest and Range Experiment Station.

Megahan, W.F. 1974b Deep-rooted plants for erosion control on granitic road fills in the Idaho batholith. Research Paper INT-161. Ogden, Utah: U.S. Department of Agriculture, Forest Service, Intermountain Forest and Range Experiment Station.

Megahan, W.F. 1977. Reducing erosional impacts of roads. In: Guidelines for watershed management. FAO Conservation Guide. Rome, Italy: Food and Agriculture Organization of the United Nations; 237-261.

Megahan, W.F., N.F. Day & T.M. Bliss 1978. Landslide occurrence in the western and central northern Rocky Mountain physiographic province in Idaho. In: Youngberg, Chester T., ed. Proceedings of 5th North American forest soils conference, Fort Collins, Colorado. Fort Collins, CO: Colorado State University; August 1978: 116-139.

Megahan, W.F., W.S. Platts & B. Kulesza 1980. Riverbed improves over time: South Fork Salmon. In: Symposium on watershed management. Vol. 1; New York: American Society of Civil Engineers: 380-395.

Megahan, W.F. 1983. Hydrologic effects of clearcutting and wildfire on steep granitic slopes in Idaho. Water Resources Research. 19(3):811-819.

Megahan, W.F., K.A. Seyedbagheri & P.C. Dodson 1983. Long term erosion on granitic roadcuts based on exposed tree roots. Earth Surface Processes and Landforms. 8: 19-28.

Megahan, W.F., K.A. Seyedbagheri, T.L. Mosko, & G.L. Ketcheson In press. Construction phase sediment budget for forest roads on granitic slopes in Idaho. In: Proceedings International symposium on drainage basin sediment delivery, Albuquerque, New Mexico, August 4-8, 1986, Published by International Association of Hydrological Sciences.

Monsen, S.B. 1974. Plant selection for erosion control on forest roads of the Idaho batholith. 1974 Winter Meeting American Society of Agricultural Engineers, Chicago, Illinois, Dec. 10-13, 1974, Paper No. 74-2559, 18 p. Amer. Soc. Ag. Engr., St. Joseph, Michigan.

Ohlander, C.A. 1964. Effects of rehabilitation treatments on the sediment production of granitic road materials. Colorado State University; M.S. thesis, Fort Collins, Colorado. 78 p.

Packer, P.E. 1967. Criteria for designing and locating logging roads to control sediment. Forest Science. 13(1):2-18.

Packer, P.E. & G.F. Christensen 1964. Guides for controlling sediment from secondary logging roads. U.S. Department of Agriculture, Forest Service, Intermountain Forest and Range Experiment Station and Northern Region, Ogden, Utah, 42 p.

Pfister, R.D. 1969. Effect of roads on growth of western white pine plantations in northern Idaho. Research Paper INT-65. Ogden, Utah: U.S. Department of Agriculture, Forest Service, Intermountain Forest and Range Experiment Station.

Prellwitz, R.W. 1975. Simplified slope design for low-standard roads in mountainous area. In: Transportation Research Board Special Report 160, Low Volume Roads, TRB-Nat. Academy of Sciences; 65-74.

Rice, R.M. & J.R. Wallis 1962. How a logging operation can affect streamflow. Forest Industries 39(11):38-40.

Rosa, J.M. & M.H. Tigerman 1951. Some methods for relating sediment production to watershed conditions. Research Paper INT-26. Ogden, Utah: U.S. Department of Agriculture, Forest Service,

Intermountain Forest and Range Experiment Station; 19 p.

Schumm, S.A. 1971. Fluvial geomorphology, channel adjustment and river metamorphosis. In: River Mechanics, Volume 1, Edited by H. W. Shen, Fort Collins, Colorado, p. 5-1 to 5-22.

Silen, R.R. & H.J. Gratkowski 1953. An estimate of the amount of road in the staggered setting system of clearcutting. Research Paper PNW-92. Portland, Oregon: U.S. Department of Agriculture, Forest Service, Pacific Northwest Forest and Range Experiment Station; 4 p.

Smith, R.B. 1979. Steep slopes logging. J. of Logging Management. 101: 1794-1796, 1821.

Swanston, D.N. 1974. Guidelines for characterizing naturally unstable or potentially unstable slopes on western National Forests. In: Proceedings, symposium on new requirements in forest road construction. Sponsored by the Association of Professional Forestry of British Columbia. University of British Columbia; Publication No. FP 2406, 126-137.

Swanston, D.N. 1981. Watershed classification based on slope stability criteria. Proceeding, interior West watershed management symposium, 8-10 April, 1980, Spokane, Washington, Washington State University press, Pullman, Washington. 43-58.

Trimble, Jr., G.R. & N.R. Tripp 1949. Some effects of fire and cutting on forest soils in the lodgepole pine forests of the northern Rocky Mountains. J. of Forestry. 47:640-642.

Trimble, G.R. & R.S. Sartz 1957. How far from a stream should a logging road be located? J. of Forestry. 55:339-341.

United States Environmental Protection Agency, Region X, Water Division 1975. Logging Roads and protection of water quality. EPA 910/9-75-007. 306 pp.

Ward, T.J., R.M. Li & D.B. Simons 1982. Mapping landslide hazards in forest watersheds. J. of Geotechnical Engineering. 108(GT2): 319-324.

Ziemer, R.R. 1981. Storm flow response to road building and partial cutting in small streams of northern California. Water Resources Research. 17(4):907-917.

Environmental Geotechnics and Problematic Soils and Rocks, Balasubramaniam et al. (eds)
© 1987 Balkema, Rotterdam. ISBN 90 6191 785 9

The Deep Creek Mudflow of April 16, 1979, Uintah County, Utah, USA

Benjamin L.Everitt
Utah Division of Water Resources, Salt Lake City, USA
Andrew E.Godfrey
Fishlake National Forest, Richfield, Utah, USA

ABSTRACT: On April 16, 1979, about 12 acres (5 ha) of gently sloping alluvial pasture near Vernal, Utah, liquefied, caved into a tributary of Deep Creek, and flowed downstream. The event lasted about 5 hours, involved an estimated 300,000 yards (230,000 m^3) of earth, and resulted in a dendritically branched arroyo system one-half mile long and up to 21 feet (6.4 m) deep. The liquefying layer appears to have been sand with interbedded gravel. Liquefaction appears to have been initiated by high pore pressure caused by unusually high groundwater levels. Although the ground was unusually wet, there is no evidence to suggest that groundwater conditions were unique. This event and the resultant landform indicate that liquefaction-sapping is a viable process for rapid arroyo formation.

CHRONOLOGY

In the mid-afternoon of April 16, 1979, ranchers began to notice an unusual surge of muddy water flowing in Deep Creek, above Lapoint, Uintah County, Utah (Fig. 1). The day had been warm and sunny, with no precipitation. Although streams were already high from snowmelt runoff, the flow was unusual enough to prompt an investigation upstream. About 3 p.m. the owner of the site went to investigate a loud noise in the lower end of his pasture, and found what he described as a large sinkhole in the process of formation (Vernal Express, April 19, 1979). Witnesses describe a constant rumbling as slabs of alluvium caved from the sides of the expanding gully and settled into the soupy mud which filled the bottom (UGMS, 1979). Water gushed from animal burrows exposed by the caving walls. Erosion continued through the evening and into the night (Vernal Express, 19 April 1979).

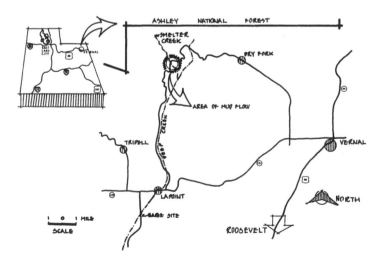

Figure 1 Location Map

The sediment eroded from the site flowed down Smelter Creek as a slurry, becoming progressively diluted downstream, until it was incorporated as part of the water-borne sediment load of Deep Creek. The Deep Creek stream gage at State Highway 246, near Lapoint, Utah, about fifteen and a quarter miles (25 km) downstream from the site (Fig. 1), recorded a peak discharge of 310 cubic feet per second (cfs, 8.8 m 3/s) the year's highest, on April 17 (USGS, 1980, p. 129). On April 15 the average daily flow was 17 cfs (0.5 m 3/s), and the average for the month was 16.1 cfs (0.45 m 3/s). The site was inspected on April 18th by the Utah Geological and Mineral Survey (UGMS); its description appears in the UGMS Survey Notes (UGMS, 1979). The authors visited the site on April 19th and May 26th.

DESCRIPTION

The site is in the northeast quarter of Section 22, T3S, R19E, Salt Lake Base and Meridian, in the drainage of Smelter Creek, tributary to Deep Creek (Fig. 1). It is a valley-bottom pasture, underlain by alluvium, which slopes southeastward at 2 to 5 percent (Fig. 2). It includes about 100 acres (41 ha) set between hills of Navajo Sandstone. The pasture is under-lain by the Jurassic Carmel Formation (Kinney, 1955, plate 1). Upstream from the pasture Smelter Creek drains 1¼ square miles (325 ha) of the Navajo, Chinle, Shinarump, and Moenkopi Formations. An alluvial fill underlies the pasture at least to the depth exposed by the out-flowing mud. The alluvium consists mostly of interbedded fine sand, and silty sand, with beds of sandy gravel increasing in number and grain size toward the downstream end of the arroyo (Fig. 3). Judging from its color and texture this alluvium is derived from the Triassic and Jurassic rock exposed in the Smelter Creek drainage.

Previous to the 1952 aerial photograph (Fig. 2), Smelter Creek had been diverted along the road on the west side of the pasture. By 1952, it had begun to cut a gully, which by 1979 had vertical walls 20 feet high in places. The prior course of Smelter Creek appears on the aerial photo along the base of the hill to the east of the pasture. The 1979 event began in the north wall of the pre-existing gully, and followed an independent course headward, inter-secting the old course of Smelter Creek only along its upstream half.

On April 16, 1979, the spring thaw

Figure 2 Aerial photo dated August 18, 1952, before the mudflow.

STRATIGRAPHIC SECTIONS, SMELTER CREEK ARROYO

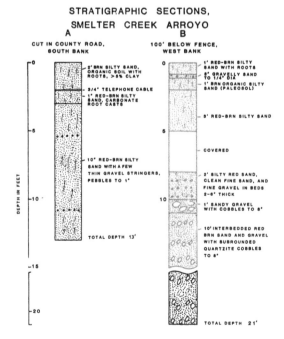

Figure 3 Stratigraphic sections of alluvium exposed in arroyo walls.

Figure 4   Distant view of the new arroyo, looking upstream.

was well underway; the snow line had receded to about 100 feet (30 m) above the elevation of the pasture. The ground was saturated, both from melting snow and by infiltration from Smelter Creek.

The nearest weather stations are at Vernal, Utah, 15 miles (25 km) to the southeast, and at Roosevelt, Utah, 19 miles (30 km) to the southwest. For the period of October 1978 through April 1979 the precipitation at both stations had been 150 percent of normal. Normal October through April precipitation at the elevation of the mudflow site is 8 inches (20 cm), and average annual precipitation is 14 inches (36 cm) (Jeppson et al., 1968). The only earthquake in Utah for the 24 hour period preceding the start of the mudflow was a micro-earthquake of magnitude 0.6 at 1:22 p.m. MST near Snowville in Box Elder County about 180 miles northwest of the site (Richins et al., 1981, p. 60).

When we visited the site on May 26, 1979, we found a large dendritically branched arroyo one-half mile (800 m) long and up to 300 feet (90 m) wide (Fig. 4). The headcut had vertical walls 13 feet (4 m) high where it had severed the county road. A three-fourth inch (2 cm) telephone cable, originally buried at a depth of 2 feet in the borrow pit along the west edge of the highway, as well as several fences, were hanging unbroken across the gulch, suggesting that the soil disaggregated and

collapsed downward before horizontal displacement began.

The arroyo has an estimated volume of 300,000 yards (230,000 m^3). To move this volume of material in 5 hours requires an average discharge of 450 cfs (13 m^3/s). This estimate of discharge is not incompatible with the instantaneous peak of 310 cfs (8.8 m^3/s) measured downstream at the Lapoint stream gage, with allowances for precipitation of sediment and attenuation of the flood wave in the intervening 15 miles.

To move 450 cfs (13 m^3/s) through the channel of Smelter Creek, whose estimated cross-sectional area just below the new arroyo is 300 square feet (28 m^2), requires an average velocity of 1.5 feet per second (0.5 m/s) in this reach; again a not unreasonable figure.

Flat-topped remnants of the original pasture remain between anastomosing branches (Fig. 5). At mid-depth is a prominent terrace with a hummocky surface formed of pieces of sod let down from the pasture but not transported very far laterally (Fig. 6). Many of these terraces are buttressed by large rotated blocks of sodded alluvium. This morphology strongly suggests that the sodded upper alluvium retained its strength, but broke into blocks and settled while an underlying layer liquefied and flowed out from underneath. In places, sod is draped over the edge of the arroyo (Fig. 7), suggesting that only the upper foot or two of sediment did not liquefy.

Figure 5   View downstream along the arroyo from near its midpoint.

Figure 6   View upstream in the arroyo from near its midpoint to show rotated blocks of sod.

The large slabs of sod in Fig. 7 also suggest that the sediment underlying the sod patches liquefied as a unit and then drained slowly away, letting the floating sod down gently like ice stranded by a waning river.

The alluvial fill exposed by the arroyo is mostly sand, interbedded with silt and gravel, generally coarser at depth and toward the downstream end of the exposure (Fig. 3). At a depth of 10 to 20 feet below the original surface, interbedded sandy gravel and gravelly sand with boulders to 1 foot in length are exposed at the downstream end of the new arroyo (Fig. 8). On May 26, water was still flowing from beds of openwork gravel exposed just above the bottom of the arroyo. The sediment which we describe, of course, is that which did not liquefy. The sediment which did liquefy is gone, and may have had different properties. We infer that it is the sand interbedded with or adjacent to the gravel which liquefied.

On May 26, dried mudcake still coated the sides and bottom of the channel of Smelter Creek immediately downstream from the new arroyo, its upper limit forming a mud-line from three to five feet above the existing channel bottom. The mud-line consisted of a 1 to 3 inch thick coating of poorly sorted silty gravelly sand (Fig. 9). Mud marks on streamside vegetation show a steep gradient away from the main channel, suggesting a viscous flow which tended to solidify as it came to rest away from the channel. Water was ponded in side drainages behind the natural levees of the mud flow, but there seems to have been little overall aggradation of the channel of Smelter Creek.

The arroyo formed by the Deep Creek mud-flow contains most of the characteristics that Higgins (1982, p. 151) describes for

Figure 7   East wall of arroyo with draped sod.

Figure 8   Base of the right wall of the arroyo near its lower end.
Largest boulders are about 1 foot long.

drainage systems formed by sapping. It is relatively steep-sided and blunt-ended (Fig. 4). The valley sides and head meet the floor in a sharp angle. The floor is fairly broad but its width varies irregularly along its length (Fig. 5). The floor of the arroyo is flat and in the center is veneered with sediment (Fig. 6). Many of the short stubby tributaries of the arroyo lack catchments above their heads (Fig. 7). The main arroyo has an angular course and tributaries meet the main arroyo at sharp angles, probably representing subsurface control. The course of the main arroyo does not follow the old channel of Smelter Creek, but proceeds more nearly up the mid-line of the valley southwest of it (Fig. 2), probably reflecting some structure in the underlying sediment, possibly older buried channels of Smelter Creek.

These criteria serve to distinguish a drainage formed by sapping from one formed by overland flow. To distinguish one formed by sapping from one formed by slumping or other forms of coherent failure we would add the criterion that only a thin veneer of sediment was preserved downstream from the valley.

The site was again visited on March 12, 1983. By that time Smelter Creek had been rerouted in a man made channel to the west and south of the mudflow site. The property owner had graded the banks of

353

Figure 9   Remnant mud in the channel of Smelter Creek downstream from the new arroyo.

part of the arroyo to make them useful as pasture again and had constructed a stock watering pond in the bottom of the arroyo. Water was flowing from the western walls of the arroyo, where they had not been modified, apparently receiving recharge from infiltration from Smelter Creek. Cattails and other phreatic vegetation had established themselves in the bottom of the arroyo indicating a permanent water table.

RECONSTRUCTION

Our reconstruction of the event and its causes are as follows: Due to the rapid thaw, water from melting snow and Smelter Creek saturated the ground, and thoroughly charged permeable beds at the base of the alluvial section. An interconnecting network of animal burrows may have contributed to the permeability. The 80-foot difference in elevation between the upper and lower ends of the pasture and the confining cap of silty topsoil permitted the development of artesian pressure within the permeable beds. We imagine that pore pressure within the confined sand layers continued to increase until it sufficiently exceeded overburden pressure to cause a blowout along the bank of the pre-existing arroyo, where the sand and gravel would have been near the surface and under the greatest hydrostatic head. Blowouts caused by hydrostatic pressures in porous soil capped by sod have been described by Eisenlohr (1952), and by Hack and Goodlett (1960, p. 45).

Once erosion began to reduce overburden pressure, the saturated sand continued to liquefy as long as the interbedded gravel or the animal burrows supplied the water to maintain high pore pressures. We further imagine the entire dendritic arroyo expanding headward by sapping and bank caving up-gradient from the first blowout. Each slide from the bank served to reduce the overburden pressure on an adjacent slice of confined sand which then liquefied, by now set in vibration by the caving of nearby banks. We presume the process either stopped when headward erosion had progressed upstream to a point of lower hydrostatic head, or when so much of the confined gravel aquifer was unroofed that pore pressure could no longer be maintained at a high level.

During erosion, the production of sediment probably kept the arroyo partly full of flowing mud, but upon cessation of bank caving, the mud continued to drain away, emptying the arroyo to its full depth. We imagine that blocks caved from banks were buttressed by the mud in the main channel until their underlying sediment re-solidified, thereby producing the mid-level terrace.

CONCLUSIONS

Several conclusions can be drawn from the Deep Creek mudflow which we believe contribute to the understanding of landscape evolution:

1. Although Smelter Creek has now been re-channeled around the site of the mud-

flow to prevent further headcutting, if left to itself the headcuts could have further eroded and adjusted to produce an ordinary-looking arroyo after a few years. It is therefore possible that similar events have gone unnoticed elsewhere, and that the phenomenon of liquefaction-sapping is more often involved in arroyo formation than has been recognized.

2. We were surprised at the grain size distribution of the mud which flowed, particularly its low proportion of material smaller than fine sand, and its very low clay content. The initial mobilization of the Deep Creek Mud Flow may be similar to lateral spreading landslides described in Lake Bonneville clay, silt, and sand elsewhere in Utah, which Van Horn (1975) attributes to liquefaction of interbedded sand layers.

3. Both liquefaction and groundwater sapping are processes requiring a high water table. The Deep Creek Mudflow occurred during unusually wet conditions, and it is reasonable to assume that these wet conditions contributed to the event. However, no conclusions can be drawn from this example regarding the correspondence of permanent high water tables or long term climate wetness and the operation of sapping as a geomorphic agent. Indeed, Eisenlohr's (1952) "blowouts", a related hydrostatic phenomenon, are clearly associated with rare and ephemeral ground-water conditions.

4. Higgins (1982, p. 152) appears to have imagined a similar process operating on Mars when he wrote of sapping causing "great slumps, block slides, and debris flows". The Deep Creek arroyo described here may serve as an analog for the Martian tributaries of Ius Chasma or similar large drainage-ways in arid places. We propose the process of liquefaction-sapping as a good way to mobilize large amounts of relatively coarse-grained sediment, and to initiate movement over moderately low gradients with a minimum of water.

5. Most documented cases of liquefaction are associated with a triggering event, such as an earthquake, which provided the dynamic loading to momentarily elevate pore pressure above overburden pressure. The documentation of spontaneous liquefaction, that is, liquefaction in the absence of dynamic loading, serves to show that evidence of liquefaction in the geologic record does not by itself constitute proof of dynamic loading.

ACKNOWLEDGEMENTS:

We wish to thank John E. Costa, U. S. Geological Survey; Emery T. Cleaves, Maryland Geological Survey; S. Bryce Montgomery, Utah Division of Water Resources; and Charles B. Hunt, for their thoughtful reviews and comments on the manuscript.

REFERENCES

Eisenlohr, W. S. 1952. Floods of July 18, 1942, in northcentral Pennsylvania. U.S. Geological Survey Water Supply Paper 1134-B, 158 p.

Hack, J. T., and Goodlett, J. C. 1960. Geomorphology and forest ecology of a mountain region in the central Appalachians. U.S. Geological Survey Prof. Paper 347, 64 p.

Higgins, Charles C. 1982. Drainage systems developed by sapping on Earth and Mars Geology, 10, 147-152.

Jeppson, R. W., Ashcroft, G. L., Huber, A. L., Skogerboe, G. V., and Bagley, J. M. 1968. Hydrologic atlas of Utah, Utah Water Research Laboratory PRWG 35-1, Logan, Utah.

Kinney, Douglas M. 1955. Geology of the Uinta River-Brush Creek area, Duchesne and Uintah Counties, Utah. U.S. Geological Survey Bulletin 1007, 185 p.

Richins, W. D., Arabasz, W. J., Hathaway, G. M., Oehmich, P. J., Sells, L. L., and Zandt, G. 1981. Earthquake data for the Utah Region, July 1, 1978 to December 31, 1980. Univ. of Utah Seismograph Stations Special Publication, 125 p.

U.G.M.S., 1979. Wet spring triggers awesome earthflow. Utah Geological and Mineral Survey, Survey notes, 13, No. 2.

U.S. Geological Survey, 1980. Water Resources data for Utah, Water year 1979. U.S. Geological Survey Water Data Report UT-79-1, p. 604.

Van Horn, Richard, 1975. A failure by lateral spreading in Davis County, Utah. Utah Geology, 2, No. 1, p. 83-88.

Vernal Express, April 19, 1979, Vernal, Utah.

Environmental Geotechnics and Problematic Soils and Rocks, Balasubramaniam et al. (eds)
© 1987 Balkema, Rotterdam. ISBN 90 6191 785 9

A landslide in clayshales: Environmental damages and measures for restoration

P.Jagannatha Rao
Geotechnical Engineering Division, Central Road Research Institute, New Delhi, India

ABSTRACT : The paper presents the case history of a landslide in the Sivalik formations of the Himalayan ranges. The failure originated as a movement along a circular arc in a relatively thick bed of clayshale. However, the slide was aggravated due to secondary mass movements, and ultimately assumed the shape of the debris flow. Considerable damage was caused to the major roadways near the crown and the toe of the slide area. Due to the retrogression of the slide, forest cover on the hill suffered extensive damage. The paper discusses various features of the landslide and presents economical remedial measures to improve surface drainage over the slide area and restore the damaged forest/vegetative cover. Measures to improve the stability of the area are described.

1. INTRODUCTION

A major landslide took place in September 1982, in one of the foothill ranges of the Himalayas. Figure 1 shows a general view of the landslide, which extends from the local ridge at the top, to a point called "10 M span Bridge" below. An initial survey conducted for evaluating the landslide hazard potential along the alignment of this new road indicated that the failed zone has a high probability of sliding (Rao, 1979). However, due to the lack of an alternative route at this point, the road was built along the ridge. In late September 1982, the hillside, involving a part of the road and the adjoining area, slid out. Minor slope failures and mass movements had been noted previously as well, especially in 1978. Mass movements, in general coincided with those monsoon seasons in which rainfall was heavy.

2. GENERAL GEOLOGY OF THE AREA

The hill ranges in which the slide took place forms a part of the Sivaliks. The Sivaliks were formed when the water-worn debris of the cental Himalayan ranges started being deposited in the "Indogangetic foredeep", a shallow lagoon of water formed immediately to the south of the newly risen Himalayan ranges (Krishnan 1960). The deposition started about 12 million years ago and ended about 1 million years ago. In general, the rock formations in the Sivalik ranges consist of different types of sandstone, shales, clayshales and limestones. The Sivaliks form the outer ranges of the Himalayas and extend from the Brahmaputra valley in the east to the Potwar plateau in the west.

The slide area was found to be made up of thick beds of clayshale underlain by sandstone.

3. DESCRIPTION OF THE SLIDE AREA

Figure 2 shows a profile along the affected slope. Together with Fig.1, this serves to give an overall picture of the slide area, which extends for about 165 m (at its widest point in the direction along the road). The distance from crown to toe is about 310 m and the slope of the hill face varies from 30° to 34°.

Referring to Fig.1, it can be seen that the main slide mass is located at the crown of the slide area. The material involved in the slide is purple coloured clayshale. As shown in

Fig.1 Panoramic view of the landslide.

Fig. 2, the formation in this zone consits of thick beds of clalyshales. The lower part of the slide area consists of variegated shales, red shales with sandstone, and micaceous sandstone.

4. TYPES OF FAILURE

The primary slope failure consists of a rotational failure in the thick bed of reddish purple clayshale, with the crown of the slide located near the ridge. Figure 3 shows the probable slip circle along which the initial failure occured. It was estimated that the undrained shear strength of the clayshale at the incipient state of failure was of the order of 150 KN/m², as the clayshale became saturated due to infilatration. The strength in the dry condition was of the order of 1000 KN/m². Numerous tension cracks were observed in the crown area, and are shown in Fig. 4. The width of the tension cracks ranged from 2 cm to 6 cm. Figure 5 shows the shear striations produced by the moving mass during the slide movement.

However, in late September 1982 and October 1982, as rains continued, the wet debris from the slided-out mass started flowing down the slope as the water content increased. The mass movements assumed the character of a debris flow. As may be seen from Fig. 1, the topography of the area was such that the debris flow channels cover a little distance above the point marked '10 m span bridge'. Since the clearance at the bridge was narrow, the debris overflowed on to the road, and so, both the upper and lower roads were rendered impossible for traffic for about a week. Further, some of the exposed hillside faces in the slide area were found to be susceptible to erosion, leading to the formation of gullies from fast flowing water. Failure of these slopes was found to occur in the form of minor slumps. Such debris material gets added to the main body of the slide.

Reconnaissance of the area affected by the landslide has shown that while at the top relatively thick beds of clayshale were present, in the lower

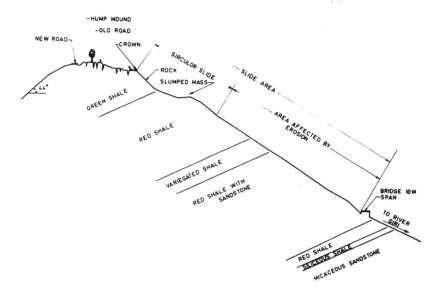

Fig. 2 Profile of the slide area

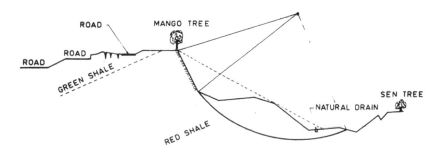

Fig. 3 Profile section with slip circle.

layers there was frequent interbedding of thin layers of clayshale and sandstone, and these materials had greater strength, tending to be more rocklike. These formations have been studied by means of stereonet diagrams so as to evaluate the possibility of mass movements. As a result, it was found that under proper conditions, both sandstone and clayshale beds can develop a tendency to be detached from the rockmass in wedge-shaped blocks formed by the intersection of the planes of discontinuities. The most probable direction of movement of these slided-out blocks was found to be NE at the top of the rock beds and towards SW at the lower part. Figure 6 shows the stereonet diagram for the lower part of the slide area.

5. NATURE OF MATERIALS INVOLVED IN THE SLIDE

The main material involved in the instability is reddish purple clayshale. Clayshales are the predominant material in the Sivalik ranges. These are essentially sedimentary deposits that have been grouped into three categories: (i) shales, which are rocklike and do not lose strength on soaking, (ii) clayshales, which lose a significant part of their strength on saturation with water and (iii) overconsolidated clays, where no additional strength effects of diagenetic bonds is experienced, and the material exhibits only the mechanical effects of the stress history. Figure 7 shows diagramatically the formation of the

Fig. 4 Tension cracks near the crown of the slide.

Fig. 5 Shear striations on the failure plane.

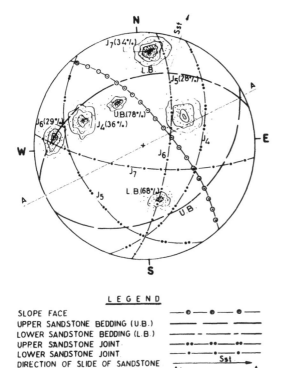

Fig. 6 Polar concentration diagram of the south western part of the slide.

above three gradations of materials, (Rao, 1976).

6. DAMAGES TO THE HILLSIDE ECOLOGY

The landslide presently under discussion is the biggest of the slides that have taken place along the newly constructed 9 km long road. Few other minor slides have occurred at other locations during the years 1978-82. As a result of all such landslides, a considerable amount of forest cover was destroyed and also erosion of the top soil was accentuated. Besides the areas that have suffered loss of vegetative cover, other areas have also got denuded due to natural factors. One such extensive area with depleted vegetative cover may be seen to the left of the slide area in Fig. 1.

Observations at a number of such landslide locations have shown that if no efforts are made to improve the conditions in an area affected by the

landslide, the damage to the hillside ecology increases at an accelerated pace. Landslide exposes fresh mass, and the slided-out debris is in a loose state. Both these are highly vulnerable to erosion from rain water. In the present slide area, large quantities of material from the top portion of the slide area got softened under the influence of continuing rains and flowed down the narrow channel down below, ultimately ending up as a silt load in the River Giri, further downhill of the slide area. Hillslopes just below the upper road level used to be cultivated and as result of the slide, cultivation could not be carried out.

The hillside restoration steps planned have deliberately discarded the conventional approach of using a number of check walls and retaining walls. Cognisance was taken of the fact that the hillside may undergo further minor slope movements and under the influence of such movements rigid or even partially rigid masonry walls are likely to experience distress or even failure. It was considered that restoration of the slope by providing improved vegetation cover and drainage would be an economical as well as an effective approach. Accordingly remedial measures, as described in the succeeding paragraphs, were adopted.

7. RESTORATION MEASURES

The remedial measures for improving the stability of the slide area were quite simple and consisted of the following:

(a) sealing of tension cracks;
(b) Installation of contour drains made up of inexpensive materials;
(c) planting of selected species of grasses and bushes over the denuded slope areas.

The remedial techniques, briefly described, are as follows:

(a) Sealing or tension cracks

As shown in Fig. 4, a number of tension cracks were observed, near the crown of the slide. Tension cracks were also noted at other points along the boundary of the slide area. Tension cracks are signs of incipient

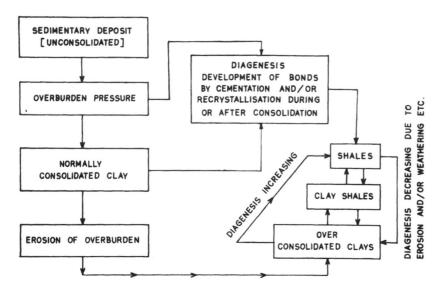

Fig. 7 Process of formation of clayshales.

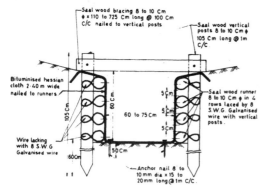

Fig. 8 Cross section of drain using Hessian Cloth.

instability. Further, they serve as sources for the ingress of water into the slide mass thereby accelerating the possibility of occurrence of sliding. Earlier field trials carried out at landslide locations in Simla and Jammu areas have shown that sealing of tension cracks serves effectively to reduce the infiltration, and significantly improves the stability of the slide area.

The procedure adopted for sealing the tension cracks is described in Appendix 1.

(b)   Installation of contour drains:

Contour drains are proposed for intercepting surface runoff water and lead the same away to a suitable nearby drainage channel. Lined masonry drains, besides being costly, are not ideal for the present slope, as minor movements are likely to occur. In such drains, the masonry lining develops cracks as the slope experiences minor movements and these cracks serve as points through which water can seep into the slope, thereby reducing the effectiveness of the drains.

Hence lightweight, flexible impervious flow-channels were devised and a typical one is shown in Fig. 8. The drain consists of hessian cloth treated with bitumen. A set of short wooden posts are driven along both the widthwise edges of the drain. The cross section of the drain is formed by packing the bed with rubble and covering it with the bitumen impregnated hessian cloth. The vertical side of the drain is formed by nailing the cloth to wooden posts. The wooden posts are braced widthwise as well as longitudinally to give the structure adequate stability. Such drains can accommodate moderate settlements without undergoing any functional disruption.

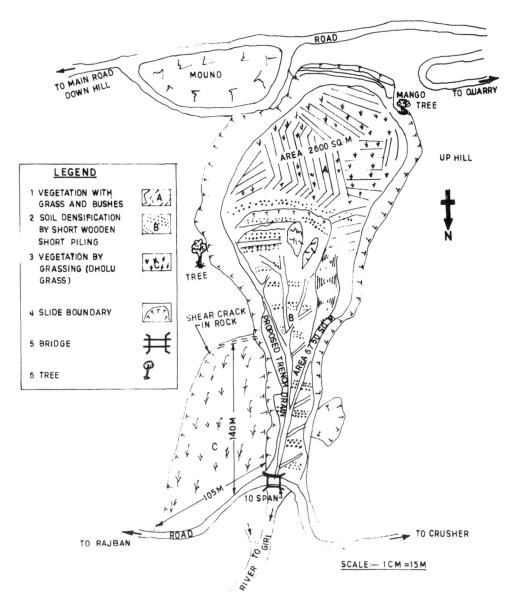

Fig. 9 Plan showing restoration of ecological improvement by vegetation

(c) Restoration of vegetative cover

Slope faces exposed by fresh sliding and the debris material from the slide have been found to be especially vulnerable to erosion from flowing rain water. If such erosion is allowed to continue unchecked, the formation of deep gullies was found to result, leading to a secondary cycle of instability. It is necessary that such slopes be given an extensive cover of vegetation to protect the slopes from deterioration and to ensure stability. This aim is achieved by a combination of the following methods:

(a) plantation of selected species of grasses;

(b) plantation of a variety of quick sprouting bushes;

(c) plantation of short stems of species of trees which take root and grow.

363

By adopting the above three types of plantations, an erosion-resistant cover is developed over the slope. Grass cover serves primarily to check surface erosion while roots of shrubs and plants also contribute their share by holding the soil together. The bushes and plants form a canopy of two levels on which rain drops fall at first so that the full destructive impact of the falling drops is not felt by the loose soil.

Figure 9 shows the plan adopted for such a plantation. Local species of grasses and bushes are preferred because of their adaptability to the local soil and environmental conditions. A variety of grass called "Dholu" grass was found to be hardy and capable of taking root even in very steep slopes. Similarly, suitable types of bushes identified were : Linea Grandis (local name: Maddaar), Ipomea (Sadabahar) and Agaue Americanna (Ramban).

8. SUMMARY

The landslide described occurred in a thick bed of clayshale as a typical circular slide. As a result of this major slide and other minor slides along the 9 km road, considerable denudation of the forest and vegetative cover had occurred over the hillside slopes.

Effective and economical measures to improve the stability of the area are described. These consisted of sealing the tension cracks, and restoring the vegetative cover by planting suitable species of grass, shrubs and plants. A flexible type of bitumen impregnated hessian cloth was provided for improving surface drainage.

ACKNOWLEDGEMENTS

The author wishes to acknowledge the help rendered by Shri Ajaib Singh, Shri R.K. Jain, and Shri D. Mukherjee, Scientists, in the field investigations. Thanks are due to Shri T.K. Natarajan, Deputy Director and Head, Geotechnical Engineering Division, for encouragement in carrying out the work. Thanks are to Shri O. Mascernhas, Scientist, for carefully proofreading the manuscript.

The paper is being published with the kind permission of the Director, Central Road Research Institute, New Delhi.

REFERENCES

Krishnan M.S. (1960) "Geology of India and Burma"

Higginbothams, 4th ed, 604 pp. Rao, P.J. (1976) "Stability of cuts in clayshales",

Symposium on foundations and excavations in weal soils.

Vol. I, Indian Geotechnical Society, 1976.

Rao, P.J. (1976) Unpublished report on "Susceptibility of landslide of the road near Rajban".

APPENDIX - I

Sealing of Tension Cracks

All round the periphery of the landslide, numerous tension cracks have been observed, with widths ranging from a few millimetres to as much as 30 cm. These constitute a source of instability of alarming magnitude. Further, the open cracks are a direct source of water infiltration during the rains.

Water saturates the body of the unstable mass and also lubricates the shear plane, both factors working to decrease the factor of safety of the hillside. It is therefore of foremost importance to seal these cracks effectively to prevent any ingress of water. Such a treatment had been of immense help in restoring the stability of unstable slopes in Simla and Jammu areas where the technique had been tried earlier.

All along the length of the cracks, a trench of about 30 cm depth and 60 cm width should be cut. The trench width should normally extend 30 cm beyond either edge of the crack. The excavated material, if predominantly soil, should be mixed with water to give it the optimum moisture content for compaction purposes. The consistency of the soil at this moisture content would normally be such that the soil can easily be manipulated to form a lump when pressed in the hand, but does not wet the hand.

The deeper part of the cracks should be filled up first by tamping thoroughly with crow-bars or suitable wooden poles. Filling of the trench should be done by spreading the wet soil (at O.M.C.) to about 15 cm depth and then compacting the same hand rammers. The top layer should be made such that the original shape of the slope is retained to the maximum extent possible. The surface over the filled up tension crack should then be made waterproof by spraying it evenly with bituminous cutback of the following composition:

Bitumen	1 kg
Kerosene	0.6 kg
Parafin Wax	0.01 kg

The bitumen should first be heated to about 100° C, and then the other two components added. The mixture may be sprayed with a garden sprayer, at an average rate of 0.2 kg/m^2.

Environmental Geotechnics and Problematic Soils and Rocks, Balasubramaniam et al. (eds)
© 1987 Balkema, Rotterdam. ISBN 90 6191 785 9

Role of remote sensing in the detection of potential sites for landslides/rockfalls in the Deccan Trap lava terrain of Western India

V.V.Peshwa & Vivek S.Kale
Department of Geology, University of Poona, Pune, India

ABSTRACT : Photointerpretation was carried out in the Deccan Trap (Cretaceous - Eocene) terrain in the vicinity of the crest of the Western Ghats. The Ghats divide the relatively planar coastal area in the west from the Plateau in the east. The processes of mass-wasting of the bedrock (sometimes leading to rockfalls/landslides) is particularly active in this terrain, accentuated by the seasonal heavy Monsoonic precipitation (of the order of 600-700 cms/year). The basaltic flows along the crest of the Ghats are highly fractured and traversed by basic intrusive dykes. The weathering and denudation in these rocks is initiated along the cooling joints and structural fractures, and results in the loosening of large massive slabs of the jointed basalts and intrusives.

Multiband LANDSAT imageries are capable of locating linear features like fracture-zones and intrusive dykes, on a regional scale. The high-scale aerial photographs can supply detailed local information regarding these features, and help identify the locales of accentuated mass-wasting on possible failure-prone slopes. Field investigations and monitoring of such sites, particularly those slopes which are in the vicinity of human settlements, railways, roadways, etc., could help prevent disasters.

1 INTRODUCTION

The parts of Western India which are underlain by the Deccan Trap basaltic flows can broadly be divided into three physiographic regions, namely (i) the Western Konkan coastal plains, (ii) the Eastern Deccan plateau and (iii) the Western Ghats dividing (i) and (ii). Increasing socio-economic interaction between the Konkan plains and the Deccan plateau has resulted in a phenomenal growth in the density of the lines of communication (e.g. railways, roads, telecommunication and other cables, etc.) between these two parts, obviously traversing the Ghats. This has resulted in the growing realisation that the precipitous escarpments of the Ghats are succeptible to failure, leading to rock-falls and landslides.

Though essentially quite small in magnitude and localised in extent, these slope failures have become an ever-increasing hazard to the lines of communication. In some sections, road-blocks due to rock-falls during the rainy season, have become common and loss of life and property is not unknown.

Case studies of earlier rock-falls and landslides, identifiable on aerial photographs, are summarily discussed to develop a qualitative model of these slope failures. The role which remotely sensed data could play in identifying and locating such vulnerable locations is highlighted in this paper.

2 WESTERN GHATS

Geographically, the term Western Ghats includes the roughly North-South trending chain of hill ranges and steep escarpments which run parallel to the western coast-line of India (see Inset Fig. 1). Steep, high-rise west-facing escarpments, varying in hieght between 500 m. and 1500 m. and having regional slopes of well over 17° (Fig. 1), comprise its northern parts, where it is underlain by Deccan Trap basalts. The main escarpment of the Ghats commonly has an inclination of over 35°, with local subvertical faces being fairly frequent.

The rock-falls and landslides are particularly localised in those parts of the Ghats with slopes exceeding 35°. The main risk of these slope failures is in those parts of the Konkan plains which are present directly along the base of the scarp.

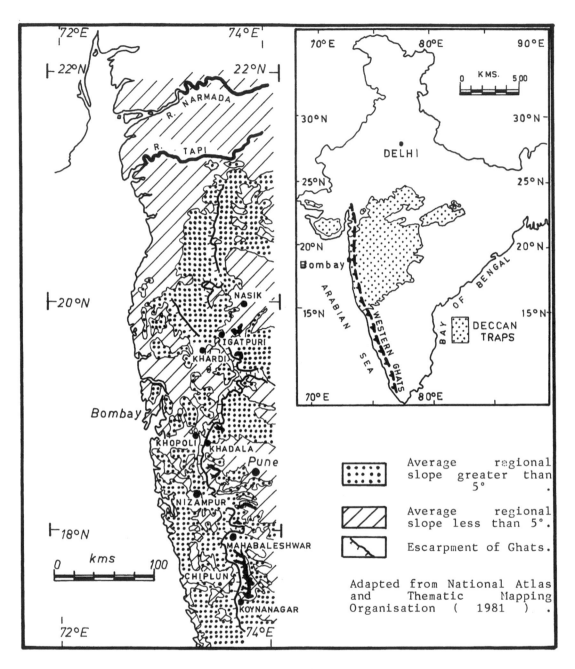

Fig   1   :   SLOPE MAP OF THE WESTERN GHAT REGION UNDERLAIN BY THE
DECCAN TRAPS . (Showing the locations mentioned in the text
and the position of the Western Ghat escarpment)

3 DECCAN TRAPS

The exposures of horizontal to subhorizontal basaltic flows and associated effusives and pyroclastics encountered along the Ghats, comprise the products of the extensive and voluminous outpouring of flood-basalts of the Deccan Volcanic episode, ranging in age from about 83 Ma. to 31 Ma. - Cretaceous to Eocene (Alexander, 1981).

These exposures of the Deccan Traps show a variety of landforms, including spectacular multiple scarps, radiating spurs and conical hills (Subramanyan, 1981) along the Western Ghats. Figure 2 depicts the generalised topographic profile, characterised by steep slopes, narrow deep valleys and subvertical escarpments along the Ghats.

The Deccan Traps exposed along the crest of the Western Ghats were earlier considered to represent parts of the undisturbed sections from this volcanic province. Recent work, (Mulay, 1980; Peshwa and Mulay, 1983; Peshwa et. al., 1984) based on aerial photointerpretation using high and medium scale photographs has shown that these flows are imprinted with records of subsequent structural disturbances, in the form of a major N-S to NNE-SSW trending fracture system, and a minor ENE-WSW to E-W trending system of fractures. The intrusive dykes in this region show characteristic polygonal cooling joints, with close-spaced joints near the contact, and wider spacing in the central parts of the dyke.

The humid, subtropical climate, heavy Monsoon precipitation of around 600-700 cms/year (saturated between the months of June and December), exalting the predominance of chemical degradation over physical disintigration in the process of weathering of these rocks. As a result, the rock commonly yields a lateritic-clayey admixture as a weathering product.

This mass-wasting of the Trappean rock on a local scale, is innitiated and accelerated (as shown in Fig 3) along the fractures and joints, as these planes permit the seepage of water in the otherwise almost impervious rock. Under these spatially controlled processes of mass-wasting of the basaltic bedrock, therefore, the profile of the products is governed by the geometry of the inherent joint/superimposed fracture systems present in the Trappean rocks. Spheroidal weathering and exfoliation are very common in the exposures of the Deccan Traps.

On a regional scale, this controlled denudational activity is manifested in the form of fracture-controlled drainage development and the geomorphic evolution of the Trappean country (Peshwa and Mulay, 1978; Mulay, 1980). The generalised denudational history of the Western Ghats is depicted in Fig 4, which also indicates the locations of possible slope failure and the positions in which the debris of such failures in the past is observably accumulated.

4 CASE HISTORIES

In course of the photogeological investigations along the Western Ghats, some instances of the past slope-failures could be observed and studied in the field. Each of these cases represents an example of sites which have past histories of repitative slope-failures. They have been used primarily to decipher the cause-effect relations for the observed failures of slopes along the Ghats.

4.1 Khardi - Bhatsanagar area

Near Khardi, about 200 m. southwest of the Bhatsa dam, a major NE-SW trending regional fracture-zone is exposed in the Bhatsa river channel. This fracture-zone is marked in the field by the presence of very close-spaced NE-SW trending fractures all along its length of over 7-8 km., and it can be delineated as a photolineament of the negative relief type on the aerospace imageries. The river channel is flanked by a number of steep cliffs, with individual heights of 10 m. to 30 m., many of which have records of repetitive failure. The river channel is controlled by the follows this fracture zone, with intensely weathered basalt exposed all along its length and somewhat lesser weathered basalt along its flanks. The barren scars produced due to the cliff-dislocations are vividly shown on the aerial photographs (scale : 1:30,000) due to their barren nature, which imparts a lighter tone to them. The debris of the failures has accumulated at the base of the cliffs and shows light tone and granular texture on the aerial photographs.

4.2 Khopoli - Khandala area

A subvertical cliff (delineated as a dark-toned photolineament on aerial photographs) produced by erosion of a dyke is seen about 3-4 kms. west of Khopoli, at the hamlet of Atkargaon. This N-S trending dyke, with its inherent close-spaced jointing in contrast to the compact host basalts,

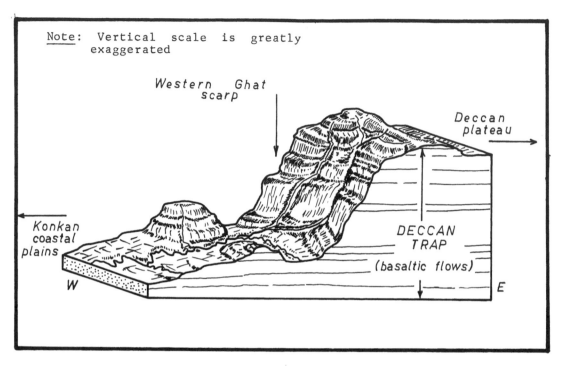

Fig 2 : IDEALISED SKETCH SHOWING THE GEOMORPHIC PROFILE ACROSS THE
WESTERN GHATS .

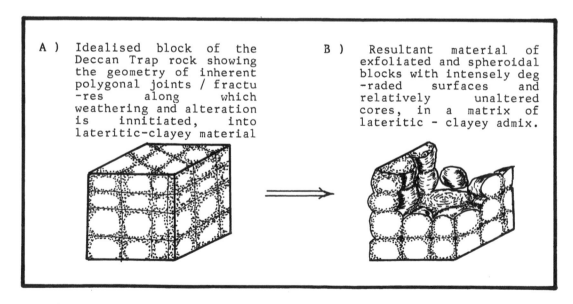

A) Idealised block of the
Deccan Trap rock showing
the geometry of inherent
polygonal joints / fractu
-res along which
weathering and alteration
is innitiated, into
lateritic-clayey material

B) Resultant material of
exfoliated and spheroidal
blocks with intensely deg
-raded surfaces and
relatively unaltered
cores, in a matrix of
lateritic - clayey admix.

Fig 3 : TYPICAL PATTERN OF WEATHERING AND EROSION OF THE DECCAN
TRAPS (Depicted for an idealised cubical block) .

is vulnerable to vertical erosion along its length, giving rise to a narrow linear valley. The weathering of the host basalts in this cliff-face receeds in intensity, away from contact with the dyke. The resulting cliff on the eastern face of this valley is a site of continual rock-falls during the rainy season. The accumulated debris at the base of this slope, the intrusive dyke and the steep escarpment, are all clearly seen on the (scale : 1:25,000) aerial photographs.

In the Bor Ghat section of National Highway No. 4 (between Khopoli and Khandala) are subvertical escarpments with heights of over 200 m. These escarpments, in complete contrast to the ones at Atkargaon, have no known record of failure, and tower above the Khopoli Hydroelectric Power Station. Aerial photointerpretation reveals a marked absence of any fracture zone or dyke traversing these cliffs, which are underlain by considerably lesser weathered basalts than those encountered at Atkargaon.

4.3 Nizampur - Kumbhe area

Near the village of Kumbhe, east of Nizampur, a 5-6 m. wide dyke is exposed along a steep narrow valley which has a singular drop of over 300 m. The valley which has evolved along the dyke is cluttered all along its length of about a kilometre by the weathered rubble of dyke material. The small rhomboidal fragments, which results due to the polygonal jointing of the dyke, are known to roll "en-masse" down the valley, during the rainy season, causing a serious hazard to the settlements and fields at the base of the valley.

The dyke, much of whose exposed material is highly weathered, appears as a light-toned photolineament on aerial photographs (of scale 1:5,000). It is also responsible for a case of river-capture at the top of the valley.

4.4 Mahabaleshwar area

The Ghat scarp at Mahabaleshwar (a hill-station, with a record of being one of the locations receiving the highest precipitation in this region) is known for its picturesque beauty. The escarpments, most with singular drops exceeding a couple of hundred metres, are known to be stable, except at a few locations where failure has been reported to have caused considerable damage. The "Arthur's Seat point", located on an isolated spur, is one such

example. The spur has evolved due to erosion along a regional fracture-zone which can be seen on aerial photographs. The earlier viewing gallery at this point had collapsed into the deep gorge below. Field studies incidate that the earlier gallery was probably constructed exactly atop this fracture-zone. The entire gorge and the base of the valley are lined by the rubble which collapses episodically from this overhang.

4.5 Koyananagar area

Sathe, et. al. (1968) report the development of fissures and cracks as well as slope failures in Koyananagar area, in sequel to the Koyna earthquake of 11 December 1967 (Richter scale : 6.5). However, they point out that the fractures were restricted to the soil cover, and did not have alarming width. Neither were they observed penetrating into the (basaltic) bedrock. Aerial photographs of the Ghats in this region indicate the presence of regional fracture zones west of the Koyna reservoir, cutting across the scarp-line. Local failures are noted along these fracture zones, and rubble recorded by aerial photographs has accumulated at the base of the Ghat-scarp.

5 SUCCEPTIBILITY OF SLOPE FAILURE

Slope stability studies by various leading workers in this field for example Terzagi (1950), Morgenstern and Sangrey (1978), Tein and Sangrey (1978) among numerous others, indicate that the stability of any slope is primarily dependent upon the following factors :

(a) Slope angle;
(b) Physical properties of the material underlying the slope, e.g. shear strength(s); cohesive strength (c) and the angle of internal friction ($\emptyset$) of the material; and
(c) the trigger mechanism.

Barton (1973) shows that compact basalts have $\emptyset$ values varying between 31° and 38°, and that wet, saturated samples yield weaker strength characteristics while the dry, unaltered samples are considerably stronger. The compressive strengths of the Deccan Trap basalts are recorded as varying between 1000 and 2600 kg/sq.cm. by Parthasarathy and Shah (1981). Their observations again confirm that the saturated and/or weathered samples are considerably weaker. Simple estimates of slope stability would therefore suggest that the horizontally

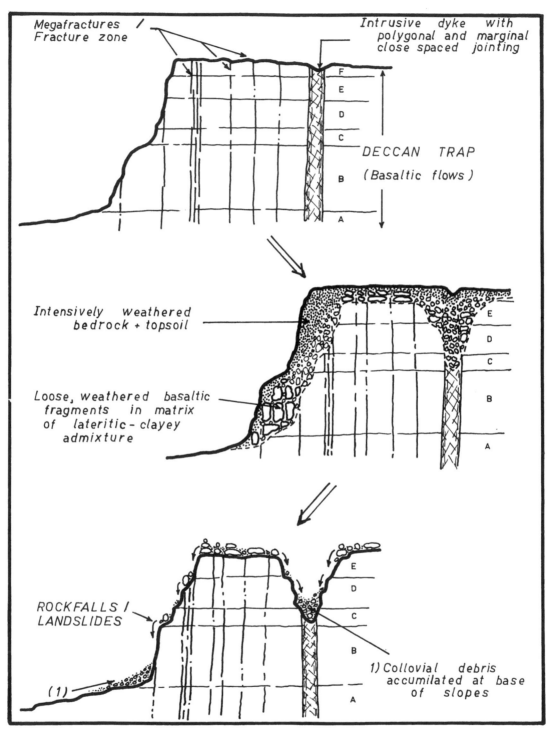

Fig 4 : IDEALISED SEQUENCE OF WEATHERING AND EROSION SHOWING EVOLUTION OF UNSTABLE BLOCKS / SLOPES WHICH ARE SUCCEPTIBLE TO FAILURE

stratified, completely unaltered, compact basaltic flows of the Traps should sustain perfectly stable slopes with inclinations exceeding 30°.

The earlier described field studies show primarily that the basaltic flows exposed along the Ghats are neither homogeneously compact (being traversed at numerous locations by fracture zones/dykes and obviously cooling joints not to mention the presence of amygdaloidal flows, nor dry and unaltered in nature. The slope failures noted in the Ghats invariably fall in the catagories of "rock-falls" or "rock-topples" in the classification of Varnes (1978). The lateritic-clayey admixture which is produced as a product of the weathering of the Trappean rocks is considerably weaker than the basaltic rocks themselves.

It may thus be concluded that the slope failures observable in the Western Ghats are due to a combination of factors which are induced as a consequence of the mass-wasting of the Trappean rocks.

The weathered rocks which have developed extensively along the crest of the Ghats are the primary cause of the slope instability, because of their inherent weak nature as compared to unweathered rocks. Their location vis-a-vis the pre-existing fracture and joint surfaces along which failure of these slopes takes place, further supports this. The relatively incohesive and low " $\emptyset$ " material of lateritic - clayey composition, which is particularly profusely generated along these parting planes in the Trappean rocks, probably loses its strength on being saturated during the Monsoon season. Its role as a sort of lubricating medium between the individual rock-blocks after being saturated is also indicated. All these features together are ultimately responsible for the slope instability and could finally culminate in failures.

This also explains the apparent constraint on the rock-falls/land slides in the Ghats, known to occur only during the monsoon season. The unstabilised slope may sustain during the dry season. With the onset of the Monsoonic rainfall, the water table rises due to the percolating influx, in turn saturating the material of lateritic-clayey composition, present along the joint/fracture planes. The weakened material consequently fails to sustain the load of overburden, leading to rock-falls/landslides. Thus this model of slope failure in the Ghats, does not primarily require a separate trigger mechanism in the form of, say, earthquakes, as is commonly supposed. In fact, as observed in the Koyananagar area, even intense earthquakes are incapable of producing any major and noticiable failures in the bedrock basalts.

In this model explaining, broadly, the causative factors of slope failures in the Western Ghats the following crucial conditions can therefore be considered to identify a failure-prone slope :

(a) Slope inclinations exceeding 35°.
(b) Imprint of some joint or fracture system or the presence of a dyke along the crest of the slope.
(c) Deep weathering profile along the slope, particularly with very intensly-weathered material present along the parting surfaces, at the crest of the slope.
(d) Significantly shallow to medium depth of the groundwater table, which is sensitive to rainfall induced infiltration.

6 APPLICATIONS OF REMOTELY SENSED DATA

The LANDSAT multiband imageries are capable of making it possible to locate the steep slopes by virtue of their relatively barren nature, or simple dwarf-grass cover, in contrast to the denser vegetation which is present on the gentler slopes. As a result of the geometry of the slope, also with reference to the LANDSAT MSS system, steeper slopes are highlighted as low-reflectance areas yielding relatively lighter tone. A detailed interpretation of the multiband imageries also yields information regarding the presence of highly-weathered lateritic/sub-lateritic basalts due to their tonal and textural contrast with the relatively unaltered bedrock exposures. Similarly, megafracture zones and major dykes can be located from these imageries on a regional scale. Thus by mapping these features on a regional scale, it should be possible to identify areas where there is a critical coincidence of the three factors, namely (a) steep slope, (b) highly weathered bedrock exposures and (c) presence of a fracture zone or dyke intersecting the slope profile.

Detailed interpretation of high-scale aerial photographs covering these areas will provide local information on these factors in detail and thereby help locate accurately the actual sites which would represent failure susceptible slopes.

Field verification of the interpreted data would aim at establishing the degree and approximate depth of the weathering profiles at these critical sites. Groundwater table observations could provide conclusive evidence regarding the actual possibilities of slope failures at these sites.

7 CONCLUSIONS

It may be concluded that investigations using remotely sensed data on two scales (low scale for regional studies and high scale for local identifications), could provide considerable information regarding the failure susceptibility of the escarpment faces in the Western Ghats. Mehta and Patel (1980) and Gupta and Bhandari (1980) have earlier discussed case histories highlighting the applicability of the interpretations of remotely - sensed data in landslide investigations, which also help a broad understanding of cause-effect relations of slope-failures in particular terrains. These relations can then be confirmed for specific locations by detailed field studies. These confirmed results can refer back to the interpreted data so as to locate hetherto unknown locations of failure-prone slopes. Corrective measures can then be applied to this slopes to prevent future failures and damage.

ACKNOWLEDGEMENTS

The help and invaluable comments on this work by Shri K. R. Datye, Director, Dubon Project Engrr., Bombay are gratefully acknowledged. The second author (VSK) is thankful to C.S.I.R., New Delhi for the financial support in the form of a Senior Research Fellowship.

REFERENCES

ALEXANDER, P.O. 1981. Age and duration of the Deccan Volcanism : K - Ar evidence. **Mem. Geol. Soc. India : 3 : 244-258.**

BARTON, N. 1973. Review of a new shear strength criterion for rock joints. **Engrr. Geol. 7 (4) : 287-332.**

FUJITA, T. 1980. Slope analysis of Landslides in Shikoku, Japan. **Proc. Intern. Symp. Landslides; New Delhi : 1 : 169-174.**

GUPTA, S.K. and BHANDARI, R.C. 1980. The role of aerial photointerpretation for Landslide studies in a part of Himachal Pradesh. **Proc. Intern. Symp. Landslides; New Delhi : 1 : 99-101.**

GUPTA, V.J. 1976. **Indian Cainozoic Stratigraphy;** Hind. Publ. Corp., New Delhi : 144 p.

MEHTA, H.S. and PATEL, A.N. 1980. Applications of digital techniques in identifying landslide - prone area. **Proc. Intern. Symp. Landslides; New Delhi : 1 : 25-28.**

MULAY, J.G. 1980. Geology of the area around Khandala (Maharashtra) India : with special reference to remote sensing. **Unpubl. Ph.D. Thesis, Poona Univ. :** 203 p.

MULAY, J.G. and PESHWA, V.V. 1978. Fracture control on the drainage pattern of the area south of Lonavala and Khandala (Maharashtra). **Proc. Symp. Morphology & Evolution of Landforms; New Delhi :** 222-225.

PARTHASARATHY, A. and SHAH, S.D. 1981. Deccan Volcanics : Rock material and rock mass characteristics and their significance in Engineering Geology. **Mem. Geol. Soc. India : 3 : 233-243.**

PESHWA, V.V., MULAY, J.G. and KALE, V. S. 1984. Fracture zones in the Deccan Traps of Western and Central India : A study based on Remote Sensing techniques. **Proc. 5th A.C.R.S., Kathmandu, Nepal : E1.**

RIB, H.T. and LIANG, T.A. 1978. Recognition and identification (in **Eds** : Schuster and Krizek) **Landslides : Analysis and Control :** 34-80.

SATHE, R. V., PHADKE, A.V., PESHWA, V.V. and SUKHATANKAR, R.K. 1968. On the development of fissures and cracks in the region around Koyananagar affected area. **Jour. Univ. Poona (S. & T. Section) : 34 : 15-19.**

SCHUSTER, R.L. and KRIZEK, R.J. **(Eds)** 1978. **Landslides : Analysis and Control. (Sp. Rep. Transport Res. Board) N.R.C.,** N.A.S., Washington, D.C. : 234 p.

SUBRAMANYAN, V. 1981. Geomorphology of the Deccan Volcanic Province. **Mem. Geol. Soc. India : 3 : 101-116.**

SUKHESHWALA, R. N., 1981. Deccan Basalt Volcanism, **Mem. Geol. Soc. India : 3 : 8-18.**

TERZAGHI, K. 1950. Mechanisms of Landslides. **Engrr. Geol. (Barkey Vol) Geol. Soc. America :** 84-121.

VARNES, D. J. 1978. Slope Movement : Types and Processes **(Eds.** Schuster and Krizek) **Landslides : Analysis and Control :** 11-33.

WU TEIN, H. and SANGREY, D.A. 1978. Strength properties of material and their measurements (in **Eds.** Schuster and Krizek) **Landslides : Analysis and Control :** 139-154.

Environmental Geotechnics and Problematic Soils and Rocks, Balasubramaniam et al. (eds)
© 1987 Balkema, Rotterdam. ISBN 90 6191 785 9

Mechanics of deposition of sand (and dust formation in sand storms)

A.V.Jalota
Civil Engineering Department, Motilal Nehru Regional Engineering College, Allahabad, India

ABSTRACT : Mechanics of deposition of Sand has been studied using a stereo-microscope, and slow motion film technique. A single Sand particle, after striking deposition surface, rebounds in random direction - spinning with high speed which has been quantified for Certain heights of fall (of the particle). The events taking place on the surface of deposition are also studied. The study is extended to mechanics of deposition of group of sand particles, deposited by Controlled process of vertical deposition. The analysis of the results bring out some of the causes of formation of Macropores which form an inherent property of packings of granular soils. The study suggests mechanics of formation of dust in Sand Storms.

INTRODUCTION

The study deals with deposition of granular soils only. A pioneer work in the field of deposition is the study made by Kolbuszewski (1948). That study (Kolbuszewski 1948) gives "Laws of Deposition" which indicate the influence of velocity with which a particle arrives for deposition (V), and Intensity of deposition (I) in terms of weight of particles deposited in a unit time over a unit area (e.g. gms/cm^2/sec) on porosity of a deposit formed, for a controlled process of vertical deposition. The work on deposition reported in the study (Kolbuszewski 1948) has been, Later, extended to airblown deposition (Kolbuszewski 1950, 1953). These studies deal with the effect of the factors I & V on overall porosity (N) of the sand deposits formed, as well as on relative porosity (Nr) used to describe the state of a granular mass.

Further study (Amirsoleymani 1965) brings out that deposits of same sand having same porosity (and thus same relative porosity), formed by different processes of deposotion, have different macro properties. For example the study (Amirsoleymani 1965) presents that the contours of porosity in deposits formed by diffrent processes of deposition such as (a) air -blown (wind tunnel) deposition, (b) deposition in water, a-

nd (c) vibration method, are different, as shown in Fig. 1.

It is further contended that the machanical behaviour of these deposits are different. It has, also, been demonstrated quantitatively (Aminsoleymani 1966) that mechanical properties of samples (such as stress -strain behaviour in Triaxial shear strength test) of same porosity but produced by different methods such as by (i) spooning and tamping, and (ii) deposition of sand by a funnel, are different.

These studies bring out clearly that the process of deposition as well as v-ariation in attributes of a deposition process, such as I and V, effect the state ot a packing of sands, and Relative porosity, by itself, is inadequate to describe state of grannular soil mass.

A study of two dimensional packing (Kolbuszewski 1963) of sand size spheres has illustrated the existance of different patterns of pores.

Fig.2.(a & b) shows these patterns for a dense and a loose packing. The zones of loosely packed particles quite apparent in these figures are defined as Macropores (Kolbuszewski 1965). The presence of a few zones of voids large enough to house a particle (similar to macropores) has been indicated in studies (Bernal and Mason 1960, Scott 196-2). Also, an indirect reference about t-

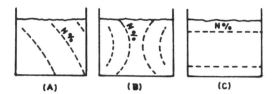

Fig. 1. Contours of porosity in various types of depositions (After Amirsoleymani 1965)

Fig. 2. Two dimensional packing of Sand size spherical beads. a) A dense packing. b) A loose packing. (After Kalbuszewski - 1965)

he significance of a 'Canal-System' consisting of pores of different sizes (description viewed as quite close to that of macropores) has been made in another study (Jong and Veruijt 1965).

The author has made a study of state of grannular soil mass as a system (Jalota 1972). A number of studies (Amirsoleymani 1966, Jalota 1972, Kolbuszewski 1965) have intiated a new school of thought and has brought out that Macropores from a fundamental property of grannular packings. The author has demonstrated that relative porosity and two parameters based on quantification of macropores can describe the state of grannular soil mass uniquely. These are discussed in more details elsewhere

(Jalota 1972).

The author felt that a detailed study of mechanics of deposition was necessary to understand formation of depositions and to study the effect of various attributes (such as I and V) of a process of deposition.

This paper presents the experimental investigations and results regarding mechanics of controlled process of vertical deposition; sand particles falling through air.

2. EQUIPMENT USED

2.1 Controlled process of vertical deposition equipment

In order to study mechanics of deposition it was decided to study a process of deposition with a wide range of control of intensity of deposition (I), and velocity of deposition (V), and giving reproducible state of packings.

An equipment was fabricated to achieve controlled process of deposition with a wide range of control of I, and V.Special emphasis was laid to achieve a very low value of I so as to approch state packing having minimum porosity. The apparatus produced a thin layer of sand on a belt-conveyor which was mounted on a trolley. This produced a curtain of sand moving to and fro, giving vertical deposition of sand particles. The speed of the belt conveyor and the trolley could be controlled separately to provide a large variation of I. Controlling various parameters, the equipment was used to produce packings of sand of minimum porosity. (porosity lower than the one that could be achieved by vibration method). This is discussed separately (Jalota 1972). Assembly of the equipment used is shown in fig.3.

This apparatus could produce packings of a wide range of porosity, and the packings were closely reproducible.

2.2 Stereo microscope

A stereomicroscope was used to visually examine the process of deposition. However the particles striking the surface of deposit already formed, stormed out of view quickly and the field of view was very small.

Fig. 3. Vertical controlled depositions apparatus.

Fig. 4. Single particle deposition apparatus.

2.3 Slow-motion film examination

Therefore to examine the mechanics of deposition comprehensively, a slow-motion film examination was planned. A high speed cine-camera which provided a wide range of speed was used, along with a timer unit and a voltage regulator to set out the desired speed and the event timing.

For filming the mechanics of particle deposition, it was necessary to switch on the camera at such a time that sand curtain reached (and was within) the field of view when the camera had reached the desired steady speed. However the time avilable to film the event and the field of view so was small

that it would have been very difficult to start the camere precisely at such a time manually. So a microswitch arrangment was fabricated so that the camera was switched on automatically when the sand curtain reached a pre-determined point. The timer unit of the camera produced a mark on the film after every 1/1000th of a second.

Initially, it was planned to study the mechanics of deposition when a single particle falls at a time and a small equipment shown in Fig.no. 4 was developed for the purpose. Subsequently, a study was made of deposition of group of particles falling as a curtain of sand.

3. RESULTS/OBSERVATIONS

3.1 Microscopic observations

The observations made by stereo – microscope may be summarized below:

(i) The particles fall vertically (nearly) and hit the surface of the deposit. In general, after hitting the surface, the particles bounce up as projectiles, hit the surface again and settle in some position (or may repeat the sequence again). The falling particles invariably, cause disturbance to the particles already deposited.

(ii) In depositions with a low I, the bouncing effect is more apparent than in high I depositions. In the latter case, the trajectories possess either low heights or are along the surface, i.e. the particles roll along the surface.

(iii) The bouncing reaction is more pronounced when V (or height of fall of particles - H) is more.

(iv) The direction of the path followed by the bouncing (or the rising) particles is random.

(v) Some particles have been observed to be clinging to the container walls indicating the electrostatic effect.

However the above events take place in such a short interval of time that little details of the mechanics of the particle deposition can be apprehended.

In order to promote a better understanding of the deposition process a study is made of an idealized case of two dimensional packing to start with. A two-dimensional-packing apparatus (after Kolbuszewski 1965) has been used, after some slight midifications. Spherical beads are dropped into this apparatus one by one. It has been observed that the falling particle either stays in position (if it falls into a hollow space big enough to accommodate it), or jumps over to the side and/or rolls along the surface before coming to rest. In general, the pore is small in the first case, and is big in the latter case (perhaps due to frictional resistance to its movement and/or electrostatic forces which may be responsible for the settlment of the particle in the adverse position). Quite often, the big pores propagated into formation of macropores. Sometimes the impact of the oncoming particle causes the settled particles to move out of their position or causes a re-arrangement of the particles below the surface, resulting in collapse of some of the less stable macropores and formation of the new ones. The effect of the reduction in the height of fall of the particles invariably affected a formation of a large number of macropores. State of packing of low and high porosity are produced as shown in Fig.2. These depositions indicate the formation of macropores such as marked as x - x in Fig.2 (a & b).

3.2 Slow-motion film observations

A number of slow motion films have been produced for a comprehensive study of the mechanics of (a) a single particle of sand falling at a time, and (b) vertical controlled deposition. Camera speeds varying from 4000 to 7000 frames per second (fps) have been used. The observations made may be sub-classified as :

The films have been produced by keeping the camera tube (i) horizontal (films B1, B2, B3), and (ii) inclined at different inclinations (films B4, B5 and B6), to record the events caused during the deposition of a single particle falling from diffrent heights (of 46,29,and 10cms) on to a loose, or a dense bed.

For studying the vertical controlled deposition, films have been produced to record deposition of group of particles. i) for low I (film A2, A3, A4, A6), ii) for high I (films A5, A7), and iii) for a low value of V (film A3).

1. The films B1, B2 and B3 show the bouncing, spinning, etc. of a falling particle; whereas films B4 and B5 illustrate the events caused (by the falling particle) at the surface of the deposit. It is observed that the velocity of the particle (for H=46cms) arriving for deposition, is 3m/sec nearly (films B1, B2, B3). The particle hits the surface and rebounds to a height more than 6 mm, spinning at a high speed (of nearly 1000 revolutions per second, rps). When it falls again, it may rebound again though to a smaller height (such as < 3 mm), and the speed of spinning is 300 to 500 rps. The length of the trajectory followed by the particle varies in a wide range. Normally the particle, after bouncing once or more number of times, settles in a hollow, that is an optimum position. Sometimes, especially for small H, the particle spins along the surface, hits some other particle, and rises up. The spinning is much slower in this case (200 to 300 rps).

Film B4 (H of 46 cms) shows that the falling particle, invariably, knocks up two or more particles from the surface. The latter ones bounce up by 3 mm or less and spin at a lower speed (200 to 300 rps). All the particles, whether rising up or grazing along the surface, are spinning during their state of motion. In general, the particle spinning at a high speed settles in an optimum position, that is in a hollow cavity; wheras the particle rolling along the surface, or rising a very small height may settle in an adverse position resulting in a big void or 'arching action'. Furthermore, the particles falling on a loose bed (film B5) have low height of rebound and slow spinning.

2. Films A2, A4 and A6 show that the behaviour of the particles, when the intensity of deposition is very low is similar to that of the particles deposited individually. This results in the formation of small pores (and there are less chance of formation of macropores). When I is high)films A5 and A7), the particles do not have much opportunity to rebound, or the bouncing particles are interferred with by the falling particles. Generally, the particles roll along the surface. This results generally in big voids and formatiom of a large number of macropores. The effect of reduction in H (film A3) is similar to the observations given above for film B5.

A few extracts from the films, two groups of four consecutive frames, are given in the Fig.Nos. 5 and 6. Fig. No.6 demonstrates the high speed of spinning of a falling particle.

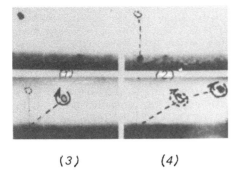

(3)   (4)

Fig. 5. A particle hitting deposition surface and rebounding.

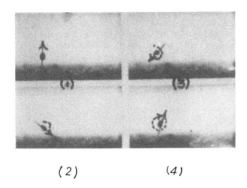

(2)   (4)

Fig. 6. Spinning of a particle.

4. DISCUSSION OF RESULTS :

4.1 Mechanics of deposition of single particle

The present investigations discover, for the first time, the high degree of the speed of spinning of a particle when it hits a surface of deposited particles and rebounds from it. This is illustrated in Fig.6. However the speed of the particle, after the second rebound, is less. This is due to the partial loss of its kinetic energy due to the subsequent impact. When the particle falls after the first or the second rebound, it tends to occupy the position of the least potential energy and thus settles in a hollow space between the deposited particles. This results in formation of small voids and thus a dense packing.

Particles which roll along the surface, often occupy an adverse position and result in the formation of big voids which may further propagate to from macropores. The reason for this may be that : (i) these particles possess low kinetic energy (e.g. due to a small height of fall – film B5) and low speed of spinning, and (ii) due to the low momentum, the electrostatic forces may become more affective and cause the formation of chains of particles if it is a coarse sand, or column of particles if it is a fine sand (Kolbuszewski 1948). The argument that the particles rolling along the surface contribute more in the formation of macropores is supported by the observation of the two dimensional packing.

The result of an examination of the film B5 indicates the effect of the deposition of a particle on a loose bed. The low height of rebound, and slow spinning may be due to the cushioning effect of the loose bed. This results generally in the deposition of the particle in an adverse position. The low heigh of fall of the particle results in low momentum and similar deposition as above.

4.2 Deposition of group of particles

The analysis of the films of group 'A' for vertical controlled deposition indicates the effect of the interaction of the particles. When I is very low, the number of particles arriving over a Unit area is so small that their deposition is similar to the deposition of a single particle. For example, for low I (and H of 46 cms) the falling particles rebound, spin with high speed (nearly 1000 rps) and tend to settle in preferred positions. The loose or relatively less stable particles on the surface of the deposit are knocked around to get more stable position. The particles have enough time to come to rest before the arrival of fresh particles. The chances of the formation of arches and the effective role of the electrostatic forces is small (due to the high momentum possessed by the particles and little interference between the particles). This results in small void and the number of macropores formed is small.

However when I is high, the height of the rebound of the particles is small due to the interference of the falling particles. Particularly a large number of particles simply roll along the surface. In the present process of deposition, the sloping surface at the end of the layer of the sand being deposited also causes the rolling of the particles down its slope in the direction of the travel of the sand-curtain. These particles possess low momentum. Thus the electrostatic forces and frictional resitance may become quite effective to cause the formation of a honeycomb or loose structure. All these resons contribute to the formation of a large number of macropores.

The film A3 indicates that the reduction in H results in lower height of rebound, and slower spinning. This is due to the low value of momentum because of the low V. This results in the formation of big pores. However the formation of big pores resulting in loose state of packing of the bed further causes reduction in the momentum (and spinning speed etc. of the particles) due to the cushinoning effect discussed above (film B5). The big void may propagate to form macropores.

Some of the particles settle in such a position that arches are formed. These arches may be stable ones causing the formation of large macropores or may collapse due to the tamping action of the falling particles and may result in smaller macropores (Jalota 1972). Further based on the observations of two dimensional aparatus, it may be said that the container walls cause arching effect and thus cause a loose packing of the particles near them (Lees 1970). Hence the boundary conditions help in the formation and propagation of macropores. A reason for the formation of macropores in depositions formed by settlement of particles in water has been given in another study (Kolbuszewski 1948).

4.3 Formation of dust in sand storms

The sand storms contain dust. Various reasons put forth by different studies are reviewed by Bagnold (1941). It is concluded that dust is formed due to the impact of sand particles falling on the ones lying as a deposit, though this conclusion has been found unconvincing by Bagnold also (Bagnold 1941). This is so because the mass of a sand particle is very very small and the momentum is too small to cause disintegration of sand particles deposited as the hardness of the quartz, the main constituent mineral of sand, is quite high.

The study made by the author (Jalota 1972) for the first time discovers that a particle when it strikes the surface of a sand deposit, rebounds from the surface and is spinning with a high speed. A speed of spinning as high as 1000 revolutions per second (rps) has been observed (This phenomenon is illustrated in fig.6). This action is due to the high velocity of the falling particles, angularity, and low radius of gyration of the particles. Thus these particles act as high speed drill/grinder and produce dust. This phenomenon is discussed in detail elsewhere (Jalota 1985).

5. CONCLUSIONS

The following conclusions can be drawn as a result of the above discussion :

Mechanics of a single particle deposition

1. When a particle falls and hits a particle/particles deposited already, it rebounds, spinning with a considerably high speed. However the heights of the ttrajectory followed by it, and its speed of spinning depends upon its velocity of fall, and density of the surface of deposition.
2. The particles on the surface of the deposite hit by the falling particle may also rise up following a low trajectory and spinning at a smaller speed.
3. The particle, if it rebounds after hitting the deposition surface again, follows a lower trajectory and spins with a slower speed.
4. If the particle falls on a loose bed (i.e. having a high porosity) the bouncing and spinning action is relatively small.

Vertical controlled deposition

5. For low values of I, the number of particles deposited over a unit area in a unit time, are so small that their depositional behaviour corresponds to that of a single particle deposition. The particles generally occupy optimum positions, if the value of V is large. Therefore the number of the macropores formed is small.
6. For high values of I, the bouncing and spinning action of a particle is

restricted due to the interaction of a large number of particles. This results in the formation of a large number of macropores.

7. For lower values of height of fall of the particles, H (and thus of V) the number of macropores formed is large.

8. For low V or high I, the formation of loose deposit further helps in the propagation of macropores.

Causes for formation of Macropores

9. The crowding and the low momentum of the particles, the electrostatic forces, and arching action, all contribute to the formation of a large number of macropores.

10. The container walls help to bring about the formation of macropores in their vicinity. These macropores may be propogated further across the deposit.

A summary of the above indicates that the state of dispersion and the kinetic energy possessed by the particles, arriving for the deposition has the most significant effect on the formation of macropores (and thus on the state of packing).

Dust formation

11. High speed of spinning of the particles imparts an action of a drill/grinder to the falling particles resulting in formation of dust.

ACKNOWLEDGEMENT

The author is grateful to Prof. J.Kolbuszewski, Ex-Head of the Department of Transportation and Environment planning, University of Birmingham,Birmingham (U.K.) for acting as the guide for the main study, of which this presentation is only a part. Further the help extended by the faculty members of the University by way of discussions is also acknowledged with thanks.

REFERENCES :

Amirsoleymani, T. 1965. Packing of granular materials- Effects of deposition. Report to the Department of Transportation and Environmental Planning,University of Birmingham. (U.K.)

Amirsoleymani, T. 1966. Packing of granular materials with special reference to Triaxial testing. Thesis for the Degree of Ph.D., University of Birmingham. (U.K.)

Bagnold, R.A. 1941. The physics of blown sand and desert sand. Matheun Ltd., London.

Bernal, J.D. and Mason, J. 1960. Packing of spheres-coordination of randomly packed spheres. Nature 188.

Jalota A.V. 1985. Dust in sand storms. Proc. Asia Pacific Symposium on Wind Engineering, 85, Roorkee (India)

Jalota A.V. 1972. "Mechanism and Laws Governing Formation of Macropores in Granular Soils". Ph.D. Thesis University of Birmingham (U.K.)

Josselin De Jong and Veruijit. 1965. Primary and secondary consolidation of a spherical clay sample. Proceeding 6th International Conference on Soil Mechanics and Foundation Engineering. Volume I.

Kolbuszewski, J. 1948. An experimental study of maximum and minimum porosity of sands. Proceeding 2nd International Conference Soil Mechanics and Foundation Engineering Volume I.

Kolbuszewski, J. 1950. Notes on deposition of sand. Research. Vol.3.

Kolbuszewski, J. 1953. Note on factors governing the porosity of wind deposited sands. Geological Magazine. Volume No. 1.

Kolbuszewski, J. 1963. "Quicksand". New Scientist. Vol. 20.

Kolbuszweski, J. 1965. Sand particles and their densities. Lectures given to the material Science Club's symposium on Densification of particulate materials. London 26th February, 1965.

Lees, G. 1970. Studies of Inter-particle void characteristics. Journal of Engineering Geology. Vol. 2. No. 4.

Scott, G.D. 1962. Radial distribution of random close packing of equal spheres. Nature 194.

Environmental Geotechnics and Problematic Soils and Rocks, Balasubramaniam et al. (eds)
© 1987 Balkema, Rotterdam. ISBN 90 6191 785 9

Reducing landslide hazards*

William J.Kockelman
United States Geological Survey, Menlo Park, Calif.

ABSTRACT: Landslides continue to cause costly damage and loss of lives. Many techniques are available for reducing landslide hazards; 27 are described in this paper. An overview of these techniques is useful to planners who implement hazard-reduction programs, to engineers who serve as advisors to local or state governments, and to decisionmakers who select the most appropriate technique for a given situation. Prerequisites for the successful use of these techniques are hazard information understandable to nongeologists and adequate communication of this information to those who will, or are required to, use it. It is concluded that certain factors needed to ensure the lasting effectiveness of these techniques are usually beyond the control of the public planner, engineer, and decisionmaker.

1 INTRODUCTION

According to the U.S. Geological Survey (1982) "Ground failures caused by landslides, subsidence, swelling clays, and construction-induced rock deformation cause billions of dollars in property losses in the United States each year. Together they have ... exceeded the annual combined losses from floods, earthquakes, hurricanes, and tornadoes by many times--those from landslides and subsidence alone amounting to at least $1.5 billion per year over the past 50 years (Jahns, 1978)...."

The Appalachian, Rocky Mountain, and Pacific Coast regions have suffered the greatest landslide damage; recent examples include: over $4 million per year in Allegheny County, Pennsylvania; over $22 million in Cincinnati, Ohio, between 1974 and 1980; $250 million in Utah in 1983; over $3 million per year in Colorado; $66 million and 25 lives lost in the nine-county San Francisco Bay area in 1982; over $40 million in Orange County, California, between 1966 and 1981; and an average $500 million in the Los Angeles area during each year of heavy storms.

---

*Original version of this paper was published in the Bulletin of the Association of Engineering Geologists 1986, v. 23, no. 1, College Station, TX.

2 LANDSLIDE-HAZARD-REDUCTION TECHNIQUES

Numerous techniques for reducing landslide hazards and their costs are available to planners, engineers, and decisionmakers. Some of these techniques, such as public acquisition of hazardous areas, are well known to the planning profession. Others, such as control structures, are commonly used by engineers. Still others, such as warning signs and regulations, are obvious and practical, but require maintenance and enforcement. Still others are innovative and untested when applied to landslides, but have been successfully used in solving flood and soil problems. These and other techniques are shown in List 1 under the general headings of (1) discouraging new development in hazardous areas, (2) removing or converting existing development, (3) providing financial incentives or disincentives, (4) regulating new development, and (5) protecting existing development.

The techniques may be used in a variety of combinations to help solve both existing and potential landslide problems. For example, Stokes and Cilweck (1974) described special techniques used for the guidance and control of land development in Ventura County, California. These techniques consist of subdivision and grading ordinances, and various guidelines and standards to

List 1. Hazard-reduction techniques

Discouraging new development in hazardous areas by:

Disclosing the hazard to real-estate buyers
Posting warnings of potential hazards
Adopting utility and public-facility service-area policies
Informing and educating the public
Making a public record of hazards

Removing or converting existing development through:

Acquiring or exchanging hazardous properties
Discontinuing nonconforming uses
Reconstructing damaged areas after landslides
Removing unsafe structures
Clearing and redeveloping blighted areas before landslides

Providing financial incentives or disincentives by:

Conditioning federal and state financial assistance
Clarifying the legal liability of property owners
Adopting lending policies that reflect risk of loss
Requiring insurance related to level of hazard
Providing tax credits or lower assessments to property owners

Regulating new development in hazardous areas by:

Enacting grading ordinances
Adopting hillside-development regulations
Amending land-use zoning districts and regulations
Enacting sanitary ordinances
Creating special hazard-reduction zones and regulations
Enacting subdivision ordinances
Placing moratoriums on rebuilding

Protecting existing development by:

Controlling landslides and slumps
Controlling mudflows and debris flows
Controlling rockfalls
Creating improvement districts that assess costs to beneficiaries
Operating monitoring, warning, and evacuating systems

assist developers in meeting the County's policies and procedures. In the case of flood problems, Kusler (1982) described 75 case studies of hazard-reduction programs, many of which are applicable to landslides. Most include various combinations of two or more techniques, such as, restrictive rebuilding regulations combined with acquisition for a mudflow area; rebuilding moratoriums and more detailed regulations combined with monitoring and warning systems, posting of signs, and partial acquisition; and ample setbacks from bluffs subject to erosion combined with acquisition and citizen education.

The techniques are generally applicable to all types of surface-ground failures, including flows, slides, and falls whether triggered by stream or coastal erosion, rainfall, flooding, earthquakes, volcanoes, or human activity. Some of the triggering mechanisms for these failures are discussed by Coates (1977: 7-13).

3 PREREQUISITES FOR THE USE OF HAZARD-REDUCTION TECHNIQUES

Prerequisites for the successful use of any of the hazard-reduction techniques are the availability of adequate, detailed hazard information that is in a form usable by nongeologists; and its communication to those planners, engineers, and decisionmakers who will, or who are required to, make use of such information. A comprehensive national program for landslide-hazard reduction, developed by the United States Geological Survey (1982), sets forth goals and tasks for making landslide studies, evaluating and mapping the hazard (sometimes called "zonation"), disseminating the information to potential users, and subsequently evaluating the use of the information. Examples of the types of maps to be developed under such a program, and lists of typical users and communication techniques are included in the program.

Landslides are not simple processes, nor are they easily described. For example, the National Research Council (1982) Committee on Methodologies for Predicting Mudflow Areas addressed the need for an unambiguous and technically acceptable definition of "mudslide" used by the Federal Emergency Management Agency in administering the National Flood Insurance Program. The work of Varnes (1978: 11-33) in describing and diagramming slope-movement types and in providing a review of the principles and practice of zonation (Varnes and the International Association of Engineering Geology Commission on Landslides and Other Mass Movements on Slopes, 1984) is very helpful.

Recognition and identification of landslides have been discussed by Rib and Liang (1978) and by Slosson (1983), but the preparation of reliable landslide-hazard (susceptibility) maps can be expensive and is at a relatively early stage of development and testing. Brabb (1984) describes and gives examples of various innovative types of landslide maps --reconnaissance-inventory, complex-inventory, susceptibility, computer-assisted, loss-evaluation, and risk-determination maps.

Examples of the type and detail of hazard maps adequate for county (1:24,000) and regional (1:125,000) planning and decisionmaking include the slope-stability maps prepared by Nilsen and others (1979) for the nine-county San Francisco Bay region. These maps and their relevance and application to land-use planning are described in a report by Vlasic and Spangle (1979). Other hazard maps used in the United States at scales ranging from 1:12,000 to 1:100,000 show:

1. Mudflow hazards and paths (Los Angeles County Flood Control District, 1984)

2. Potential rockfalls and areas susceptible to sliding in Jefferson County, Colorado (Glenn Scott, 1972)

3. "Mudslide"-risk zones in southern Ventura County, California (Evans and Gray, 1971)

4. Slope stability during earthquakes in San Mateo County, California (Wieczorek and others, 1985)

5. Landslide-prone areas in the State of West Virginia (Lessing and others, 1976)

6. Landslide-hazard areas in King County, Washington (King County Department of Planning and Community Development, 1980)

7. Mudflow inundation for the Cowlitz and Toutle Rivers, Cowlitz County, Washington (Swift and Kresch, 1983)

8. Debris-flow hazards for the Las Vegas Southeast Quadrangle, Nevada (Katzer, 1981)

9. Slope stability in Fairfax County, Virginia (Obermeier, 1979)

10. Areas most susceptible to land-sliding in Beaver County, Pennsylvania (Pomeroy, 1979)

11. Susceptibility to shallow land-sliding in Marin and Sonoma counties, California (Ellen and others, 1982)

12. Potential for debris flows along the Wasatch Front, Utah (Wieczorek and others, 1983)

A recent statute enacted by the California Legislature (1983) requires the State Geologist to develop maps of landslide hazards in urban and urbanizing areas. This Landslide Hazard Identification Act provides that the California State Mining and Geology Board, in consultation with the State Geologist, develop guidelines for the preparation of the maps, and priorities for those areas to be mapped. The California Legislature (1983, sec. 2686) further specifies that the maps shall be at scales suitable for local planning purposes. The Act defines "landslide hazard" to mean "an identified recent activity of, or high potential for, landslide, slope failure, creep, or debris-flow hazards...." According to Gray (1984), about $300,000 in new funding annually became available on July 1, 1984, and the California State Division of Mines and Geology began mapping five high-priority areas at a scale of 1:24,000.

Planners and engineers prepare various planning studies, such as environmental impact assessments and multi-hazard inventories, and various development plans, such as utility-service plans and subdivision layouts. The availability of landslide-hazard information that is usable by nongeologists is also a prerequisite for the preparation and adoption of such studies and plans. Once these and other studies and plans-- transportation, land-use, and community facilities--are prepared, any combination of the techniques discussed in the following sections can be used to implement the plans in order to reduce landslide hazards.

4 DISCOURAGING NEW DEVELOPMENT IN HAZARDOUS AREAS

There are several methods to discourage development in hazardous areas. These include disclosing hazards to real-estate buyers, posting warning signs, adopting appropriate utility and public-facility service-area policies, informing and educating the public, and recording the hazard in public documents. Any of these may be used to discourage responsible developers and prudent home buyers from locating projects or purchasing homes in landslide areas.

4.1 Disclosure

Preparing landslide-hazard information for nongeologists and enacting hazard-disclosure laws can make people aware of landslide hazards. Furthermore, disclosing these hazards at the time of purchase can alert property owners to potential danger and loss. For example, to provide for protection against flood losses through a federally subsidized flood-insurance program, the United States Congress (1974b) requires lenders to notify prospective borrowers that the real estate being mortgaged is located in a flood-hazard area, as identified by the Federal Insurance Administrator. To provide for public safety from earthquakes, the California Legislature (1972) requires a seller or his agent to tell the prospective buyer that the real estate is located within an official fault-rupture zone, as delineated by the State Geologist (Hart, 1985).

Disclosure can also be required at the local level. For example, in the ordinance enforcing on-site geologic investigations before construction, the Santa Clara County (California) Board of Supervisors (1978) requires that all sellers of real estate lying partly or wholly within the county's flood, landslide, and fault-rupture zones provide the buyer with a written statement of the geologic risk.

To help Realtors (members of the National Association of Real Estate Boards) comply with these federal, state, and county disclosure laws, five local boards of Realtors in the San Francisco Bay area prepared street maps showing some or all of the flood, landslide, and fault-rupture zones. The five maps cover one entire county and parts of three others, and include more than 50 cities. In addition, the San Jose (California) Board of Realtors (1978) designed a form to be attached to a real-estate contract to comply with the county disclosure ordinance. This example of disclosure is described in greater detail by Brown and Kockelman (1983: 84-91).

4.2 Posted warnings

Warning signs that are posted and readily visible to buyers, developers, and the public help alert prospective developers and purchasers to the hazard. Warnings should be based upon adequate data and be posted where the landslide areas intersect or abut public rights-of-way, such as "slide area" signs placed along highways. In a similar way, extensive flood-warning signs have been used by the United States Bureau of Reclamation on the Sacramento River near Redding, California and, according to Kusler (1982: 291), flood elevations with special labels prepared by the United States Army Corps of Engineers have been posted in the City of Crookston, Minnesota.

Warnings can also take the form of rubber-stamped notations placed on subdivision plats or on building and zoning permits. These types of warnings alert potential property buyers that sites are in hazardous areas.

4.3 Public-facility service-area policies

Local governments and utilities can adopt policies that prohibit the financing and construction of public facilities or utilities in potential landslide areas to promote safety and reduce public losses. If such areas lack streets, bridges, water-supply and sewer systems, and other community facilities, they will be far less likely to attract new development or to tempt developers to extend existing development. These community facilities are not only subject to costly damage, but they influence the location and density of private development. This technique for prohibiting the construction of public utilities that will serve hazardous areas can be used in conjunction with regulatory techniques, such as subdivision and zoning ordinances. However, to be fair to prospective developers, the policies need to be adopted long before any development plans are made.

4.4 Public information and education

Public information programs are essential for bringing landslide information to the attention of the public. Responsible developers and prudent citizens, when told of landslide hazards, may not wish to risk property losses or expose their clients or families to the dangers and trauma caused by landslides. Since any hazard-reduction program depends on the understanding and support of an informed public, education is vitally needed. Preparing, announcing, and disseminating information on landslide damage, susceptibility, and hazard-reduction techniques can be accomplished through conferences, legislative hearings, workshops, newsletters, press releases, bulletins, and letters to key officials.

Conferences or hearings convened soon after disasters usually have key participants, attentive audiences, and media coverage. Examples of such conferences are the 1983 Governor's Conference on Geologic Hazards (Utah Geological and Mineral Survey, 1983), the 1982 National Research Council and United States Geological Survey conference on debris-flows, landslides, and floods (Brown, 1984), and the Governor's Conference on Environmental Geology (Colorado Geological Survey, 1969). A report on problems expressed and recommendations made during a hearing was prepared and disseminated by the California State Assembly Select Committee on Landslide Prevention (1980).

Guidebooks or model ordinances written for nongeologists by federal, state, or local units of government or professional associations can be helpful; examples include Use of Slope-Stability Information in Land-Use Planning (Vlasic and Spangle, 1979), Facing Geologic and Hydrologic Hazards--Earth-Science Considerations (Hays, editor, 1981), Geologic Principles for Prudent Land Use --A Decisionmaker's Guide for the San Francisco Bay Region (Brown and Kockelman, 1983), Reducing Landslide Hazards --A Guide for Planners (Erley and Kockelman, 1981), the Colorado Geological Survey's Model Geologic Hazard Area Control Regulations (in Lessing and others, 1976, Appendix C), and Recommended Guidelines for Preparing Engineering Geologic Reports (California Division of Mines and Geology, 1975).

Another way to disseminate information on landslide damage and site instability is through the use of maps, for example, the maps that show five zones of instability for most of West Virginia. A bulletin accompanying these maps (Lessing and others, 1976) informs buyers, builders, and homeowners about "danger signals" and landslide-correction methods, and explains the responsibilities of state agencies concerned with landslides.

4.5 Public records

Public records of land ownership also can help alert land purchasers, local assessors, and lenders to potential landslide hazards if such hazards are made part of a public record. Such records might include the filing of detailed maps of landslide areas with the appropriate recorder of deeds. A request for entry of the hazard maps into tract indexes would result in the abstracts of titles and subsequent conveyances containing entries referring to the hazards. Subdivision ordinances requiring identification of landslide areas result in such information being filed automatically with a recorder of deeds.

The Santa Clara County (California) Board of Supervisors (1978) requires geologic reports for hazardous zones. If the report indicates unusually severe geologic conditions, including landslides, the property owner must sign a statement before any development takes place. This statement acknowledges that the owner is aware of the hazards, accepts the risks of development, and relieves the County of liability.

The statement is then recorded in the County Recorder's Office and can be expunged only if subsequent information-- approved by the County Geologist-- indicates that the hazard no longer exists or has been reduced. However, no new structures for human occupancy can be located in active landslide areas until the areas have been stabilized according to accepted engineering practices.

5 REMOVING OR CONVERTING DEVELOPMENT

Recurrent damage from landslides can be avoided by permanently evacuating areas that continue to have slope failures. Structures may be removed or converted to a use which is less vulnerable to landslide damage. The feasibility of such action depends on the value of the structures, their potential for contributing to unstable conditions, whether they can be successfully reinforced, and the level of citizen concern. Techniques for removal or conversion include the acquiring or exchanging of hazardous areas and relocating their occupants, incorporating nonconforming-use provisions into zoning ordinances, clearing and redeveloping damaged areas, removing unsafe structures or public nuisances, and urban redevelopment.

5.1 Acquisition, exchange, and relocation

A government agency may acquire hazardous lands through purchase, condemnation, tax-foreclosure proceedings, dedication, devise (will), or donation. The agency can then control development, using it in the public interest. It might choose to sell or lease part or all of it, stipulating that no structure be built that would be vulnerable to landslides. It might lease the land for crops or grazing, thus recovering part of the acquisition costs.

In some cases, simple purchase may be more economical than extending and maintaining public services or utilities in landslide areas and perhaps being liable for post-landslide repairs. It is important that the government agency have a detailed geologic map showing the distribution of landslides. Some park districts in the San Francisco Bay area have acquired landslide terrain for a park, only to find out later that they are also responsible for lateral support where a property line cuts across a landslide mass.

Exchange of privately owned hazardous areas for publicly held stable areas is possible if such public lands are available and close to the hazardous area and if the occupants are willing to relocate. Arizona is relocating residents in 10 communities subject to flooding, pursuant to a program authorized by the Arizona Legislature in 1978. This program includes exchange of public lands for private lands subject to flooding, and state financial aid for relocation (Kusler, 1982: 71).

A public agency may also acquire less-than-fee interest in hazardous areas. This method costs the public less than purchase, because only certain property rights are purchased. Such interest may be in the form of an easement to protect scenic views, conveyance of development rights to assure the continuation of existing private parks and open spaces, or grants of public access and development rights for construction and use of park facilities. The purchase of easement or development rights can limit development, and the owner receives fair compensation for the release of these rights. Easement lines do not require accurate landslide-hazard boundaries (as do zoning districts, for instance) because the boundaries of the area to be acquired can be determined by agreement. Easements should be obtained in perpetuity or for as long as the landslide hazard exists, however periodic inspections and enforcement of the permitted land use by the agency holding the easement is necessary.

Generally, easements have been found most effective in areas without heavy development pressures. Although their studies are directed to flood plains and wetlands, Field Associates (1981) have provided an excellent handbook on the use

of acquisition techniques that could be applied to landslide areas, including selecting properties, setting priorities, and acquiring less-than-fee interests; sources of assistance and case histories are also given.

Numerous federal and state financial aid programs are available for acquiring land for park and recreational uses that can be compatible with some landslide hazards. One example is the Federal Land and Water Conservation Fund Act described in the Catalog of Federal Domestic Assistance by the United States Office of Management and Budget (1983: 414, 415). In the case of flood-hazard areas (Kusler, 1982: 292), the Ramsey County Open-Space Planning Office developed an acquisition and relocation plan for most of the Town of Lilydale, Minnesota:

About 97% has been implemented. It called for county acquisition of approximately 320 acres, including eight businesses and 113 households (mostly mobile homes) with funds from the state and the Minnesota Metropolitan Council. Total project costs are estimated at $4,750,000 not counting development costs estimated to be an additional $1,931,000. The town site is being converted to use as a park.

5.2 Discontinuance of nonconforming uses

When new or amended land-use zoning regulations are enacted, they often result in some existing land uses becoming nonconforming. Nonconforming uses are defined as those that exist at the time a zoning ordinance is adopted or amended, and do not conform to the new ordinance. For example, if an ordinance prohibits new residences within a landslide area, those residences existing prior to the ordinance become nonconforming.

Zoning or other land-use ordinances may provide that nonconforming uses may be continued but cannot be extended or enlarged. They may stipulate that, if the current uses are discontinued for a designated period, any future use must conform to the ordinance. Regulations may also require that total structural repairs or alterations over the lifetime of a nonconforming building be limited to a percentage of the assessed or market value.

Legislation may also eliminate nonconforming uses by providing for their amortization over a reasonable period of time. This practice of eliminating non-conforming uses is described in detail by R.L. Scott (1972). Model ordinance provisions and comments on the discontinuance of existing land uses are described in A Model Land Development Code by the American Law Institute (1975: 142-166).

5.3 Post-disaster redevelopment

Disasters which generate great public and political awareness can often provide an opportunity to redevelop in a safer manner by the joint use of techniques such as rebuilding moratoriums, acquisition, relocation, regulations, slope stabilization, and financial incentives and disincentives. William Spangle and Associates and others (1980) describe reconstruction plans and actions taken after several disasters—some of which included landslides—in the San Fernando Valley, California; Anchorage, Alaska; Bluebird Canyon, Laguna Beach, California; and Valdez, Alaska.

In the case of destruction of shoreline development by storms and associated landsliding, geologists Pilkey and Howard (1981: 10) are emphatic:

There is immediate need for measures to prevent redevelopment after storms. No longer should local, state, and federal governments expend public funds in redevelopment; on the contrary, these governments should assume responsibility for the protection of the public (as opposed to the private) interest, not to re-create an untenable situation or to guarantee recurrent destruction of such properties.

5.4 Removal of unsafe structures

Government agencies can remove structures damaged by landslides by applying their public-nuisance-abatement powers. Section 203 of the Uniform Building Code by the International Conference of Building Officials (1985), adopted by many cities and counties, indicates how this can be done:

All buildings or structures regulated by this code which are structurally unsafe or not provided with adequate egress, or which constitute a fire hazard, or are otherwise dangerous to human life are ... unsafe. Any use of buldings or structures constituting a hazard to safety, health, or public welfare by reason of inadequate maintenance, dilapidation, obsolescence, fire hazard, disaster, damage, or abandonment is ... an unsafe use.

All such unsafe buildings ... are hereby declared to be public nuisances and shall be abated by repair, rehabilitation, demolition, or removal ....

This power gives many communities that have adopted similar provisions in their building codes the authority to remove any structures damaged by landslides. For example, the 1978 Bluebird Canyon (southern California) landslide required timely demolition and removal of 22 damaged homes; the owners refused to sign agreements allowing their homes to be demolished. According to William Spangle and Associates and others (1980), the Laguna Beach City Council "felt that appeals by landslide victims could prolong actions needed to protect habitable properties surrounding the landslide area from the effects of approaching winter rains.... The City Manager, with confirmation from the City Council, declared a nuisance to exist...and ordered the demolition.... Demolition began two weeks later ...."

5.5 Urban redevelopment

Sometimes landslide hazards can be reduced by purchasing, and then converting or redeveloping, land that has been adversely affected by landslides. State laws authorizing the creation of public agencies usually provide for: the preparation and adoption of redevelopment plans; the acquisition, clearance, disposal, reconstruction, and rehabilitation of blighted (including damaged) areas; and the relocation of those persons displaced by a redevelopment project.

Redevelopment agencies usually are empowered to issue bonds, receive part of the taxes levied on property in the project, and use federal grants or loans available under various programs of the Federal Housing and Community Development Act of 1974. These programs are described in the Catalog of Federal Domestic Assistance by the United States Office of Management and Budget (1983: 360-362, 365, 366).

Reconstructing public facilities located in landslide areas before a disaster also affords an opportunity to reduce the risk of damage. Such facilities could include roads, bridges, utilities, and community facilities that are subject to rebuilding by reason of functional or structural obsolescence. This end can be achieved by reinforcing the structure, designing it to accommodate displacement, relocating in less hazardous areas, or bridging. The term "bridging" refers to the construction of spans over hazardous areas--a technique primarily used for highways; it is expensive and consequently only used as a last resort (Gedney and Weber, 1978: 175).

6 FINANCIAL INCENTIVES OR DISINCENTIVES

Financial techniques of encouraging or discouraging development in landslide areas include amending government assistance programs, increasing awareness of legal liability, adopting appropriate lending policies, requiring insurance, and providing tax credits or lower assessments.

6.1 Federal and state financial assistance

Federal and state programs that provide grants, loans, loan guarantees, tax credits, tax deductions, depreciation allowances, insurance, revenue sharing, or other financial assistance have a tremendous effect on public and private development. Obviously, the enabling legislation for these programs can be amended by the United States Congress or state legislatures to provide for site investigations in landslide areas, avoidance of hazardous areas, or stabilization of slopes.

Less popular among elected officials, but equally effective, are financial "disincentives" which act as deterrents to the use and development of hazardous areas. A "disincentive" could reduce the federal share of a grant if the facility to be funded were to be located in a landslide area. For example, the United States Congress (1973) introduced provisions into the Flood Disaster Protection Act of 1973 for withholding federal benefits from flood-prone communities that chose not to participate in the National Flood Insurance Program. In providing loans and grants for disaster recovery, the United States Congress (1974a) requires local and state governments to evaluate and mitigate hazards.

Other examples of existing federal programs or legislation that could be amended to discourage development in landslide areas include: (1) farm, business, and industrial grants and loans; (2) assistance for economic development; (3) public works projects; (4) urban mass transportation grants; (5) housing subsidies; (6) school and hospital construction; (7) assistance for sewage treatment, water treatment, and solid-waste disposal facilities; (8) dam, reservoir, and impoundment projects; and (9) highway and road construction.

6.2 Legal liability

American jurisprudence recognizes civil liability for death, bodily injury, property damage, emotional distress, and economic loss, any of which may be caused by a landslide. Those potentially liable include the former owners or their agents, present owner, developer, tract engineer, soils or geotechnical engineers, consulting engineering geologists, architect or building designer, contractor, governmental entities, adjoining property owners, or any combination of these parties. The potential liability and adverse judgments should serve as a disincentive to those who may develop landslide areas. Sutter and Hecht (1974), supplemented by McGuire in 1985, discussed the types of lawsuits that injured parties may bring against those alleged responsible for their losses:

1. Fraud --a former owner advises the purchaser that a house is "in perfect condition" when cracks (caused by recent ground failure) had been repaired and repainted.
2. Negligence--a neighbor changes the natural drainage, thus causing a landslide.
3. Strict Liability--a mass producer and seller of lots improperly cuts, fills, and compacts earth to create a building site.
4. Breach of Warranty--parties to a real estate sales agreement insert an express guarantee of soil and geologic stability, which is otherwise.
5. Failure to Comply with Regulations-- a developer or subdivider fails to perform the geologic investigations required by a state statute or local ordinance.
6. Public Negligence--a city grading or building inspector fails to perform periodic inspections of the lot grading or building construction to ensure that the work complies with the municipal code.

The results of work by the Association of Bay Area Governments (1984) on the liability of businesses and industries for earthquake hazards and losses are applicable to landslides, particularly those triggered by a seismic event. In its Executive Summary, the Association concludes that the legal defense that an earthquake is an "act of God" may only work in two very limited situations where the event:

1. was of such type or size as to be unforeseeable and the business did not act negligently with respect to dealing with a foreseeable event, or

2. was foreseeable, and the defendant took all reasonable actions to prevent harm, but nonetheless damage still occurred.

It can be argued that landslides (other than those triggered by a seismic event) are less an "act of God" than are earthquakes, and therefore such defense in a lawsuit is less likely. Tank (1983) concludes the landslide and subsidence chapter in his Legal Aspects of Geology:

Recent court decisions have identified the developer or his consultants as primarily responsible for damage due to land failure. The prudent developer will avoid landslide-prone areas or institute control measures before movement begins, where it appears that these measures will be adequate in stabilizing the area.

In a recent massive-earth-movement damage case, the California Court of Appeals First District (1984) has held that it is the duty of a real-estate broker selling a house to conduct a reasonably competent and diligent inspection of the property and disclose to the buyer any defects revealed by the inspection.

The California State Constitution (1879) as interpreted in various court decisions guarantees that the appropriate unit of government pays for what amounts to an after-the-fact condemnation of private property that it either took or damaged for a public purpose. This theory or remedy of inverse condemnation was substantially broadened by the California Supreme Court (1965) when it held that it was irrelevant whether Los Angeles County could have been expected to foresee a disaster after it did road grading at Portuguese Bend in Palos Verdes; homeowners won $7 million in damages.

Recent court decisions are placing increased responsibility for landslide losses on public agencies in Palos Verdes, Malibu, Pacific Palisades, and other southern California communities. Cox (1984) reported that one of the plaintiffs' attorneys in a multimillion dollar series of lawsuits in Malibu's Big Rock Mesa is arguing that "the county should have designed a water system that would have withstood earth movements, rather than breaking ... and dumping water into already saturated clay and sand."

This trend has serious implications not only for government but also for professionals working in the fields of land use and development. Slosson and Havens (1984) suggested that legal liability is becoming an incentive for more careful work by professionals and listed some steps taken by the American Society of Civil Engineers to familiarize its members with the implications of recent case law.

6.3 Lending policies

Private lenders and government lending agencies can help reduce landslide hazards by denying loans or requiring nonsubsidized insurance for development in landslide areas. Almost all construction today involves loans or mortgages by private lenders. Many of these loans are insured by government agencies that specify minimum requirements for eligibility. For example, the United States Department of Housing and Urban Development (1973) sets minimum requirements for construction and design, and insists that natural hazards, including landslides, be taken into consideration.

Private lenders also can require consideration of landslide hazards. From their surveys of home-mortgage originators in the states of Washington and California, Palm and others (1983: xi) observed that the response of those California lenders recognizing seismic hazards, including landslides, "is usually to avoid loans in particularly hazardous areas or to require that homeowners purchase earthquake insurance." Marston (1984: 16) noted that at least two of the 12 largest residential lenders used landslide-hazard information in their evaluations of applications for loans.

6.4 Nonsubsidized insurance

The cost of landslide insurance can discourage development in hazardous areas or encourage land uses that are less subject to damage. Landslide insurance from private sources is costly for areas known to be susceptible to landslides, because losses due to landslides lack the random nature necessary for a sound insurance program. In this respect, landslide areas are comparable to flood areas, and the American Insurance Association (1956) statement on flood insurance may be applicable:

> Flood insurance covering fixed-location properties in areas subject to recurrent floods cannot feasibly be written because of the virtual certainty of loss, its catastrophic nature, and the reluctance or inability of the public to pay the premium charge required to make the insurance self-sustaining.

The federal government attempts to discourage development in "mudslide" areas by means of the National Flood Insurance Program (NFIP). As amended by the United States Congress (1974b), this program requires that some kind of flood insurance must be purchased before any form of federal financial assistance will be given for construction or acquisition in areas where "mudslide" hazards have been identified However, there is no funded program for identifying and mapping "mudslide" hazard areas for the NFIP. The word "mudslide" indicates a legislative term rather than a geologic term.

Sound insurance programs at reasonable rates cannot be made available in known fault-rupture, flood, and landslide areas unless the premium costs are subsidized. Government subsidies to property owners and their mortgagers who suffer damage may lead to development in hazardous areas because the potential loss is indemnified. According to Miller (1977), after national flood insurance became available, lending institutions in four of the 15 communities studied reversed earlier restrictions on mortgages in hazardous coastal areas.

Based on a survey of 1,203 local governments, Burby and French (1981: 294) concluded that "It often appears that the NFIP induces in-creased flood plain development...." State and local officials interviewed by Kusler (1982: 36) argued that "bank financing would not have been available for much of the new development without flood insurance."

6.5 Tax credits or lower assessments

Real-property tax credits or assessment policies that encourage open-space uses or resource protection can also be used to discourage certain types of vulnerable development in landslide areas. In agricultural areas, state farmland-preservation laws provide encouragement to leave undeveloped lands that may be susceptible to landsliding. For example, the Wisconsin Legislature (1977) provides for a state income-tax credit of up to $4,200 annually as an incentive to farmers to preserve farmland. The credit doubled in 1982 if a county created an exclusive agricultural zone.

According to the United States Council on Environmental Quality (1976: 67) at least 42 states had adopted preferential tax policies to encourage farmers and other owners of open space to retain their holdings. For example, the Williamson Act passed by the California Legislature (1965) enables cities and counties to establish agricultural preserves, wherein land uses are restricted to agricultural and other open-space uses. An owner receives a tax assessment based on the restricted use rather than its market value; in return, the owner makes at least a 10-year commitment to retain the land as open space. The effectiveness of these laws in reducing development in landslide areas depends upon their being combined with the regulatory techniques discussed in the following section.

7 REGULATING DEVELOPMENT IN HAZARDOUS AREAS

Various types of land-use and land-development regulations can be used to reduce landslide hazards; they are often the most economical and the most effective means available to a local government. It is unrealistic to assume that development can be indirectly discouraged for a long period, and other techniques, such as protecting existing development or purchasing hazardous areas. can be very costly.

A community, through its development regulations, can prohibit, restrict, or regulate development in landslide areas. It can restrict landslide areas to open-space uses such as parks, grazing, or certain types of agriculture. Or, if more vulnerable development is allowed in hazardous areas, the density of development can be kept to a minimum--both to reduce its effects and the potential for damage. Grading, hillside development, zoning, sanitary, special-zone, and sub-division regulations as well as mora-toriums on rebuilding can be used to meet these objectives.

7.1 Grading ordinances

Grading ordinances can be used to ensure that excavating, cutting, and filling of landslide areas are designed and con-ducted in such a way as to avoid cutting into the toe of a landslide, removing lateral support, surcharging the land-slide head, or otherwise reducing its stability. Such goals can be obtained by grading regulations designed to:

1. Require a permit prior to scraping, excavating, filling, or cutting any lands.
2. Prohibit, minimize, or carefully regulate the excavating, cutting, and filling activities in landslide areas.
3. Provide for the proper design, con-struction, and periodic inspection and maintenance of weeps, drains, and drain-ageways, including culverts, ditches, gutters, and diversions.
4. Regulate the disrupting of vegeta-tion and drainage patterns.
5. Provide for proper engineering de-sign, placement, and drainage of fills, including periodic inspection and main-tenance.

The Grading and Excavation Code, Chapter 70 of the Uniform Building Code published by the International Conference of Building Officials (1985) and used as a model or adopted by many local govern-ments, requires a permit for most types of grading; section 7006(c) states that:

When required by the building official, each application for a grading permit shall be accompanied by two sets of plans and specifications, and supporting data consisting of a soils engineering report and engineering geology report. The plans and specifications shall be prepared and signed by a civil engineer when required by the building official.

A former grading supervisor and county geologist for Orange County, California, (Scullin, 1983) has written an excellent guidebook on the administration, inspec-tion, and enforcement connected with excavation and grading codes. It in-cludes a discussion of the grading-plan review process, sample enforcement-control forms, and a checklist for the proper maintenance of hillside lots by homeowners.

7.2 Hillside-development regulations

Some communities have adopted regulations to limit the amount of development (building sites per acre) that may take place in hillside areas; sometimes these regulations are incorporated into the grading or zoning ordinances. The regu-lations usually include a formula for determining the density of development that will be permitted for a given steepness of slope.

Hillside-development regulations can be used both for safety and aesthetic purposes. Controlled development con-serves the views, and lower densities re-duce exposure to hazards. Proper grading procedures, such as correct excavation and fill practices, are imperative to assure slope stability. Without them, the potential for landslides remains the same for individual sites; with them, the number of adversely affected sites is reduced. An engineering geologist is needed to assist the local government planner in making an accurate judgment about where density restrictions and other regulations can be useful. The steepness of the slope is not solely the determinant of slope stability; lithology and geologic structure also must be considered.

In Performance Controls for Sensitive Lands, Thurow and others (1975) discuss, and give examples of, three principal types of hillside-development regula-tions:

1. Provisions that decrease allowable development densities as the slope in-creases.
2. Provisions that assign use and dens-ity on the basis of soil characteristics in slope areas.
3. A "principles approach" that is re-latively free of precise standards, but emphasizes case-by-case evaluation on the basis of a number of specific policies.

Other examples include hillside-development standards (Simi Valley City Council, 1978) which are applicable to lands in residential zoning districts having slopes of 10 percent or greater, and environmental hillside districts (Cincinnati City Council, 1979) which contain at least four of the following characteristics:

1. Slopes of 20 percent or greater.
2. Existence of Kope formation.
3. Prominent hillsides which are readi-ly viewable from a public thoroughfare

located in a valley below a hillside identified in the Cincinnati Hillside System.

4. Hillsides which provide views of a major stream or valley.

5. Hillsides functioning as community separators or community boundaries as identified in a community plan accepted and approved by the city planning commission.

6. Hillsides which support a substantial natural wooded cover.

7.3 Land-use zoning regulations

Land-use regulations are an accepted method for controlling development. They provide direct benefits by restricting future development in hazardous areas and by limiting the expansion of existing development already in those areas. The types of zoning districts that best suit landslide areas include agricultural, open-space, conservancy, park, and recreational land-use zones. Such districts can be developed to permit grazing, woodlands, wildlife refuges, and public or private recreation. Provisions can be incorporated into district regulations to prohibit specific uses that could trigger landslides or that would be vulnerable to landslide damage. Prohibited uses might include clear cutting of trees, road construction, dwellings, off-road vehicles, irrigation, liquid-waste disposal, and the permanent sheltering or confining of animals.

In Colorado, for example, geologic hazards, including landslides, mudflows, and unstable or potentially unstable slopes, have been declared by the state legislature to be matters of state interest. To assist communities in designing land-use regulations, the Colorado Geological Survey prepared model Geologic Hazard Area Control Regulations (in Lessing and others, 1976: Appendix C) for adoption by counties and municipalities. The model regulations permit only the following "open" uses in designated geologically hazardous areas:

1. Agricultural uses such as general farming, grazing, truck farming, forestry, sod farming, and wild-crop harvesting.

2. Industrial-commercial uses such as loading areas, parking areas not requiring extensive grading or impervious paving, and storage yards for equipment or machinery easily moved or not subject to geologic-hazard damage.

3. Public and private recreational uses not requiring permanent structures designed for human habitation such as parks, natural swimming areas, golf courses, driving ranges, picnic grounds, wildlife and nature preserves, game farms, shooting preserves, target ranges, trap and skeet ranges, and hunting, fishing, skiing, and hiking areas if such uses do not cause concentrations of people in areas during periods of high-hazard probability.

These uses are permitted only to the extent that they do not conflict with the local municipal or county zoning ordinance.

District regulations also can be designed to reduce the density of development. For example, the San Mateo County (California) Board of Supervisors (1973) created a resource-management zoning district to carry out the objectives and policies of their open-space and resource-conservation plans. The district regulations limit the number of dwellings in zones with unstable slopes to one unit per 16 hectares (40 acres) and require geologic site investigations to ensure that the reduced development is located in safe areas. The lower net number of dwellings permitted may then be clustered at a higher density in the nonhazardous areas.

7.4 Sanitary ordinances

On-site sewage-disposal systems (septic tanks, absorption fields, and seepage beds and pits) can trigger landslides; they are also likely to malfunction when slope movements occur, creating serious health hazards. Sanitary ordinances can help eliminate some of these problems and regulations can be designed to:

1. Require a permit from the Building Official before an on-site sewage-disposal system is installed, and require that the permit application show the boundaries of landslide areas.

2. Prohibit on-site soil-absorption sewage-disposal systems and private water-supply systems in landslide areas.

3. Require on-site soil-absorption sewage-disposal systems in landslide areas to be replaced with alternate systems, such as public sanitary sewers or holding tanks.

The model ordinance by the Southeastern Wisconsin Regional Planning Commission (1969: Appendix H) can be adapted to landslide hazards and incorporated into sanitary, health, and plumbing codes.

7.5 Special hazard-reduction regulations

Some communities put overlays of hazard zones on their basic land-use zoning-district maps. The hazards shown on the overlays are then used to supplement the basic use and site regulations found in the zoning ordinance. Such regulations can be designed to:

1. Preserve vegetation, maintain drainage, control off-road vehicles, avoid the most hazardous areas, require clustering of dwellings, and reduce development densities.

2. Prohibit certain operations that increase loads, reduce support, or otherwise cause instability of slopes—operations such as filling, irrigating, disposing of solid and liquid wastes, and removing the toe of a landslide.

3. Prohibit certain uses -- storage of radioactive, toxic, flammable, and explosive materials -- that could cause serious health and safety hazards if released by landslide movement.

The Portola Valley (California) Town Council (1974) has developed criteria for permissible land uses in the least stable and most stable areas shown on its map of potential land movement. For example, roads, houses, utilities, and water tanks are prohibited in certain zones. The Town Council also requires use of the same criteria in administering the town's zoning, subdivision, site development, and building ordinances. The Portola Valley Town Council (1979) subsequently adopted additional regulations that reduce the maximum number of dwellings permitted in areas of potential land movement.

The King County (Washington) Department of Planning and Community Development (1980) has mapped landslide areas. The King County Council (1979) has provided in its zoning ordinance relating to these hazard areas that:

No building permit, grading permit, shoreline substantial development permit, conditional use permit, unclassified use permit, variance, rezone, planned unit development, subdivision or short subdivision shall be granted for development on any Class III landslide hazard area unless King County determines, upon review of a soils study completed by a qualified soils engineer or engineering geologist, that the proposed development together with any required conditions to mitigate

adverse environmental impacts can be safely accommodated on the site and is consistent with the purposes of this ordinance. The soils study shall include specific recommendations for mitigating measures which should be required as a condition of any approval for such development.

Other examples are provided by Kusler (1982: 331, 347, 349)—regulations for high-risk erosion along the Lake Michigan shoreline (Lincoln Township, Michigan), for mudflow hazards (City of Burbank, California), and for flood-related erosion and debris flows (City of Palm Desert, California).

7.6 Subdivision ordinances

Regulating the design and improvement of subdivisions is another way to prevent or control the development of landslide areas. A dilemma occurs when government officials approve a subdivision, accept public rights-of-way, extend utilities, and then attempt to impose zoning and other regulations that would prohibit further development. This dilemma can be avoided by adopting a subdivision ordinance designed to:

1. Require the delineation and designation of landslide areas on subdivision plats and certified survey maps.

2. Require dedication or reservation of landslide areas for public or private parks or other community purposes; and require dedication of, or easements along, those waterways necessary for adequate drainage.

3. Require that public and private roads, bridges, utilities, and other facilities be located or designed and constructed to avoid landslide areas or to withstand anticipated movement.

4. Select road and utility alignments and grades to minimize cuts and fills.

5. Prohibit the creation and improvement of building sites in landslide areas.

The Southeastern Wisconsin Regional Planning Commission (1969: Appendix F) has prepared a model subdivision ordinance relating to various flooding, soil, and geologic problems. It can be adapted to landslide hazards. The model ordinance has been adopted by many local units of government; these ordinances have been in effect for over 10 years without successful legal assault.

7.7 Rebuilding moratoriums

Temporarily prohibiting rebuilding or construction of new development in areas after a landslide or severe ground failure may be used as an interim measure. This technique provides the local unit of government with time to prepare plans for removing unsafe structures, constructing retaining buttresses, adopting grading or zoning ordinances, or acquiring unstable land and relocating its former occupants. Examples of rebuilding moratoriums include one by the Rancho Palos Verdes (California) City Council (1978) for an area shown on their "landslide moratorium map"; one for the City of San Bernardino, California, for an area damaged three times by mudflows; and another for Cowlitz County, Washington, for the mudflow area of the Toutle River (Kusler, 1982: 268, 269, 308, 309).

8 PROTECTING EXISTING DEVELOPMENT

Because many landslide areas offer highly desirable views, they are subject to development pressures; such development has taken place in landslide areas and will probably continue. Property damage from landslides often leads to a demand for costly public works to provide protection for existing development. This demand is usually limited to smaller landslide areas because of the costs involved and the necessity for careful and accurate engineering design, construction, and maintenance.

There are several ways in which the property owners in landslide areas can be protected. Those discussed in this section are: controlling slides and slumps; controlling mudflows and debris flows; controlling rockfalls; creating special-assessment districts; and operating monitoring, warning, and evacuating systems. Many of these are more thoroughly discussed in Schuster and Krizek (1978).

8.1 Slide and slump controls

Once a slope has begun to move, several types of controls are available to avoid or correct the movement, depending on the type of failure, kinds of materials involved, and location of the slope. Gedney and Weber (1978) separated these into three categories: (1) avoiding the problem, (2) reducing the driving force, and (3) increasing the resisting forces; they also provided examples with cross sections from several state highway or transportation departments. These controls also are applicable to mudflows and debris flows. Examples of these and other types of controls used in Canada, Austria, West Germany, Turkey, Italy, Switzerland, Japan, and Hungary are discussed and illustrated by Veder (1981: 73-180).

Bukovansky (1977) classified slope-stability improvements into four groups: change of slope shape, change of drainage, addition of retaining forces, and other special treatment methods. In addition to providing diagrams of several of these methods, he described and commented on each group:

> The change of the slope shape may include a total excavation of the slide, a partial excavation of the upper landslide portion, loading the landslide toe, or a combination of both. Excavation is typically an economic and safe way to treat unstable areas. It becomes difficult and expensive with the increasing size of the area.

> Drainage, very effective and frequently used, is limited to cases where ground water exists within the area. Surface drainage is efficient in only a few cases of small slides; usually, subsurface drainage is required. It can be achieved with the help of trenches filled with pervious gravel, horizontal drainage holes, drainage wells, and, in extreme cases, subsurface drainage tunnels.

> Retaining forces include buttresses, piles, retaining walls of various types, bolts, and anchors. Their applications are limited to slides or areas of a limited extent. Buttresses are by far the most popular method because of low costs and simple installation.

> Special treatment methods include freezing, thermal methods, grouting, electroosmosis, and blasting. All of these methods have been used in special cases where more common methods were not feasible. They are expensive and of limited applicability for common stability problems.

Types of controls also can be divided into drainage, removal, restraint, or relocation. Royster (1979) described each and provided examples of their application by the Tennessee Department of Transportation; he concluded:

> Whatever the measure chosen, and whatever the level of effectiveness required, the engineering geologist or geotechnical engineer must also consider what might have been done to prevent the problem in the first place. Hindsight, while of little benefit in solving the immediate problem, can be of great benefit in avoiding similar occurrences in the future.

These controls can be used in various combinations. For example, the California Division of Mines and Geology Staff (1982: 64-77) described, illustrated, and recommended various methods of repair and reconstruction of slopes that failed in the Baldwin Hills (Los Angeles County, California). The methods included replacing failed and weakened materials with compacted engineered fill, covering erodible materials on steep slopes with gunite or clear plastic sheeting, flattening unstable slopes when sufficient space exists downhill, constructing revetment systems and retaining walls, installing surface and subsurface drainage devices, planting proper vegetation, and using various other methods for modifying slopes that have not failed.

8.2 Mudflow and debris-flow controls

Mudflows and debris flows can be retained, diverted, or channeled; some structures in their paths can be protected by bracing and reinforcing their upslope walls to deflect or resist pressure or impact. Hollingsworth and Kovacs (1981: 23-27) described retaining walls, deflecting walls, debris-catchment basins, and debris fences specifically for mudflows and debris flows.

Large reservoirs also entrap mud and debris by reducing flow velocities and thereby causing sedimentation. Both debris basins and reservoirs require removal of accumulated sediment; they are more effective when soil and water conservation is practiced throughout the watershed. For example, the Los Angeles County Flood Control District now operates and maintains over 120 debris basins and over 220 stabilization structures in 47 watersheds. The 1969 storms in southern California deposited over 13 million m^3 (17 million yd^3) of debris in basins and reservoirs, necessitating a large and costly emergency cleanout (Alexander, 1984).

It is sometimes difficult to separate floods and the techniques to reduce their hazards from mudflows and debris flows and the techniques to reduce their hazards. For example, much of the more than 2 billion m^3 (3 billion yd^3) of material displaced by the 1980 Mount St. Helens volcanic eruption remains in the upper reaches of the watershed. Some of this easily erodible material formed unstable, natural dams holding back Spirit Lake and other lakes. According to the Cowlitz County Board of Commissioners (1983), one-third of this debris is expected to erode away by the year 2001; there is also the possibility that a rapid failure of the avalanche-formed dam at Spirit Lake could cause catastrophic flooding and debris flows within the next few years.

The County's watershed-management plan for the Toutle and Cowlitz rivers provides for combinations of flood and debris-flow hazard-reduction techniques such as: controlling sediment, dredging, raising existing levees, restricting future development, maintaining debris-retention structures, stabilizing river banks, relocating dredged soils, continuing a rebuilding moratorium, practicing soil and water conservation, lowering Spirit Lake, and installing a flood-warning system.

8.3 Rockfall controls

Loose or falling rocks can be retained, intercepted, or directed away from development. Control methods identified by Piteau and Peckover (1978: 219-223) include relocation, intercepting ditches and berms, wire-mesh anchors, wire-mesh catch nets and fences, catch walls, rock sheds, and tunnels. They discussed and illustrated each and noted that:

> Relocation is most effective where rocks are free falling from steep rock faces in proximity to the roadway or where stabilization or some other protection measure is not feasible. Relocation is feasible when space is available and the design criteria are not affected. When combined with proper ditch shaping and possibly with other ditch-level protection, relocation can be the most economical solution and, in some cases, one of the few solutions to the problem. Unless relocation practically eliminates the

possibility of accidents, other measures also should be included along with relocation.

Rock slopes need to be carefully mapped by the engineering geologist using the stereographic methods presented by Hoek and Bray (1981). Practical remedial measures on rock slopes are described by Schach and others (1979), and Crimmins and others (1972).

8.4 Assessment districts

If it becomes necessary to stabilize landslide areas, construct control or protection structures, or to repair damaged public facilities, the costs can be assessed wholly or in part against the owners of the lands that will benefit. For instance, the City of San Jose, California, anticipated subsidence damage to public roads and utilities in an area within a new development. The City Council passed an ordinance and resolution establishing a 7.2-hectare (17.9-acre) maintenance district before the area was developed; the City's Director of Public Works prepared a budget, an assessment formula, and a proposed assessment. These were adopted by the San Jose City Council (1979) after public hearings and validation by the Superior Court.

The California Legislature (1979) passed the Beverly Act which provided for the formation of Geologic Hazard Abatement Districts. The California Division of Mines and Geology Staff (1982: 77) has suggested that one large Geologic Hazard Abatement District be created for the Baldwin Hills area to encompass all of the damaged properties within Los Angeles County and the cities of Los Angeles and Culver, or several smaller districts ranging in size from less than 15 to more than 35 properties. They concluded:

The participants in such districts could not only make sure that slopes are stabilized the most efficient and economical way, but also that they are maintained properly after stabilization. The districts would also help to resolve the problems of change in ownership or inability on the part of individual homeowners to maintain their portions of a slope. The lack of maintenance on a single property, such as the failure to clean a drain, can cause slope instability problems for several nearby properties.

When created, these districts are authorized to acquire real estate; acquire, construct, operate, manage, or maintain improvements; and pay for such improvements. Improvements are defined as "any activity necessary or incidental to the prevention, mitigation, abatement, or control of a geologic hazard." Such hazard is defined as "an actual or threatened landslide, land subsidence, soil erosion, or any other natural or unnatural movement of land or earth." A district consists of an area specifically benefited by, and therefore subject to, a special assessment to pay for the improvements. The first Geologic Hazard Abatement District in California, Abalone Cove Landslide Abatement District, was formed by the Rancho Palos Verdes City Council (1981).

A detailed and equitable cost-assessment formula was devised by Lung (1981) for a proposed Geologic Hazard Abatement District in the Mount Washington district of Los Angeles; it serves as a model for other agencies presently forming Geologic Hazard Abatement Districts in California.

8.5 Warning and evacuating systems

Potentially unstable land can be monitored so that residents can be warned and, if necessary, evacuated. The most common forms of monitoring at present are field observations, inclinometers, extensometers, and electrical fences or trip wires. An excellent brochure by Kaliser (undated) on observing and reporting landslides (including debris flows and mudflows) has been prepared for local government officials and emergency preparedness officers. According to Piteau and Peckover (1978: 224-225), methods for monitoring rockfalls currently being tested include: vibration meters, television observation, guided radar, and laser beams.

Under the Disaster Relief Act (United States Congress, 1974a) the United States Geological Survey (1977) developed a three-category procedural warning system for geologic hazards including landslides. As a result, the appropriate government officials in Billings, Montana, Kodiak, Alaska, and Wrightwood, California, have been warned of potential rockfalls, landslides, and cyclic landsliding, respectively. The United States Geological Survey (1983) revised its criteria for the three hazard-warning categories: (1) a degree of risk greater than normal for the area; (2) a hazardous condition that has recently developed or has only recently been recognized; or (3) a threat that warrants consideration of a near-term public response.

Although the time, place, and magnitude of landslides can be predicted only in relatively small areas in which detailed geological and engineering studies have been conducted, susceptible areas can be identified on a larger scale. For example, in 1982 United States Geological Survey scientists warned of mudflow and debris-flow hazards throughout northern California by means of a news release and provided a list of signs that might indicate deeper-seated landslides (King, 1982). Monitoring of high-intensity rainfall areas and relating them to steep slopes can also provide a means for recognizing debris-flow hazards over large areas. Campbell (1975: 30-31) commented on the value of warning and evacuation in the case of the 1969 life-taking debris flows in the Santa Monica Mountains (California):

> Residents who are notified that storm conditions have reached a point where debris flows may be generated by soil slips if high-intensity rainfall continues should be alerted, should be prepared to recognize approaching danger, and should move quickly out of harm's way. Small children, invalids, and elderly people might be evacuated at such a time, but general evacuation of whole neighborhoods should not be necessary. The records show that even without advance planning, many people were able to react in ways that saved them from injury. Obviously, advance planning would provide for quicker and better protective response.

Immediate relay of information is vital in areas where landslides (such as rockfalls and debris flows) happen rapidly. For example, the 20 flood-warning sirens connected to the monitoring system of the Spirit Lake blockage which have been installed along approximately 80 km (50 mi) of the Toutle and Cowlitz Rivers are downstream of about 2 billion m^3 (3 billion yd^3) of volcanic debris--a mudflow and debris-flow hazard that could be catastrophic.

9 EVALUATION OF REDUCTION TECHNIQUES

The effectiveness of each hazard-reduction technique varies with the time, place, and persons involved in the planning and implementing of the program for reducing the hazard. The public, as well as planners, engineers, and decision-makers, live and work in a complex geologic environment. This geologic environment, however, is just one aspect of the surroundings that affect their lives and work; other aspects are social, economic, political, and aesthetic. Some of these are more apparent or more important than others to the individual planner, engineer, decisionmaker, or their constituents.

Very few systematic evaluations have been made of hazard-reduction techniques, even fewer for landslides specifically. No rigorous benefit-cost studies have been conducted, although a few intensive evaluations have been made for flooding and some other hazards which may be applicable to landslides.

Analyses by Slosson (1969) and Slosson and Krohn (1979) have shown substantial reductions in landslide failures and site damage because of the adoption and enforcement of more modern grading codes in the City of Los Angeles. The success of the City of Los Angeles' landslide-reduction program is discussed by Fleming and others (1979: 434-437).

A.W. Martin and Associates (1975: 50) performed an analysis of 13 techniques for protecting development in landslide areas. Their study, made for Allegheny County, Pennsylvania, found that "the cost of preventing and controlling landslides is often prohibitive.... Avoidance is highly recommended for individual dwelling lots."

After an investigation of the damages and losses from the January 1982 mudflows in northern California, the California State Board of Registration for Geologists and Geophysicists (1982) generally concluded that:

> ...a large degree of responsibility for slope failures rests with units of the state and local governments .... A significant step forward would be requiring local governments to use the services of a licensed professional, either as a staff person or consultant, in the review and approval of geologic reports. The responsibility for reducing future loss of property and life from mudflows requires the full cooperation of the city government, professional geologic community, and the State Board of Registration.

In the case of flooding, Sheaffer and Roland, Inc. (1978: 10) concluded that "experience to date indicates that the current approach to correcting non-conforming uses through zoning mechanisms is not effective. Nonconforming uses, particularly residences, are allowed to continue even when they are substantially damaged unless they are purchased." Surveys of state and local programs conducted as part of a study by Kusler (1982: appendices I and III) support their conclusion.

Evaluations of those techniques used for disclosing fault-rupture zones and the processing of loans may be relevant to landslide hazards. For example, prerequisites for effective disclosure of hazards by real-estate sellers include a seller's or real-estate agent's knowledge and integrity, a buyer's awareness of the potential danger or financial loss before making the commitment to purchase, and a buyer's concern about hazards as related to his or her other priorities. According to a study by Palm (1981) of disclosure of official fault-rupture zones by real-estate agents in Berkeley and part of Contra Costa County, California, these prerequisites are often lacking.

Marston (1984: 16-17), after her analysis of interviews with officers of the 12 largest home-mortgage lenders in California, concluded that "of the five lenders using earthquake information, only two had formal policies, and only the first of these actually enforced a limited policy of refusing to make loans on commercial property in the San Francisco Bay Area astride the San Andreas Fault"; and that for 11 of the 12 lenders interviewed, the lenders' awareness of the hazard was not translated into enforced policies.

Remedial public works to protect existing development is costly and can be self-defeating. It may encourage development by leading the public to believe that landslide problems have been eliminated, not simply reduced. Intelligent landslide-area management and regulation are still needed for effective hazard reduction. With regard to cost, Brooks (1982, pp. 18, 19) notes that:

> The cost of public works has increased sharply in the last decade. Not only has the cost of construction increased by a factor of about three over the last decade, but the cost of borrowing (expressed as the interest rate paid by government) has also

tripled. Therefore, the annual cost could have increased between three and nine times, depending on the length of the repayment period. Thus, there are strong economic incentives to consider and use other approaches.

CONCLUSIONS

Numerous techniques are available to planners, engineers, and decisionmakers for reducing landslide hazards. However, even if adequate landslide hazard information is available, presented in a language understandable by nongeologists, effectively communicated, and properly used, the lasting effectiveness of any technique for landslide-hazard reduction depends upon many other factors, usually outside the control of the public planner, engineer, or decisionmaker. For example:

1. Continued awareness and interest by the public.

2. Careful revision of enabling legislation (if needed) by state legislatures.

3. Accurate site investigations by registered geologists and geotechnical engineers.

4. Conscientious administration of codes and regulations by grading inspectors and building officials.

5. Sustained support of inspection and enforcement officials by political leaders.

6. Consistent enforcement by government attorneys.

7. Judicious adjustment of regulations by administrative appeal bodies.

8. Skillful advocacy (if challenged) and proper interpretation by the courts.

9. Concern for individual, family, and community safety by real-estate buyers, developers, insurers, and financiers.

Rossi and others (1982: 21, 22) summarized their Natural Hazards and Public Choice--The State and Local Politics of Hazard Mitigation as follows:

> It is now clear that for most local communities natural disasters take a back seat to more pressing problems.... The other problems faced by communities are more urgent, and more predictably so, than the possible consequences of low-probability events. Furthermore, the low importance accorded to natural disasters means that the politics surrounding them is difficult to forecast and that there is little political dis-

400

advantage or gain accruing to officials because of their opinions on natural-hazard mitigation.

ACKNOWLEDGMENTS

Robert Schuster, engineering geologist; Earl Brabb, research geologist; Robert Sydnor, engineering geologist; Jon Kusler, attorney; James Slosson, former California State Geologist and former President, California State Board of Registration for Geologists and Geophysicists; and Theodore Smith, engineering geologist, provided many valuable suggestions and critical comments that have improved this paper. Special thanks are owed Catherine Campbell for editing, Cynthia Ramseyer for word processing, and the United States Geological Survey Western Region headquarters' library staff for help in obtaining reference materials.

REFERENCES

Alexander, T. M., 1984. personal communication, Supervising Civil Engineer, Los Angeles County Flood Control District: Los Angeles, CA, 90051.

American Insurance Association, 1956. Studies of Floods and Flood Damages, 1952-1955: New York, NY, 296 p.

American Law Institute, 1975. A Model Land Development Code--Complete Text and Commentary: Washington, DC, 524 p.

Association of Bay Area Governments, 1984. The Liability of Businesses and Industries for Earthquake Hazards and Losses--Executive Summary: Oakland, CA, 8 p.

A. W. Martin and Associates, Inc., 1975. Review of Construction Methods for Use in Landslide-Prone Areas, prepared for the County of Allegheny Department of Planning and Development, Pittsburgh, Pennsylvania: King of Prussia, PA, 57 p.

Brabb, Earl E., 1984. Innovative approaches to landslide hazard and risk mapping. In Proceedings, 4th International Symposium on Landslides: Toronto, Canada, September, 1984, Vol. 1, p. 307-323.

Brooks, Norman H., 1982. Storms, Floods, and Debris Flows in Southern California and Arizona, 1978 and 1980--Overview and Summary of a Symposium, September 17-18, 1980: National Academy Press, Washington, DC, 47 p.

Brown, Robert D., Jr. and Kockelman, William J., 1983. Geologic Principles for Prudent Land Use--A Decisionmaker's Guide for the San Francisco Bay Region, U.S. Geological Survey Professional Paper 946: U.S. Government Printing Office, Washington, DC, 97 p.

Brown, William M., III , 1984. Debris Flows, Landslides, and Floods in the San Francisco Bay Region, January 1982 --Overview and Summary of a Conference held at Stanford University, August 23-26, 1982: National Academy Press, Washington, DC, 83 p.

Bukovansky, Michal, 1977. Mitigation of unstable slope hazards. In Shelton, David C. (ed.), Governor's Third Conference on Environmental Geology: Colorado Geological Survey, Special Publication 8, Denver, CO, p. 7-20.

Burby, Raymond J. and French, Steven P., 1981. Coping with Floods--the Land Use Management Paradox: Journal of the American Planning Association, Vol. 47, No. 3, p. 289-300.

California Court of Appeals First District, 1984. Easton v. Strassburger: Division 2, A010566, CIV. 53113, 199 California Reporter 383.

California Division of Mines and Geology, 1975. Recommended Guidelines for Preparing Engineering Geologic Reports: Note 44, Sacramento, CA, 2 p.

California Division of Mines and Geology Staff, 1982. Slope Stability and Geology of the Baldwin Hills, Los Angeles County, Calif., Special Rept. 152: Sacramento, CA, 93 p., 2 plates, scale 1:4,800.

California Legislature, 1965. Williamson Act, as amended: California Government Code, section 51200 and following, West's Annotated Codes.

California Legislature, 1972, Alquist-Priolo Special Studies Zone Act, as amended: California Public Resources Code, section 2621 and following, West's Annotated Codes.

California Legislature, 1979. Geologic hazard abatement districts: California Public Resources Code, section 26500 and following, West's Annotated Codes.

California Legislature, 1983, Landslide Hazard Identification Act: California Public Resources Code, section 2670 and following, West's Annotated Codes.

California State Assembly Select Committee on landslide prevention, 1980. A Report on Committee Hearing, Sacramento, California, April 18, 1980: Assembly Publications Office, Sacramento, CA, 22 p.

California State Board of Registration for Geologists and Geophysicists, 1982. April 19, 1982 Minutes of a Public Meeting: Sacramento, CA, Attachment A.

California State Constitution, 1879. as amended: Art. 1, Sec. 19, West's Annotated Codes.

California Supreme Court, 1965. Albers v. County of Los Angeles: 62 California Reports 2d 250, 42 California Reporter 89, 398 Pacific Reporter 2d 129.

Campbell, Russell H., 1975. Soil Slips, Debris Flows, and Rainstorms in the Santa Monica Mountains and Vicinity, Southern California, U.S. Geological Survey Professional Paper 851: U.S. Government Printing Office, Washington, DC, 51 p.

Cincinnati City Council, 1979. Ordinance 552-1979 amending environmental quality hillside districts: Cincinnati Zoning Code, sections 3403.2 and following, Cincinnati, OH.

Coates, D.R. (ed.), 1977. Landslides--Reviews in Engineering Geology: Geological Society of America, Vol. III, Boulder, CO, 278 p.

Colorado Geological Survey, 1969. Governor's Conference on Environmental Geology: Special Publication 1, Denver, CO, 78 p.

Cowlitz County Board of Commissioners, 1983. Toutle-Cowlitz Watershed Management Plan--Map, Executive Summary, and Policy Position Statement: Cowlitz County Department of Community Development, Kelso, WA.

Cox, Gail Diane, 1984. Local government losing ground to landslide lawsuits--Hillside homeowners seek compensation for property damage--Courts sympathetic: Los Angeles Daily Journal, Vol. 73, Los Angeles, CA, April 9, 1984.

Crimmins, Robert, Samuels, Reuben, and Monahan, Bernard, 1972. Construction rock work guide: Wiley-Interscience, New York, 241 p.

Ellen, Stephen; Peterson, David M.; and Reid, George O., 1982. Map Showing Areas Susceptible to Different Hazards from Shallow Landsliding, Marin County and Adjacent Parts of Sonoma County, California, U.S. Geological Survey Miscellaneous Field Studies Map MF-1406: Reston, VA, scale 1:62,500.

Erley, Duncan and Kockelman, William J., 1981. Reducing Landslide Hazards--A Guide for Planners: Planning Advisory Service Report No. 359, American Planning Assoc., Chicago, IL, 29 p.

Evans, J. R. and Gray, C. H., Jr., (eds.), 1971. Analysis of Mudslide Risk in Southern Ventura County, California, California State Division of Mines and Geology, Sacramento, CA, 22 maps, scale 1:24,000.

Field Associates, Inc., 1981. State and Local Acquisitions of Floodplains and Wetlands--A Handbook on the Use of Acquisition in Floodplain Management: U.S. Water Resources Council, Washington, DC, 137 p.

Fleming, Robert W.; Varnes, David J.; and Schuster, Robert L., 1979. Landslide hazards and their reduction: American Planning Association Journal, Vol. 45, No. 4, Chicago, IL, p. 428-439.

Gedney, David S. and Weber, William G., 1978. Design and construction of soil slopes. In Schuster, Robert L. and Krizek, Raymond J., (eds.), Landslides--Analysis and Control: National Academy of Sciences, Transportation Research Board Special Report 176, Washington, DC, p. 172-191.

Gray, Cliffton H., Jr., 1984. personal communication, Geologic Hazards Officer, California State Division of Mines and Geology, Los Angeles, CA.

Hart, Earl W., 1985. Fault-rupture hazard zones in California--Alquist-Priolo Special Studies Zones Act of 1972 with Index to Special Studies Zones Maps: California Division of Mines and Geology Special Publication 42, Revised 1985.

Hays, W. W., (ed.), 1981. Facing Geologic and Hydrologic Hazards--Earth-Science Considerations, U.S. Geological Survey Professional Paper 1240-B: U.S. Government Printing Office, Washington, DC, 109 p.

Hoek, Evert, and Bray, John, 1981. Rock slope engineering: London, Institution of Mining and Metallurgy, 3rd edition, 358 p.

Hollingsworth, Robert and Kovacs, G.S., 1981. Soil slumps and debris flows--prediction and protection: Bulletin of the Association of Engineering Geologists, Vol. XVIII, No. 1, p. 17-28.

International Conference of Building Officials, 1985. Uniform Building Code: Whittier, CA, 817 p.

Jahns, Richard H., 1978. Landslides. In Geophysical Predicitons--Studies in Geophysics: National Academy of Sciences, Washington, DC, p. 58-65.

Kaliser, Bruce N., undated. Observing and Reporting Landslides and Debris Flows (Mudflows): Utah Geological and Mineral Survey, Salt Lake City, UT, brochure.

Katzer, Terry, 1981. Flood and Related Debris Flow Hazards, Nevada Bureau of Mines and Geology, Las Vegas Area Map 3A1: Reno, NV, scale 1:24,000.

King County Council, 1979, Ordinance No. 4365 on zoning: Seattle, WA, 11 p.

402

King County Department of Planning and Community Development, 1980. as amended, Sensitive Areas Map Folio: Seattle, WA, 58 maps, original scale 1:24,000.

King, Edna, 1982. Northern California Landslide Hazard Will Persist into Summer, U.S. Geological Survey press release mailed March 1, 1982: Public Affairs Office, Menlo Park, CA, 3 p.

Kusler, John A., 1982. Regulation of Flood Hazard Areas to Reduce Flood Losses, U.S. Water Resources Council: U.S. Government Printing Office, Washington, DC, Vol. 3, 357 p.

Lessing, Peter; Kulander, Byron R.; Wilson, Bruce D.; Dean, Stuart L.; and Woodring, Stanley M., 1976. West Virginia Landslides and Slide-Prone Areas, West Virginia Geological and Economic Survey Environmental Geology Bulletin No. 15: Morgantown, WV, 64 p., 28 maps, scale 1:24,000.

Los Angeles County Flood Control District, 1984. Crest Fire, January 26-28, 1984, Burned Area Flood Hazard Locations: Los Angeles, CA, 3 maps, scale 1:12,000.

Lung, Richard, 1981. Mount Washington Geologic Hazard Abatement District, Geotechnical Investigation Report: Leighton and Associates, Irvine, California, unpublished consulting report prepared for the Department of Building and Safety of the City of Los Angeles, Report No. 1800632-01, dated July 15, 1981, 6 pl., 4 app., 36 p.

Marston, Sallie A., 1984. A Political Economy Approach to Hazards--A Case Study of California Lenders and the Earthquake Threat, University of Colorado, Department of Geography Working Paper 49: Boulder, CO, 27 p.

Miller, H. C., 1977. Barrier Islands, Barrier Beaches, and the National Flood Insurance Program--Some Problems and a Rationale for Special Attention. In Barrier Islands and Beaches--Technical Proceedings of the 1976 Barrier Islands Workshop, Annapolis, MD.

National Research Council, 1982. Selecting a Methodology for Delineating Mudslide Hazard Areas for the National Flood Insurance Program, Committee on Methodologies for Predicting Mudflow Areas: National Academy Press, Washington, DC, 35 p.

Nilsen, Tor H.; Wright, Richard H.; Vlasic, Thomas C.; and Spangle, William E., 1979. Relative Slope Stability and Land-use Planning in the San Francisco Bay Region, California, U.S. Geological Survey Professional Paper 944:

U.S. Government Printing Office, Washington, DC, 96 p., map scale 1:125,000.

Obermeier, Stephen F., 1979. Slope Stability Map of Fairfax County, Virginia, U.S. Geological Survey Miscellaneous Field Studies Map MF-1072: Reston, VA, scale 1:48,000.

Palm, Risa, 1981. Real estate agents and the dissemination of natural hazards information in the urban area: Valuation--American Society of Appraisers Journal, Vol. 26, p. 172-182.

Palm, Risa; Marston, Sallie; Keller, Patricia; Smith, David; and Budetti, Maureen, 1983. Home Mortgage Lenders, Real Property Appraisers, and Earthquake Hazards, Program on Environment and Behavior Monograph 38: University of Colorado Institute of Behavioral Science, Boulder, CO, 152 p.

Pilkey, Orrin H., Jr., and Howard, James D. (conveners), 1981. Saving the American Beach--A Position Paper by Concerned Coastal Geologists. In Results of the Skidaway Institute of Oceanography Conference on America's Eroding Shoreline--The Need for Geologic Input into Shoreline Management, Decisions, and Strategy, Savannah, Georgia, March 25-27, 1981: Duke Univ., Durham, NC.

Piteau, Douglas R. and Peckover, F. Lionel, 1978. Engineering of rock slopes. In Schuster, R. L. and Krizek, R.J., (eds.), 1978. Landslides-Analysis and Control: National Academy of Sciences, Transportation Research Board, Special Report No. 176, Washington, DC, p. 192-227.

Pomeroy, John S., 1979. Map Showing Landslides and Areas Most Susceptible to Sliding in Beaver County, Pennsylvania, U.S. Geological Survey Miscellaneous Investigations Series Map I-1160: Reston, VA, scale 1:50,000.

Portola Valley Town Council, 1974. A resolution approving and adopting "geologic map" and "movement potential of undisturbed ground (map)" and establishing land-use policies for lands shown on said map: Portola Valley, CA, 5 p.

Portola Valley Town Council, 1979. Planned unit developments; parcel area requirement in areas of land movement potential: Zoning Ordinance, section 6201.3, Portola Valley, CA, p. 22-23.

Rancho Palos Verdes City Council, 1978. Ordinance No. 108U establishing a moratorium on certain permits, on the processing or approval of tentative maps or parcel maps in certain areas of the City, and declaring the urgency thereof: Rancho Palos Verdes, CA.

Rancho Palos Verdes City Council, 1981.
Resolution No. 81-4 ordering the
formation of the Abalone Cove Landslide
Abatement District: Rancho Palos
Verdes, CA.

Rib, Harold T. and Liang, Ta, 1978.
Recognition and identification. In
Schuster, R.L. and Krizek, R.J.,
(eds.), 1978, Landslides--Analysis and
Control: National Academy of Sciences,
Transportation Research Board, Special
Report No.176, Washington, DC, p.34-80.

Rossi, Peter H.; Wright, James D.; and
Weber-Burdin, Eleanor, 1982. Natural
Hazards and Public Choice--The State
and Local Politics of Hazard Miti-
gation: Academic Press, New York, New
York, 337 p.

Royster, David L., 1979. Landslide remedial measures: Bulletin of the Association of Engineering Geologists, Vol.
XVI, No. 2, College Station, TX, p.
301-352.

San Jose Board of Realtors, 1978. San
Jose Real Estate Board Standard Form,
Addendum A: San Jose, CA.

San Jose City Council, 1979. Ordinance
No. 19651 establishing maintenance districts and assessment procedures, May
15, 1979, and a resolution establishing
Maintenance District No. 3 and confirming the assessment, July 3, 1979: San
Jose, CA.

San Mateo County Board of Supervisors,
1973. Ordinance No. 2229 adding a
resource-management district and regulations to the county zoning ordinance:
Redwood City, CA, 24 p.

Santa Clara County Board of Supervisors,
1978. Geological ordinance No. NS-
1205.35: Santa Clara County Code, section C-12-600 and following, San Jose,
CA.

Schuster, R.L. and Krizek, R.J., (eds.),
1978. Landslides--Analysis and Control:
National Academy of Sciences, Transportation Research Board, Special
Report No. 176, Washington, DC, 234 p.

Scott, Glenn R., 1972. Map Showing Landslides and Areas Susceptible to Land-
sliding in the Morrison Quadrangle,
Jefferson County, Colorado, U.S. Geological Survey Map I-790-B: Reston, VA,
scale 1:24,000.

Scott, R.L., 1972. The Effect of Nonconforming Land-use Amortization:
American Society of Planning Officials,
Planning Advisory Service Report No.
280, Chicago, IL, 22 p.

Schach, R., Garsnoi, K., and Heltzen, A.
M., 1979. Rock bolting--a practical
handbook: Pergamon Press, NY, 84 p.

Scullin, C.M., 1983. Excavation and Grading Code Administration, Inspection,
and Enforcement: Prentice-Hall, Inc.,
Englewood Cliffs, NJ, 405 p.

Sheaffer and Roland, Inc., 1978. Alternatives for Implementing Substantial
Improvements Definitions, Report to the
U.S. Department of Housing and Urban
Development: Federal Emergency Management Agency, Federal Insurance Administration, FIA-1, April 1980, Washington DC, 39 p., 23 tables.

Simi Valley City Council, 1978. Ordinance
376 establishing hillside performance
standards, Simi Valley Municipal Code,
section 8175 and following: Simi Valley, CA, 24 p.

Slosson, James E., 1969. The role of
engineering geology in urban planning.
In The Governor's Conference on Environmental Geology: Colorado Geological
Survey, Special Pub. No. 1, Denver, CO,
p. 8-15.

Slosson, James E., 1983. Recognition of
Landslides. In Legal and Legislative
Approaches to Western States Geologic
Hazards: Continuing Legal Education,
Utah State Bar, Salt Lake City, UT, p.
117-148.

Slosson, James E. and Havens, Gay W.,
1984. Legal Liability--An Incentive for
Mitigation. In Federal Emergency
Management Agency, Proceedings, Legal
Issues in Emergency Management:
Emmitsburg, MD, 11 p.

Slosson, James E. and Krohn, James P.,
1979. Mudflow and debris flow damage
February 1978 storm--Los Angeles area:
California Geology, California State
Division of Mines and Geology, Sacra-
mento, CA, Vol. 32, No. 1, p. 8-11.

Southeastern Wisconsin Regional Planning
Commission, 1969. Soils Development
Guide: Planning Guide No. 6, Waukesha,
WI, 247 p.

Stokes, A.P. and Cilweck, Blase A., 1974.
Geology and land development in Ventura
County: California Geology, California
State Division of Mines and Geology,
Sacramento, CA, Vol. 27, No. 11, p.
243-251.

Sutter, John H. and Hecht, Mervyn L.,
1974. Landslide and Subsidence Liability, California Continuing Education of
the Bar, California Practice Book No.
65; Supplement by John F. McGuire,
March 1985: The Regents of the University of California, Berkeley, CA, 240
p, 147 p.

Swift, Charles H., III and Kresch, David
L., 1983. Mudflow Hazards Along the
Toutle and Cowlitz Rivers from a Hypothetical Failure of Spirit Lake Blockage: U.S. Geological Survey Water-

Resources Investigations Report 82-4125: Reston, VA, 10 p., 10 maps, scale 1:24,000.

Tank, Ronald W., 1983. Legal Aspects of Geology: Plenum Press, New York, NY, 583 p.

Thurow, Charles; Toner, William; and Erley, Duncan, 1975. Performance Controls for Sensitive Lands—A Practical Guide for Local Administrators: American Planning Association (formerly American Society of Planning Officials), Planning Advisory Service Report No.307 and No.308, Chicago, IL, 155 p.

United States Congress, 1973. Flood Disaster Protection Act amending the Housing and Urban Development Act of 1968: Public Law 92-234, 87 Statutes 975, 42 U.S. Code 4001-4128.

United States Congress, 1974a. Disaster Relief Act of 1974: Public Law 93-288, 88 Statutes 143, 42 U.S. Code 5121-5202.

United States Congress, 1974b. National Flood Insurance Act of 1968, as amended: Public Law 93-383, 88 Statutes 739, 42 U.S. Code 4104a.

United States Council on Environmental Quality, 1976. Seventh Annual Report: U.S. Government Printing Office, Washington, DC, 378 p.

United States Department of Housing and Urban Development, 1973. Housing Production and Mortgage Credit Minimum Property Standards: U.S. Government Printing Office, Washington, DC, 4 volumes.

United States Geological Survey, 1977. Warning and preparedness for geologic-related hazards—proposed procedures, Federal Register Vol. 42, No. 70, p. 19292-19296: U.S. Government Printing Office, Washington, DC.

United States Geological Survey, 1982. Goals and Tasks of the Landslide Part of a Ground-failure Hazards Reduction Program, U.S. Geological Survey Circular 880: U.S. Government Printing Office, Washington, DC, 49 p.

United States Geological Survey, 1983. Revision of terminology for geologic hazard warnings, Federal Register Vol. 48, No. 197, p. 46104: U.S. Government Printing Office, Washington, DC.

United States Office of Management and Budget, 1983, Catalog of Federal Domestic Assistance: U.S. Government Printing Office, Washington, DC, 872 p., 6 appendices.

United States Office of Science and Technology Policy, 1978. Earthquake Hazards Reduction—Issues for an Implementation Plan: Washington, DC, 231 p.

Utah Geological and Mineral Survey, 1983. Governor's Conference on Geologic Hazards: Circular 74, Salt Lake City, UT, 99 p.

Varnes, David J., 1978. Slope movement types and processes. In Schuster, R.L. and Krizek, R.J., (eds.), Landslides—Analysis and Control: National Academy of Sciences, Transportation Research Board, Special Report 176, p. 11-33.

Varnes, David J. and the International Association of Engineering Geology Commission on Landslides and Other Mass Movements on Slopes, 1984. Landslide hazard zonation—A review of the principles and practice. In Natural Hazards: U.N. Economic, Scientific and Cultural Organization, Vol. 3, 63 p.

Veder, Christian, 1981. Landslides and Their Stabilization: Springer-Verlag, New York, 247 p.

Vlasic, Thomas C. and Spangle, William E., 1979. Use of slope-stability information in land-use planning. In Nilsen, Tor H.; Wright, Richard H.; Vlasic, Thomas C.; and Spangle, William E., 1979, Relative Slope Stability and Land-use Planning in the San Francisco Bay Region, California, U.S. Geological Survey Professional Paper 944: U.S. Government Printing Office, Washington, DC, p. 55-87.

Wieczorek, Gerald F.; Ellen, Stephen; Lips, Elliott W.; Cannon, Susan H.; and Short, Dan N., 1983, Potential for Debris Flow and Debris Flood Along the Wasatch Front Between Salt Lake City and Willard, Utah, and Measures for their Mitigation: U.S. Geological Survey Open-File Report 83-635, 42 p., map scale 1:100,000.

Wieczorek, Gerald F.; Wilson, Raymond C.; and Harp, Edwin L., 1985. Map showing slope stability during earthquakes in San Mateo County, California: U.S. Geological Survey Miscellaneous Investigations Series Map I-1257-E, scale 1:62,500.

William Spangle and Associates, Inc.; Earth Sciences Associates; Degenkolb, H.J., and Associates; Duggar, George S.; and Williams, Norman, Jr., 1980. Land Use Planning After Earthquakes: William Spangle and Associates, Inc., Portola Valley, CA, 158 p.

Wisconsin Legislature, 1977. Farmland Preservation Act: Wisconsin Statutes, sections 71.09(11) and 91.01 and following, West's Annotated Codes.

Section D
Wastes and their use

Environmental Geotechnics and Problematic Soils and Rocks, Balasubramaniam et al. (eds)
© *1987 Balkema, Rotterdam. ISBN 90 6191 785 9*

New theoretical models for waste disposal sites with clay liners

R. Kerry Rowe
Geotechnical Research Centre, University of Western Ontario, London, Canada

John R. Booker
School of Civil Engineering, University of Sydney, Australia

ABSTRACT: New theoretical solutions for modelling contaminant migration through clay liners and into surrounding soil are reviewed. Analytic solutions for one dimensional migration are outlined. Consideration is then given to modelling one, two and three dimensional migration using a finite layer technique. The use of these techniques for determining appropriate soil-leachate parameters and for modelling field situations is illustrated by means of several hypothetical examples.

1 INTRODUCTION

In the design of clay barriers for waste disposal sites, consideration should be given to the potential contamination of groundwater due to contaminant migration through the clay barrier. Historically, modelling of contaminant migration has been primarily performed using Finite Element and Finite Difference techniques. While these approaches have met with some success, serious numerical problems can arise when attempting to model predominantly diffusion controlled contaminant migration through a clay liner and into an aquifer where contaminant transport is primarily by advection. The use of these techniques requires extensive experience and training in numerical analysis if reliable results are to be obtained. In addition, a large computer and significant computer time and manpower resources may be required because the solutions must be obtained at a large number of points and times even if the designer is only interested in what happens at a few specific points and times. For these reasons, it has often been considered impractical to use numerical modelling in design situations.

Recently, Rowe and Booker (1984; 1985a,b; 1986a,b) have developed a number of new theoretical solutions which allow convenient and accurate analysis of contaminant transport through clay liners for a range of different practical situations. These analyses can be performed on a microcomputer and can be used without an extensive knowledge of numerical analysis. Thus, the objective of this present paper is to review these new theoretical solutions for 1D, 2D and 3D contaminant migration and to illustrate some potential applications.

2 GENERAL GOVERNING EQUATIONS

The transport of substances through a saturated clay can often be approximated by a Fickian-type law having the form:

$$\underset{\sim}{f} = nc\underset{\sim}{v} - nM_D \, \underset{\sim}{\nabla}c$$
$$M_D = \text{Diag}(D_{xx}, D_{yy}, D_{zz}) \tag{1}$$

where $\underset{\sim}{f}$ is the flux and c the concentration of the dispersing substance, n is the effective porosity of the clay, M_D is the matrix of "coefficients of hydrodynamic" dispersion (incorporating the effects of molecular diffusion and mechanical dispersion) and $\underset{\sim}{v}$ is the seepage velocity. The advection velocity $\underset{\sim}{v}_a = n\underset{\sim}{v}$ will be used to describe advection when discussing the results. Consideration of mass balance shows that

$$\nabla^T f + n \frac{\partial c}{\partial t} + \rho K \frac{\partial c}{\partial t} = 0 \tag{2}$$

The last term takes account of the possibility of some adsorption onto the clay skeleton for the simplest case of equilibrium controlled ion exchange at low concentrations, where ρ is the bulk density of solid and K is the distribution coefficient. The distribution coefficient K may often be estimated from the results of a laboratory column test, as indicated in a later section.

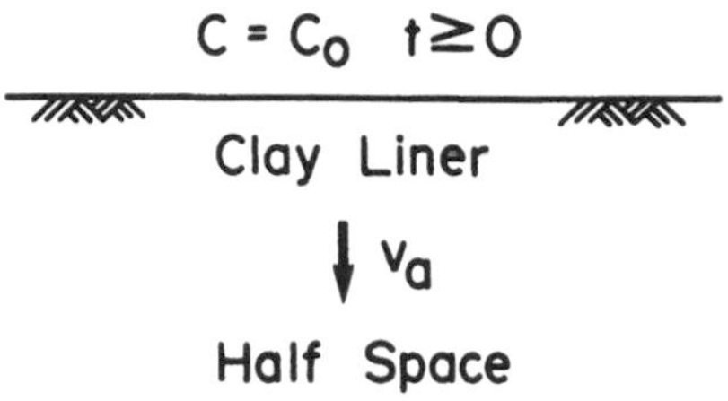

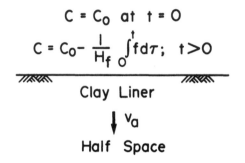

$$C = C_0 \quad t \geq 0$$

Clay Liner

↓ v_a

Half Space

Fig. 1. Problem considered by Ogata (1970).

$$C = C_0 \text{ at } t = 0$$

$$C = C_0 - \frac{1}{H_f} \int_0^t f d\tau; \quad t > 0$$

Clay Liner

↓ v_a

Half Space

Fig. 2. Problem considered by Booker & Rowe (1986b): Concentration of contaminant in source fluid varies with time. H_f=volume of leachate/unit area.

3 ANALYTIC SOLUTIONS - 1D MIGRATION

Ogata (1970) has published an analytic solution for one-dimensional advective-diffusive contaminant transport in a semi-infinite homogeneous layer subject to a constant surface concentration c_0. The situation considered is shown schematically in Fig. 1, and the solution for the concentration c at any depth z and time t is given by

$$c = J[z,t] \tag{3a}$$

where

$$J[z,t] = \frac{c_0}{2} \left\{ \exp\left(\frac{vz}{2D}\right) \exp(-ab) \, \mathrm{Erfc}\left(\frac{a}{2\sqrt{t}} - b\sqrt{t}\right) \right.$$

$$\left. + \exp(ab) \, \mathrm{Erfc}\left(\frac{a}{2\sqrt{t}} + b\sqrt{t}\right) \right\} \tag{3b}$$

$$a = z\sqrt{\frac{(n+\rho K)}{nD}}$$

$$b = v\sqrt{\frac{n}{4D(n+\rho K)}}$$

and $v = v_z$ = average linearized seepage velocity in the z direction,

$D = D_{zz}$ = effective diffusion coefficient in the z direction.

This solution may be expected to provide a reasonable model for contaminant migration at "small times". In this context "small time" implies that there is no significant change in contaminant concentration within the landfill and also there is no significant transport of contaminant into any underlying aquifer. What constitutes a "small time" will depend on the mass of contaminant within the landfill and the thickness of the clay liner.

Landfills are of finite extent and have a limited active life. Typically, decomposition of waste will commence shortly after placement and the concentration of contaminants within the leachate will increase until a maximum value is reached. This process, which may take several years, can be modelled using the principle of superposition and an analytic solution for a linear increase in concentration with time has been developed by Booker and Rowe (1986).

If the landfill is constructed on a clay liner, the time to reach the peak concentration in the leachate is often small compared with the time scale imposed by slow pollutant migration through the clay. In these cases, it can be assumed that the contaminants of interest attain their maximum value, c_0, in the landfill shortly after construction (time zero) and that these concentrations will then decrease with time as contaminant is transported into the soil. The rate at which the concentration varies will depend on the volume of leachate within the landfill and this can be expressed in terms of the quantity H_f (the equivalent height of leachate) which is defined as the volume of leachate divided by the plan area of the landfill.

Booker and Rowe (1986) have derived an analytic solution for one-dimensional advective-diffusive contaminant transport in a semi-infinite homogeneous layer for this case of a variable concentration of contaminant within the landfill (see Fig. 2). The solution for concentration at any depth z and time t is given by:

$$c = c_0 \exp(ab-b^2 t) \, (bf(b,t)$$

$$-df(d,t))/(b-d) \tag{4a}$$

where

$$f(b,t) = \exp(ab+b^2 t) \, \mathrm{erfc}\left(\frac{a}{2\sqrt{t}} + b\sqrt{t}\right) \tag{4b}$$

$$f(d,t) = \exp(ad+d^2 t) \, \mathrm{erfc}\left(\frac{a}{2\sqrt{t}} + d\sqrt{t}\right) \tag{4c}$$

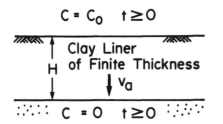

Fig. 3. Problem considered by Rowe & Booker (1985a): clay liner of finite thickness.

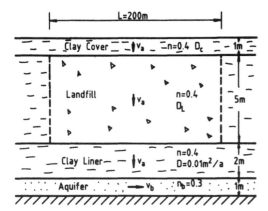

Fig. 4. Example problem analyzed using Finite Layer Technique.

$$d = \frac{nD}{H_f} \sqrt{\left(\frac{n+\rho K}{nD}\right)} - b$$

and all other terms are as previously defined.

If we allow $H_f \to \infty$ so that the surface concentration is constant, we find that $d=-b$ and Eq. 4 reduces to Eq. 3 above.

The solution given in Eq. 4 may be expected to provide a reasonable model of contaminant migration for both small and large times provided the clay barrier is relatively thick in the direction of contaminant transport. This situation may arise when the landfill is located in a deep natural clay or clay till deposit. However, it is also common for the clay barrier to be relatively thin (often only 1-2 m), being underlain by a permeable aquifer in which there will be horizontal flow. In this case, some washing of the clay barrier will occur at the interface with the aquifer and Eqs. (3) and (4) will only be valid for "small times" (i.e. before the contaminant

reaches the aquifer). The simplest idealization of this case assumes that the velocity of groundwater flow in the aquifer is sufficient to provide complete washing of the base of the clay barrier (i.e. the concentration at the base $c_b=0$ as indicated in Fig. 3. Rowe and Booker (1985) have developed an analytic solution for this case of 1D advective-diffusive transport through a clay barrier of thickness H and concentration at any depth z ($0 \le z \le H$) and time t; viz.

$$c = -\{ \sum_{p=0}^{\infty} \exp(-\frac{V}{D}[(p+1)H-z])$$

$$J[2(p+1)H-z,t]$$

$$- \sum_{p=0}^{\infty} \exp(-\frac{V}{D}pH)J(2pH + z,t)\} \quad (5)$$

where $J[Z,t]$ given by Eq. 3b

is evaluated for $Z=2(p+1)H-z$
and $Z=2pH+z$
for the first and second summations respectively; and all other terms are as previously defined.

Equations 3-5 provide useful closed form solutions for a number of limiting cases. Clearly, however, situations frequently arise where the closed form solutions only provide a crude approximation to the actual situation because of more complicated boundary conditions (e.g. partial washing of the base of a clay barrier), non-homogeneity of the soil deposits, and 2D and 3D contaminant transport. To obtain solutions for these more complicated situations, it is necessary to adopt a semi-analytic approach, as outlined in the following sections.

4 A FINITE LAYER APPROACH

Rowe and Booker (1984; 1985a,b; 1986a,b) have developed a finite layer procedure for solving equations (1) and (2) under one, two and three-dimensional conditions for a wide range of practical situations where the soil deposit can be idealized as being horizontally layered.

The finite layer technique involves splitting the soil deposit into separate layers. For example, separate layers may be used to model surface runoff, the clay cover, the waste, the clay liner, and any underlying aquifers, or other soil layers. The approach is semi-analytic. The governing equations and boundary conditions are simplified by introducing a Laplace Transform and, in the case of 2D and 3D analyses, a Fourier Transform. The transformed equations can be solved analytically to give the variation in concentration and flux throughout each layer of the deposit. However, in general, it is not possible to invert the transformed solutions analytically and a computer is required to perform a number of integrations (specifically, to invert the Laplace Transform and, any Fourier Transforms). This approach is easily programmed (eg. Rowe, Booker & Caers, 1983; 1985) and the calculations can be performed on a micro-computer. (Readers are referred to the references for mathematical details and discussion concerning implementation).

This approach allows consideration of
(1) coupled diffusive (dispersive)-advective transport in any of the layers
(2) geochemical reactions between the clay and the contaminant (this may be particularly important for organic contaminants)
(3) depletion of contaminant within the landfill with time due to mass transport into the clay cover and liner
(4) contamination of both surface runoff and groundwater due to diffusive-advective contaminant transport
(5) modelling different degrees of washing of the liner due to different velocity of flow in an underlying aquifer; and
(6) 1D, 2D, or 3D contaminant transport.

Since the solution is semi-analytic, the concentration of contaminant can be determined at specific points and times without determining the entire concentration field at previous times. In many applications, it is really only necessary to determine the magnitude and time of occurrence of the maximum expected base concentration at a few specific points which will usually be specified as monitoring points in the aquifer (e.g. beneath the landfill and at a few points outside the landfill).

Using a "binary chop" algorithm (which can be readily incorporated into the computer program), the peak concentration expected to occur at the point can usually be determined to a precision of .1% or better, from relatively few trials (typically less than 10). Clearly, even fewer trials would be required if one has a reasonable initial estimate of the time at which the peak will occur. Since the data preparation time is also quite small, sensitivity studies can be easily performed to determine the effect of uncertainty regarding input parameters upon the calculated peak concentrations at the specified monitoring points.

The major computational effort associated with the finite layer approach results from the numerical inversion of the Laplace

and Fourier Transforms. For one-dimensional problems, it is only necessary to invert the Laplace Transform. This can be done very efficiently and the computation time is trivial. For two-dimensional problems, it is also necessary to invert a Fourier Transform. The computation time for a two-dimensional problem is typically two orders of magnitude greater than for 1D problems. This computation time is not trivial but in the authors' experience it is still substantially less than the computation time required to obtain equivalent precision using 2D finite element algorithms. The advantages of the finite layer approach are most evident when considering the concentrations at points away from the landfill at modest to large times (although the procedure is computationally superior and easier to use for all cases examined by the authors).

Three-dimensional calculations require significant computer time, and this generally makes parametric studies prohibitively expensive in terms of computer resources. However, the effect of parameters which would influence a full three-dimensional analysis can usually be illustrated by considering a typical two-dimensional cross-section such as that illustrated in Fig. 4. In fact, in many practical situations, it may be appropriate to perform preliminary calculations using a one-dimensional analysis. Two-dimensional and three-dimensional analyses can then be performed for more detailed studies. This will be discussed in more detail by Rowe and Booker (1985b, 1986b).

5 USE OF 1D THEORY FOR PARAMETER DETERMINATION

One of the key problems in any geotechnical analysis is the selection of appropriate soil parameters. For analyses performed using the theory outlined in the previous sections, the key soil-related quantities are the advective velocity, the diffusion coefficient (coefficient of hydrodynamic dispersion) and the sorption potential (in terms of ρK) for each layer.

The advective velocity may be calculated from seepage considerations provided that the hydraulic conductivity of the clay is known. Geotechnical engineers have considerable experience in estimating the hydraulic conductivity of clay permeated by leachate, as discussed by Quigley, Rowe and Fernandez (1986).

Usually the most feasible method of determining the diffusion coefficient and the sorption potential of particular contami-

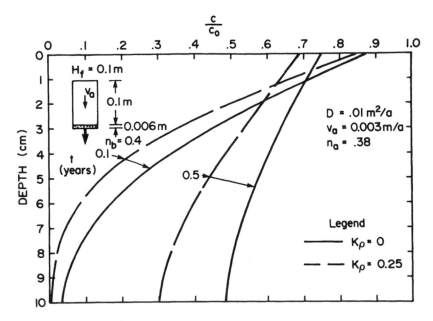

Fig. 5. Determination of soil parameters: Sensitivity of results to the distribution coefficient K

nants in a clayey soil is to perform laboratory column tests. This approach involves taking a 0.05-0.15 m thick sample of the soil of interest and placing it in a column as described by Rowe and Caers (1986). The leachate of interest is then placed above the soil with a height of leachate H_f typically in the range from 0.05 to 1 m. Contaminant is then permitted to migrate through the specimen under the prescribed head and is collected at the bottom of the sample. The concentration of the species of interest in the source fluid and the effluent should be monitored with time. By adjusting the parameters ρK and D, the calculated variation in contaminant with time can be matched to these observed values to obtain an estimate of the two parameters. When adopting this procedure, an initial estimate of D can be obtained by first determining the value of D for a non-reactive reference species (e.g. Chloride) and then multiplying this value by the ratio of the diffusion coefficient of the species of interest in pure water to the diffusion coefficient of the reference species in pure water. By terminating the test at some time t and then determining the variation in concentration with depth in the sample, the values of D and ρK determined as described above can be checked by comparing the calculated and observed concentration profiles throughout the specimen at the end of the test.

The accuracy of the values of D and ρK deduced from this procedure will depend on the variation in concentration in the source fluid which occurs during the test. This in turn will depend on the length of the test and the height of leachate H_f. In general, the greater the value of H_f, the longer the test required to obtain a reasonable estimate of the soil parameters. If results are required for height of leachate H_f of the order of 1 m, it may be desirable to perform several identical tests which are terminated at different times. The value of ρK and D may be deduced from the concentration profile obtained in any one of these tests and may be checked by comparing calculated and observed concentration profiles at other times. This procedure allows more accurate determination of parameters ρK and D for cases where the variation in concentrate with time in the source fluid is relatively small (i.e. less than 15%) over the period of the test. The use of values of H_f less than 1 m is normally recommended.

The sensitivity of concentration profiles to the values of ρK is illustrated in Fig. 5, which shows the profile at t=0.1 and 0.5 years as calculated for $\rho K = 0$ and 0.25. The effect of ρK on the concentration calculated at the surface is evident in Fig. 5 for this case where H_f=0.1 m. There is also a substantial difference in the variation in concentration with depth at a given

413

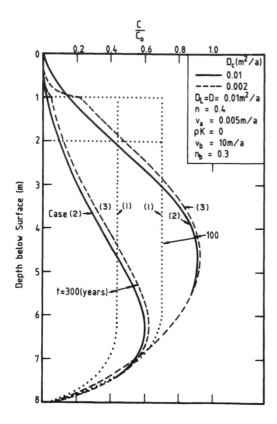

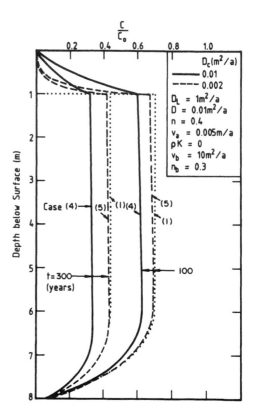

Fig. 6. Concentration profile in the clay
cover, landfill and clay liner. Effect of
assumptions regarding D_C in the clay cover;
D_L=0.01 m^2/a in the landfill. Modified 1D
analysis.

Fig. 7. Concentration profile in the clay
cover, landfill and clay liner. Effect of
assumption regarding D_C in the clay cover;
D_L=1 m^2/a in the landfill. Modified 1D
analysis.

time t greater than 0.1 years, even for
this quite small difference in sorption
potential (i.e. ρK). Thus, this procedure
provides quite a sensitive means of estimating both the diffusion and distribution
coefficients.

The application of this technique to obtain diffusion coefficients and distribution coefficients for a number of inorganic
chemicals in a natural clay till from
Sarnia Ontario has been described by Rowe
and Caers (1986). Parameters determined
from this study have been successfully used
for predicting the actual observed field
behaviour at a landfill in Sarnia Ontario
(Rowe et al., 1985; Quigley and Rowe,1985).

6 USE OF 1D APPROACH FOR FIELD SITUATIONS

To illustrate potential applications for
the proposed theory, consider a situation
such as that shown in Fig. 4. The problem

shown is two-dimensional, with primarily
vertical advective-diffusive transport downwards through the clay barrier and predominantly horizontal advective transport in
the aquifer. This situation where the clay
barrier is thin relative to the dimensions
of the landfill and is underlain by a thin
aquifer in which flow is horizontal can be
approximately analysed by assuming 1D transport in the clay cover, landfill and clay
liner and modelling the horizontal transport in the base aquifer by a boundary condition, as described by Rowe and Booker
(1985). [A comparison of results for the
modified 1D and a full 2D analysis will be
given in the next section.]

The leachate-clay liner properties (i.e.
diffusion and distribution coefficients and
hydraulic conductivity) are critical when
considering the effectiveness of the clay
barrier, however,adopting approaches such
as those discussed in the previous section
for determining D, ρK, and those discussed

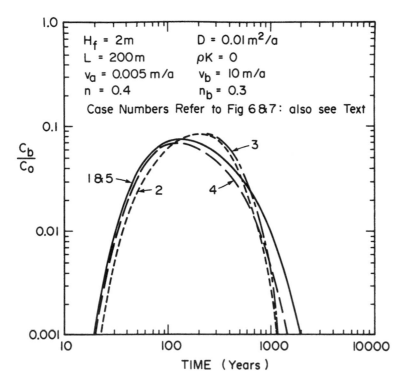

Fig. 8. Effect of assumption concerning D_C and D_L in the clay cover and landfill on the variation in contaminant concentration within an aquifer beneath the clay liner.

by Fernandez and Quigley (1985) and Quigley, Rowe and Fernandez (1986) for determining hydraulic conductivity, it is possible to obtain quite reasonable engineering estimates of these parameters. Much greater uncertainty exists with regard to the effective diffusion/dispersion coefficient in the landfill and clay cover, and the velocity of flow within the aquifer.

The effect of uncertainty regarding the diffusion/dispersion coefficients D_C and D_L in the clay cover and landfill (respectively) has been discussed in some detail by Rowe and Booker (1986b). To illustrate some of these effects, consider the situation shown in Fig. 4 and take v_a=0.005m/a, v_b=10m/a, ρK=0. Figures 6 and 7 show the variation in contaminant concentration with depth at two times (t=100,300 years) as calculated from a modified 1D analysis assuming an initial contaminant concentration c_0 in the landfill at time t=0. The results presented in Fig. 6 were obtained assuming that the diffusion coefficient D_L in the landfill is the same as in the underlying clay (i.e., D_L=D=0.01m²/a). The results in Fig. 7 were obtained assuming a very high diffusion/dispersion coefficient in the landfill (D_L=100D=1 m²/a).

In both figures, the concentration profile is shown for three cases:

Case 1 - the landfill and clay cover are modelled as a boundary condition (Rowe and Booker, 1985a) rather than as distinct layers. This approach implicitly assumes perfect mixing of contaminant within the landfill and a perfect clay cover (i.e., no contaminant can escape upwards from the landfill). In this case, the volume of leachate is expressed in terms of the equivalent height of leachate H_f, where H_f=volume of leachate/area= 200x5x0.4/200=2 m.

Cases 2 and 4 - the clay cover has a diffusion coefficient equal to that in the clay barrier (i.e. D_C=D=0.01 m²a). This might correspond to a saturated clay liner.

Cases 3 and 5 - the clay cover has a diffusion coefficient one fifth of that in the clay barrier (D_C=0.2D=0.002 m²/a). This might correspond to a partially saturated clay liner.

Inspection of Figs. 6 and 7 shows that:
- some contaminant escapes to the surface through the clay barrier for D_C=0.01 m²/a as evidenced by the concentration gradients;

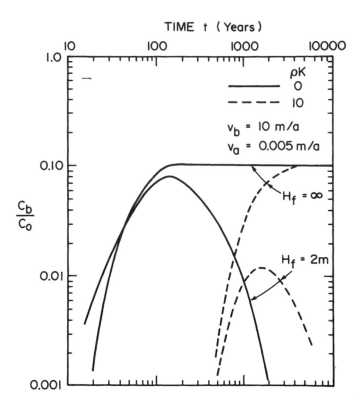

TIME t (Years)

Fig. 9. Effect of sorption and the height of leachate on the variation in contaminant
concentration with time within a base aquifer beneath the clay liner.

- for D_C=0.002 m^2/a, a negligible mass of contaminant escapes to the surface above the clay cover;
- a value of diffusion/dispersion coefficient D_L=1 m^2/a gives an almost uniform concentration within the landfill (there is in fact a small concentration gradient); and
- given D_L=1 m^2/a, the concentration profile for D_C=0.002 m^2/a (case 5) is practically the same as that obtained by treating the landfill as being perfectly mixed with a perfect cover (case 1).

The assumption regarding the values of D_L and D_C in the landfill and cover does have an appreciable effect on the concentration profile in the landfill and cover. However the most important observation that can be drawn from these results is that the concentrate profile in the clay liner, and particularly the concentration in the underlying aquifer, is relatively insensitive to the values of D_C and D_L adopted. This is illustrated in Fig. 8, which shows the variation in the concentration of contaminant within the base aquifer with time for the five cases considered. From this it is clear that for this problem, even a varia-

tion of several orders of magnitude in the diffusion coefficient in the landfill D_L had a relatively small effect on the maximum concentration of contaminant in the aquifer. This finding has important practical applications in that it implies that in similar cases reasonable estimates of the contamination of underlying aquifers can be made even if there is considerable uncertainty regarding the parameters D_L and D_C in the landfill and clay cover. In these cases it seems reasonable to ignore the details of the landfill and cover in preliminary calculations by treating the landfill and cover as an equivalent boundary condition defined in terms of the equivalent height of leachate H_f and the initial concentration c_o, as in case 1 considered above (details regarding the formulation of this boundary condition for 1D and 2D cases are given by Rowe and Booker, 1985a,b).

Clearly, these findings should not be overextrapolated, and in final designs it is essential to perform sensitivity studies similar to those reported here. The finite layer approach is particularly suitable for performing these sensitivity studies.

For a given clay liner, the potential

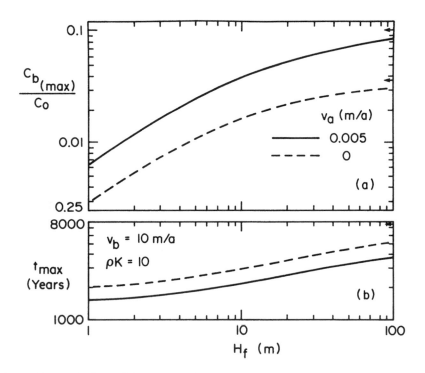

Fig. 10. Effect of the height of leachate H_f on the (a) maximum base concentration; and (b) time to reach the maximum base concentration.

contamination of groundwater is dependent on the mass of contaminant within the landfill. This may be related to the concentration c_o and the volume of leachate (which can be expressed in terms of the equivalent height of leachate H_f). The effect of this mass of contaminant can be illustrated for a given value of c_o by varying the height of leachate H_f. Typically, H_f lies in the range from 1 to 5 m (although values outside this range do occur).

Fig. 9 shows the variation in concentration within the aquifer with time for $H_f=2m$ and $H_f=\infty$ ($H_f=\infty$ implies that the surface concentration remains constant at c_o for all time). For a constant surface concentration, the degree of sorption onto the soil (i.e., ρK) has no effect on the maximum concentration of contaminant in the aquifer, although it does increase the time required to reach the maximum. Consideration of the finite mass of contaminant(e.g. the results for $H_f=2m$) gives rise to lower peak concentrations. This is particularly true when there is interaction between the migrating chemical species and the soil as indicated by Fig. 10, which shows the peak base concentration for a range of H_f values ($1m \le H_f \le 100m$) and two advective velocity

v_a, assuming a moderate level of sorption ($\rho K=10$).

The thickness H of the clay barrier may be expected to have a significant effect on the concentrations of contaminant within any underlying aquifer as shown in Fig. 11. From results such as these, the thickness of a clay liner can be selected to satisfy environmental limits on the maximum contaminant concentration permitted within the aquifer beneath the clay liner.

7 2D ANALYSES FOR FIELD SITUATIONS

The results presented in the preceding section were obtained using the modified 1D analysis. Since the situation being modelled is really two-dimensional, this raises the question as to what error is involved in adopting the approximate 1D approach. The answer to this question depends on the problem being considered. For situations where the liner and aquifer are relatively thin, it may be expected that the modified 1D analysis would provide a reasonable approximation. That this is so is illustrated by Fig. 12, which compares the concentration of contaminant in the aquifer beneath the downstream edge of

417

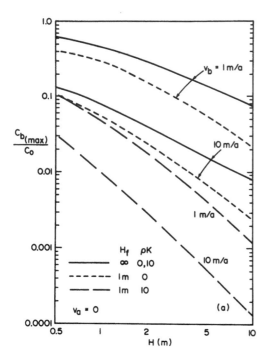

Fig. 11. Effect of the clay liner thickness on the maximum contaminant concentration within a base aquifer.

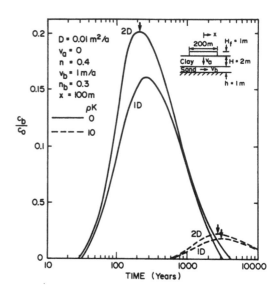

Fig. 12. Variation in contaminant concentration with time within a base aquifer beneath the edge of the landfill using a modified 1D and a full 2D analysis (Modified from Rowe & Booker, 1985).

landfill calculated from a full 2D analysis with the value calculated from the 1D analysis for a range of times and two values of ρK. The simple 1D approach captures the form of the variation in concentration with time and the maximum 25% difference in peak concentration would be acceptable in preliminary calculations for many practical problems.

Clearly, if a designer is concerned with the concentrations of contaminant at points outside the landfill it is necessary to perform a two-dimensional analysis. Figure 13 shows the maximum concentration in a sand-aquifer as calculated from a 2D analysis for a point beneath the downstream edge of the landfill (x=100 m) and at a point 300 m away from the landfill (x=400 m). These results were calculated for the situation shown in the insert to Fig. 13 for a range of base velocities within the aquifer, assuming here that the dispersion coefficient in the aquifer D_b is directly related to the base velocity v_b (i.e., $D_b=15\,v_b$ m^2/a). Beneath the landfill (i.e., at x=100 m) the maximum concentration in the sand $c_{b(max)}$ decreases monotonically with increasing base velocity because of the higher dilution of contaminant that occurs with increasing flow in the aquifer.

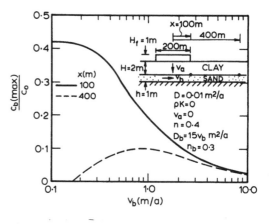

Fig. 13. Maximum concentration of contaminant ever attained at two points (x=100m, x=400m) assuming different flow velocities v_b in the underlying aquifer.

Rowe and Booker (1985b) have shown that two quite different mechanisms compete to control the maximum base concentrations at points in the aquifer outside the landfill; namely, vertical diffusion into adjacent clay and dilution. Generally speaking, the longer it takes for contaminant to move from beneath the landfill to the point of interest, x, the more diffusion that can

418

occur into the clay and hence the lower the concentration at point x. The time interval between when the contaminant is released into the aquifer and when it arrives at a point outside the landfill is primarily dependent on the base velocity v_b. Thus the lower the value of v_b, the greater the amount of attenuation that will occur before the contaminant reaches point x. However, as noted above, increasing v_b will also decrease the concentration of contaminant beneath the landfill due to dilution. Because of these conflicting mechanisms, there is a critical velocity at which the maximum peak concentration at any particular point will be reached. For the example case considered here, this critical velocity is about 0.8 m/a. Values of v_b either larger or smaller than 0.8 m/a will result in a smaller peak concentration at point x in the aquifer. This has important practical consequences since it implies that the design for either the maximum or minimum expected base velocity in the aquifer is not necessarily conservative.

8 CONCLUSIONS

New solution techniques have recently been developed for calculating contaminant migration from waste disposal sites. In many practical situations where the soil deposit can be idealized as being horizontally layered, these approaches are more efficient, numerically stable and easier to use than traditional finite element or finite difference approaches. This paper has attempted to outline these new solutions and to illustrate some practical situations where the approaches could be applied.

9 ACKNOWLEDGEMENT

The work described in this paper was supported by grant No. A1007 from the Natural Sciences and Engineering Research Council of Canada. Additional funding was provided by NSERC Strategic Grant G0921. The authors also wish to thank Mr. C.J. Caers for his conscientious assistance with the computations reported in this paper.

10 REFERENCES

Booker, J.R. and Rowe, R.K. 1986. One dimensional advective-dispersive transport into a deep layer having a variable surface concentration. Faculty of Engineering Science, University of Western Ontario Research Report GEOT-2-86.

Fernandez, F. and Quigley, R.M. 1985. Hydraulic conductivity of natural clays permeated with simple liquid hydrocarbon. Cdn. Geot. J., V. 22, No. 2, pp. 205-214.

Ogata. 1970. Theory of dispersion in a granular medium. US Geol. Survey Prof. Paper 411-1.

Quigley, R.M. and Rowe, R.K. 1985. Leachate migration through clay below a domestic waste landfill, Sarnia, Ontario, Canada: Chemical interpretation and modelling philosophies. Proc. Int. Symp. on Indust. and Hazardous Waste, Alexandria, (accepted for publication ASTM).

Quigley, R.M., Rowe, R.K. and Fernandez, F. 1986. Engineered clayey barriers to control groundwater contamination. Water Pollution Res. J. of Can. (in press).

Rowe, R.K. and Booker, J.R. 1984. The analysis of pollutant migration in a non-homogeneous soil. Geotechnique, V. 34, No. 4, pp. 601-612.

Rowe, R.K. and Booker, J.R. 1985a. 1-D pollutant migration in soils of finite depth. J. Geot. Engrg. ASCE, V. 111, GT4, pp. 479-499.

Rowe, R.K. and Booker, J.R. 1985b. 2D pollutant migration in soils of finite depth. Cdn. Geot. J., V. 22, No. 4, pp. 429-436.

Rowe, R.K. and Booker, J.R. 1985c. The analysis of multiple-contaminant migration through clayey soils. Proc. of 5th Int. Conf. on Num. Methods in Geomechanics, Nagoya, V. 1, pp. 581-588.

Rowe, R.K. and Booker, J.R. 1986a. A finite layer technique for calculating 3D pollutant migration in soil. Geotechnique, V. 36, No. 1 (in press).

Rowe, R.K. and Booker, J.R. 1986b. Prediction of diffusion-controlled contaminant transport in landfill-clay liner systems. Faculty of Engineering Science, University of Western Ontario Research Report GEOT-1-86.

Rowe, R.K., Booker, J.R. and Caers, C.J. 1984. POLLUTE - 1D pollutant migration through a non-homogeneous soil: Users Manual. Report No. SACDA 84-13.

Rowe, R.K., Booker, J.R., Caers, C.J. 1985. Migrate-2-D pollutant migration through a non-homogeneous soil: Users Manual. Available through SACDA, Faculty of Engineering Science, University of Western Ontario.

Rowe, R.K. and Caers, C.J. 1986. Use of a laboratory column test to determine the diffusion and distribution coefficients of contaminant through undisturbed soil (in preparation).

Rowe, R.K., Caers, C.J., Booker, J.R. and Crooks, V.E. 1985. Pollutant migration through clayey soils. Proceedings XI Int. Conf. on Soil Mech. and Foundation Engrg., San Francisco, V. 3, pp. 1293-1298.

Environmental Geotechnics and Problematic Soils and Rocks, Balasubramaniam et al. (eds)
© *1987 Balkema, Rotterdam. ISBN 90 6191 785 9*

Land waste disposal systems

T.Viraraghavan
Faculty of Engineering, University of Regina, Canada

ABSTRACT: Some issues concerning the land disposal of potash and uranium mining wastes are discussed, and some general requirements of hazardous waste impoundments are reviewed. A brief survey of natural and artificial barriers is included.

1 INTRODUCTION

The decision to build a land waste disposal facility such as a sanitary landfill, a wastewater irrigation system, an on-site wastewater disposal system, or water treatment impoundments at any particular site and its design must be based on soil, geologic, groundwater, and topographic conditions at the site as well as the nature of the waste requiring storage and/or treatment. The location must not be based on considerations of economics alone or the availability of a cheap parcel of land for which there is no immediate intended use. A land waste disposal system should form an integral part of the particular land environment for which it is intended, and not be a facility imposed on it. The system must renovate wastes and be designed so as not to cause contamination or pollution of groundwater or surfacae water. All land waste disposal operations must include provisions for monitoring groundwater quality upgradient and downgradient of the site, to evaluate the performance and to take corrective actions when necessary.

This paper will briefly review land treatment of certain mining wastes (potash and uranium), general aspects of hazardous waste impoundments, and natural and artificial barriers.

2 WASTE MANAGEMENT IN THE POTASH INDUSTRY

Large volumes of both solid and liquid wastes are generated by the potash industry in Saskatchewan. The waste, consisting primarily of NaCl, is stored on the surface near the plants in waste disposal basins which are artificially constructed lagoons and/or natural depressions. Several of the mines use deep well injection to dispose of some of the excess brine in conjunction with surface storage.

The waste disposal basins are underlain by Pleistocene sediments. Infiltration of brine from the waste disposal basins into the underlying sediments and subsequent incorporation into groundwater flow systems present a potential hazard of contamination of the groundwater resources in the vicinity of the mines.

The companies operating in Saskatchewan must retain the waste within the boundaries of the mine property (i.e. waste disposal basins) or dispose of part of the waste by other means, such as backfill of solid waste in mined-out areas (not practiced at this time), or disposal of brine by deep-well injection. The biggest disadvantage of the present waste disposal method, if no other techniques such as deep-well injection are used to reduce the waste volume, is the constant need for increased storage capacity. This means that existing lagoons will have to be expanded constantly in future years to supply the necessary storage space.

The bulk of the waste is directly related to the type and trade of ore. The ore that is present in Saskatchewan is composed of a mixture of sylvite and/or carnallite and halite. Thus far, all mining activity in Saskatchewan has concentrated on those areas where the ore is primarily sylvinite and has no or only a low carnallite content. In addition to a much lower K_2O equivalent than sylvite, carnallite ore has another disadvantage because it contains $MgCl_2$, which is undesirable in the finished end product and must be removed. This is readily achieved by selective solution, but as a result larger volumes of waste brine are produced than during the processing of sylvinitic ore.

Another source of waste determined by the original depositional environment, is the insolubles. The insolubles consist primarily of clays and minor amounts of anhydrite, gypsum, quartz, and carbonate minerals. The clays are either interspersed with the ore or they occur as discrete thin layers in the ore bed. The other minerals generally occur interspersed with the ore. If the location of the mines in Saskatchewan is considered with respect to the type of ore and insoluble content, then it can be seen that the ore in southeastern Saskatchewan contains approximately 1–5 percent carnallite, whereas the ore mined by the various companies in central Saskatchewan is essentially free from carnallite but contains a higher percentage of insolubles.

In addition to the volume of waste generated during the milling and refining of the potash ore, three other sources can be recognized. These are "wash" water in the mill, shaft water, and spent mill chemicals. All wastes from these various sources are discharged in a waste disposal basin. The waste, both liquid and solid, from the mill and other sources is pumped as a slurry into the waste disposal basins. The waste disposal basins consist of artificially constructed lagoons and/or natural depressions. The solid particles in the slurry build "alluvial" fans above and deltas below the brine level from the point of discharge in the waste disposal basin. Table 1 shows a typical chemical analysis of brine.

Waste volume production is in the order of 1.5 tons of solid waste per ton of finished product, and 20 to 40 gallons of brine per ton of ore milled. The size of the individual disposal lagoons ranges from 400 to 1000 acres (160–400 ha).

In addition to the storage of waste in disposal lagoons, all but one mine use deep-well injection to reduce the volume of liquid waste. The disposal formulations are all located below the Prairie Evaporite Formation. Injection rates range from 150 to 1200 (US) gpm ($0.5-4.5$ m^3/min).

Tests using class A evaporation pans have shown that evaporation from brine surfaces (pans) during the summer season exceeds the annual precipitation by a factor of 1.2–1.5. Unfortunately, results obtained from evaporation pans are not really compatible with evaporation from large surfaces of water or brine. Similar experience has been gained by the potash industry. Originally it was thought that evaporation would be an important factor in brine volume reduction; but surface roughness, turbulence, unfavourable temperature gradients (too great a depth in most brine ponds), and the very high salt content of the brine resulted in far less evaporation than expected. Several years of operating experience have shown that evaporation from brine ponds approximately equals the annual precipitation; therefore evapoaration is no longer regarded as a factor in the reduction of brine volume.

Table 1. Typical chemical composition of brine (Cleven 1984)

Parameter	Concentration
Specific conductance (μmhos/cm)	264 000
pH	7.6
Density (g/cm^3)	1.19
Ca (ppm)	1 980
Mg	7 940
Na	88 500
K	39 500
HCO$_3$	120
SO$_4$	2 100
Cl	200 000
Fe	<0.02
Mn	8.4
Pb	0.018
Cu	0.012
Zn	0.045
F	0.1

The dykes of the brine ponds undergo severe erosion. The dykes are built from till and are compacted during construction. In most cases special measures, such as riprap covers or wave breaker booms, were not taken initially to protect the dykes from wave erosion. The combination of strong winds (wind speeds of 30-50 km/h are common), wide open prairies with little or no protective forest cover in the vicinity of the lagoons, the relatively large brine storage areas, and the absence of a protective wave-breaking cover near the dykes caused severe erosion by wave action. This problem is further aggravated by the brine itself. The brine has an average specific gravity of 1.19, which in comparison with water not only results in greater impact energy of the waves on the dykes, but also tends to keep more material in suspension, thereby further increasing the impact energy. This combined effect has resulted in accelerated erosion of the unprotected dykes by wave action. The rates of erosion observed over a period of one year are as high as 3 ft in places.

To alleviate some of the problems of dyke erosion, several methods have been used, such as lining the dykes with scrap iron and/or old cars, discarded conveyor belts, straw bales booms, and waste salt (Cleven 1984). The most successful of these methods is the deposition of waste salt against the dyke. Although the exposed salt is subjected to minor karst type erosion by rainfall, intermittent spraying of the surface with brine results in precipitation of halite in the interstices of the granular waste salt. The "cemented" waste salt forms a hard crust at the surface and wave action has little effect on the stability of this surface. Another beneficial effect of the waste salt on the inside of the dykes is that it reduces the surface area subjected to brine infiltration in the immediate vicinity of the dyke.

The long-term effect of potassium-rich NaCl brines on the montmorillonitic clays may be more serious than dyke erosion. Montmorillonitic clays are abundant in the sediments underlying the brine and in the fill used for dyke construction. The effect of the saturated, potassium-rich NaCl brines on these clays is twofold. Initially it will cause dispersion of the clay minerals and subsequently it will induce mineralogical changes from montmorillonites to illites. In the first instance this will result in a decrease in the permeability, whereas in the second instance an increase in permeability can be expected. The increase in permeability in the latter case is caused by the reduction in the original "mineral" volume that occurs during the transformation from montmorillonite to illite.

Apparently there have been no detailed studies conducted on the rate of change in the mineralogy, permeability, and porosity of the sediments subjected to brine infiltration. This area warrants study because a progressive increase in the permeability of the sediments would result in higher brine infiltration rates, which may become environmentally unacceptable.

Even though deep well injection is used, the waste disposal ponds will continue to grow in size because of the brine runoff from the waste salt piles. As the size of the salt piles grow, precipitation on the piles results in a saturated barine which is conveyed into the waste disposal ponds.

A partial solution to this problem is backfilling. Most, but not all, of the waste salt could be returned to the mine because after the ore is mined and processed, the waste salt occupies more volume than the original ore itself. Therefore the waste salt piles would continue to grow, but at a reduced rate, even if backfilling was employed. Backfilling is, for the most part, a matter of economics. It would no doubat cost more to operate a mine which employs backfilling since the distances, both vertical and horizontal, travelled in removing the ore and returning the salt can be as large as 6 miles for a round trip.

Detailed seepage and stability investigations were undertaken in 1979 (Menely, 1979). These investigations concluded that the problems were serious in three of the mines, PCS Lanigan, IMCK2 and PCS Rocanville. In the case of PCS Lanigan, brine contamination in subsurfacae strata was evident as far away as 3 miles. It was concluded that this resulted from

an overland transport process rather
than direct infiltration at the mine.
The overland transport process could
have been by air pollution, and the
overtopping of dykes by wave action and
dyke seepage. After this problem was
identified, a major interceptor ditch
system was constructed by PCS to
prevent further contamination. It was
a fairly major undertaking and costed
approximately one million dollars.

3 WASTE MANAGEMENT IN URANIUM MINING INDUSTRY

Approximately 110 million tonnes of
tailings are currently stored in a
variety of surface impoundments in
Ontario and Saskatchewan, having a
cumulative surface area of
approximately 635 hectares (Dance
1985). Approximately 17 million tonnes
are stored in impoundments in northern
Saskatchewan.
    The waste management systems at
three operating mills in Saskatchewan,
located at Cluff Lake, Key Lake, and
Rabbit Lake, have utilized different
waste management concepts, as briefly
described below (Robertson and Clifton
1985).

3.1 Cluff Lake

A conventional hydraulic waste
management system is adopted at this
location. Environmental loadings are
kept at acceptable levels through
control of permeability, head and
gradient. This is accomplished through
the utilization of low head ponds
confined with a dyke that incorporates
a soil bentonite cutoff penetrating the
fracatured bedrock, and the
incorporation of toe drains in portions
of the dyke. The liquid waste stream
is subjected to chemical treatment for
removal of radioactive species.
Secondary treatment consists of
precipitation of radium from solution
and filtration prior to discharge.

3.2 Key Lake

Key Lake waste management system
incorporates thickened discharge and
subaerial deposition to increase
tailings density to the maximum
possible amount, thereby reducing the
total area required for the deposition
of tailings. The deposition in thin
layers reduces the vertical
permeability, decreasing the mass of
contaminants reaching the underdrainage
system. The underdrainage system
consists of a drained sand blanket
overlying an improved soil bentonite
sealing layer. Drainage in the blanket
reduces the head applied to the sealing
layer, reducing the seepage to
foundation soil and groundwater
systems. Water that must be discharged
to the environment is subjected to
chemical treatment for removal of
radionuclides. Key Lake waste
management system incorporates the
concepts of reduction of permeability,
reduction of gradients, minimization of
liquid effluents, and chemical
treatment.

3.3 Rabbit Lake

The scheme entails placing filtered
tailings in the abandoned Rabbit Lake
pit. The tailings will be surrounded
by a highly pervious zone that will
allow drainage radially and towards the
base of the pit from which the leachate
will be collected and pumped out for
treatment during operations.
Underdrainage will be maintained during
operations to achieve consolidation,
thereby minimizing the amount of
leachate produced in the post-
operational phase. This scheme
incorporates a dewatered tailings and
hydraulic balancing concept.

4 DESIGN AND CONSTRUCTION OF HAZARDOUS WASTE IMPOUNDMENTS - GENERAL CONCEPTS

Hazardous wastes pose unique problems;
they are placed in impoundments over
long periods and pose the potential for
chemical action with the materials used
in construction or geologic materials
at the site (Shaw and D'Appolonia
1985). If the wastes are not contained
properly, leakage through the
embankment, unplanned seepage into
groundwater, and/or the embankment
failure could affect public health and
safety and create an environmental
disaster.
    In the design and operation of
hazardous waste impoundments, emphasis
should be laid on the following two
aspects:

i) the chemical influence of the waste on the soil and rock forming the reservoir and embankment and any liners made of natural or artificial materials; and

ii) the effect of contaminant seepage and transport on groundwater and streams.

A knowledge of waste chemistry as it relates to its effect on the soil and rock at the site and the materials used for construction is of the utmost importance. This information is also important to understand and estimate contaminant attenuation (through adsorption/ion exchange by soil or liner) and contaminant transport. The important factor lies in the control of seepage relative to the chemical adsorption and aquifer characteristics of the site. It is necessary that design and construction practices do not result in unplanned seepage, or that unplanned seepage does not interfere with the structural integrity of the impoundment.

5 NATURAL AND SYNTHETIC BARRIERS

The design and installation of natural and artificial barriers is one of the most important steps in the construction of any waste management system (Haug, 1985). Some of the common natural and synthetic materials for liners (barriers) are indicated below. Natural (i) compacted earth (clay) (ii) concrete (iii) soil-cement (iv) asphalt concrete (v) asphalt emulsions (vi) bentonite membranes (vii) sand-bentonite mixtures. Synthetic (i) chlorosulfonated polyethylene (Hypalon) (ii) polyvinyl chloride (PVC) (iii) polyethylene (PE) (iv) chlorinated polyethylene (v) butyl rubber. A number of factors shown below influence the choice of a particular liner: 1) the intended usage, 2) the physical environment, 3) groundwater and leachate chemistry, 4) design life, 5) seepage rate, 6) physical constraints, and 7) cost. Compacted earth liners are the most common of all liner materials and usually the most cost effective. There are a large number of different synthetic materials available for use as liners; because of this, selection of any one synthetic liner by a design engineer often poses great difficulty. A study of case histories (Haxo et al.,

1984; Fuller, 1981; Ball et al., 1985; Schwope, 1986) on how different liners have behaved in the past is frequently used by engineers to predict how similar liners would act in the proposed facility. It is sometimes necessary to conduct the tests specified by ASTM to develop a reliable estimate of a liner material's performance, keeping in perspective the type of waste impoundment. For example, traditional permeability tests using water alone qualified four clay soils for lining hazardous waste facilities on the basis of low permeabilities (1×10^{-7} cm sec^{-1}). But these same clays underwent large permeability increases when tested with basic, neutral polar and neutral nonpolar oraganic fluids (Brown and Anderson, 1983).

6 SUMMARY AND CONCLUSIONS

The poor performance of a number of hazardous and other industrial waste impoundments is generally attributed to the lack of good project planning during the design and construction phase. Failure to execute proper quality assurance and control procedures during the project also contributes to the failure of these facilities. Deviations from desired liner specifications and poor liner/waste compatibility testing also results in poor performance. Rigorous site-specific investigations to develop the basis for proper design and construction, including liner/waste permeability studies, the use of competent consultants and contractors for design, construction and inspection, close scrutiny of all phases of design, construction and operation by the owner/operator, and efficient project management, are essential factors in obtaining a successful performance of a surface waste impoundment facility.

REFERENCES

Bass, J.M, Lyman, W.J. and J.P. Tratnyek 1985. Assessment of synthetic membrane successes and failures at waste storage and disposal sites. EPA/600/S2-85/100, United States Environmental Protection Agency, Cincinnati, Ohio.

Brown, K.W. & D.C. Anderson 1983.
Effects of organic solvents on the
permeability of clay soils. EPA-
600/S2-83-016, United States
Environmental Protection Agency,
Cincinnati, Ohio.

Cleven, L.O. 1984. Environmental
overview: Saskatchewan potash
mining, Regina.

Dance, J.T. 1985. Physical and
chemical characteristics of uranium
tailings. Proc. Seminar on
Engineering of Waste Management
Systems, Dept. of Civil Engineering,
University of Saskatchewan,
Saskatoon, Canada.

Fuller, W.H. 1981. Liners of natural
porous materials to minimize
pollutant migration. EPA-600/S2-81-
122, United States Environmental
Protection Agency, Cincinnati, Ohio.

Haug, M.D. 1985. Natural and synthetic
barriers. Proc. Seminar on
Engineering of Waste Management
Systems, Dept. of Civil Engineering,
Uniersity of Saskatchewan,
Saskatoon, Canada.

Haxo, Jr., H.E., White, R.M., Haxo,
P.D. & M.A. Fong. 1983. Liner
materials exposed to municipal solid
waste leachate. EPA-600/S2-82-097,
United States Environmental
Protection Agency, Cincinnati, Ohio.

Menely, W.A. 1979. Seepage and
stability analysis – Saskatchewan
potash mines. Regina.

Robertson, A.M. & A.W. Clifton. 1985.
Design concepts in uranium waste
management. Proc. Seminar on
Engineering of Waste Management
Systems, Dept. of Civil Engineering,
University of Saskatchewan,
Saskatoon, Canada.

Schwope, A.D., Costas, P.O. & W.J.
Lyman. 1986. Resistance of
flexible membrane liners to
chemicals and waste. EPA/600/S2-
85/127, United States Environmental
Protection Agency, Cincinnati, Ohio.

Shaw, D.E. & E. D'Appolonia. 1985.
Design and construction of hazardous
waste impoundments. Proc. Seminar
on Engineering of Waste Management
Systems, Dept. of Civil Engineering,
University of Saskatchewan,
Saskatoon, Canada.

Environmental Geotechnics and Problematic Soils and Rocks, Balasubramaniam et al. (eds)
© *1987 Balkema, Rotterdam. ISBN 90 6191 785 9*

Sanitary landfill treatment, Interstate Highway 84

Philip Keene
Consultant, Middletown, Conn., USA

ABSTRACT: A large quantity of old refuse fill, including layers of earth cover, was transported to a new location, where it was placed and compacted in layers. Measurements were made of the volume decrease of the material due to the above. During the placing, settlement platforms were installed at various elevations and settlement readings taken frequently. These indicate that "primary" settlement was extremely rapid and "secondary" settlement has proceeded continuously, at rates comparable to peats. Samples of the material, when separated in the laboratory, indicate the fill was approximately 48% refuse and 52% earth cover by volume.

1 INTRODUCTION

Refuse fills are widespread, being found over the entire world. When the refuse is covered with earth, usually over each layer of refuse, it is called sanitary landfill. Because of its increasingly large volume and the increasing restrictions on land use, better methods of refuse disposal are being studied vigorously in many countries. Recent examples of this in southern California are described (Egan et al 1975, Forsyth & Egan 1976, E N-R 1975). Nevertheless, there are countless sanitary landfills which will remain and on which we must build. Their importance can be judged by the increasing amount of research and publications on their properties and behavior. Even the urbane "New Yorker" magazine (1975) had a lead article on a huge sanitary landfill in the Bronx.

The compressibility of sanitary landfills is usually great and is due chiefly to three reasons: (a) consolidation due to its own weight and that of the overlying material; (b) raveling - i.e. the movement of material, chiefly soil, into the voids below; (c) chemical and biological decomposition of the material. To date, there are relatively few quantitative data on the compressibility of refuse fills and their behavior has been assumed to be similar to that of peats (Keene & Zawodniak 1968) (Moran et al 1958) (Rao et al 1977).

This paper is concerned with a sanitary landfill material that was moved to a new location where it was spread and compacted in layers. Data are given on the resulting volume decrease and on the subsequent compressions during the following five years. The approximate percentages of the ingredients in the material are given, based on samples taken on the job.

2 CONSTRUCTION OPERATIONS

An Interstate Route 84 highway project in Newtown, a rural town in western Connecticut, included the moving of a large quantity of sanitary landfill to a new location. The material, totaling 144,007 cu yd (110,200 m^3) in its original location, was known as the "town dump" and consisted of refuse, chiefly paper products, and an excessive amount of sand, gravel and silt cover placed at intervals for sanitary reasons. All the material had a low moisture content; it could be described as "damp". There are indications that the "dump" began 40 or 50 years ago. It was located in a large area having deep deposits of sand, gravel and silt. The Town's operators used very generous amounts of earth cover over the refuse, as the refuse was dumped off the top of a steep slope and the earth cover was also dumped in that manner.

As this sanitary landfill was in the path of the proposed west-bound lanes of

Figure 1. Dumping sanitary landfill in pit

the highway and was very compressible and unstable, it was necessary to move it. Since the State Department of Transportation could find no "takers" for this landfill, it was decided to widen the median strip slightly at the proposed interchange nearby and bury the refuse in the median area between east-bound and west-bound lanes, placing the material in compacted layers. Here the natural granular deposits extend 70 ft (21 m) below highway grade.

In the Spring of 1971, the highway contractor excavated a large pit, approximately 550 ft long, 200 ft wide and 50 ft deep (168 x 61 x 15 m) in the median area down to Elevation 312 (95 m). This elevation is six ft (2 m) above ground water table, per State Department of Health requirements. On May 18, 1971, removal of the Town's sanitary landfill and placement in the pit was begun. This operation was continued without interruption until completion on June 22. The mixture of refuse and earth was placed in 30 in (0.76 m) layers in the pit.

Each layer was compacted by the 12 cu yd (9 m^3) WABCO haul trucks and a Caterpillar D-8 bulldozer that spread and compacted the material (Fig. 1).

Over each layer was placed a 6 in (0.15 m) cover of fresh sand, thoroughly compacted,

per health requirements. To complete this filling required 14 layers, the final one reaching Elev 354 (108 m).

To measure the compression of the fill, nine settlement platforms, each with a 2 in (5 cm) pipe riser, and four concrete monuments, of the type used for property boundary markers, were installed in the middle portion of the fill. Fig. 2 shows a plan and profile of the fill, platforms and monuments. Three platforms, Nos. 1, 5, and 8, were installed on June 3, when the fill was up to about Elev 336, by digging a pit for each platform to Elev 330, placing the platforms and backfilling the pits. Because of the rather unstable fill material, it was found that placing the platforms in such pits reduced the possibility of their being damaged by construction equipment. Similarly, four platforms were installed at Elev 345 and two platforms at Elev 350 after the fill was finished at Elev 354 (108 m). In spite of this, a few of the pipe risers attached to the platforms were disturbed by construction equipment during the filling and had to be reset. For purposes of highway aesthetics and drainage, two layers of earth, chiefly sand, were added above the top fill layer to final grade, Elev 360 (110 m), in late July. Four monuments were installed at

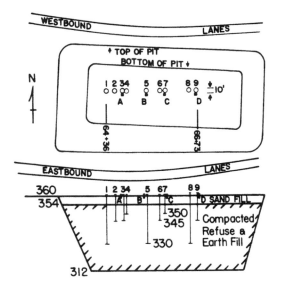

Figure 2. Plan and profile of fill and
platforms #1 to #9 and monuments
A to D (1 ft = 0.305 m)

Elev 360 (110 m) in April 1972 to measure
settlements at the top. Fig. 3 shows the
tops of pipe risers Nos. 3 and 4 and monument A.

3 INGREDIENTS OF FILL

In order to analyze the data more realis-
tically, six samples were taken of the
mixture of refuse and old earth cover from
the Town Dump. Three were taken in the
median location at about Elev 330 and three
at about Elev 346. The samples weighed
about 30 lbs (14 kg) each. The material in
each sample was separated into its ingredients -- large gravel (over 19.0 mm), small
gravel (over 2.36 mm), sand, silt, paper
products, rags, wood, plastics, glass and
metal. Each ingredient was dried and
weighed separately and its solid volume
calculated from its weight and specific
gravity. Table 1 shows the maximum and
minimum percentage, by weight, of each
ingredient in the six samples.

Figure 3. Completed fill showing pipe risers
numbers 3 and 4 and monument A

Table 1. Ingredients in samples, town dump

Max. and min. percentage, by dry weight

	Max.	Min.
Large gravel	13.5%	2.0%
Small gravel	10.3	7.0
Sand	63.0	50.0
Silt	6.3	5.2
Paper products	16.5	10.5
Rags	6.3	0.5
Wood	1.4	0.2
Plastics	1.8	0.5
Glass	2.8	1.5
Metal	8.5	1.0

As stated previously, the percentage of soil is considerably higher than found in "typical" sanitary landfills, due to the abundance of available soil and the method of dumping at the Town Dump.

Putrescible material, chiefly food waste, was negligible in the samples, but observations on the project indicate it was originally less than 5% of the refuse. Since World War II, food waste brought to municipal disposal areas is very small, due to the extensive practice of packaging food for sales. In the early days of the work, some rats were seen, but these were exterminated. In addition, there were some large materials in the Town's sanitary landfill, including about 100 refrigerators and TV sets, 700 rubber tires and some tree stumps and limbs. These bulky materials were taken off the project limits and disposed of.

Included in the cooperation by the State Department of Health were eight visits to the site by the Principal Sanitarian of the Southwest Regional Office before and during construction. The Sanitarian took a representative sample of garbage in the fill for analysis. This showed a moderate quantity of undecomposed organic material, including portions of fruit and other garbage, as well as a large quantity of paper; much of the latter was intact, readable and not decomposed.

4 CALCULATIONS OF VOLUME DECREASE BY COMPACTION

The material in the Town Dump was measured by surveying and found to be 144,007 cu yd (110,200 m^3) as mentioned previously.

Based on the total (dry) weights in the samples, the mass volumes of the earth portion and of the refuse portion were calculated. For the earth portion, the weight was converted to mass volume by a density factor of 100 pcf (1.60 gm/cc). For the refuse portion, the conversion factor was 30 pcf (0.48 gm/cc); this value was selected after reviewing various references (Bromwell, 1971; Merz & Stone, 1962; Sowers, 1968; Yen & Scanlon, 1975; Zoino, 1973). The result indicates the earth portion was 52% or 74,800 cu yd (57,300 m^3) and the refuse portion was 48% or 69,200 cu yd (52,900 m^3).

In order to measure the volume decrease of the Town's sanitary landfill material when placed in its new location, the volume of this excavated material, 144,007 cu yd (110,200 m^3), was compared with its volume of 119,371 cu yd (91,300 m^3) after placement and compaction in the median area. The latter volume is after making the small correction due to compression while placing the fill; it does not include the fourteen 6 in (0.15 m) layers of fresh sand. The volume decrease due to compaction was 24,636 cu yd (18,900 m^3) or 17.1% of its original volume.

In an effort to calculate approximately the volume decrease of the refuse portion of the landfill material, it was assumed that its earth and refuse portions remained in separate layers; this, of course, is not true. Nevertheless, it was assumed to be correct, to simplify the following calculations. It was assumed that the transfer and compaction of the landfill resulted in a 15% volume decrease of the soil portion or 11,220 cu yd (8,600 m^3). This is based on measured volume decrease of granular soils due to thorough compaction on various highway projects. The remainder of the volume decrease, 13,416 cu yd (10,300 m^3), represents a volume decrease of 19.4% in the refuse portion due to compaction. Thus, the refuse portion was compacted to a mass density of about 37 pcf (0.59 gm/cc). Obviously the above calculations are very approximate, but may be useful in presenting the picture to the reader.

5 CALCULATIONS OF COMPRESSION (SETTLEMENT)

As described previously, settlement platforms were placed at three different elevations in the fill: Elev 330, 345 and 350 or 18, 33, and 38 ft (5, 10 and 11 m) above bottom of the pit. Settlement readings were taken frequently until September 1971; thereafter they were taken once per month and then at longer intervals. Figs.

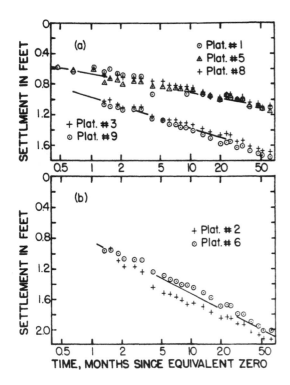

Figure 4. Time - settlement
observations under platforms
(1 ft = 0.305 m)

4 and 5 show the results of these readings.
It should be noted that some compression of
the material under each platform occurred
before the platform was installed. To
judge the amount of this compression, it
was estimated by extrapolation of the data
that this would be 0.3 ft (0.09 m) for the
material under the deepest platforms, Nos.
1, 5 and 8, placed at Elev 330. Under the
other six platforms, this would be approximately 0.6 ft (0.18 m) plus the measured
compression below Elev 330 of 0.11 ft
(0.03 m) which occurred while the fill from
the lower platforms to the upper platforms
was being placed. These amounts are included in Figs. 4 and 5.

The settlement readings for the platforms
were plotted in groups of two or three, on
the basis of the same platform elevation
and similar compressibility. The "equivalent zero" time (t=0) was taken as June 12,
1971, the date on which approximately one-
half the load had been placed.

Table 2 gives a tabulation of the primary
and secondary compressions. For simplicity,
an average was used for each group of two
or three platforms. As the platforms were
placed at intermediate elevations in the

fill, it is possible to get a good idea of
the compression at various depths. Table
2 shows the compression of the bottom 18
ft, middle 15 ft, and top 9 ft (5.5, 4.6,
and 2.7 m). For simplicity, all of the
compression is assumed to be in the Town
Dump material (refuse and old soil cover)
and none in the fourteen 6 in (0.15 m)
fresh sand layers.

It is difficult to determine from the
data the amount of "primary" compression
and at what time it was essentially completed. The earliest compression was
estimated by extrapolation, as mentioned
before; also a few of the pipe risers were
disturbed and then reset. However, it
seems clear that primary compression was
very rapid and was virtually completed at
one-half to one month after the refuse
filling was finished on June 22, 1971.
The total amount of primary compression
was about 3% of the thickness of the Town
Dump material. The coefficient of secondary compression, C_α, is based on the
thickness of Town Dump material before
primary compression. If based on thickness
at the end of primary compression, C_α would
be larger by about 3%. The large variations in values of C_α are not surprising
in view of the heterogeneous nature of tne
material and the complex mechanisms causing
these compressions.

It should be remembered that these compressions are smaller than for the usual
refuse fills because of the large earth
content of the material and the heavy compaction when placed in its final location.
The secondary coefficient for the refuse
portion alone could be estimated at approximately twice the coefficient for the entire material, as the earth portion is
about 52% of the total volume and its
secondary compression is very small. However, this simple assumption might be found
quite erroneous when better knowledge is
gained regarding the complicated mechanics
of long-term settlements in such fills.

It is of interest to compare the secondary coefficients with those reported by
Sowers (1973). He shows values from plots
for two extremes: conditions favorable
(0.08) and unfavorable (0.03) to decomposition. The secondary coefficients on
this project plot close to his "unfavorable"
plot, but if adjusted upward because of the
unusually large percentage of earth in the
fill, they would plot rather close to his
"favorable" plot. The conclusion might be
reached that this project is located in an
intermediate environment, which is moist
with 45 in (1.15 m) of annual precipitation
but has a rather cold climate.

The secondary coefficients on this proj-

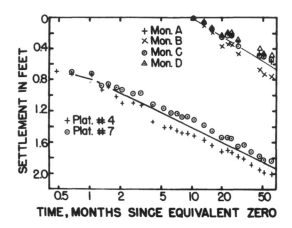

TIME, MONTHS SINCE EQUIVALENT ZERO

Figure 5. Time - settlement observations under platforms and monuments (1 ft = 0.305 m)

ect were compared with those of three projects reported by Zoino (1973) and found to be approximately the same. This agreement may be due to two compensating differences: more earth in the landfill and more favorable condition for decomposition in the subject project.

It should be mentioned that settlement data in Fig. 5 for the four monuments has been used in Table 2 to show compressions in the upper fill. However, as the monuments are in the thick 6 ft (1.8 m) earth fill placed at the top for drainage and highway aesthetics and they were installed ten months after the fill was completed, the usefulness of the data is limited.

6 SUMMARY AND CONCLUSIONS

1) A large amount of sanitary landfill,

Table 2. Primary and secondary compression

Between	Elevs.	Total* Thick. ft	Town* Matl. ft	Pri-* mary Comp. ft	Secondary∅ Compress. ft ft/cycle		c_α
Bottom & Pls. 1, 5, 8	312–330	18.3	15.3	0.6	0.50	0.24	0.016
Bottom & Pls. 3, 9	312–345	33.7	28.2	0.8	0.86	0.42	0.015
Pls. 1, 5, 8 & 3, 9	330–345	15.4	12.9	0.2	0.36	0.18	0.014
Bottom & Pls. 2, 6	312–345	33.7	28.2	0.8	1.28	0.68	0.024
Pls. 1, 5, 8 & 2, 6	330–345	15.4	12.9	0.2	0.78	0.44	0.034
Bottom & Pls. 4, 7	312–350	38.7	32.7	0.9	1.02	0.64	0.020
Pls. 1, 5, 8 & 4, 7	330–350	20.4	17.4	0.3	0.52	0.40	0.023
Pls. 4, 7 & Monuments	350–360	10.8	4.1	0.2+	0.3+	0.16+	0.04
Totals (approx.)	312–360	49.5	36.8	1.0	1.4		

* Includes compression before platforms installed.

∅ To August 24, 1976   + Probably includes appreciable compression in thick earth fill at top.

$$c_\alpha = \frac{\Delta H}{H} \bigg/ \log \frac{t_2}{t_1}$$

where ΔH is compression occurring between t_1 and t_2; H is thickness of Town Material before compression; t_1 and t_2 are elapsed times on secondary compression phase (1 ft = 0.305 m)

432

composed of refuse and earth in approximately equal parts, by volume, was moved to a new location where it was spread and compacted in layers.

2) This operation resulted in a shrinkage of 17%.

3) Compression of the material in its new location was measured at various levels for five years. Compression appears to be similar to that of peats. "Primary" compression was about 3%, occurring rapidly, and was essentially completed at one-half to one month after completion of filling. "Secondary" compression varied, with C_α varying between 0.014 and 0.034.

4) If the landfill had contained a smaller amount of earth, the percent compression of the landfill would have increased. The amount of increase is not known, as the effects of mechanisms such as raveling by movement of earth into voids of the refuse is not known quantitatively.

The information from this project appears to be similar, qualitatively, to that reported by authors of recent papers. Their approach to the analyses is to consider that these settlements may be somewhat analogous to those of very compressible soils such as peats. Their studies have used the concepts of initial-primary and secondary compression, but have emphasized that raveling, physico-chemical and bio-chemical decay and combustion will greatly affect the results. For these reasons, settlements in the later years may be unexpectedly more severe than would be experienced in soils such as peats; also, using temporary overloads to reduce long-term settlements may not be as effective as in the case of natural soils.

Research is needed to learn more of the complicated processes that contribute to settlements of sanitary landfills.

7 ACKNOWLEDGEMENTS

During the design and construction of the project, the Environmental Health Section of the State Health Department, headed in the Southwest Regional Office by Robert Mundy, cooperated fully with the State Department of Transportation. This resulted in a highly satisfactory solution to the problem, one that received strong praise in the local newspapers.

The Soils and Foundations Division of DOT made the subsurface investigations and analyses and furnished recommendations to the designers for the refuse disposal. During the construction this Division made further investigations to assist the Dis-

trict IV construction personnel, headed by Arthur Carlson. Mention should be made of the numerous and careful settlement readings made by Robert L. Hock of District IV.

REFERENCES

Bromwell, L.G. 1971, Construction on refuse fills, Special Summer Program, Dept. Civil Engg., Mass. Inst. of Technology.

Egan, J.P., Leech, L.R. and Campbell, J. 1975, Embankment construction utilizing sanitary landfill material, Division of Construction and Research, California DOT, Final Report.

Forsyth, R.A. and Egan, J.P. 1976, Use of waste materials in embankment construction, Office of Transportation Laboratory, California DOT, presented at Annual Meeting, Transp. Research Board.

Keene, P. and Zawodniak, C.D. 1968, Embankment construction on peat utilizing hydraulic fill, Third International Peat Congress, Quebec, Canada.

Merz, R.C. and Stone, R. 1962, Landfill settlement rates, Public Works, Vol. 93.

Moran, Proctor, Mueser, and Rutledge 1958, Study of deep soil stabilization by vertical sand drains, Bureau of Yards and Docks, Dept. of the Navy, Washington, D. C., Publication NOy 88812.

Rao, S.K., Moulton, L.K. and Seals, R.K. 1977, Settlement of refuse landfills, Geotechnical practice for disposal of solid waste materials, Specialty Conf., Am. Soc. C. E.

Recycled trash as road base is gold as state saves $1 million, Engineering News-Record March 13, 1975.

Sowers, G.F. 1968, Foundation problems in sanitary landfills, Journal of Sanitary Engg. Div., ASCE, Vol. 94, No. SA1, Proc. Paper 5811.

Sowers, G.F. 1973, Settlement of waste disposal fills, Proceedings, Eighth International Conference on Soil Mechanics and Foundation Engineering, Moscow, USSR.

The Talk of the Town, The New Yorker, April 28, 1975.

Yen, B.C. and Scanlon, B. 1975, Sanitary landfill settlement rates, Journal of Geotechnical Engg. Div., ASCE, Vol. 101, No. GT5.

Zoino, W.S. 1973, Stabilizing landfills with surcharge, Confer. Session No. 44, Hway. Research Bd. Annual Meeting.

Environmental Geotechnics and Problematic Soils and Rocks, Balasubramaniam et al. (eds)
© 1987 Balkema, Rotterdam. ISBN 90 6191 785 9

Colliery wastes in the United Kingdom – Findings and implications: 1966-1985

R.K.Taylor
University of Durham, UK

ABSTRACT: A major programme of site investigations and research projects was established by the National Coal Board following the Aberfan flowslide disaster in 1966. By 1981 the major part of the programme had been completed, so test results were statistically analysed. Accordingly, the composition and engineering properties of colliery discards were subsequently reviewed (Taylor 1984a).

This paper considers some of the main findings of the above review, and in the light of present-day construction practices. Colliery discards are dominantly composed of mudrocks with the addition of some waste coal. In this respect 'coarse discards' are similar in character to other non-marine 'shale' fills. Those fine discards which are pumped to lagoons have a high coal content and span a considerably wider particle size range than metal mine tailings.

1 INTRODUCTION

On 21 October 1966 122330m^3 of colliery discard from Tip 7, Aberfan (South Wales) was involved in a complex failure which developed as a flowslide, moving down the 12.5 degree hillside at a velocity of between 16 to 32 kph. One hundred and forty-four persons were killed, of whom 116 were school children. Background geological, geotechnical and mining details of this failure, which was probably triggered by an adjustment slip in loose discard at the toe of the heap, are given in Woodland (1968), Welsh Office (1969) and Bishop (1973).

The tragedy led to the longest disaster inquiry in British legal history; the loose tipping of coarse discard in spoil heaps and the use of upstream methods in the construction of colliery lagoon banks were discontinued (National Coal Board 1970; McKechnie Thomson and Rodin 1972). Strictly regulated civil engineering methods are now used by the mining industry (National Coal Board 1971), in that

tips are constructed by earth-moving equipment in horizontal layers not exceeding 5m in thickness. Colliery lagoon embankments are constructed by the downstream method which ensures much greater control on the stability of the structure.

As previously reported at A.I.T (Taylor et al, 1980), pressures on land use have led to the increased development of overtipping procedures, whereby lagoons may be incorporated within a tip. At these sites lagoon banks are initially designed to become an integral part of the flanks of the tip disposal structure (Fig. 1).

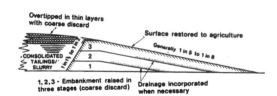

Figure 1. Composite disposal structure. Waste heap incorporating lagoon.

Post-Aberfan National Coal Board sponsored research into the engineering and mineralogical properties of colliery discards, including their susceptibility to liquefaction, had largely been completed by 1981. Consequently it was decided to collate and statistically analyse the results from 3338 samples taken from 149 spoil heaps and 1973 samples from 25 lagoons, in order to evaluate trends and regional variations. Some of the findings from this review (Taylor 1984a) are the subject of this presentation.

2 COMPOSITION OF DISCARDS

Colliery discards are mainly the mudrocks from the roof and floor measures originating from coalface cutting operations, although other rock types may be conveyed to the surface from underground development work. The bulk of the discard is conveyed with the coal as 'run-of-mine' (ROM). From the screening and washing of ROM in the coal preparation process two products emerge: coarse discard separated from the raw coal, and fine discard ($< 0.5mm$ diameter) which is carried in suspension.

Present production of coarse discard is about 59Mta of which about 3Mta is used underground, whilst fine discards make up about 6Mta of the total. Approximately 45Mta are largely disposed of in tips and about 50% of fine discards (ie 3Mta) are deposited in lagoons — the cheapest method of disposal. The remaining 5Mt of discards are disposed of at sea. Fine discards are of two types: coal-rich slurry from the coal-washing process and clay-rich tailings from the froth flotation process used at many collieries to clean coal of less than 0.5mm size.

2.1 Mineralogy

Table 1 shows that there are major mineralogical differences between coarse and fine discards. The first point to note is that the average coal content is exceedingly high in lagoons. The ratio of clay minerals to quartz indicates that in terms of

detrital (or transported) minerals clay minerals are relatively more abundant in lagoon sediments, as might be expected. The high coal content in lagoons is matched by a higher content of non-detrital minerals associated with coal, namely, pyrite (ferrous sulphide) and ankerite (an Fe Ca Mg carbonate) found in the cleat (small-scale joints) of coal. These findings are borne out by the geochemistry which is not shown here.

An important feature of UK mudrocks of Coal Measures age (eg. coarse discards) is that they are largely non-marine in origin, whereas mudrocks world-wide are dominantly marine (Potter et al, 1980). The latter mudrocks tend to contain both calcite (compared with siderite in the Coal Measures) and pyrite in greater proportions. Pyrite oxidation leading to acid attack on calcite, which is more susceptible to weathering than other carbonates, is considered by the writer to be a more important chemical weathering process in marine mudrocks than in non-marine types. It is of interest to consider the hydrogen ion concentration (pH) and the water and acid soluble sulphate levels in UK coarse discards from old colliery tips (Table 1). Although discards are customarily alkaline or neutral when first emplaced, they become progressively more acid in the outer-zone ($< 1m$ depth). Nevertheless, of the 224 unburnt coarse discard samples analysed in Table 1, 71% were acceptable for construction use both in terms of TRRL sulphate recommendations (Sherwood and Ryley (1970) and Department of Transport (1976) specifications ie. water and acid soluble sulphate contents were less than 2.0g/1 and 1%, respectively).

In unburnt sections of old loose tips, chemical weathering does not appear to have affected discard to any degree below 3m depth (see for example Spears et al, 1971). In the 50-year-old Yorkshire Main tip comparisons with fabricated triaxial specimens of fresh discard from underground did not reveal any decrease in effective shear strength parameters due to physical or chemical weathering: composite

TABLE 1. Average mineralogy of coarse and fine discards
(% by wt.)

		Coarse Discards (unburnt) 74 samples 15 tips	Fine Discards 47 samples 10 lagoons
	Quartz	17.5	6.0
	Illite	31.5	34.0
Clay	Mixed-layer clay	26.0	
minerals	Kaolinite	10.5	8.5
	Chlorite	0.5	-(b)
	Carbonates(a)	1.0	2.0
	Pyrite	-(b)	2.0
	Organic carbon	13.0	47.5
	(mainly coal)	100.0	100.0

a) Dominantly siderite and some ankerite in tips.
    Ferroan dolomite (ankerite) in fine discards
b) Small quantities in some samples

Note: in all discards trace amounts of sulphates, feldspars,
       rutile and phosphate total less than 2%.

Coarse Discards (unburnt)

1. Average hydrogen ion concentration (pH), = 6.78 ± 1.31
    138 samples
2. Average acid soluble sulphates (wt %),   = 0.20
    224 samples
    Ninety-five per cent of distribution
    between: 0.05 and 2.80
3. Average water soluble sulphates (wt %),   = 0.10
    224 samples
    Ninety-five per cent of distribution
    between: 0.01 and 0.60

strength parameters of discard in the tip — $c'=0$, $\phi'=35°$ (49 specimens); fabricated fresh discard — $c'=0$, $0'=33.5°$ (19 specimens). The largest recorded decrease in effective shear strength parameters due to weathering or comminution by earth-moving plant was 21%, although the more commonly noted drop was 10%.

2.2 Fundamental properties

Particle size distributions of discards from tips and lagoons are illustrated in Fig. 2 and summary statistics are given in Table 2. Bulk samples from tips are likely to be more representative than U100 driven samples. Perhaps the most surprising feature of the fine discards (Fig. 2) is the amount of extraneous discard in lagoons which is greater than 0.5mm in grain size. Moreover, the size range (band width) of fine discards is much broader than that of coarse discards, although the former are finer-grained. Mean size distributions indicate that slurry is coarser-grained than tailings which are, of course, finer-grained than coarse discards. It is of interest that metal mine tailings span a much reduced range compared with fine discards from UK coal-mining operations (Taylor 1984a).

The high coal contents of slurry and tailings means that there are big differences between fundamental properties of coarse and fine

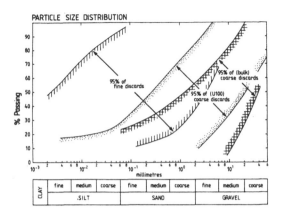

PARTICLE SIZE DISTRIBUTION

Figure 2. Showing grain size distributions of coarse and fine discards.

discards (Table 2), although they both classify mainly as silts and inorganic clays of low to medium plasticity on the Casagrande Plasticity Chart. However, about 50% of all discard samples are non-plastic in character. Specific gravities of fine discards range from 1.4 to 2.6, with a mean of 1.94, whilst coarse discards have a mean of 2.31 and a range of 1.6 to 2.8. The content of (waste) coal is an important control on all properties in which weight is involved.

Natural water content is, of course, greatly different in the two discard types (mean 11.0% tips, 34.8% lagoons) and so too is bulk density - mean $1.902 Mg/m^3$ tips, $1.457 Mg/m^3$ lagoons. It is of interest that compared with an average voids ratio of 0.82 in lagoon sediments, the old, loose tips compacted under self-weight to an average void ratio of 0.34. Indeed, the average compaction achieved in this manner was about 95-96% of the average obtained in the BS compaction test (2.5kg rammer).

Earlier work (Taylor 1975) showed that bulk density correlated with the clay minerals content, and to a lesser extent with carbonates (siderite). Density shows a strongly antipathetic relationship with coal as might be expected. Natural water content was found to

correlate with 10Å clays in particular, and unexpectedly with the coal component. It is suspected that this latter correlation could reflect the presence of coal of low diagenetic rank (higher moisture content) which was unsaleable in the past. Plasticity index correlated positvely with the clay minerals component as would be expected.

3 SHEARING RESISTANCE

Effective shear strength parameters are of greater import than undrained strength in the construction of tips and embankments, because the rate of construction is customarily slow enough for excess pore water pressures to dissipate. Laboratory investigations have shown that drained shear strengths are practically independent of initial density in all but clay-rich coarse discards. Initially high moisture contents had no discernable effect on the shear strength attained. Hence, for construction purposes drained shear strength can be regarded as being independent of layer thickness so long as the particle size distribution does not change significantly during construction. It is during transportation from the washery and during spreading on a heap that appreciable breakdown takes place, not after burial. Moreover, apart from certain clayey discards, a high rate of brittleness does not appear to be a feature of these non-marine mudrock fills.

The importance of coal content is illustrated by the histograms in Fig. 3 which indicate that effective shear strength parameter, ϕ', of fine-grained, coal-rich slurry is higher than that of coarse discard, which in turn is higher than clay-rich tailings. Statistically, there is also a good case for adopting a curved shear strength envelope passing through the origin, ie c'=0. For all discards a power fit has been adopted, viz:

$$\tau = m(\sigma')^z$$

where: τ = shear stress
   σ' = effective stress normal to the plane of failure
   m and z = constants (z=1 signifies a linear envelope).

438

TABLE 2. Fundamental properties of coarse and fine discards

	COARSE DISCARDS (UNBURNT)					FINE DISCARDS (TAILINGS PLUS SLURRY)				
	Distribution type*	No. of tests	Mean or median	Std. Dev.	Percentiles 2.5% 97.5%	Distribution type*	No. of tests	Mean or median	Std. Dev.	Percentiles 2.5% 97.5%
Standard Penetration Test N	Nl	1179	22	13.13	- -	-	-	-	-	- -
Liquid Limit %	Nl	720	36.8	6.92	- -	Sk	373	38.0 (38.8)	(7.63)	25.0 59.0
Plastic Limit %	Sk	720	21.0 (21.4)	(4.0)	16.0 31.0	Sk	362	23.0 (23.5)	(4.5)	16.0 34.0
Specific Gravity	Nl	683	2.31	0.20	- -	Nl	409	1.94	0.24	- -
Natural Moisture Content %	Sk	1267	11.0 (12.0)	(6.0)	4.7 27.0	Nl	390	34.8	12.9	- -
Bulk Density Mg/m³	Sk	1102	1.902 (1.863)	(0.219)	1.356 2.209	Sk	321	1.457 (1.492)	(0.233)	1.112 2.057
Void Ratio	Sk	547	0.34 (0.39)	(0.19)	0.14 0.90	Nl	221	0.82	0.27	- -
Ash Content %	Sk	324	75.8 (73.1)	(11.2)	47.6 88.3	Nl	154	52.0	16.4	- -
Clay Size %	Sk	82	5.0 (6.7)	(6.7)	1.0 18.0	Nl	259	16.0	10.6	- -

* Sk = Skew distribution, Nl = normal distribution

NOTE THAT MEANS AND STANDARD DEVIATIONS ARE SHOWN IN BRACKETS WHEN MEDIAN IS APPROPRIATE MEASURE OF CENTRAL TENDENCY.

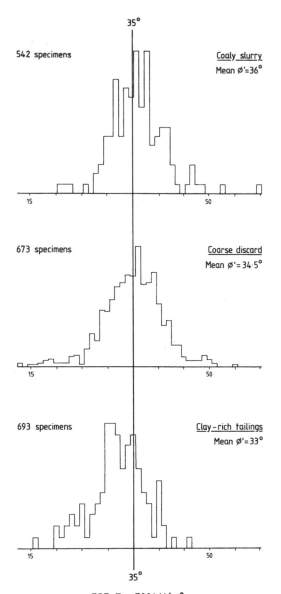

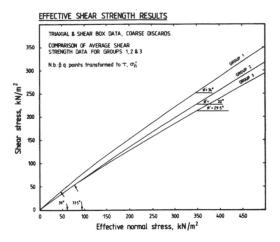

Figure 4. Statistical groups for UK coarse discards - peak effective shear strength (average Mohr envelopes shown).

Figure 3. Statistically valid decrease in shear stength parameter, ϕ' ($c'=0$), from coal-rich slurry to coarse discard to clay-rich tailings.

Coarse discards fall into 3 statistically valid groupings, the means for which are shown in Fig. 4. These groups are undoubtedly a reflection of the diagenetic rank of the associated coal seams (depth of burial) and certain similarities in mineralogy between coalfields. For example, the stronger group 1 materials include 'high rank' (deeply buried) seams in the Durham, Northumberland and South Wales coalfields, together with finer-grained, lower rank discards from Scotland. The latter is related on a mineralogical basis since Durham, Northumberland and Scottish discards are very similar; they contain an abundance of the geotechnically inert clay mineral, kaolinite. At the other extreme, the group 3 (weaker) discards are made up almost entirely of low rank materials (coal rank code numbers generally 700-900). These rocks and coals were generally under less overburden cover than the aforementioned during their geological history.

All fine discards have been grouped together in Fig. 5; there is no significant difference between the shear strengths of coarse and undifferentiated fine discards. It is only when fine discards are divided into slurry and tailings that the ranking order of shear strength emerges: slurry (strongest), coarse discard, tailings (weakest) - Fig. 3.
The residual shear strength of coarse discards is concluded to be a function of mineralogy, geological history (depth of burial), the normal stress range used in the

TABLE 3.   Residual shear strength parameters coarse discards
----

	Number of test sets	Minimum value	Mean value	Maximum value
C'_r, kN/m^2	28*	0	20.1	35
ϕ'_r degrees	30*	9	17.5	37

----

   *   Excluding extremes

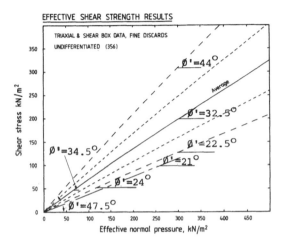

Figure 5. Peak effective shear strength of all UK fine discards (tailings plus slurry).

test, the shear displacement attained, and possibly the type of equipment used ie. reversing box as opposed to ring shear (see Section 4.9 in Taylor 1984a). Curvature is a feature of the residual Mohr envelope of coarse discards, but because of computer storage restrictions it was not allowed for in data processing. Consequently, a linear shear strength envelope has resulted in a high cohesion intercept shown in Table 3; it is unlikely to be valid. The range of apparent ϕ'_r values applicable to most stability problems ranges from 9 to 37 degrees (Table 3). It is representative of a wide spectrum of discards varying from weak, clayey (under-clay) types, to indurated highly siliceous siltstones commonly containing disseminated clay-ironstone (siderite). Values of ϕ'_r in Table 3 generally tend to cluster

between 10 to 21 degrees.

One hundred and seventy-eight coarse discard and 318 fine discard specimens were processed in a total stress analysis using purely linear envelopes. The requisite shear strength parameters are given in Table 4. It will be observed that both types of discard indicate a largely frictional behaviour, with mean parameters as follows:

$\phi_u = 21°$,
$c_u = 12.4$kN/m^2
(coarse discard)

$\phi_u = 24.5°$,
$c_u = 2.1$kN/m^2
(fine discard)

Undrained shear strength parameters can be used in stability analyses to test the sensitivity of total stress conditions, in the possible event of excess pore water pressures failing to dissipate during construction, which is customarily slow.

4   LAGOON SEDIMENT PARAMETERS

A noticeable difference was found between tailings and slurry in terms of piezometer response. Artesian and elevated water tables close to the lagoon surface were recorded in lagoons containing tailings, whereas in better-drained coal-rich slurry, water tables grouped within the 2m to 6m depth range below the surface. Lagoon investigations were an important component of the research programme in view of overtipping implications and the excavation of some coal-rich slurries (for sale) by machines standing on the floor of a lagoon.

TABLE 4. Total stress shear strength parameter

	No. Specimens	Apparent Cohesion (kN/m^2)	Angle of Shearing resistance (degrees)
COARSE DISCARDS			
1. Up 2nd stan. error est.		19.4	38.5
2. Probable maximum envelope		c.25.0	33 - 34
3. Up 1st stan. error est.		13.8	29.5
4. Mean	178	13.3	21.0
5. Lr, 1st. stan. error est,		11.2	13.5
6. Lr. 2nd. stan. error est.		9.6	6.0
FINE DISCARDS			
1. Up. 2nd stan. error est		66.8	38.0
2. Up. 1st stan. error est.		32.0	31.0
3. Mean	318	2.3	24.5
4. Lr. 1st stan. error est.		(-24.4)	(18)
5. Lr. 2nd stan. error est.		(-49.6)	(12)

Note: ( ) negative cohesion intercepts theoretically incorrect

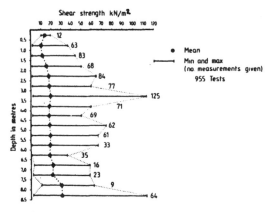

Vane shear strengths in colliery lagoons

H = 2D vanes

Figure 6. Peak vane shear strength distribution at 0.5m intervals, taken from 11 different colliery lagoons.

4.1 Vane shear strengths

More than 1500 vane tests were carried out in lagoons of which 955 measurements are shown as a composite plot in Fig. 6 at 0.5m depth intervals. Down to a depth of about 6.5m the minimum shear strength measured at all depths can be very low indeed (1 to $2kN/m^2$). Near to the surface of a lagoon, however, both minimum and average strengths increase, which is probably a reflection of the desiccated crust that usually forms on most lagoons in dry weather. Below a depth of about 3m, Fig. 6 indicates that the average shear strength remains remarkably constant down to a depth of about 7m where it begins to increase. Maximum peak vane strengths are more erratic because they commonly involve buried desiccated surfaces. Nevertheless, Fig. 6 shows an increase from $20kN/m^2$ at the surface to over $60kN/m^2$ at 2.5m depth. At greater depths the maximum shear strength range is 40 to $75kN/m^2$. Considering that the data in Fig. 6 were taken from 11 different lagoons, strength variations do not appear to be as great as individual vane profiles had suggested.

Bearing in mind the minimum strengths in Fig. 6 and indications of under-consolidation signified by elevated piezometer levels (especially in tailings), it was necessary to devote considerable time to the analysis of vane profiles (see Cobb 1977, Kirby 1980). Three useful trend lines have been adopted for diagnostic purposes. Peak vane strengths of the more argillaceous sediments can lie on the $c/p' = 0.3 (\pm0.1)$ line, where c is the measured vane shear strength (cohesion) and p' is the vertical effective pressure at test depth (see Lambe and Whitman 1969). In contrast, remoulded shear strengths tend to be distributed around Skempton's (1957) $c/p' = 0.11 + 0.0037$ PI line, where PI is the plasticity index. Peak strengths falling below this line may well signify under-consolidated sediments. As a third alternative, granular slurry of high permeability and densified crusts from previous filling levels have been found to follow an $s = p' \tan \phi'$ trend, where s is the shearing resistance of the sediment and ϕ' the effective stress angle of shearing resistance.

Vane tests and the monitoring of excess pore water pressures are important guides used in overtipping operations. Field trials on this aspect were reported at AIT by Taylor et al, 1980. Subsequent to these field trials at Maltby and Silverhill Collieries it has been possible to develop stability curves as a guide to overtipping operations (Kirby 1980; also A.R. Bacon in Taylor 1984a). Based on a high water table (0.75m below the surface) stability analyses and field trials indicated that for a lagoon to be overtipped, the minimum peak shear strength (H = 2D vane) within any layer must exceed 4.5kN/m^2. Vane profiles invariably dictate that earthmoving plant of low bearing capacity must be used in overtipping operations (for example, a Caterpillar D6 dozer).

For the purpose of excavating sediments by a machine standing on the floor of a lagoon a number of stability assessments have been made, once again using field vane tests. The limiting height of a free-standing temporary face, z, is about:

$$z = \frac{(\pi+2)}{\gamma} \quad , \quad \text{where } \gamma$$

is the unit weight of the sediment.

This simplifies to: $z \simeq c_{min}/3$

Because the profile being excavated is usually concave towards its base, the minimum horizontal shear strength is more significant because failure surfaces can translate into a planar movement along the interface between strong and weak laminations near the base of the excavation. We have found that for lamination thicknesses typical of UK lagoons an H = $D/4(14^o)$ diamond-shaped vane is convenient for measuring the peak shear strength in a horizontal direction. An appropriate equation of torque for this vane using a parabolic stress distribution was found to be as follows (Kirby 1980):

$$T = \frac{\pi D^3 c}{7} \sec \alpha \quad \text{where:}$$

α = 14° in the case of H = D/4 diamond-shaped vane

D = diameter of vane
c = horizontal shear strength
T = torque

4.2 Consolidation and permeability

The relatively high level of self-weight compaction attained in old tips has already been commented upon. Even so, it was rather surprising to find that 40 to 50% of the coefficients of (triaxial) consolidation (c_{vt}) were less than 10m^2/year. Ninety-five per cent of the values ranged from 1.3 to 650m^2/year, with the median being 6.3m^2/year. Only about 25% of values were greater than 100m^2/year. In contrast, specimens made up in a loose state for liquefaction tests had c_{vt} values ranging from 83 to 1128m^2/year (average 379m^2/year), within the confining pressure range 30 to 250kN/m^2.

Fine discards tested mainly in standard oedometers gave a median value of c_v (25.30m^2/year) which was four times that of coarse discards from old tips. Moreover, the range encompassing 95% of all values is much greater (2.87 - 1432m^2/year). Nevertheless, only 22.5% of the c_v values were in excess of 100m^2/year.

Research projects in which a large (254mm) Rowe Cell was used have indicated significantly higher c_v

values than suggested by standard 20mm thick oedometer specimens. On a small-scale it has been recognised that individual laminations tested in an oedometer can exhibit c_v values spanning the range $2m^2/year$ (clay bands) to $2000m^2/year$ (sand size coal). Rowe Cell comparisons for sediments from different parts of a lagoon show c_v values of 200 to $1000m^2/year$ (more typically 400-$600m^2/year$) at the inlet end, which compares with $2 - 40m^2/year$ (typically $2-10m^2/year$) for clay-rich sediments at the outlet. No indication of anisotropy in lagoon sediments was recorded in the Rowe Cell tests, which is probably an indication of the discontinuous nature of individual laminations in the sediments on the scale of a tailings lagoon.

Coefficients of volume compressibility (m_v) of lagoon deposits are remarkably constant across an individual lagoon. Some 80 tests were analysed (within a vertical stress range up to $1000kN/m^2$) and were found to range from $m_v = 0.002 - 2.0m^2/MN$, with a median value of $0.4m^2/MN$. In conventional soils this would encompass stiff clays and normally consolidated alluvial clays.

The distribution of permeability in tips has been found to be bimodal with two distinct peaks at $k = 10^{-3}$ to $10^{-4}m/s$ and $k = 10^{-7}$ to $10^{-8}m/s$. The overall range for lagoons is not greatly different to that of tips ($k = 1$ to $10^{-9}m/s$ tips, 10^{-1} to $1^{-8}m/s$ lagoons). Most of the lagoon permeabilities however, group within the range 10^{-2} to $10^{-6}m/s$. All the values quoted were field determinations, although it was generally found that laboratory constant head tests only underestimate field values by about one order of magnitude. Values calculated from standard oedometer tests are up to four orders of magnitude lower than field tests.

Information on horizontal and vertical permeabilities of fine discards on the scale of a lagoon is incomplete. Tentative data from electrical analogue modelling suggest that lateral drainage to the banks is not inconsequential, despite predominant upwards drainage, and in some lagoons, drainage through the floor into permeable bedrock or sandy (Drift) deposits.

It is intended to conduct further analysis of flow through embankments in order to investigate more rigorously the interrelationship between the sediments and embankment materials. In particular the interface between the two is important and very low permeabilities have been measured at this junction (Cobb 1977).

Experience within the industry has indicated that the types of embankment detailed in the Technical Handbook (National Coal Board 1970) can probably be reduced to a single model. This is largely because the flatter slopes resulting from planning controls raises fewer problems in respect of stability.

5. LIQUEFACTION INVESTIGATIONS AND FURTHER WORK ON POREWATER CHEMISTRY

No mention has been made of laboratory work or field trials concerned with liquefaction. Details have previously been reported at an AIT symposium (Taylor et al, (1980), and fuller accounts are given elsewhere (Taylor 1984 a and b).

Monotonic controlled load triaxial tests indicated that weaker coarse discards under very low effective stresses can exhibit partial liquefaction behaviour. However, the equivalent depth of discard in a tip affected would only be about 3m. More importantly, the densities used in tests in which the phenomenon was observed were significantly less than those achieved by the modern construction methods referred to in the introduction.

Tailings and slurry specimens formed from pillar and U100 samples did not liquefy in cyclic loading triaxial tests designed to simulate a 200-year return period British earthquake. Because a few remoulded coal slurry samples did, field trials were carried out. These involved the construction of overtip embankments, one being on tailings of high sensitivity (=4.2). Explosives were also used to generate ground motions (Taylor et al, 1980). The results of these trials indicated that liquefaction as a consequence of overtipping operations, or earth tremors, is extremely unlikely. It was also estimated that lagoon sediments with a plasticity index greater than 10 are unlikely to liquefy, even when thoroughly remoulded.

Work is still progressing at Durham University in connection with the restoration of active tips. A project nearing completion is an assessment of the rate of leaching of chlorides from coarse discard embankments. At some collieries, particularly in the eastern coalfields, deep strata waters can be very high in chlorides (e.g. > 100 000mg/l as Cl). Since residual brines in a colliery embankment discard can induce physiological drought in young vegetation it is important to have some knowledge of their distribution and leaching rates in an active tip or embankment under restoration. Continuous monitoring indicates that chlorides increase in the surface layers in dry summer weather when soil suctions in excess of $80kN/m^2$ can be generated. In wet winter periods the profiles reverse with chlorides increasing downwards from the surface into the tip to a depth of 0.5m or so. There is however, an overall decay in Cl concentration with time and preliminary analyses suggest that in many tips low concentrations can be attained by natural leaching in about 5 years.

A much more severe problem concerning the reclamation of the older generation of 'loose tips' was the development of acidity and sulphates as a consequence of pyrite oxidation. Consequently, botanical investigations of tolerant species and neutralising treatments were carried out in some coalfields (see, for example, Bradshaw and Chadwick 1980).

6 CONCLUSIONS

This brief review of some of the more important engineering properties of colliery discards in the United Kingdom indicates that these materials are much more systematic in behaviour than might have been expected 20 years ago.

The frictional character of coarse discards under both effective and total stress test conditions is notable. Moreover the influence of (waste) coal in enhancing shear strength is such that coaly slurry from lagoons is stronger than equivalent coarse discard of sand and gravel sizes. On a regional basis the influence of the geological history of burial (diagenesis), and probably some clay mineral control, has been revealed by statistical treatment of shear strength data.

Given good quality control, coarse discards are little different from other 'shale fill' construction materials.

The greater understanding of fine discards (tailings and slurry), which detailed analyses have helped to elucidate, are of import in both excavation and lagoon overtipping operations. In situ vane testing in association with push-in piezometers are a useful guide in assessing the feasibility of these operations.

Flatter (environmental) slopes with an emphasis on land use economy and progressive restoration favour the development of composite disposal structures in the future.

ACKNOWLEDGEMENTS

The work described was carried out under research contracts awarded to the writer by the National Coal Board. The views expressed in the paper are not necessarily those of the board in all cases.

REFERENCES

Bishop, A.W. 1973. The stability of tips and spoil heaps. Q.Jl. Engng. Geol. London., 6: 335-387.

Bradshaw, A.D. & Chadwick, M.J. 1980. The restortion of land, London: Blackwell Scientific Publications.

Cobb, A.E. 1977. Stability and degradation of colliery shale embankments and properties of tailings lagoon deposits. Unpublished Ph.D. thesis, University of Durham, England.

Department of Transport, 1976. Specification for road and bridge works, London: HMSO.

Kirby, J.M. 1980. Shear strength, consolidation and drainage of colliery tailings lagoons. Unpublished Ph.D thesis, University of Durham, England.

Lambe, T.W. & Whitman, R.V. 1969. Soil mechanics. New York, John Wiley & Sons.

McKechnie Thomson, G. & Rodin, S. 1972. Colliery spoil tips - after Aberfan. London: Instn. Civ. Engrs., (Paper 7522).

National Coal Board Technical Handbook, 1970. Spoil heaps and lagoons, Second draft, London: National Coal Board.

National Coal Board, 1971. NCB (Production) Codes and Rules, Tips. First Draft, London: National Coal Board.

Potter, P.E., Maynard, J.B. & Pryor, W.A. 1980. Sedimentology of shale, New York: Springer - Verlag.

Sherwood, P.T. & Ryley, M.D. 1970. The effect of sulphates in colliery shale on its use in roadmaking. RRL Report LR 324, Road Research Laboratory, Ministry of Transport.

Skempton, A.W. 1957. The planning and design of the New Hong Kong Airport. Discussion. Proc. Instn. Civ. Engrs., 7: 305-7.

Spears, D.A., Taylor, R.K. & Till, R 1971. A mineralogical investig- ation of a spoil heap at Yorkshire Main Colliery. Q.Jl. Engng. Geol., London. 3: 239-252.

Taylor, R.K. 1975. English and Welsh spoil heaps - mineralogical and mechanical interrelationships. Engineering Geology, 9: 39-52.

Taylor, R.K. 1984a. Composition and engineering properties of British colliery discards, London: Mining Department, National Coal Board, London, 244pp.

Taylor, R.K. 1984b. Liquefaction of seabed sediments: triaxial test simulations. Proc. Symposium, Seabed Mechanics IUTAM, Newcastle upon-Tyne, Sept. 1983 (Ed. B. Denness), 131-38.

Taylor, R.K., Kirby, J.M. & Lucas, J.M. 1980. An investigation of overtipping a colliery lagoon. Proc. Internat. Conf. Engineering for Protection from Natural Disasters, Bangkok. (Eds. P. Karasudhi, A.S. Balasubramaniam and W. Kanok-Nukulchai), 629-641.

Welsh Office, 1969. A selection of technical reports submitted to the Aberfan Tribunal. London: HMSO.

Woodland, A.W. 1968. Field geology and the civil engineer, Proc. Yorks. Geol. Soc., 36: 531-576.

Environmental Geotechnics and Problematic Soils and Rocks, Balasubramaniam et al. (eds)
© 1987 Balkema, Rotterdam. ISBN 90 6191 785 9

New process of sewage sludge treatment and utilization of slag for construction materials

S. Iwai & Y. Miura
Nihon University, Funabashi, Japan

T. Kawakatsu
Osaka Gas Co. Ltd, Japan

ABSTRACT: The molten slag made from sewage sludge by the coke bed type melting method has properties suitable for construction materials. It is known that interlocking blocks using the molten slag can produce good pavement.

1 INTRODUCTION

In Japan, efficient sewage systems prevail widely in the large cities. On the other hand, large amounts of sewage sludge are produced and now their disposal is becoming the serious problem.

Sewage sludge has been disposed of by means of their utilization for fertilizer, filling land or, after burning, as ash. However, more effective measures of sewage sludge disposal are now required due to recent increased use of chemical fertilizers, difficulty in securing reclaimed land around large city areas and protection against secondary environmental pollution.

Not only simply treating and disposing of the sewage sludge, but also it is necessary to use suitable treatment measures in order to utilize sewage sludge as construction materials. Simply burning the sewage sludge results in a powder which is difficult to handle. Also, the chemical composition of the ash means it is difficult to utilize as it is.

Accordingly, a coke bed type melting process, which burns and melts the sewage sludge and solidifies the molten ash was developed and a demonstration plant has been built to examine the economic and operational performance of the process. It became clear that stabilized treatment of the sewage sludge can be carried out by this process. The molten ash is cooled for solidification after it flows out of the furnace. This solidified molten ash is called molten slag. Depending on the method of cooling, two types of slag are obtainable, one being air-cooled slag and the other water-crushed slag.

Both have been used as construction materials in tests, and it has been decided to utilize them for the surface course aggregate of interlocking blocks for a new paving material of unusual and elegant appearance. The interlocking block paving has the advantage of easy repair and producing almost no waste material during repair work. Thus, its use as paving materials is on the increase. In order to confirm the merits of modified interlocking block in which molten slag is used, its production and trial pavement was made.

2 MELTING THE SEWAGE SLUDGE

There are the following advantages if sewage sludge is treated by the melting method, rather than the burning method;

1. The reduction in weight and reduct volume ratio are large.

2. It is easy to handle because the molten slag is in grain or lump form.

3. Although sewage sludge contains some harmful heavy metal, this is settled in the molten inorganinc components (SiO, CaO etc.), and will never melt out.

4. As the inorganic components in the sewage sludge are molten and solidified, they can be used as aggregate and/or paving material.

Furthermore, there is no need to put a great deal of work for securing reclamation area for disposal inland and the disposal cost of sewage sludge can be eliminated, if molten slag can be utilized efficiently.

Table 1. The melting method for the sewage sludge treatment (After Okada (1983)).

Type of furnace	Object of melting	Description	Major energy
Reverberatory furnace	Dried sludge (Water content: About 10%)	The dried sludges are placed around the core of furnace in the shape of the doughnut, then the furnace is fired. As the upper portion of the wall of inside of the furnace is heated, it functions as some type of reverberatory furnace and keeps the materials in melting . The materials are heated by the burner until the furnace gets constant operating condition. Since then, the temperature is controlled by means of varying the height of the top of furnace.	Fuel oil (+ Electric power)
	Burnt ashes	With fixing the burnt ashes inside of furnace, the furnace is heated by the burner. As the inside upper portion of furnace is heated and functions as the reverberatory furnace, it maintains the materials in melting condition. Because the burnt ashes have no heat, this systems depend on the fuel from outside source for supplying energy.	
Arc discharge furnace	Dried sludge	If the electric power is charged in between the carbon electrode and the base material, the arc discharge occures with generating the heat. Thus this system maintains the materials in melting condition.	Electric power
	Burnt ashes	– ditto –	
Whirl flow furnace	Dried sludge (Water content: About 5%)	Inducing the whirled gas flow with the sludge and the air along the outside of cylindrical furnace, and making the furnace burn completely by means of extending the flow retention time so that the inside wall of furnace being heated to remain the materials in melting condition.	Electric power (For drying)
Microwave irradiation furnace	Burnt ashes	When microwave is irradiated to the burnt ashes (a dielectric material), the wet ashes are caused the rapid oscillating for self-generating heat to be molten. The temperature can be controlled by adjusting the output of microwave.	Electric power
Coke bed furnace	Dried sludge (Water content: About 40 – 50%)	Charging the sludge and coke alternately into furnace while securing the draft and supplying the energy to maintain the materials in melting condition. Melting the burnt ashes can also be performed by this mean.	Coke

Table 2. Production costs for sewage sludge slag (After Japan Sewage Ass'n (1983)).

Melting method	Surved organization	Sludge to be treated		Scale of treatment		Melting temperature (°C)	Cooling method	Utility cost		Maintenance cost (yen/t)	Depreciation cost (yen/t)
		Type of sludge	Water content (%)	Sludge charging (t/yen)	Slag production (t/yen)			Cost of sludge charged in melting furnace per dry weight (yen/t)	Cost of slag produced per unit weight (yen/t)		
Microwave irradiation furnace	Yokohama city	Burnt ashes	0	3.8	3.6	1,450	Slow cooling	16,800	17,700	14,900	5,100
Arc discharge furnace	Kawasaki city	Dried sludge	18.1	17.6	4.3	1,450 – 1,550	Water cooling	85,500	286,600	12,800	
Coke bed furnace	Osaka city	Dried sludge	55	14.6	45.4	1,500	Water cooling	17,100	24,700	4,100	7,700
Reverberatory furnace	Kobe city	Dried sludge	20	2.1	0.33	1,400	Water cooling	15,200	77,400	34,800	16,900
Reverberatory furnace	Kobe city	Burnt ashes	0 – 0.04	2.0 – 2.9	1.9 – 2.9	1,320 – 1,440	Slow cooling	21,500 – 41,300	21,800 – 47,000	19,000	42,900

449

2.1 Sewage sludge metling method

Several melting methods can be used for sewage sludge: the reverberatory furnace method, arc discharge furnace method, whirl flow furnace method, microwave irradiation method and coke bed method, as shown in Table 1 (see Okada (1983)). Table 2 shows the scale and cost of the treatment methods in Japan (see Japan Sewage Ass'n (1983)). As compared the production cost with the other methods, the cost of molten slag production by coke bed melting method is the lowest, especially the maintenance cost. It can be said that sewage sludge treatment using the coke bed type melting method will be nearly economical overall.

2.2 Coke bed type melting method

The sewage sludge melting process with coke as the fuel as a substitute for petroleum fuel was developed as new process for treating sewage sludge. A primary demonstration plant was built to evaluate this process and has a sewage sludge treatment capacity of 20ton/day. After a thorough investigation of operational performance and the characteristics of the molten slag, it was confirmed that stabilized processing had been achieved. A second demonstration plant with a higher capacity (sludge treatment capacity 50 ton/day) was built to investigate operational efficiency and to study the effec-

tive methods for utilizing molten slag at the sewage treatment plant. The flow diagram of this process is shown in Figure 1. A full view of the secondary demonstration plant is shown in Figure 2.

As shown in Figure 1, the sewage sludge melting system consists of a sludge feeding and drying system, sludge melting system, slag treating system, heat recovery system and exhaust gas treating system.

The sludge to be treated (dehydrated cake, water content 75-80%) is pressurized and fed into the drier by feeding pump. The dried sludge (water content 40-50%) is charged into the melting furnace with specific amounts of coke and flux added. The combustibles contained within the sludge become exhaust gas by burning and exhaust to the atmosphere after recovering the heat, filtering dusts, and desulfurizing. The inorganic matter contained in the sludge is melted by the heat from the coke and continuously flow from the bottom of the furnace to become either air-cooled or water-crushed slag.

3 THE PROPERTIES OF MOLTEN SLAG

The molten slag is the molten-solidified ash of the molten sewage sludge. Two types of molten slag can be produced. When the molten ash flowing from the furnace is cooled in the air, it becomes rock shaped lumps. This is called air-cooled slag. When cooled by water, as it is cooled

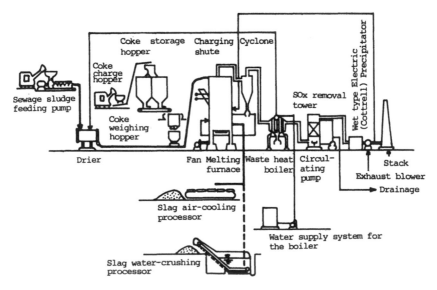

Figure 1. Flow diagram of the coke bed type melting process.

Figure 2. Coke bed type melting plant.

Table 3. Type of the sewage sludge for treatment.

		Dehydrated cake		Sludge treatment method
		Type	Water content (%)	
A		Mixed raw lime-sludge cake	79.4	Mixed raw sludge → Added lime & chloride iron agent → Vacuum dehydration
B		Mixed raw polymeric-sludge cake		Mixed raw sludge → Added polymeric agent → Centrifugal dehydration

Table4. Chemical composition of molten slag.

Chemical component	SiO_2	CaO	Al_2O_2	FeO	MgO	P_2O_5	MnO	TiO_2	K_2O
Content (weight %)	39.4	32.2	14.7	4.1	3.5	1.4	1.1	0.74	0.54

rapidly and crushed during the cooling down, the molten ash solidifies into a grain-like form called water-crushed slag. The color of the molten slag is black or dark brown and looks like obsidian.

The chemical composition of the molten slag varies depending on the composition of the inorganic components which are sewage sludge, flux, and coke ashes. The chemical composition of the sewage sludge varies largely depending on the method of sludge treatment in the sewage treatment plant. The sludge treated by this demonstration plant is, as shown in Table 3,

made of two types of sludge mixed in equal quantity. The chemical composition of the molten slag from this melting plant is shown in Table 4.

3.1 The properties of air-cooled slag

When the air-cooled slag is to be used as the construction materials, it should be controlled by the process shown in Figure 3.

After the molten ash is cooled and solidified by the air slow-cooling device,

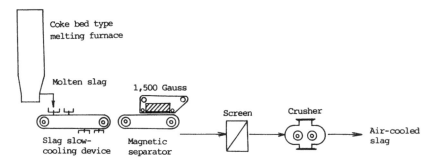

Figure 3. Flow diagram of air-cooled slag proportioning process.

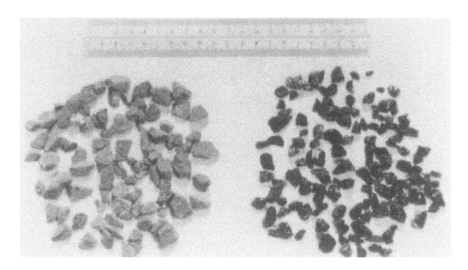

Figure 4. Crushed stone (2.5-5mm) (left) and air-cooled slag (right).

Table 5. Comparison of test results of air-cooled slag on aggregate.

Test \ Material	Crushed stone (from Ako)	Air-cooled slag
Fineness modulus (F.M.)	6.60	6.59
Decantation test (%)	0.61	0.07
Specific gravity (surface dried)	2.62	2.87
Absorption ratio (%)	0.852	0.241
Unit weight (kg/l)	1.53	1.71
Solid volume percentage (%)	58.9	57.2
Soundness test (%)	3.9	14.6
Abrasion test (%)	16.1	32.2

it becomes molten slag, after which large-sized pieces of metal in the molten slag are removed by the magnetic separator. After sieving slag particles sized 10-25mm by screen, crushing them in the crusher, the slag can be used as construction materials. Figure 4 shows the controlled air-cooled slag.

Table 5 shows the comparative results of the air-cooled slag and crushed stone brought from Ako. The gradation of the air-cooled slag is proportioned according

Table 6. Gradation of JIS-A-5011 air-cooled iron-blast-furnace slag aggregate and air-cooled slag.

Sieve size (mm)	25	20	10	5	2.5
Test result	100	96.0	31.6	9.1	3.9
Allowable range of gradation (%)	100	90 - 100	20 - 5	0 - 10	0 - 5

Table 7. Modified CBR test results.

	Plasticity index (PI)	Max. dry density (g/cm^3)	Opt. water content (%)	Degree of compaction (%)	Modified CBR (%)
Air-cooled slag	N.P.	2.38	4.8	95	106
Mechanical stabilized crushed stone	N.P.	2.16	7.0	95	89
JIS-A-5011 crushed stone for road construction	——	——	——	——	Over 80[*]

[*] Manual for design and construction of asphalt pavement (Japan Road Ass'n (1980)).

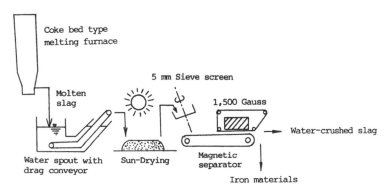

Figure 5. Flow diagram of water-crushed slag proportioning process.

to the gradation of air-cooled iron-blast-furnace slag aggregate for concrete specified by JIS-A-5011 as shown in Table 6. The crushed stone is also proportioned to the same gradation. Table 7 shows the results of the modified CBR test for air-cooled slag. As the above data make clear, the air-cooled slag can stand comparison with the crushed aggregate.

3.2 The properties of water-crushed slag

Water-crushed slag for construction mate-rials should be processed as shown in Figure 5.

The molten ash flowing from the furnace is charged into the water spout and, in the process of cooling, is solidified to molten slag which is crushed to grain. After drying, the impurities and foreign matter are removed by 5mm mesh screen, then, iron pieces in the slag are removed by magnetic separator. Figure 6 shows the water-crushed slag.

The aggregate properties of the water-crushed slag are shown in Table 8. Regarding particle size distribution, although

Figure 6. Water-crushed slag (>5mm).

Table 8(a). Gradation of water-crushed slag.

Sieve size (mm)	Passing (%)
10	100
5	100
2.5	90.3
1.2	49.5
0.6	15.3
0.3	5.5
0.15	1.7
Fineness modulus	3.39

Table 8(b). Test results of water-crushed slag.

Test		Value
Decantation test	(%)	0.22
Specific gravity (surface dtied)		2.84
Absorption ratio	(%)	0.93
Unit weight	(kg/l)	1.60
Solid volume percentage	(%)	56.7
Soundness test	(%)	8.7
Chloride test	(%)	0
PH test (H_2O and KCl solution)		9.1 & 8.7

these are large size particles in comparison with ordinary fine aggregate, it seems that the slag has properties similar to ordinary aggregate.

4 INTERLOCKING BLOCKS USING WATER-CRUSHED SLAG

As mentioned above, it is now realized that water-crushed slag has properties similar to ordinary aggregate, and that it has a distinctive black or dark brown color. To make use of these colors, using water-crushed slag for the aggregate of surface-course mortar for interlocking blocks (hereinafter called ILB), newly designed ILB were produced and tested.

4.1 Interlocking blocks(ILB) and their advantages

An ILB is one with irregularities at the sides providing an interlocking function with the blocks arranged side by side, with a resultant load distribution effect. Thus, ILB are used not only for sidewalks but for roadways and the paving in factory yards of under heavy traffic loads (see Proc. of CBP (1984)). It appears that the range of applications of ILB paving with continue to expand.

The ILB have some advantages in terms of production, repair work and use of water-crushed slag as aggregate.

1. ILB have a simple shape, are easy mass-produce. Thus, costs will be low.

Table 9(a). Mix proportions for surface-course mortar.

Slag content (%)	Water cement ratio (%)	White cement (kg/m³)	Calcium carbonate (kg/m³)	Water-crushed slag (kg/m³)	Crushed white granite (kg/m³)	Water (kg/m³)	Total (kg/m³)
10				225	1223		2290
30	29.4	510	182	675	815	150	2332
50				1126	408		2376
70				1576	—		2418

Table 9(b). Mix proportions for base-course concrete.

Water cement ratio (%)	Cement (kg/m³)	Fine aggregate (kg/m³)	Coarse aggregate (kg/m³)	Water (kg/m³)	Total (kg/m³)
30.0	400	1139	871	120	2530

Table 10. Results of test on the materials for the ILB.

	Type of material	Material	Specific gravity	Finess modulus (FM)	Max. particle size (mm)	Absorption (%)
Surface course	Cement	White cement	3.07	—	—	—
	Fine aggregate	Calcium carbonate	2.56	—	0.3	—
		Water crushed slag	2.85	3.39	5.0	0.93
		Crushed white granite	2.58	5.40	10.0	0.7
Base course	Cement	Ordinal grade portland cement	3.15	—	—	—
	Fine aggregate	Sand	2.71	2.50	5.0	Less 3.0
	Coarse aggregate	Mechanical stabilized crushed stone	2.50	4.00	15.0	Less 3.0

2. As ILB are produced by hydraulic press equipped with a vibrator, sufficient strength is obtained.

3. When underground laid facilities such as gas pipelines, electric power cables, telephone cables, waterworks, sewage pipelines and so on under ILB pavement have to be maintained, the repair work of pavement is easy with almost no waste materials, because the ILB at the maintenance location can be taken up and, after completion of the repair of the underground facilities, replaced.

4. The water-crushed slag can be used for aggregate as it is, immediately. It does not need secondary treatment as the burnt ash and air-cooled slag do. Also the black or dark brown slag can be used as colored aggregate as artificially colored aggregates.

5. By changing the proportions in the water-crushed slag to be mixed, ILB of different color lightnesses can be produced. Thus, the color design for pavement can be varied. If necessary, it is also possible to color the ILB by using pigments.

4.2 Producing ILB used water-crushed slag

The newly designed ILB using water-crushed

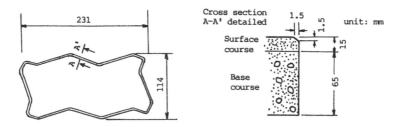

Figure 7. The shape and dimensions of the ILB.

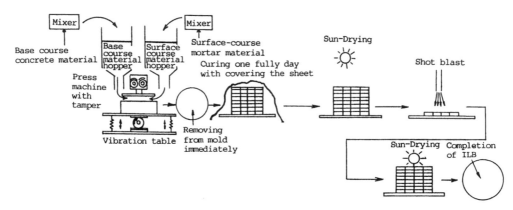

Figure 8. Flow diagram of the ILB production process.

Figure 9. Block-making machine.

slag was produced to comfirm above advantages. The mix proportions of the ILB are as shown in Table 9. The ILB consist of two parts, surface course and base course.

The aggregate of surface-course mortar is made by mixing the water-crushed slag with crushed granite, paying attentions to the color so that the ILB can look like natural stone. Calcium carbonate is added to compensate the fine aggregate and prevent shrinkage cracks in the surface-course mortar. Also by means of varying the quantity of water-crushed slag to be mixed, different ILB lightnesses have been produced. The targeted compressive strength of the ILB is 600kgf/cm^2 at full area loading. Table 10 shows the characteristics of the materials. The shape and

456

dimensions of the ILB are shown in Figure 7. The thickness of the surface-course mortar and the base course are 1.5cm and 6.5cm, respectively.

Figure 8 shows the production process of the ILB. The block-making machine with a hydraulic press equippped with vibratory tamper and vibration table is used to form the blocks. The formed blocks are removed from the mould immediately and cured by hydration heat for one full day under a covering sheet. After that, they are cured by sun-drying in the open yard. After 2 weeks curing in the yard, the ILB are shot-blasted to get a natural appearance. Figure 9 gives a side view of the block-making machine.

4.3 The properties of the ILB

A cylindrical specimen is taken from the newly designed ILB by core boring machine to perform the compressive strength test. As Figure 10 shows, thin plywood and teflon sheets coated with silicone oil are attached to both ends of the specimen to remove the edge friction, after attach, the compressive strength test is performed by unversal testing machine.

Figure 11 shows the relationship between the compressive strength and dry density. It is noticed that the compressive strength increases in proportion to increasing the dry density. Also it exceeds the targeted compressive strength which is converted to the compressive strength of the specimen.

The relationship between the compressive strength and water-crushed slag content is

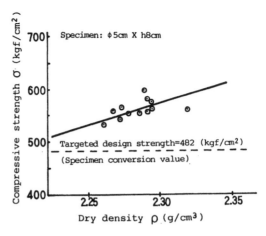

Figure 11. Relationship between compressive strength and dry density.

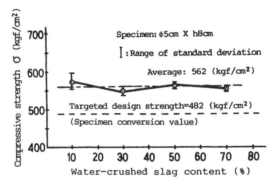

Figure 12. Relationship between compressive strength and water-crushed slag content.

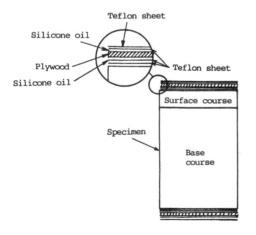

Figure 10. Edge treatment for the specimen (Plywood capping).

shown in Figure 12. The compressive strength is expressed in terms of average value, and the standard deviation is also illustrated in the figure. In general, it seems that compressive strength does not depend on the quantity of water-crushed slag content, and that the dispersion of compressive strength is small.

Figure 13 shows the relationship between the water absorption ratio and dry density. As the dry density increases, the water absorption ratio decreases. Thus, it is assumed that as the water absorption decreases, the compressive strength will increase.

These test results meet the quality standards of the ILB generally used in Japan (ILB Ass'n (1984)). It is clear that these ILB can be put to practical use.

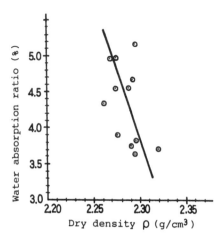

Figure 13. Relationship between water sorption ratio and dry density.

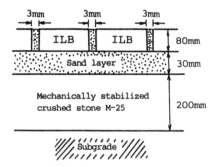

Figure 14. Pavement structure.

Figure 15. ILB with different lightness.

Figure 16. Repair work (removing the ILB).

4.4 Trial pavement using the new ILB

Trial paving was made using the new ILB with water-crushed slag as the aggregate of the surface-course mortar. Color-conditioning was investigated for appearance and ease of repair was also investigated. A structure of the trial pavement is shown in Figure 14.

By means of arranging ILB with different lightnesses, the contrasts in the road surface was inspected. Figure 15 clarifies that the paving surface can be accented sufficiently by using different lightnesses of color only. In comparison with conventionally-colored ILB, it has a subtler appearance. Especially when the new ILB are wet, the contrast between white and black are set off to provide pleasant elegance. Such ILB contains about 30% to 50% water-crushed slag.

Figures 16 and 17 show the ease of removal and replacement of ILB pavement when underground repair work is required.

Until now, when repair work has been performed under conventional pavement, waste materials such as removed lumps of concrete or asphalt concrete have been produced, but with the ILB pavement there is almost no waste material. Other advantages are that the existing paving materials at the repair site are reused economically and there are no unsightly indications of repair left at the repaired section, after repair works are completed.

Figure 17. Completion of repair work (re-placing the ILB).

5 CONCLUSION

A coke bed type melting process to burn and melt sewage sludge and solidify the molten ash was developed for the treatment of sewage sludge. A demonstration plant was built to investigate economic and operating performance of the process at the sewage treatment plant. It was confirmed that the process can be operated stably with treated sewage sludge at maximum rate 50ton/day.

Air-cooled slag (in lump state) and water-crushed slag (in grain state) can be obtained. These slag properties stand comparison with generally-used aggregate.

The new ILB with water-crushed slag used as aggregate for the surface-course mortar has almost the same strength as ILB in general use. In addition, it became clear that different colors of ILB can be produced by varying the quantity of the slag to be mixed. It may become possible to design good color-conditioning for roads and to supply materials which can be developed more favorable pavement for road user. It is clear that the some of these advantages can be added to the original merits of interlocking blocks in general use. Thus, it is clear that molten slag, especially water-crushed slag has higher value added for surface-course mortar of ILB, and that sewage sludge can be utilized as a resouce for construction materials.

ACKNOWLEDGEMENT

The authors gratefully acknowledge the kind cooperation and assistance in the development of the coke bed type melting process from members of the Osaka Prefecture, Sewage Technology Improvement Research Association (Chairman, Prof. Masakatsu Hiraoka, Kyoto University).

REFERENCE

Okada, K. 1983. Treatment and disposal of sewage sludge. Enviroment-Pollution News (in Japanese).
Japan Sewage Association 1983. Treatment of sewage sludge for construction materials (in Japanese).
Proceedings of Concrete Block Paving 1984. 2nd International conference on Concrete Block Paving.
Japan Road Association 1980. Manual for design and construction of asphalt pavement.
Interlocking Block Association in Japan 1984. Outline of interlocking block marketing (in Japanese).

Section E
Problematic soils and rocks

Environmental Geotechnics and Problematic Soils and Rocks, Balasubramaniam et al. (eds)
© 1987 Balkema, Rotterdam. ISBN 90 6191 785 9

Strength of a compacted mudstone

R.H.Chen & J.J.Wang
National Taiwan University, Taipei

ABSTRACT: The mudstone in southern Taiwan is a weak rock. However, it seems likely
that compaction of the excavated mudstone will improve its engineering behavior, and
then it can be used. The tested sample is unweathered mudstone obtained from a new cut
along a road. The engineering properties measured from the tests are strength, compressibility, and swelling pontential. In order to simulate short-term and long-term conditions, both as-compacted and soaked tests are performed. Statistical correlation between the engineering properties and testing parameters are then established. Hence,
the prediction of engineering properties of compacted mudstone can be made from the
laboratory results.

1 INTRODUCTION

The area of mudstone in southern Taiwan
is more than 100 Km^2. This mudstone is
very stiff when dry but easy to be slaked
in the water. Because of its poor engineering characteristics and bad land use,
erosion problem in the mudstone area is
very serious. Furthermore, due to the high
salinity of this sedimentary weak rock,
agricultural development in this area is
unsuccessful. On the other hand, it is
expected that there will be large amount of
excavations in mudstone for residential
area in the future. It seems practicle to
compact the excavated mudstone to improve
its engineering behavior so that it can be
used as the subgrade of the road or as an
embankment material.

The purpose of this study is to find out
the feasibility of using crushed mudstone
as an embankment fill. Hence, in considering the stability of an embankment, two
critical conditions are concerned, i.e.,
end-of-construction (short-term condition)
and long-term situation. It is hoped, that
the relationship between the behaviors at
two critical conditions and the initial
compactive parameters can be developed from
the laboratory results.

2 BEHAVIOR OF A COMPACTED EARTH DAM

The most important factors for a material

to be considered as the fill of a compacted
earth dam are its durability and shear
strength. The mudstone being studied is
stiff but nondurable and hence is not a
good material for a rock-fill dam (Surendra
& Lovell, 1981). However, it is possible
to be compacted and used for constructing
the soil-fill dam. Then, its slaking
characteristic, strength, compressibility,
and swelling potential are of main concern.

For a compacted earth-fill dam, its short-
term and long-term behaviors are quite
different. In short-term condition, the
shear strength of the fill is contributed
partly by the capillary force (Lowe, 1967).
This negative pressure will decrease due
to evaporation or saturation of water.
Hence, to check short-term stability of a
dam, the undrained shear strength of compacted samples in both as-compacted and soaked conditions must be known.

In long-term condition, the dam will consolidate and settle down. The mechanism of
consolidation of a dam is a combination of
consolidation by self-weight, swelling due
to saturation, broken down and rearrangement
of aggregates, etc. The void ratio in the
fill will change due to consolidation which
will affect the long-term characteristics
such as shear strength, the amount of
swelling or settlement. These long-term
behaviors are possibly controlled by the
initial compactive factors such as initial
degree of saturation, dry density, water
content, and compactive effort. In the

following paragraphs, the test program for developing the relationship between testing parameters will be described.

3 TEST PROGRAM

The test program includes compaction tests, unsaturated unconsolidated undrained (UUU) and saturated unconsolidated undrained (SUU) triaxial tests, consolidated undrained triaxial test with pore water pressure measurement ($\overline{CU}$), and hydrostatic compression (HC) test. The preparation of the compacted samples follows the procedure suggested by Head (1982). The grain sizes of the samples are those passing #40 sieve.

3.1 Compaction test

The purpose to perform this test is to develop the relationship among compactive effort, water content, and dry density.
There are three levels of compactive efforts: modified AASHTO, standard AASHTO, and low compactive effort which is equivalent to two-fifth of standard AASHTO. The compacted samples with water contents at O.M.C. (optimum water content), O.M.C. + 2%, and O.M.C. -2% will be used in the following tests.

3.2 UUU triaxial test

From this test the short-term shear strength of as-compacted mudstone is determined. Test procedure may refer to that of Bishop & Henkel (1964). Compacted samples are tested under the confining pressures of 1.0 kg/cm^2, 2.0 kg/cm^2, and 4.0 kg/cm^2.

3.3 SUU triaxial test

This study is to find out the effect of saturation on the short-term shear strength of as-compacted samples. Test procedure is the same as that of UUU test, while method of saturation follows that of Chan (1983).

3.4 $\overline{CU}$ triaxial test

The long-term shear strength of compacted mudstone is obtained from the result of this test. Test procedure follows that of SUU test. The consolidation pressures of 1.0 kg/cm^2, 2.0 kg/cm^2, and 3.0 kg/cm^2 are applied on the compacted samples to simulate the consolidation of soil by the overburden and saturation of water.

3.5 HC test

This test will determine the swelling and consolidation characteristics of the compacted mudstone.

4 TEST MATERIAL

Test samples were obtained from a new cut near the town called Nanhwa in southern Taiwan. Samples are unweathered, dark gray, medium stiff, and no obvious joint on the surface. The contents of silt and clay are 68% and 32% by weight, respectively. The soil indices are liquid limit LL = 29.0, plastic limit PL = 12.9, and specific gravity G_s = 2.74. The mudstone is classified as CL soil according to the Unified Soil Classification System.

5 TEST RESULTS

The test results are presented below. Statistical analysis is also made on these parameters so that the engineering characteristics of mudstone can be correlated.

5.1 Compaction characteristics

The compaction curves of three energy efforts are shown in Fig. 1. It can be seen that these curves (aggregates passing #40 sieve) are very similar to typical compaction curves of soil. For constant compactive effort, dry-of-optimum sample has more water deficiency, and thinner double layer. The attractive force between particles induces the structure to be flocculated (Lambe, 1958). The flocculated structure prevents particles from moving to a more oriented structure by compaction. In the mean time, capillary force within aggregates developed at low water content restrains the aggregates from moving to a denser condition. Hence, for dry-of-optimum sample, low dry density is obtained at low water content when the compactive effort is constant. As the water content increases, the double layer expands more and the attractive force between soil particles becomes small and the particles move more easily and become more orientally. Consequently, the void ratio is reduced and the dry density increases.
The three-dimensional diagram of water content, dry density, and compactive energy ratio (compared to standard AASHTO) is shown in Fig. 2. The contours are the curves of constant water content. In order to explain the result easier, the projection of Fig.2 on the plane of dry density-energy ratio is

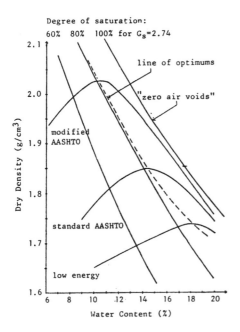

Figure 1. Compaction curves of mudstone

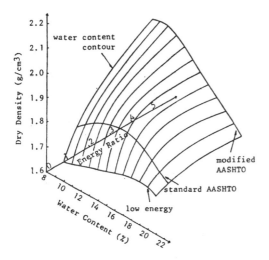

Figure 2. Relationship among dry density, water content, and compactive effort

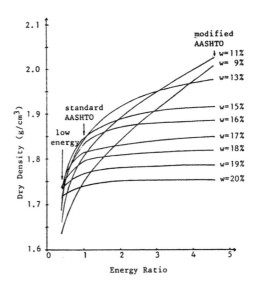

Figure 3. Dry density vs. energy ratio at constant water content

shown in Fig. 3. It is clear that, at low water contents (9% — 11%), compactive effort is very effective in increasing the dry density. As the water content gets higher (w > 13%), dry density gets lower even for the same compactive effort. This is because at high water content, the degree of saturation is so high (for example, at w = 20%, Sr = 92%, 95%, and 97% for low, standard, and modified AASHTO, respectively) that the voids in the sample may not be continuous and hence the entrapped air is difficult to be expelled out by compaction

The behaviors of compacted samples on dry-of-optimum and wet-of-optimum are quite different, the regression analyses are thus made separately. The combination of parameters were chosen from those of Liang & Hu (1984):

(a) dry-of-optimum

$$\gamma_d = 1.489 + 0.931 \ \sqrt{E_r}/w + 0.0006 \ w^2/\sqrt{E_r} \tag{1}$$

(b) wet-of-optimum

$$\gamma_d = 2.499 - 0.045 \ w + 0.0004 \ w^2 \tag{2}$$

where γ_d is the dry density in g/cm^3, w is the water content in per cent, E_r is the energy ratio. The coefficients of correlation (R^2) are 0.9175 and 0.9177 for equations (1) and (2), respectively. The above two equations can be used to predict the dry density if the water content and compactive effort are known.

5.2 Short-term shear strength

The short-term shear strength of compacted soil are functions of its water content, dry density, and soil structure. The UUU shear strength of compacted mudstone under

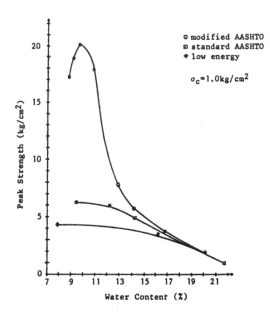

Figure 4. Peak strength vs. water content
(σ_c = 1.0 kg/cm^2)

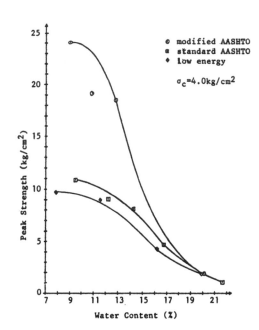

Figure 6. Peak strength vs. water content
(σ_c = 4.0 kg/cm^2)

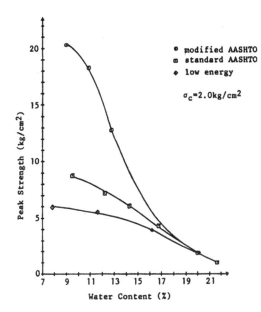

Figure 5. Peak strength vs. water content
(σ_c = 2.0 kg/cm^2)

various confining pressures are shown in Figs. 4-6. The figures show that the shear strength decreases as water content increases. This is partly because the structure of compacted mudstone is floc-culated at low water content and is dispersed at high water content. Besides, more water will destroy the capillary force, produce higher pore water pressure, and reduce the effective stress. Figs. 4-6 also show that the shear strengths for various compactive efforts are almost identical at high water content.

In order to make comparison easier, Figs. 4-6 are replotted on Figs. 7-9. These figures show that, at the same water content, shear strength increases with dry density. This phenomenon is more significant at low water contents. As the water content equal to about 20%, the phenomenon is not obvious.

Figs. 10-12 show that, for constant dry density, shear strength reduces as water content increases. Note that the same dry density implies that the contact area on the shear plane is about the same. However, as water content increases, the repulsive force between particles and the pore water pressure induced during shearing are higher and hence the shear strength drops.

Since the confining pressure (σ_c) has different effect on the shear strength of samples compacted at different compactive effort, the regression analysis is thus separated for various confining pressures. Equations (3)—(5) represent the shear strength at failure $(\sigma_1-\sigma_3)_f$ in terms of dry density, water content, and initial degree of saturation.

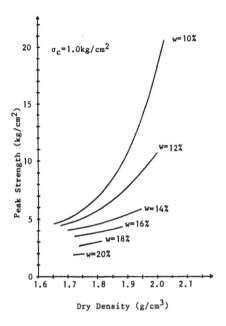

Figure 7. Peak strength vs. dry density at constant water content (σ_c = 1.0 kg/cm^2)

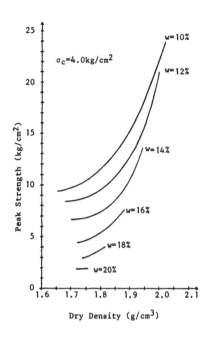

Figure 9. Peak strength vs. dry density at constant water content (σ_c = 4.0 kg/cm^2)

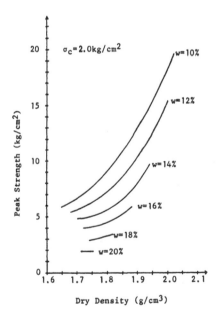

Figure 8. Peak strength vs. dry density at constant water content (σ_c = 2.0 kg/cm^2)

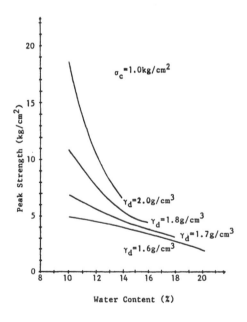

Figure 10. Peak strength vs. water content at constant dry density (σ_c = 1.0 kg/cm^2)

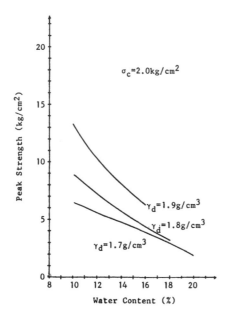

Figure 11. Peak strength vs. water content at constant dry density (σ_c = 2.0 kg/cm^2)

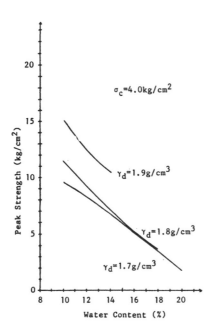

Figure 12. Peak strength vs. water content at constant dry density (σ_c = 4.0 kg/cm^2)

(a) σ_c = 1.0 kg/cm^2

$$(\sigma_1-\sigma_3)_f= 2.390+15.53\gamma_d\sqrt{S_i}/w-2.060w+0.069w^2$$
$$(R^2 = 0.88) \tag{3}$$

(b) σ_c = 2.0 kg/cm^2

$$(\sigma_1-\sigma_3)_f=-6.544+17.35\gamma_d\sqrt{S_i}/w-0.842w+0.026w^2$$
$$(R^2 = 0.97) \tag{4}$$

(c) σ_c = 4.0 kg/cm^2

$$(\sigma_1-\sigma_3)_f=-5.974+20.48\gamma_d\sqrt{S_i}/w-0.880w+0.019w^2$$
$$(R^2 = 0.96) \tag{5}$$

Where S_i is the initial degree of saturation, the shear strength is in kg/cm^2. The shear strength parameters (c,ϕ) can be obtained from the envelope of Mohr's circles constructed at various confining pressures by equations (3)—(5).

5.3 Long-term shear strength and volume change

The long-term shear strength and the volume change characteristics of compacted soil are affected by the initial compaction conditions. Fig. 13 presents the volume change of samples compacted at standard AASHTO under hydrostatic compression condition. The samples swell most under unconfined condition (σ_c = 0). The consolidation is significant for the samples under confining pressure and the critical consolidation pressure (under which has no volume change) is smallest. The water sucked by the compacted sample is partly for saturating the sample and partly for swelling. Fig. 14 shows that the dry-of-optimum sample has higher swelling potential than that of wet-of-optimum sample if the sucked water is deducted. Fig. 15 shows the comparison of shear strength at SUU condition with that at UUU condition under various confining pressures. The decrease of shear strength due to saturation is obvious. This phenomenon is due to several reasons: suction of water which increases the void ratio and reduces the contact area between particles, hydration of cations on the surface of particles, and the cementation that may be destroyed. Because the dry-of-optimum samples swell most, consequently the reduction of shear strength in them is most significant.

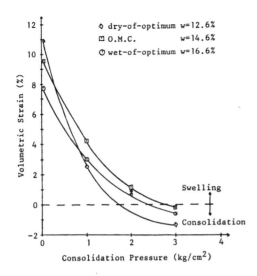

Figure 13. Volumetric strain vs. consolidation pressure for dry-of-optimum samples compacted by standard AASHTO

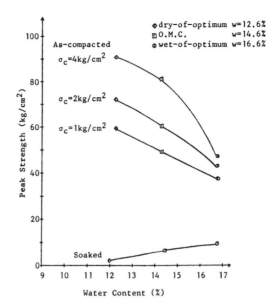

Figure 15. UUU & SUU peak strengths vs. water content

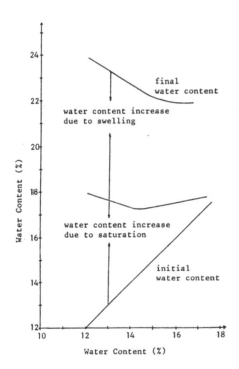

Figure 14. Water content increase due to swelling and saturation.

6 CONCLUSIONS

1) For the crushed mudstone studied (passing #40 sieve), its compaction characteristic is very much like that of soil. At low water content (w = 10%), compaction has significant effect on the increasing of dry density while at high water content (w = 20%), the effect is little.

2) From statistical analysis, the dry density of dry-of-optimum sample is functions of compactive effort and water content; while the dry density of wet-of-optimum sample has little connection with the compactive effort (see equations (1) and (2)).

3) The undrained shear strength of compacted mudstone at the same water content increases with dry density, but decreases as water content increases for samples of constant dry density.

4) The shear strength parameters (c, ϕ) of unsaturated compacted mudstone can be obtained from equations (3)—(5) with initial compaction conditions are known.

5) The unsaturated shear strength of compacted mudstone are higher on dry-of-optimum than on wet-of-optimum. However, dry-of-optimum sample after soaking has less shear strength than wet-of-optimum after soaking.

6) Sample on dry-of-optimum swells more when

soaked and the critical consolidation pressure is less.

From the above conclusions, it is suggested that the mudstone be compacted on wet-of-optimum so that the reduction of shear strength and swelling amount after soaking can be reduced.

ACKNOWLEDGEMENTS

The authors gratefully acknowledge the financial support from National Science Council, Republic of China. Thanks are also extented to my colleagues, Prof. Wu, W.T. and Mr. Tso, T.H. for their valueable suggestions and to Profs. Hung, J.J. and Yen, C.L. who initiated this program. Mr. Kok, K.N. helped a lot in the laboratory work.

REFERENCES

Bishop, A.W. and Henkel, D.J. 1964. The measurement of soil properties in the triaxial test. London: Edward Arnold.

Chan, C.K. 1983. Back pressure saturation. unpublished.

Head, K.H. 1982. Manual of soil laboratory testing. Cornwall: Engineering Laboratory Equipment Limited.

Lambe, T.W. 1958. The structure of compacted clay. JSMFD, ASCE, 84: 1654-1 to 1654-34 and 1655-1 to 1655-35.

Liang, Y. and Hu, B.L. 1984. Study of strength control of compacted laterites. Report of National Science Council, R.O.C.

Lowe, J. 1967. Stability analysis of embankments. JSMFD, ASCE, 93: 1-34.

Surendra, M. and Lovell, C.W. 1981. Additives to limit slaking in compacted shales. Proc. Int. Symp. on Weak Rock: 225-230.

Environmental Geotechnics and Problematic Soils and Rocks, Balasubramaniam et al. (eds)
© *1987 Balkema, Rotterdam. ISBN 90 6191 785 9*

Improvement of dispersive soils by mixing with Bangkok clay or bentonite

D.T.Bergado
Asian Institute of Technology, Bangkok, Thailand

K.-Y.Kang
Technical Research Institute, Daewoo Engineering Co. Ltd, Seoul, South Korea

ABSTRACT: Several dispersive soil dams in Northeast Thailand have failed through surface and piping erosion. These dispersive soils are inorganic sandy, silty clay or clayey silt of low plasticity with a high content of dissolved sodium in the pore water. The addition of bentonite has no effect on the dispersivity. On the other hand, the addition of from 20% to 30% Bangkok clay transforms the soil into non-dispersive. The crumb test initially indicates the dispersivity of the soil. However, pinhole test and chemical tests are necessary to confirm the dispersivity. The liquid and plastic limits increase linearly with increasing percentage of Bangkok clay. The maximum dry density decreases and the optimum moisture content increases with increasing percentage of Bangkok clay.

1 INTRODUCTION

Dispersive soils are clayey silty soils which have a highly erodible nature and a higher content of dissolved sodium in pore water than ordinary soils. They are eroded by a process in which the individual colloidal clay particles go into suspension in practically stilled water. In Thailand, several earth structures have encountered problems due to internal erosion of dispersive soils. Recent research has shown that chemical stabilization is used frequently to improve these soils. In this study, dispersive soils from three separate locations in Thailand, namely: Phitsanulok canal site, Lam Sam Lai dam site, and Ampun dam site were studied. Both the original dispersive soils and the soil mixed with Bangkok soft clay or bentonite were tested using dispersive identification tests, index tests, and tests for physical and engineering properties. The purpose of this study is to evaluate the dispersivity and the physical properties of the original dispersive soil, the dispersive soil-Bangkok clay mixture, and the dispersive soil-bentonite mixture.

2 DESCRIPTION AND IDENTIFICATION OF DISPERSIVE SOILS

Dispersive soils contain clay fraction which has high potential for a dispersive state when the soil mass interacts with water. The dispersion occurs when the repulsive force (electrical surface forces) between individual clay particles exceeds the attractive forces (van der Waals attraction) so that when the clay mass is in contact with water, individual clay particles are progressively detached from the surface and go into suspension (Sherard et al, 1972). Dispersive soils have sodium as the main factor that contributes to their dispersivity. Holmgreen and Flanagan (1977) describe the reason why sodium causes soil dispersivity. The effectiveness of sodium as a dispersing agent is related to its ionic charge of plus one. For the same negative charge on a clay particle, an ion swarm of univalent sodium contains twice as many ions as does a corresponding swarm of divalent ions such as calcium and magnesium ions. Osmotic potentials are thus greater in sodium systems because osmotic potential is proportional to the number of ions in the double layer. Furthermore, because of its single valence, the coulombic attraction of a sodium ion to the charged particle surface is less than that for polyvalent ions. Sodium ion swarms therefore range further from the particle surface, forming a thicker double layer. Lambe and Whitman (1969) report that as the negative charge on a clay particle is balanced by the

Table 1 Criteria for Dispersion Ratio Test

Degree of Dispersion (%)	Classification
Less than 35	Nondispersive
35 to 50	Moderately Dispersive
50 to 75	High Dispersive Potential
Greater than 75	Extremely Dispersive

cations in the double layer, the two advancing particles begin to repel each other when their double layers come into contact with each other. The repulsive force between the adjacent particles for any given spacing is therefore directly related to the sizes of the double layers. Thus, a net repulsive force is more probable for a sodium ion system. Decker and Dunnigan (1977) indicate that dispersive soils in nature commonly have alkaline pore water with pH greater than 8.5.

The dispersion test is an indicator test developed by the Soil Conservation Services (Volk, 1937) to evaluate the susceptibility of soils to erosion. A grain size curve determined by the ordinary hydrometer test is compared with the one that is determined by the hydrometer test without chemical dispersant. This test should be used only as a general guide to the identification of the erodibility of clayey soils. The dispersion ratio (D) is defined as the ratio of the percent finer than 0.005 mm diameter without and with chemical dispersant. The criteria for soil classification are tabulated in Table 1 as set by the Soil Conservation Service (Sherard et al, 1972).

Emerson (1967) describes a rapid field test for the evaluation of the dispersive nature of clays, called the crumb test. The test consists of dropping a small, moist (natural moisture) clod of soil into a clear beaker of distilled water or 0.001 normal sodium hydroxide (NaOH) solution or both. If the soil is dispersive, a colloidal cloud develops around the periphery of the clod. The tendency for the clay particles to go into colloidal suspension is observed after 5-10 minutes of immersion. Sherard et al (1976b) developed a dispersivity rating system as follows:

Grade 1 - no reaction: crumb may spread on the bottom of the beaker without a sign of cloudy water caused by colloids in suspension.

Grade 2 - slight reaction: the water gets slightly cloudy at the surface of crumb.

Grade 3 - moderate reaction: an easily recognizable cloud of colloid in suspension, usually spreading out in thin streaks on the bottom of the beaker without totally covering the bottom area.

Grade 4 - strong reaction: colloidal cloud covers approximately the whole bottom of the beaker in a very thin skin. In extreme cases, all water becomes cloudy.

The pinhole test was developed for identification of dispersive soils by Sherard et al (1976a) and modified by Coumoulos (1977) and Schafer (1978). As seen in Fig. 1, distilled water is caused to flow through a 1.0 inch (25.4 mm) long specimen with water content near the plastic limit. The water is caused to flow under a hydraulic head of 2 to 40 inches (50.8 mm to 1020 mm) through a 1.0 mm diameter hole. The classification of the soil is based on the appearance of the water, the rate of flow, and final size of the hole in the specimen. The criteria for evaluating the pinhole test result are shown in Table 2.

The soil chemical test procedure is covered in the Agriculture Handbook No. 60 edited by Richards (1954). An empirical ratio between the Exchange Sodium Percentage (ESP) and Sodium Adsorption Ratio (SAR) was found. The relative abundance of mono-and divalent ions is manifested as the SAR ratio as follows:

$$SAR = \frac{Na^+}{\sqrt{0.5\ (C_a^{++} + Mg^{++})}} \qquad (1)$$

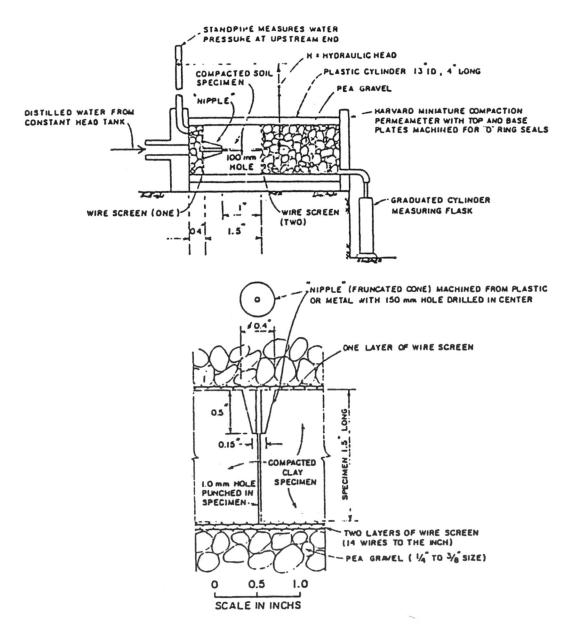

Fig. 1 Pinhole Test Apparatus and Section through Pinhole Test Specimen (After SHERARD et al, 1976a).

in which Na^+, Ca^{++}, and Mg^{++} refer to the concentrations of the designated soluble cations in the porewater expressed as milli-quivalents (meq) per liter. Richards (1954) HAS RELATED THE ESP to SAR by the following regression equation:

$$ESP = \frac{-0.0126 + 0.01475\ SAR}{1+(-0.0126 + 0.01475\ SAR)} \times 100 \quad (2)$$

The soil is classified as dispersive when it falls into the group of nonsalinealkali soil shown in Table 3.

3 ENGINEERING PROBLEMS OF DISPERSIVE SOILS

Engineering problems occur when dispersive

Table 2 Criteria for Pinhole Test

Classification	Description
D1 and D2	Dispersive soils: fail (5 x original diameter) rapidly under 2 inches head
ND4 and ND3	Intermediate soils: erode slowly under 2 inches or 7 inches head
ND2 and ND1	Nondispersive soil: no colloidal erosion under 15 inches or 40 inches head

Table 3 Classification of the Soil According to the Chemical Composition of the Soil Pore Fluid

Soil Group	ESP	EC (ms/cm)	pH	Dispersivity
Saline	< 15	> 4	< 8.5	Dispersive
Saline-Alkali	> 15	> 4	< 8.5	Nondispersive
Nonsaline-Alkali	> 15	> 4	> 8.5	Dispersive

NOTE: EC = electrical conductivity (mmhos/cm)

soils are used as construction material and/or foundation materials. Problems due to dispersive soils occur frequently on earth dams, earth embankments, irrigation canal banks, building foundations, etc. The damage could be any combination of surface erosion, piping, slaking, gully erosion, etc. as mentioned by past investigators (Sherard et al, 1972; Emerson, 1964; Nakano, 1967; Ingles and Aitchison, 1969; Cole and Lewis, 1960; Cole et al, 1977; Fernando, 1979; Shieh, 1981; Tai, 1981; Kim, 1982; Garcia,, 1983; Tai, 1983; Chen, 1984; and Kang, 1985).

Since 1970, a number of earth dams in Northeast Thailand have failed due to serious piping erosion, tunnelling erosion, and surface erosion mainly attributed to the presence of dispersive soils. The piping and tunnelling erosion occurs during the first reservoir filling. The surface erosion is caused by rainfall run-off. Cole et al (1977) summarized damage to dams in Thailand (Table 4) including the chemical and physical properties of the construction materials. The locations of these dams are given in Fig. 2. Shearard et al (1972) described how the typical dispersive soil piping failure in homogenous earth dams starts with a very small

initial leak in a narrow crack which erodes in a few hours into a tunnel of substantial diameter. The initial narrow cracks may occur due to drying and shrinkage of the plastic material, differential settlement, earthquake, hydraulic fracturing, geological weakness in the foundation or merely poor compaction. suggested that the mechanism for piping failures in earth dams could involve the dispersion of the clay fill material at the exit point of the percolating water, leading to the progressive loss of fine clay particles in suspension. Piping failures due to the action of percolating rainwater have been described as surface erosion.

Parker and Jenne (1967) report in detail the potential damage of piping to highway embankments. It was pointed out that piping damage originates from dessication cracks. Where the hydraulic gradient is sufficient, fine-grained sediments are transported in suspension along the crack to appear at an incipient pipe outlet in the embankment. This type of pipe may also occur as a result of localized subsidence due to saturation of surficial sediments forming sinks or stress cracks.

The damage due to dispersive soils also

Table 4 Summary of Test Results and Dam Behaviour

1	2	3	4	5	6	Pore Water Salt Concentration, meq/litre		9	10	11	12	13
Dam and Year of Completion	Number of Specimens	Sodium, %		Dispersion, %				Clay Size Minerals'	Compaction, Moisture	Water-tightness	Tunnel.** Erosion	Comments
		Mean	Range	Mean	Range	Mean	Range					
Lam Chieng Krai, 1972	29	84	74to95	49	5to68	40	4to93	Q I M	no tests	3 leaks, then failed	minor	piping consistent with dispersive soil, having salts above 15 meq/litre
Huay Sawal, 1972	39	71	14to96	36	0to67	6	1to17	Q K M	no tests	3 leaks, repaired	severe	2 leaks due to settlement cracks above spillway conduit
Lam Sam Lai, 1970	13	68	9to93	38	9to62	8	2to14	M(?) K	dry	failed	very severe	failure almost certainly due to dispersion piping
Ampun, 1972	16	66	19to95	29	1to64	5	1to11	Q K L	variable poor	failed at closure section	moderate	sodium high, but soil may not be dispersive
Huay Saneng, uncompleted	24	60	25to93	10	0to63	3	0to13	K70to90% M or M+V 10 to 30%	fair	water not stored yet	none	sodium high, but soil may not be dispersive
Nam Un, 1973	3	32	25to37	0	0	1	1to1	no test	good	excellent	none	not dispersive
Lam Phra Plerng, 1967	4	30	23to37	12	12to13	8	4to11	no test	good	some leakage through foundations	none	not dispersive

* M = Montmorillonite, K = Kaolinite, Q = Quartz, I = Illite, V = Vermiculite (Column 9).
** Very severe = Large deep tunnels in many places, Severe = Deep tunnels in some places,
Moderate -= Shallow tunnels, some places, Minor = Occasional Small tunnels (Column 12).

475

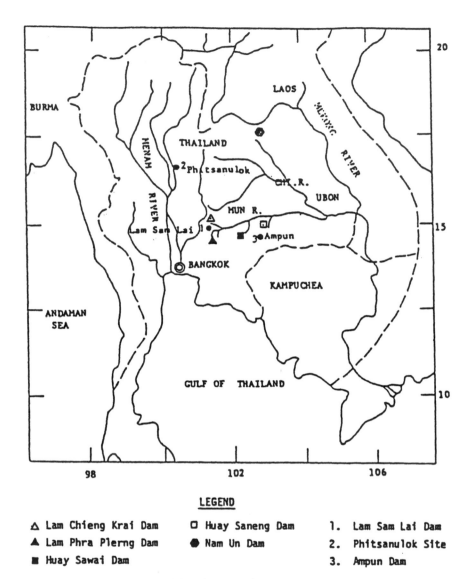

Fig. 2   Location Map

occurs in excavation cut slopes. Piping failure due to rainfall begins when rainwater enters dessication cracks in the upper parts of the slopes, percolates vertically downward, and then flows horizontally washing out dispersive clay in suspension. Irrigation canal banks made of dispersive soils are also damaged due to the interaction with the retained water. Such a situation occurred in the Lam Pao Irrigation Canal Project as reported by Tai (1983).

4 PHYSICAL, CHEMICAL, AND MINERALOGICAL PROPERTIES OF DISPERSIVE SOILS

The dispersive soils mostly encountered in Northeast Thailand have been studied previously (Cole et al, 1977; Fernando, 1979, Shieh, 1981; Kim, 1982; Tai, 1983; Garcia, 1983; Chen, 1984; and Kang, 1985). These soils have been classified predominantly as CL in the Unified Classification system, indicating sandy silt, clayey silt and silty clay of low to medium plasticity with

Table 5 Summary of Classification Test

Author	Location	Unified Classification	Description of Soil
FERNANDO (1979)	Lam Sam Lai Dam	SC-CL	Reddish brown silty with low to medium plasticity
	Lam Chieng Krai Dam	SM-CL	Yellow and brownish yellow silty clay and sandy clay
	Huay Sawai Dam	CL	low plasticity clay
SHIEH (1981)	Nakhon Rachasima	ML-CL	sandy-silty clay
(1982)	Ban Sawai	CL	reddish brown to light purple silty clay
(1983)	Lam Pao Dam	SM-SC	pinkish white nonplastic silty-clayey sand
(1984)	Khon Kaen University	CL	yellowish red silty clay

Table 6 Summary of Atterberg Limits Test

Author	Location	Liquid Limits (%)	Plastic Limits (%)
FERNANDO (1979)	Lam Sam Lai Dam	14.7-35.0	10.8-18.9
SHIEH (1981)	Nakhon Rachasima	37.3±0.5	12.5±0.8
KIM (1982)	Ban Sawai area	39.5	19.6
	Sukothai area	24.8	13.4
TAI (1983)	Lam Pao Dam	-	N.P.
CHEN (1984)	Khon Kaen University	24.3	13.8
GARCIA (1984)	Khon Kaen area	14.7-21.9	14.6-15.5

silty sand. The results of past investigations are tabulated in Tables 5 and 6. Dispersive soils have sodium as the predominant chemical in the pore water. Cole et al (1977) found an average sodium percentage of 66-80%. The pH value of dispersive soils ranges from 5.4-9.0 (Kim, 1982). In comparison, AIT clay is classified as silty clay of high plasticity (CH) with low pH of 3.9 and almost equal proportions of sodium (26%) compared to calcium and magnesium cations. The main clay-size minerals of dispersive soils are quartz and kaolinite (Chandra and Garcia,

1984; Cole et al, 1977). Shieh (1981) found that the percentages of clay minerals at Lam Sam Lai dam were 50-60% kaolinite and 20-25% for both montmorillonite and illite. AIT clay contains kaolinite and illite in almost equal amounts (33-38%). with less montmorillonite with chlorite and quartz (Chandra and Garcia, 1984).

5 METHODS OF STABILIZATION OF DISPERSIVE SOILS

Lime has been effectively used to

477

stabilize dispersive soils. The addition of hydrated lime tends to increase the total concentration of calcium cations and reduce the high sodium content controlling the dispersivity. The general order of replaceability of the cations is given by the lytropic series as $Na^+ < K^+ < Ca^{2+} < Mg^{2+} < Al^{3+}$. Any cation will tend to replace the cation to the left of it, and the monovalent cations are usually replaceable by multivalent cations. Pozzolanic reactions between the lime and clay particles producing calcium silicate hydrates may also increase the soil strength and the resulting cementation reduces soil erosion. The percentage of lime needed to render the soil non-dispersive ranges from 1 to 2 percent by weight (Fernando, 1979; Kim, 1982; Tai, 1983; Chen 1984; and Cole et al, 1977). Similar results were obtained by Haliburton et al (1975), Sherard et al (1972), and Logani and Hector (1979). Although the hydrated lime is effective, there is a disadvantage that the stabilized soil may become brittle and susceptible to cracking.

Van Olphen (1963) states that as the concentration of ions increases when sodium chloride is added, a charge reversal on the clay edges is likely to result because of anion absorption on the clay edges. Then further sodium chloride addition causes the concentration of sodium cations to increase in the clay surface as the absorbed anions are replaced and a flocculated structure results. The principal disadvantage is that, being soluble, salt may be leached out from the soil by subsequent rainfall. Chandra and Chen (1985) show that 1% sodium chloride gives maximum strength.

A 5% flyash has been found to yield optimum strength (Chandra and Chen 1985). Flyash is a waste product of burning powdered coal from thermal power plants. The effectiveness of the flyash depends on its pozzolanic activity because of its high calcium content. A 1% gypsum has been found by Chen (1985) to be most effective. Sherard et al (1976a) indicate that the additional calcium cations generated by the additions of gypsum decrease the percentage of sodium controlling dispersivity. A 0.5 percent aluminum sulfate has also been utilized to yield optimum results (Chen 1985). Apparently, the addition of aluminum ions decreases the percentage of sodium ions controlling dispersivity. Cement has been used as additive for controlling soil dispersivity (Tai, 1983). The addition of cement to dispersive soil would saturate the pore water with calcium hydroxide. Thus, the effect of lime released from cement hydration has an effect similar to adding lime to dispersive soil.

6 DISPERSIVE SOIL SAMPLE SITE DESCRIPTION

The soil samples in this study are collected from Phitsanulok irrigation canal embankment, Ampun dam site, and Lam Sam Lai dam site. These locations were suggested by the officials of the Royal Thai Irrigation Department, particularly Dr. Suphon Chirapuntu. The site locations are given in Fig. 2. The irrigation canal embankment is 176 km long with maximum height of 5 m and average width of 15 m. The embankment material consists of pale brown to yellow brown sandy silt.

The Ampun dam consists of homogenous earthfill embankment with a maximun height of 10 m, a crest length of 2.8 km and a crest width of 6 m. The classification of the embankment material is mostly clayey to silty sand. The natural soil below the base of the dam consists of brown sandy silt (ML) to silty clay of low plasticity (CL). A detailed section of the dam after repair is shown in Fig. 3. According to Cole et al (1977), during the first filling of the reservoir in 1972, a washout occurred at the location where the final closure was made. It was believed that the tunnel erosion was facilitated by the presence of undercompacted dispersive soils near the faces of the embankment.

The Lam Sam Lai dam is located at Prakong Village, Pak-Thongchai District, Nakhon Ratchasima Province. The dam has a maximum height of 14 m, crest length of 2.5 km, and a crest width of 7.0 m. The details of the dam cross-section are given in Fig. 4. The dam material consists of sand, silt, and clay particularly yellowish-brown silty to sandy clay. The dam failed by piping during the first reservoir filling when the dam embankment suffered severe tunnelling erosion.

7 SAMPLING AND PREPARATION OF TEST SPECIMENS

Prior to the collection of soil samples, crumb tests were performed to ensure that the soil is dispersive according to the guidelines in Table 1 as proposed by Sherard et al (1976b). For the laboratory tests, 50 to 70 kg of disturbed soil sample was taken from each of the three sites. The soil samples were taken at the location where severe erosion had

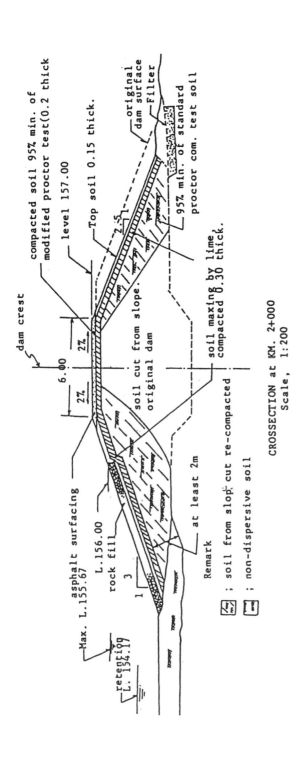

CROSSECTION at KM. 2+000
Scale, 1:200

Fig. 3 Cross Section of Improved Ampun Dam.

479

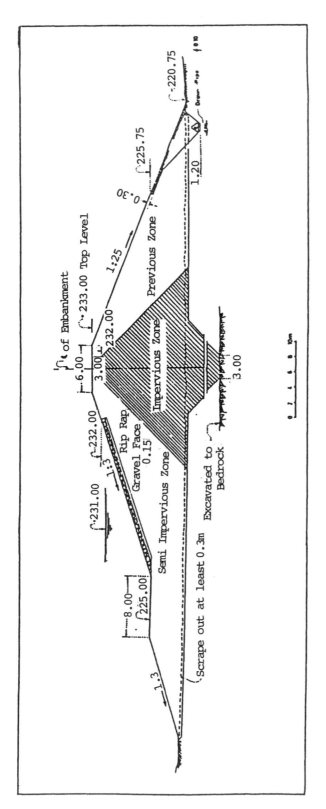

Fig. 4 Cross-section of the Lam Sam Lai Dam.

480

developed and where most dispersivity was shown according to crumb test performed with the help of officials from the Royal Irrigation Department of Thailand. At each site, about 0.80 m surface soil was first removed before samples were taken. In Phitsanulok, samples were taken from KM 40+300, Ban Ta Tong, Phitsanulok Province. In Lam Sam Lai dam, the samples were taken from the dam embankment. The sample of dispersive soil from Ampun dam was taken at KM 2+200 in the dam embankment. The samples of Bangkok clay were collected from a depth of 1.5 to 2.5 m using a helical auger at the location behind the ET Building inside AIT Campus.

The dispersive soil specimen was first air dried and pulverized by a rubber hammer. Then, the samples were seived through U.S. Standard Sieve No. 10, oven dried, and stored in plastic bags. The Bangkok clay samples were first oven dried. Then the dried clay was pulverized by placing in a rotating drum full of ball bearings. Finally, the samples were sieved through US Standard Sieve No. 10 and stored in plastic bags.

Before each test, the desired proportion of dispersive soils and Bangkok clay or dispersive soils and bentonite was mixed in dry condition. A predetermined amount of distilled water was then added gradually. Then the specimen was stored in the humidity room for 24 hours to allow for moisture equilibration. The specimens for the pinhole tests were compacted using the Harvard miniature compaction method.

8 LABORATORY TEST PROGRAM AND RESULTS

The testing program is outlined in Table 7. In the following sections, test procedures and results are presented and discussed.

9 PHYSICAL PROPERTIES

Index properties such as natural moisture content, Atterberg limits, specific gravity and grain size analysis with hydrometer analysis were established in both original and mixed soil samples. The grain size distribution of the dispersive soils from each site and of Bangkok clay are shown in Fig. 5. According to the Unified Classification System, the original soils from Ampun dam and Phitsanulok sites are both classified as inorganic clays of low to medium plasticity (CL) and the corresponding soil

from Lam Sam Lai dam is classified as well-graded silty fine sand to sandy silt (SM). The general properties of the original soil specimen are tabulated in Table 8. The effect of adding Bangkok clay and bentonite increased the Atterberg limits linearly with the proportions of the admixtures as shown in Fig. 6. Figure 7 shows that the mixture of dispersive soil with Bangkok clay lies in the CL region while the corresponding mixture with bentonite lies in the CH region.

The relationship between dry density and optimum moisture content obtained from standard Proctor compaction tests and Harvard miniature compaction test for soil samples from Phitsanulok (Soil 1) and Ampun dam (Soil 2) are shown in Fig. 8 and 9, respectively. The maximum dry density decreased while the optimum moisture content increased with increasing proportions of Bangkok clay as expected. The amount of reduction for the Ampun dam soil specimens is much higher than that for the Phitsanulok site. This phenomenon may be due to the "sandy" type of soil of the former rather than the latter.

10 TEST FOR DISPERSIVITY OF SOIL SPECIMENS

10.1 Crumb test

The original soil samples from three different sites were identified as dispersive by means of crumb test in the field. The crumb test is considered the quickest and simplest field test to identify dispersive soils. But for some reason not yet clear, this test failed to identify some kaolinite clays that were found to be dispersive by other tests (Forsythe, 1977). Sometimes, because of their qualitative nature, it is difficult to differentiate between different grades of classifications. Most of the soil samples from Phitsanulok site and Lam Sam Lai dam sites showed grade 3 to grade 4 classification which means moderate to strong reaction with water. But the Ampun dam site showed only a slight reaction (grade 2).

10.2 Pinhole test

For this test, two admixtures were used, namely: Bangkok clay and bentonite. The standard procedure suggested by Sherard et al (1976a) was followed. The results of the pinhole tests are tabulated in Table 9. It can be seen that for dispersive soil (D1, D2), the flow rate increased

Table 7 Designation of Field and Laboratory Tests

Type of Test	Mixing Material	Type of Specimen	Sample number	Test Description	Objective
Specific gravity	Nil	P-0, A-0, L-0 clay	4		Physical property
Grain size analysis	Nil	P-0, A-0, L-0 clay	4	Include hydrometer test	Physical property
Atterberg limits	Bangkok clay	P-0,10,20,30,40,50 A-0,10,20,30,40,50 L-0,10,20,30,40,50	18	Use distilled water	Physical property (LL, PL, PI)
	Bentonite	P-10,20,40 A-10,20,40 L-10,20,40	9		
Compaction	Bangkok clay	H-P0,P20,P30 S-P0,P20,P30 H-A0,A20,A30 S-A0,A20,A30 S-C0,C20,C40		Harvard miniature compaction method	1. Relationship between Yd max and Wopt 2. Permeability test
Pinhole	Bangkok clay	P-0,10,20,30,40 A-0,10,20,30,40 L-0,10,20,30,40	15	Harvard miniature compaction method	Dispersivity D1,D2, No. 4 No. 3, No. 2, No. 1
	Bentonite	P-10,20,30 A-10,20,30 L-10,20,30	9		
Chemical	Bangkok clay	P-0,40 A-0,40 L-0,40, C-)	7	Use AFE Division Laboratory	
	Bentonite	P-40 A-40 L-40, Bentonite	4	Use AFE Division Laboratory	

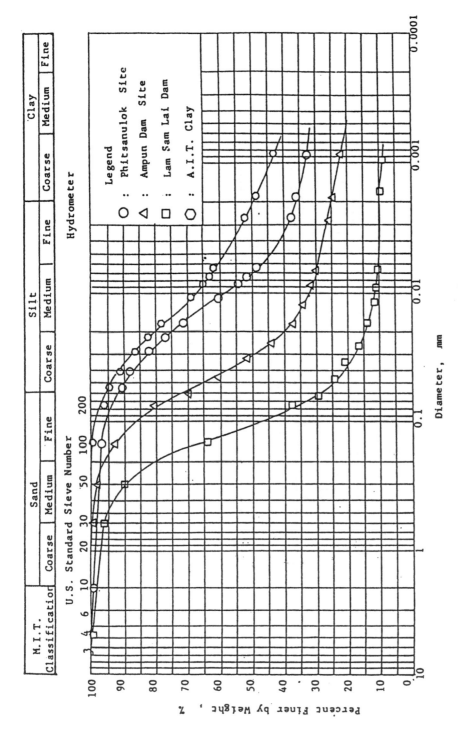

Fig. 5 Grain Size Distribution of Original Soils.

483

Table 8 Result of Atterberg Limits Test.

Site	Proportion of Mixing (%)		LL (%)	PL (%)	PI (%)	Proportion of Mixing (%)		LL (%)	PL (%)	PI (%)
Phitsanulok	Clay	10	41.8	18.5	23.3	Bentonite	0	40.5	18.0	22.5
		20	42.5	19.0	22.5		10	61.2	17.7	43.5
		30	45.5	19.7	25.8		20	92.1	19.1	73.0
		40	47.0	20.5	26.5		30	123.3	20.8	102.4
		50	48.0	21.0	27.0					
Ampun Dam	Clay	10	25.5	12.2	13.3	Bentonite	0	23.2	13.0	10.2
		20	28.4	10.7	17.4		10	46.4	19.1	73.0
		30	33.0	12.0	21.0		20	74.5	12.8	61.7
		40	39.7	14.0	25.7		30	103.5	13.0	90.5
		50	42.9	15.0	27.9					
Lam Sum Lai	Clay	10	21.1	N.P.	-	Bentonite	0	19.6	N.P.	-
		20	23.4	15.6	7.8		10	45.8	11.7	34.1
		30	27.5	11.5	16.0		20	74.4	13.2	61.2
		40	34.0	12.5	21.5		30	106.8	14.0	92.8
		50	37.0	14.0	23.0					

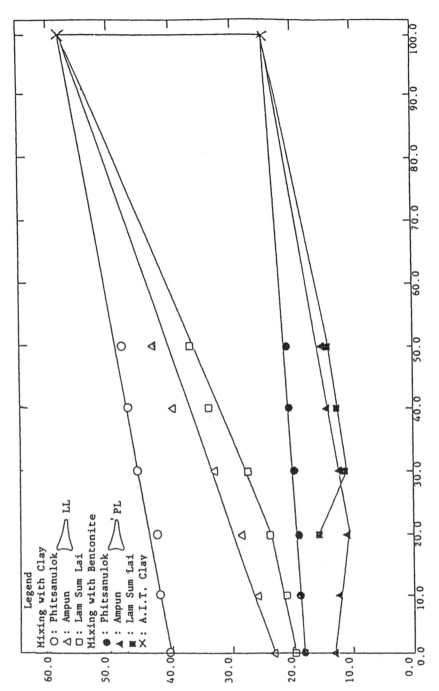

Fig. 6 Change of Atterberg Limit with Increasing Percentages of Bangkok Clay or Bentonite.

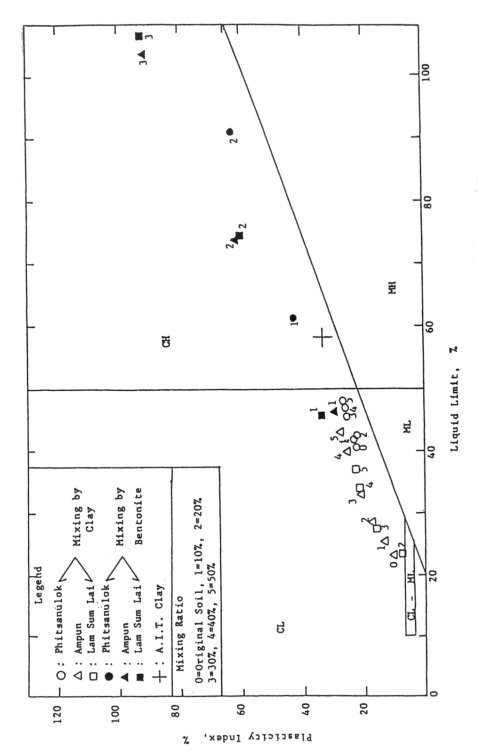

Fig. 7 Plasticity Chart of Dispersive Soils with Additive Material

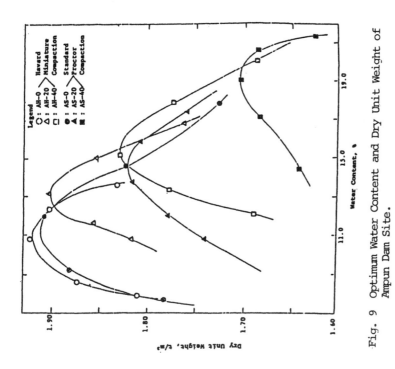

Fig. 9 Optimum Water Content and Dry Unit Weight of Ampun Dam Site.

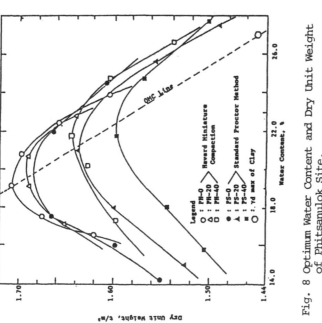

Fig. 8 Optimum Water Content and Dry Unit Weight of Phitsanulok Site.

487

Table 9 Result of Pinhole Test on the Remoulded Soil.

Site	Additives (%)		ω (%)	γd t/m³	Flow Rate (mL/sec)				Diameter of outlet (mm)	Color (side)	Particle Failing	Passing Time (head) (min)	Classification
					5cm	16cm	38cm	102cm					
Phitsanulok	nil		16.4	1.692	9.7	-	-	-	4.5	SM	few	5.0(2")	D1
	Clay	10	17.3	1.682	3.5	-	-	-	4.0	SM	few	8.5(2")	D2
		20	19.0	1.672	0.8	-	-	-	1.5	SM	few	10.0(2")	ND4
		30	19.7	1.666	0.4	4.0	-	-	3.0	BV	none	5.0(7")	ND3
		40	20.5	1.629	-	1.5	2.0	6.1	2.0	BV	none	5.0(40")	ND2
	Bento-nite	10	17.7	1.776	7.3	-	-	-	4.5	D-SM	few	4.7(2")	D1
		20	19.1	1.634	7.1	-	-	-	5.0	D-SM	few	5.0(2")	D1
		30	20.8	1.582	7.7	-	-	-	5.0	D-SM	none	4.4(2")	D1
Ampun Dam	nil		10.0	1.892	4.0	-	-	-	4.5	SM	few	2.0(2")	D1
	Clay	10	11.4	1.884	0.6	6.0	-	-	3.0	BV-CC	f.-n.	5.0(7")	ND4
		20	12.7	1.871	0.5	2.0	2.7	5.8	2.0	BV-CC	none	5.0(40")	ND2
		30	13.8	1.840	-	1.0	1.4	3.5	1.5	CC	none	5.0(40")	ND1
		40	14.9	1.784	-	-	1.2	2.9	1.0	CC	none	5.0(40")	ND1
	Bento-nite	10	11.5	1.793	8.5	-	-	-	5.5	SM	few	3.5(2")	D1
		20		1.741	7.9	-	-	-	5.0	SM	few	4.0(2")	D1
		30	13.0	1.691	9.3	-	-	-	5.0	SM	few	4.0(2")	D1
Lam Sam Lai Dam	nil		10.4	1.808	8.8	-	-	-	5.0	SM	few	3.5(2")	D1
	Clay	10	10.4	1.759	8.7	-	-	-	5.0	SM	few	5.0(2")	D1
		20	15.6	1.734	3.9	-	-	-	4.0	BV	few	8.7(2")	D2
		30	12.1	1.704	0.8	6.2	-	-	4.0	BV	f.-n.	3.5(7")	ND4
		40	14.6	1.683	0.6	6.3	-	-	4.0	BV	none	4.2(7")	ND4
	Bento-nite	10	11.7	1.722	8.4	-	-	-	5.0	SM	few	3.7(2")	D1
		20	12.8	1.706	7.3	-	-	-	5.0	SM	few	4.4(2")	D1
		30	14.0	1.632	9.8	-	-	-	5.0	SM	few	3.4(2")	D1

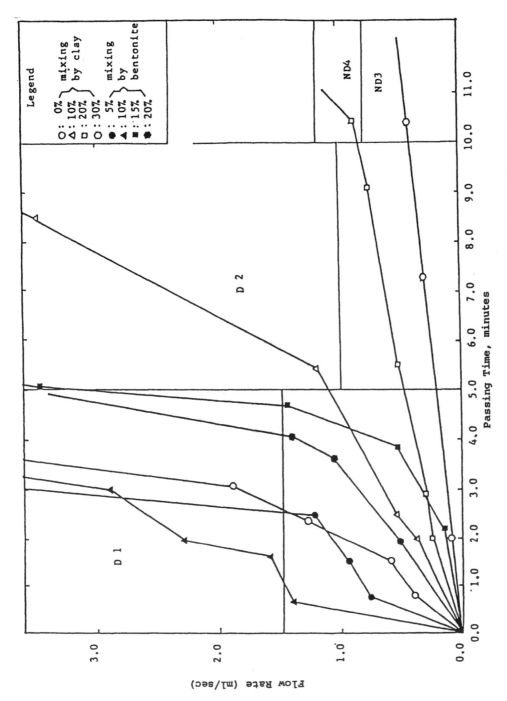

Fig. 10 Flow Rate and Passing Time of Phitsanulik Site Specimen in 50 mm Head.

489

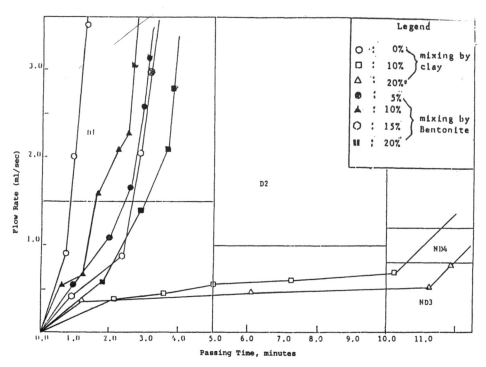

Fig. 11 Flow Rate and Passing Time of Ampun Dam Specimen in
50 mm Head

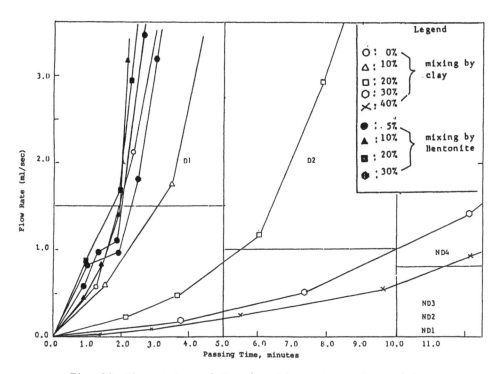

Fig. 12 Flow Rate and Passing Time of Lam Sum Lai Dam
Specimen in 50 mm Head

490

Table 10 Result of Chemical Test on Saturated Extract.

Site	Additives (%)		Saturation (%)	Saturated Paste		Saturated Extract									Classification
				pH	EC	pH	EC	Na meq/l	K meq/l	Ca + Mg meq/l	SAR	ESP %	TDS meq/l	Na %	
Ampun dam	Clay	0	36.01	3.99	0.12	7.5	0.29	2.5	0.0	0.65	4.39	4.96	3.20	78.3	1
	Bento- nite	40	59.29	4.08	1.43	5.4	2.87	16.0	0.92	11.8	6.59	7.80	28.7	55.8	4
	Bento- nite	40	221.58	8.11	1.68	8.2	2.19	18.5	0.03	0.38	36.9	34.7	16.5	97.6	2
Lam Sam Lai Dam	Clay	0	25.42	6.28	0.34	7.7	0.58	5.5	0.06	0.70	9.30	11.1	6.26	87.81	1
	Bento- nite	40	30.17	4.11	1.75	6.9	3.76	23.5	0.12	11.7	9.72	11.6	35.3	66.51	4
	Bento- nite	40	191.74	8.11	1.68	8.3	1.63	16.0	0.06	1.89	16.5	18.7	18.0	89.1	2
Phitsanu- lok	Clay	0	38.61	6.15	0.37	7.6	0.73	7.5	0.04	0.92	11.1	13.1	8.46	88.7	1
	Bento- nite	40	59.67	4.61	2.13	5.6	3.01	16.3	0.31	12.1	9.00	10.7	30.7	63.4	4
	Bento- nite	40	240.52	7.91	1.68	7.9	1.61	16.1	0.03	0.38	36.9	34.7	16.5	97.5	2
A.I.T.	Clay		82.73	3.90	3.14	4.4	4.89	26.0	0.80	24.0	7.51	8.94	50.8	51.2	3
Bentonite	Nil		377.46	8.66	2.30	7.4	2.5	21.0	0.08	0.54	40.4	35.0	21.6	97.1	2

Note: EC Unit = mm hos/cm

491

rapidly from 0.50 ml/sec to 3.5 ml/sec in the first 5 min period and tended to be steady in the next 5 minute period. The hole after the test was enlarged to 4 to 5 times the original size. The shape of the hole was quite uniform. As shown in Figs. 10 and 11, the samples from Phitsanulok and Ampun dam sites, respectively, become nondispersive with 20% of Bangkok clay. While the samples from Lam Sam Lai dam needed 30% of the admixture to stabilize, as shown in Fig. 12. It can be seen in Figs. 10 to 12 that the use of bentonite as admixture was not effective in stabilizing dispersive soils.

10.3 Chemical tests

The results of chemical tests on saturation extracts of soils are tabulated in Table 10 and plotted in Fig. 13 and 14. The SAR and ESP were determined by using Equations 1 and 2. According to the relationship between the percent sodium and total dissolved salts in saturation extract (Fig. 13), the original dispersive soils from each site fell in zone 1, which may have potential for breaching and tunnel erosion. When 40% of Bangkok clay was added, the relationship changed to zone 4 in Fig. 13, which is the transition state. The Bangkok clay in itself falls in zone 3, which is erosion resistant. In Fig. 14, the original dispersive soils in each site fall under zone A, which means dispersive soil. But the results show a tendency towards zone B, which is the intermediate zone, after mixing with 40% Bangkok clay. Bangkok clay in itself falls under zone B. But using bentonite as an additive was not effective. Even

bentonite itself is found to be dispersive.

The pH values were determined by means of a pH meter. The pH values of the original Bangkok clay and bentonite, which were used as admixtures, had values of 4.4 and 7.8, respectively. The pH values of the dispersive soils were very high than the Bangkok clay used as additive material. At high pH, the proton concentration is slow and the net surface charge of clay minerals is negative, resulting in a greater concentration of ions in the diffuse swarm and consequently greater repulsive interaction between particles. Bentonite additive was not effective, perhaps because of its high pH value and high sodium percentage, as tabulated in Table 10.

11 CONCLUSIONS

Based on the results of the experimental investigations carried out on the dispersive soil mixed with Bangkok clay or bentonite, the following conclusions can be drawn:

1) Bentonite is not suitable as an admixture to improve dispersive soils perhaps due to its high pH value and high percentage of sodium in the saturation extract. Bentonite itself has been found to be dispersive.

2) Both dispersive soils of Phitsanulok site and Ampun dam site with Unified Soil Classification (USC) of CL changed to non-dispersive soil

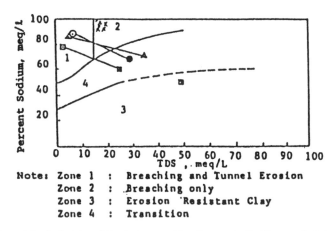

Note: Zone 1 : Breaching and Tunnel Erosion
Zone 2 : Breaching only
Zone 3 : Erosion Resistant Clay
Zone 4 : Transition

Fig. 13 Relationship between the Percent Sodium and Total Dissolved Salt in Saturation Extract.

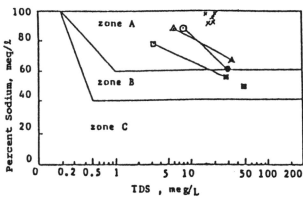

Note: Zone A : Dispersive   Percent Sodium (meq/ℓ) $= \dfrac{Na(100)}{Ca+Mg+Na+K}$

Zone B : Intermediate   Total Dissolved Salts (TDS) $= Ca+Mg+Na+K$

Zone C : Nondispersive

Fig. 14   Relationship between the Percent Sodium and Total
Dissolved Salt in Saturation Extract.

through mixing with 20% Bangkok clay. On the other hand, the dispersive soils of Lam Sam Lai dam site with USC designation of SM required a larger amount (30%) Bangkok clay.

3) Most of the dispersive soils in Northeast Thailand are inorganic silty clays, clayey silts, and sand-silt mixtures of low to medium plasticity. The liquid and plastic limits of the dispersive soil-Bangkok clay mixture increase with increasing clay percentage.

4) The maximum dry densities decreased and the corresponding optimum moisture content increased almost linearly with increasing Bangkok clay percentages, as expected, for both the standard Proctor and Harvard miniature compaction test.

5) The dispersive soils in Northeast Thailand were mainly responsible for dam failures due to piping, tunnelling, and surface erosion. The soils consist predominantly of kaolinite clay minerals and clay-size to silt-size quartz with high pH value and high content of dissolved sodium relative to the quantities of other cations such as calcium and magnesium in its saturation extract.

12 ACKNOWLEDGMENTS

The authors are gratefully indebted for the valuable advice of Prof. A.S. Balasubramaniam and Prof. Prinya Nutalaya during the research study. The authors are also very grateful to Dr. Suphon Chirapuntu, Senior Soil Engineer at the Royal Thai Irrigation Department, for providing valuable information and for providing assistance during the collection of soil samples. Finally, heartfelt gratitude is extended to Mrs. Uraiwan Singchinsuk and Miss Nualchan Sangthongsttit for their valuable help in the preparation of the manuscripts.

13 REFERENCES

Aitchison, G.D. (1960), Discussions on paper entitled Piping Failures of Earth Dams Built of Plastic Materials in Arid Climates by Cole & Lewis (1960), Proc. 3rd Australia-New Zealand Conf. on Soil Mech. and Found. Engg., pp. 230.

Chandra, S. and Chen, G.L. (1985), Improvement of Dispersive Soil By Using Different Additions, India Geotechnical J., Vol. 14, No. 3, pp. 202-216.

Chandra, S. and Garcia, E.B. (1984), Chemical and Mineralogical Study of Collapsible and Dispersive Soil, Intl. Geotechnical Conf., December, 1984, Calcutta, India.

Chen, G.L.J. (1984), Imvprovement of Dispersive Soil by Using Different Additives at Low and High pH Values, M.Eng. Thesis, AIT, Bangkok, Thailand.

Cole, D.C.H. and Lewis, J.G. (1960), Piping Failure of Earth Dams Built of

Plastic Materials in Arid Climates, Proc. 3rd Australia-New Zealand Conf. on Soil Mech and Found. Engg., pp. 93-99.

Cole, B.A., Ratanasen, C., Maiklad, P., Liggins, T.B. and Chirapuntu, S. (1977), Dispersive Clay in Irrigation Dams in Thailand, ASTM STP-623, pp. 25-41.

Coumoulos, D.G. (1977), Experience with Studies of Clay Erodibility in Greece, ASTM STP-623, pp. 42-57.

Decker, R.S. and Dunnigan, L.P. (1977), Development and Use of the Soil Conservation Service Dispersive Test, Dispersive Clays, Related Piping and Erosion in Geotechnical Projects, ASTM STP-623, pp. 94-109.

Emerson, W.W. (1964), The Slaking of Soil Crumbs as Influenced by Clay Mineral Composition, Australian J. of Soil Research, Vol. 2, pp. 221-227.

Emerson, W.W. (1967), A Classification of Soil Aggregate Based on Their Coherence in Water, Australian J. of Soil Research, Vol. 5, pp. 47-57.

Fernando, M.J. (1979), Distribution and Properties of Dispersive Soils in the Vicinity of the Lam Sam Lai Dam Site and Their Effects on Embankment Dam Construction, M.Eng. Thesis, A.I.T., Bangkok, Thailand.

Forsythe, P. (1977), Experience in the Identification and Treatment of Dispersive Soils in Mississippi Dams, ASTM STP-623, pp. 135-155.

Garcia, E.B. (1983), Chemical and Mineralogical Studies of Collapsible and Dispersive Soils in North and Northeastern Thailand, M. Eng. Thesis, AIT, Bangkok, Thailand.

Haliburton, T.A., Petry, T.M. and Hayden, M.S. (1975), Identification and Treatment of Dispersive Clay Soils, United States Bureau of Reclamation Special Report, PB-248 840.

Holmgreen, G.G.S. and Flanagan, C.P. (1977), Factors Affecting Spontaneous Dispersion of Soil Materials as Evidenced by the Crumb Tests, ASTM STP-623, pp. 218-239.

Ingles, O.G. and Aitchison, G.D. (1969), Soil-Water Disequilibrium as a Cause of Subsidence in Natural Soils and Earth Embankments, C.S.I.R.O. Div. of Soil Mech. Research, Paper No. 131, pp. 342-352.

Kang, K.Y. (1985), Stabilization of Dispersive Soils By Mixing with Bangkok Clay and Bentonite, M. Eng. Thesis, AIT, Bangkok, Thailand

Kim, G.W. (1982), Dispersive Soils in Northeastern Thailand, M. Eng. Thesis, AIT, Bangkok, Thailand.

Lambe, T.W. and Whitman, R.V. (1969), Soil Mechanics, John Wiley & Sons, Inc.

Logani, K.L. and Hector, M. (1979), Techniques Developed During Foundation Treatment of the Ullum Dam Constructed of Dispersive Soils, Proc. 13th Int. Cong. on Large Dams, Vol. 1, pp. 729-748.

Nakano, R. (1967), On Weathering and Changes of Tertiary Mudstone Related to Landslide, Soils and Foundations, Vol. 7, pp. 1-14.

Parker, G.G. and Jenne, E.A. (1967), Structural Failure of Western Highways Caused by Piping, Highway Research Record, No. 203, pp. 57-76.

Richards, L.A. (1954), Diagnosis and Improvement of Saline and Alkali Soils, Agriculture Handbook No. 60, Dept. of Agriculture, U.S.A.

Schafer, G.J. (1978), Pinhole Test for Dispersive Soil-Suggested Change, J. Geotech. Engg. Div., ASCE, No. GT6. pp. 761-765.

Sherard, J.L., Decker, R.S. and Raker, N.L. (1972), Piping in Earth Dams of Dispersive Clays, Proc. Conf. Performance of Earth and Earth-Supported Structures, ASCE, Vol. 1, pp. 589-626.

Sherard, J.L., Dunnigan, L.P. and Decker, R.S. (1976b), Indentification and Nature of Dispersive Soil, J. Geotechnical Engg. Div., ASCE, Vol. 102, No. GT4, pp. 287-302.

Sherard, J.L., Dunnigan, L.P., Decker, R.S. and Steele, E.F. (1976a), Pinhole Test for Identifying Dispersive Soils, J. Geotechnical Engg. Div., ASCE, Vol. 102, No. GT1, pp. 69-87.

Shieh, J.L. (1981), Engineering Properties of Dispersive Soils, M. Eng. Thesis, AIT, Bangkok, Thailand.

Tai, T.Y. (1983), Improvement of Dispersive Soils by Using Soil Stabilization, M. Eng. Thesis, AIT, Bangkok, Thailand.

Van Olphen, H. (1963), An Introduction To Clay Colloid Chemistry, John Wiley and Sons.

Volk, G.M. (1937), Method of Determination of Degree of Dispersion of Clay Fraction Soils, Proc. Soil Science Soc. of America, Vol. 11, pp. 561-565.

Environmental Geotechnics and Problematic Soils and Rocks, Balasubramaniam et al. (eds)
© *1987 Balkema, Rotterdam. ISBN 90 6191 785 9*

Mathematical models for saturated expansive soils with and without CNS

R.K.Katti
Department of Civil Engineering, CSRE, Indian Institute of Technology, Bombay
K.K.Moza & Dinesh R.Katti
CBIP Project, Department of Civil Engineering, Indian Institute of Technology, Bombay

ABSTRACT: Montmorillonite-based saturated expansive soils exhibit unconventional behaviour. At any given point in a saturated expansive soil, media binary stress conditions exist. One of the stress conditions is due to internal mineralogical characteristics in interaction with dipolar water and the second stress condition is due to the effect of gravity on a mass of soil. This paper describes various mathematical models for saturated expansive soil based on simplified physical models and their validity.

1 INTRODUCTION

In India expansive soil deposits are known locally as expansive black cotton soils and cover almost one third of the surfacial deposits. When these soils come in contact with water, they heave considerably, which results in loss of shear strength. If heaving is prevented, such soils undergo varying degrees of swelling pressure, depending upon their internal mineralogical constituents. This behaviours has caused considerable distress problems in various types of civil engineering structures, as shown in Fig.1. The swelling and swelling pressure characteristics of expansive soils have been attributed by various research workers to the presence of montmorillonite types of clay mineral having expanding lattices. (Finn 1958, Katti 1979).

The studies conducted in the field under all weather conditions and also in the laboratory using small and large-scale devices indicate that montmorillonite-based saturated expansive soils exhibit unconventional behaviour such as:
1. High magnitude of heave during saturation,
2. Near equal magnitude of swelling pressure in vertical and lateral directions,
3. Very high lateral pressures at a shallow depth of 1 to 1.5 m in spite of very high cohesion,
4. Increase in cohesion from a negligible value at surface to a very high value at a shallow depth of 1 to 1.5 m, which remains constant thereafter,
5. 2μ clay fraction exhibiting greater swelling pressure than the whole soil mass, whereas sand and silt fractions do not exhibit swelling pressure and
6. The self-equilibrating phenomenon at a shallow depth of 1 to 1.5 m (Katti 1969, 73, 75).

Extensive studies conducted in the field and laboratory with cohesive non-swelling soil layer indicate that swelling and swelling pressure characteristics of the order of 2 to 3 kg/cm^2 of underlying expansive soil could be counteracted by adding cohesive non-swelling soil of 1 to 1.5 m thickness (Katti 1975, 79).

On the basis of studies conducted during the past two to three decades, it is realised that in a montmorillonite-based saturated expansive soil medium, at any given point, binary stress condi-

Fig.1. Distress to structures in expansive soil area

tion exists. One of the stress conditions is generated due to mineralogical characteristics of expanding type clay mineral in interaction with the dipolar nature of water; the second stress condition is due to the effect of gravity on a mass of soil.

There is a need to relate the observed behaviour of saturated expansive soil media with the binary stress condition. This requires the identification of easily determinable internal properties of saturated expansive soil media related to both the stress conditions.

It is realised that evaluating the behaviour of saturated expansive soil with stress generated by mass of soil under gravitational influence, coupled with measured stress condition in bonded water in the saturated soil mass is quite a complex process.

In view of this, an attempt is made to develop physical and mathematical models to estimate various aspects of saturated expansive soil media by studying the inter-relations between such determinable parameters as swelling pressure, swelling pressure of an individual expanding clay particle,

496

undrained cohesion, capillary radius, grain size, compressibility index, e-p relation, stress due to gravitational effect, index properties, void ratio difference, etc. and using these relations to solve various problems of saturated expansive soil media with and without CNS.

This paper describes various mathematical models based on simplified physical models. The results from these models are compared with large-scale experimental results and field observations.

2 PROPERTIES

Extensive studies conducted on different montmorillonite-based expansive soil samples from different parts of India indicate that gravel content is less than 10%, sand 15-25% and clay 50-70%. The liquid limit, plastic limit and shrinkage limit of these soils range from 60-90%, 30-50% and 8-

12% respectively. The base exchange capacity of 2μ clay fraction ranges between 100-130 meq/100 grams, indicating the presence of montmorillonite or a combination of montmorillonite type clay minerals, which is further confirmed by X-ray and DTA analysis. These soils are inorganic in nature and contain 8 to 10% carbonate content (Katti 1979).

Swelling pressure values corresponding to no volume change condition for zero moisture content indicate that 2μ fraction exhibits swelling pressure above the whole soil mass, whereas sand and silt do not exhibit swelling pressure. Vertical and lateral swelling pressure tests conducted on expansive soil samples in modified triaxial testing equipment reveal that the ratio of lateral swelling pressure to vertical swelling pressure developed during saturation are several times higher than the weight of the soil sample. This phenomenon may be attributed to the stresses

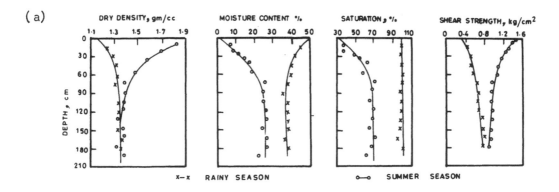

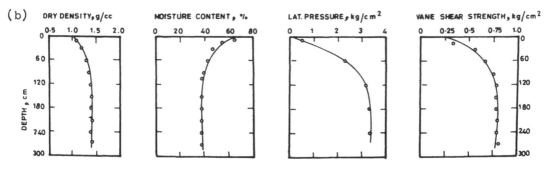

Fig.2(a) Field observations and (b) Large scale test results of Poona black cotton soil

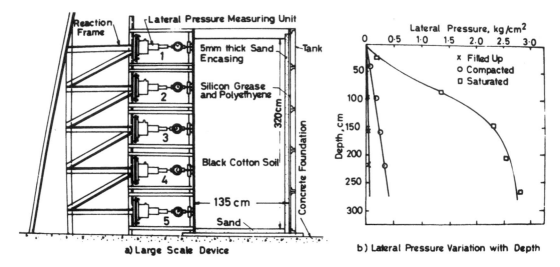

Fig.3. Large-scale device and test observations

3 FIELD AND LARGE SCALE TEST RESULTS

Density, moisture content and vane shear strength with depth in expansive soil under fully saturated condition during extreme rainy season conditions in the field show that the density and vane shear strength increase rapidly within a shallow depth of 1 to 1.5 m and remain constant thereafter (See Fig.2). The moisture content at surface is found to be close to its liquid limit value. In the case of Poona soil, surface moisture content is 75.1% and liquid limit 81.5%. The summer observations indicated that at a depth of 1 to 1.5 m although there is a change in moisture content from extreme rainy season conditions, it is not accompanied by any volume change (Katti 1973, 79).

Large scale tests conducted in the laboratory under fully saturated condition reveal similar behaviour as observed in the field during rainy season conditions. (Fig.2 (Katti 1969, 73)). Other significant observations from

generated due to interaction between montmorillonite clay mineral having expanding lattice and the dipolar nature of water in the soil mass (Katti 1979, 84, 85).

large-scale tests are :
1. The lateral swelling pressures as shown in Fig.3 increase from a negligible value at surface to near the swelling pressure value at 1-1.5 m depth, the change beyond this depth not being a marked one
2. In spite of very high cohesion, very high lateral pressures exist at a depth of 1-1.5 m
3. Cohesion increases from a negligible value at surface to a very high value at 1 to 1.5 m depth closely following the trend followed in lateral pressure.

This behaviour is unconventional and such soil systems, both in field and laboratory, exhibit self-equilibrating phenomenon at a shallow depth of 1 to 1.5 m. This has been attributed to the presence of cohesion in the soil water system (Joshi 1980, Katti 1978). Studies with CNS layer technology are based on the concept of providing a similar environment at no volume change depth in saturated expansive soil by placing a soil system having cohesion but not having clay mineral of expanding type (Katti 1975, 79).

From the studies conducted on expansive soil with CNS layer, it is observed that heaving of underlying expansive soil reduces rapidly as thickness of CNS increases

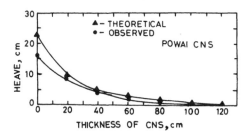

SWELLING PRESSURE OF POONA B.C.SOIL
= 3·27 kg cm^2

Fig.4. Variation of heave with thickness of CNS

(Katti 1975, 79, 85). The variation is not linear in nature, as shown in Fig.4.

With a small thickness of about 1 m of CNS on top of expansive soil, it was possible to counteract swelling pressure of the order of 2 to 2.5 kg/cm^2 of underlying expansive soil.

From this, it can be seen that in the case of a freely swollen expansive soil system, the internal characteristics and weight of soil mass above prevent expansive soil at self-equilibrating depth from swelling to a liquid limit void ratio value from a constant void ratio, indicating that the above thickness has the capacity to bring about a change in void ratio for the system from e_o to $\bar{e}$. The CNS, with its internal characteristics and weight, appears to be performing the same function.

Even though the behaviour is unconventional in nature, the whole soil system is in equilibrium at various points. This shows that in an expansive soil mass, at any given point, one has to deal with binary stress conditions, one arising from the internal mineralogical characteristics of expanding type clay mineral while interacting with the dipolar nature of water and the second due to the effect of gravity on the mass of soil.

Bearing in mind the above observations, the physical and mathematical models as developed are described below.

4 PHYSICAL AND MATHEMATICAL MODELS

4.1 Micro-particle model

One of the observations in the case of saturated montmorillonite-based expansive soil is that almost equal swelling pressures are observed in all directions. To conceptualise the phenomenon, in the year 1973 a physical model designated the micro-particle model taking into consideration the structure of montmorillonite clay mineral having expanding lattice, was developed. (Kulkarni 1973, Katti 1979). In this model, the 2μ clay fraction is idealized by assuming the particles to consist of $2\mu x 2\mu x 2\mu$, with a split in the centre having a spring exhibiting swelling and swelling pressure characteristics similar to those of montmorillonite clay mineral in 'c' direction. Such a model, as shown in Fig.5, would indicate that the swelling pressure recorded in any direction can be the same and its magnitude would be equivalent of the swelling pressure exerted by one-third the number of equivalent particles having 'c' direction perpendicular to measuring plane.

This model also underlines that although the individual clay particles are anisotropic in nature with respect to swelling and swelling pressure characteristics, the random distribution of clay particles in a soil mass gives rise to the observed phenomenon of equal pressures. Taking into consideration the random statistical distribution and mechanical analysis of soil, the following mathematical relation between the swelling pressure of an individual clay particle and measured swelling pressure is suggested :

$$q_{swi}= \frac{q_{sw} \times 1 \times 1}{1/3(\frac{p}{(2\mu)^3(1+e_i)})^{2/3} \, 2\mu \times 2\mu} \qquad ..(1)$$

Wherein q_{swi} is the swelling pressure of an individual clay particle in 'c' direction, q_{sw} is the

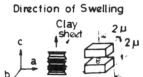

Direction of Swelling

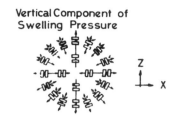

Vertical Component of
Swelling Pressure

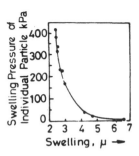

a) Model of Clay Particles
   with its Swelling

b) Orientation of Clay Particles
   and Swelling Pressure
   Components

c) Relationship Between Swelling
   and Swelling Pressure of an
   Individual Clay Particle

Fig.5. Mirco-particle model

measured swelling pressure, p the clay content and e_i the initial void ratio. By conducting a large number of swelling pressure tests under no volume change condition on a given expansive soil sample having initially zero moisture content under varying void ratios and using the above relation, the relation between swelling and swelling pressure of an individual clay particle was evaluated as shown in Fig.5c.

This model clearly indicates that the mineralogical properties of this type of clay system can contribute to exerting a distinct stress condition at a given point while intereacting with the dipolar nature of water in a saturated expansive soil system.

4.2 Cohesion model

Large-scale experiments show that high lateral pressures of the order of 2 to 3 kg/cm² are developed at a depth of 1 to 1.5 m and the soil system remains in equilibrium under fully saturated condition. This type of pressure cannot be attributed to overburden of soil, which is only between 0.2 to 0.3 kg/cm². Secondly, it was observed that cohesion also develops in a saturated expansive soil mass following the same trend as that observed for lateral pressure. Although high lateral pressures were observed on the wall, the soil at a depth

of 1 to 1.5 m and below did not show any change in volume under fully saturated condition and remained in equilibrium. Even after removal of overburden, high lateral pressures continued to remain in lateral direction.

To understand the above behaviour in saturated expansive soil, a cohesion concept was propounded (Katti 1979). In a clay water system, water which does not have any shear strength when it comes in contact with electrical charges on the surfaces of clay particles, gives rise to cohesion. It is well known that the lowering of temperature in water below $0^{\circ}C$ causes the formation of ice, which has high shear strength, the chemical composition remaining the same. This is attributed to internal electrical phenomena.

It can easily be imagined, therefore, that the dipolar nature of water in the influence zone of electrical charges can give rise to the orientation of water molecules and result in high bond strength between molecules similar to that in ice, and manifested in the form of cohesion. Such molecules may have an isosahedral structure.

A montmorillonite clay particle has very high electrical charges on its external surfaces and also in the interlayers. The water molecules entering into the interlayers due to electrical effects may cause expansion. However, the water molecules surrounding the charged surfaces, develop cohesive bonds and may prevent expansion, as shown in Fig.6b.

To understand such behaviour it is necessary to develop a physical model and mathematical relationship for the equilibrium of individual particles against swelling and equilibrium of the whole mass, as observed in the case of lateral pressures.

4.3 Equilibrium of individual particles

In a saturated soil medium, each clay particle is surrounded by dipolar water with high cohesive bonds. If a clay particle tends to swell, it is resisted by the cohesive forces around the particles. To establish the relationship between the resisting forces and the swelling pressure, the following assumptions are made :

1. A model on similar lines to Kozeny's reverse model for flow through a large number of capillaries in soils, based on principles of flow through pipes and induced internal shear (See Taylor, pp.106 for flow through soil) and

2. Water is stationary and soil particles are expanding in relation to water.

A detailed derivation is given in a paper entitled 'Lateral Pressure in Expansive Soil with Surcharges' by Joshi, R.P. and Katti, R.K., Proc. of 4th Int. Conf. on Expansive Soils, Denver, Colorado, Vol.I, pp.223-241.

However some of the more significant steps are reproduced here for convenience.

Considering equivalent cylindrical particles having the same volume as a sphere and assuming that the area distribution between solids and voids are in same ratio, the number of particles are worked out in a given area.

For equilibrium condition, the net vertical forces must be zero.

$$\left|\begin{array}{l}\text{Upward force}\\\text{due to}\\\text{swelling pressure}\end{array}\right| - \left|\begin{array}{l}\text{Resistance}\\\text{due to}\\\text{drag}\end{array}\right|$$

$$- \left|\begin{array}{l}\text{Resistance}\\\text{due to}\\\text{weight}\end{array}\right| = 0$$

$$q_{sw} = \frac{8}{3}(1-n)\,c_u' \qquad ..(1.1)$$

Wherein 'n' is the porosity and c_u' is the undrained cohesion existing without any distrubance, which is found to be around 2.5 times the c_u after removal. With the help of this equation, the following equation is derived

$$q_{sw} = \alpha\,c_u \qquad ..(2)$$

where α is the shape factor whose value ranges between 2.5 and 5. From this it is clear that there exists a mathematical relationship relating the mineralogical force system expressed in terms of swelling pressure to undrained shear strength parameter of saturated expansive soil mass.

4.4 Equilibrium of whole soil mass

For equilibrium of whole soil mass, the reaction to lateral pressure may be considered as passive pressure acting on the soil mass. To evaluate the cohesion needed to balance the passive pressure, Terzaghi's equation is used :

$$\sigma_p = \gamma.z\tan^2(45+\phi/2)+2c\,\tan(45+\phi/2)$$

$$..(2.1)$$

Assuming $\phi = 0$, the undrained cohesion needed for equilibrium is computed as 0.5 times the σ_p or $\sigma_p = 2c_u$.

4.5 Cohesion capillary radius relation

It is realized that for a given fully saturated soil system the overall electrical surface charges remain nearly constant. Thus it is expected that the change in

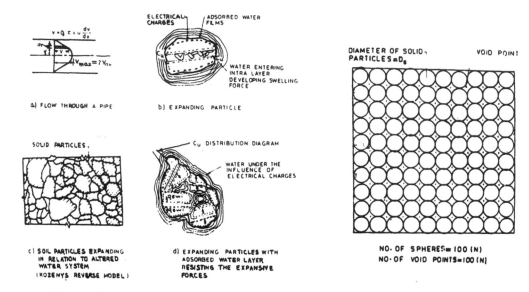

Conceptual cohesion model   Arrangement of particles

Fig.6. Cohesion model and arrangement of solid particle

cohesion would be a function of the capillary radius. With this in mind, the following equation is established between cohesion and capillary radius ;

$$c_u = c_{usl} \left[\frac{r_e}{r_{sl}} + c_1 \cdot r_e + c_2 \right]^n \quad , \quad n \neq 0$$

..(3)

where C_1, C_2 and n are constants, c_{usl} and r_{sl} the undrained cohesion and capillary radius corresponding to shrinkage limit. C_1 and C_2 are evaluated using the following boundary conditions ;

1. at $r_e = r_{11}$, $c_u = 0$

2. at $r_e = r_{sl}$, $c_u = c_{usl}$

With these boundary conditions,

$$C_1 = \frac{- r_{sl}}{r_{11}(r_{11} - r_{sl})} \quad ;$$

$$C_2 = \frac{r_{sl}^2}{r_{11}(r_{11} - r_{sl})} \quad ..(3.1)$$

The constant 'n' is determined by relating cohesion at any intermediate capillary radius between r_{sl} and r_{11}. The equation can be rewritten as :

$$C_1 r_e^2 + \left[C_2 - \left(\frac{c_u}{c_{usl}} \right)^{1/n} \right] r_e + r_{sl} = 0$$

..(3.2)

502

The following solution gives a positive value of r_e :

$$r_e = \frac{-\left\{ C_2 - \left(\dfrac{c_u}{c_{usl}}\right)^{1/n} \right\} - \sqrt{\left\{ C_2 - \left(\dfrac{c_u}{c_{usl}}\right)^{1/n} \right\}^2 - 4C_1 \, r_{sl}}}{2 \, C_1} \qquad ..(3.3)$$

4.6 Capillary radius — void ratio cum particle-size relation

Considering that the soil contains uniform fraction and the particles are distributed uniformly in rows and columns as shown in Fig.6, the number of void points will be equal to the number of solid particles. The volume of void point is given by :

$$\Delta V_e = \frac{\pi}{6} (2r_e)^3 = \frac{\text{Volume of voids}}{\text{Number of void points}} = \frac{\pi \, V_e \, D_s^3}{6 \, V_s} \qquad ..(3.4)$$

$$\therefore \; r_e = \frac{D_s}{2} \sqrt[3]{e}$$

Where D_s is the equivalent diameter of particles and e is the void ratio.

Substituting the value of r_e in equation 3.4,

$$e = \left\{ \frac{-\left\{ C_2 - \left(\dfrac{c_u}{c_{usl}}\right)^{1/n} \right\} - \sqrt{\left\{ C_2 - \left(\dfrac{c_u}{c_{usl}}\right)^{1/n} \right\}^2 - 4C_1 \, r_{sl}}}{C_1 D_s} \right\}^3 \qquad ..(4.1)$$

This shows that void ratio is related to undrained cohesion, which in turn is related to swelling pressure. The e can in turn be related to the stress condition due to gravity which is evaluated using Terzaghi's equation :

$$e = e_o - c_c \, \log_{10} \left(\frac{p}{p_o}\right) \qquad ..(5)$$

Thus the binary stresses at a given point in a fully saturated expansive soil mass are inter-related to each other.

Thus the relations, $q_{swi} - q_{sw} - c_u - r_e - D_s / e - c_c \, \log_{10} \frac{p}{p_o}$ can be considered as basic property inter-relations. Using these equations one can evaluate self-equilibrating depth and heave in expansive soils.

4.7 Equation for self-equilibrating depth

The self-equilibrating depth is evaluated by equating equation 4.1 with equation 5 and solving it for $\dfrac{p}{p_o}$:

503

$$\frac{p}{p_o} = e^{\frac{1}{0.4343c_c}\left[e_o - \left\{\frac{-\left\{C_2-(\frac{c_u}{c_{usl}})^{1/n}\right\} - \sqrt{\left\{C_2-(\frac{c_u}{c_{usl}})^{1/n}\right\}^2 - 4C_1\ r_{sl}}}{C_1\ D_s}\right\}^3\right]} \qquad ..(5.1)$$

Assuming γ is the same for p_o and p, $p = \gamma \cdot h_e$ and $p_o = \gamma \cdot h_o$

$$\therefore h_e = h_o\ e^{\frac{1}{0.4343c_c}\left[e_o - \left\{\frac{-\left\{C_2-(\frac{c_u}{c_{usl}})^{1/n}\right\} - \sqrt{\left\{C_2-(\frac{c_u}{c_{usl}})^{1/n}\right\}^2 - 4C_1\ r_{sl}}}{C_1\ D_s}\right\}^3\right]} \qquad ..(5.2)$$

where h_e is the self-equilibrating depth and h_o is a small depth corresponding to e_o.

4.8 Heave equation

Consider that with respect to initial equivalent thickness $\bar{H}$ and corresponding void ratio $\bar{e}$ as shown in Fig.7a, the heave of Δh brings uniform change in void ratio designated as e_{av}.

Volume of solids is given by ;

$$V_s = \frac{\bar{H}}{1+\bar{e}} = \frac{\bar{H} + \Delta h}{1 + e_{av}} \qquad ..(5.3)$$

$$\text{or} \Delta h = \left\{\frac{e_{av} - \bar{e}}{1 + e_{av}}\right\} h_e \qquad ..(5.4)$$

From the e-logp diagram shown in Fig.7b, $e_{av} = \frac{e_o + \bar{e}}{2}$

$$\therefore \Delta h = \left\{\frac{e_o - \bar{e}}{2+e_o+\bar{e}}\right\} h_e \qquad ..(6)$$

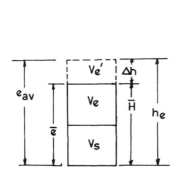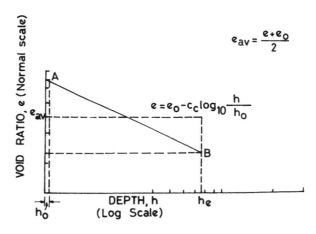

$$e_{av} = \frac{e + e_0}{2}$$

Fig.7. Void ratio heave and e-logp diagram

4.9 Interface equivalent equilibrium void ratio concept

It is observed that expansive soil itself reaches a no volume change condition below a depth of 1 to 1.5 m. In other words, expansive soil mass with void ratio at surface close to liquid limit void ratio reaches no volume change void ratio $\bar{e}$ under a height of soil h_e or equivalent intensity of stress, $p = \gamma h_e$. The change in void ratio from e_0 at surface to $\bar{e}$ at no volume change depth equal to $e_0 - \bar{e}$ results from a soil system of the order of $c_c \log_{10}(\frac{p}{p_0})$.

Utilizing this concept, equations for thickness of CNS for zero heave condition and heave under various thicknesses of CNS are developed.

4.10 Thickness of CNS

Based on interface equivalent equilibrium void ratio concept, it is realized that by placing a cohesive non-swelling soil layer on expansive soil, an equivalent of

$c_c \log_{10} (\frac{p}{p_0})$ is produced.

$$\therefore c_c \log_{10}(\frac{p}{p_0}) = \bar{c}_c \log_{10}(\frac{\bar{p}}{\bar{p}_0}) \qquad ..(6.1)$$

$$\text{or } c_c \log_{10}(\frac{\gamma \, h_e}{\gamma \, h_0}) = \bar{c}_c \log_{10}(\frac{\bar{\gamma} \, \bar{h}}{\bar{\gamma} \, \bar{h}_0}) \qquad ..(6.2)$$

$$\therefore \bar{h} = \bar{h}_0 \, (\frac{h_e}{h_0})^{c_c/\bar{c}_c} \qquad ..(7.0)$$

where $\bar{h}$ is the thickness of CNS for zero heave condition and $\bar{h}_0$ is a very small thickness.

$\bar{c}_c$ corresponds to CNS material.

4.11 Heave under various thicknesses of CNS

When thickness of CNS provided on top of expansive soil is $\bar{h}$, heave is zero. If the thickness provided on top of expansive soil is less than $\bar{h}$, the expansive soil will heave by an amount $\bar{h}_*$. Using these boundary conditions, heave under various thicknesses of CNS is evaluated from the following equation :

$$\Delta \bar{h}_* = \Delta h_0 - \bar{c}_{c_*} \log_{10} \frac{\bar{h}_*}{\bar{h}_{0_*}} \qquad ..(8.0)$$

where $c_{c_*} = \dfrac{\Delta h}{\log_{10} \dfrac{\bar{h}}{\bar{h}_{o_*}}}$, $\bar{h}_{o_*} = \bar{h}_o$

and $\Delta h_o \cong \Delta h$.

4.12 Lateral pressure equations

It was shown earlier that inter-relations exist between such determinable parameters as swelling pressure, swelling pressure of an individual clay particle, undrained cohesion, capillary radius, grain size, index properties, compressibility index, e-p relation, stress due to gravitational effect, etc. It is known that the above inter-relations can be utilized for the development of lateral pressure equations, with the proviso that the equivalent gravitational component is provided by the reaction force of the wall. The effect due to mass remains the same.

In developing lateral pressure equations for expansive soil with and without cohesive non-swelling soil layer, it is assumed that :
1. Expansive soil is free to swell in vertical direction and swelling has come to equilibrium at all points,
2. Characteristics of CNS remain the same as those in vertical direction and
3. The same amount of rotation has taken place at the vertical interface of CNS and expansive soil.

4.13 Lateral pressure equations for saturated expansive soil

A rigid rotating wall retaining saturated expansive soil is shown in Fig.8a. In developing lateral pressure equations, the stress condition generated due to internal mineralogical characteristics of montmorillonite-based expansive soil while interacting with dipolar water is considered first and subsequently the stress condition due to gravitational effect is superimposed :

$z > h_e$

At any point D for no wall movement condition e is equal to $\bar{e}$ and cohesion at D will be given by :

$$\bar{c}_u = c_{usl} \left[\frac{r_{sl}}{\bar{r}_e} + C_1 \bar{r}_e + C_2 \right]^n \quad ..(9.0)$$

in which $\bar{r}_e = \dfrac{D_s}{2} \sqrt[3]{\bar{e}}$.

Using the $q_{sw}-c_u$ relation, lateral pressure at point D for the no wall movement condition will be given by :

$$\bar{p}_1 = q_{sw} = \alpha \bar{c}_u \qquad ..(9.1)$$

or $\bar{p}_1 = \alpha c_{usl} \left[\dfrac{2r_{sl}}{D_s} \left\{ \dfrac{1}{\sqrt[3]{\bar{e}}} \right\} + \dfrac{C_1 D_s}{2} \cdot \left\{ \sqrt[3]{\bar{e}} \right\} + C_2 \right]^n \qquad ..(10)$

For rotation Θ about the bottom, the displacement at D will be $\Delta x = (1-z)\tan\Theta$. The displacement of Δx causes change in void ratio, cohesion and lateral pressure. To evaluate the void ratio, cohesion and lateral pressure corresponding to displacement Δx, the following approach is adopted. Let the displacement Δx be considered as equivalent heave causing change in void ratio in length $\bar{H}$ giving magnitude of volume change length h.

$$\therefore h = \bar{H} + \Delta x \qquad (..(10.1)$$

Assuming Δx produces uniform change in void ratio throughout h, Δx is given by using Fig.7a :

$$\Delta x = \left\{ \frac{e_{av} - \bar{e}}{1 + e_{av}} \right\} (\bar{H} + \Delta x) \qquad ..(10.2)$$

Substituting $e_{av} = \dfrac{e + \bar{e}}{2}$ from Fig.7b in the above equation

$$\therefore e = \left\{ \frac{2\,(1+\bar{e})\,\Delta x}{\bar{H}} + \bar{e} \right\} \qquad ..(11.0)$$

To check validity substitute Δx for Δh :

$$e = \left\{ \frac{2\,(1+\bar{e})\,\Delta h}{\bar{H}} \right\} + \bar{e} = \left\{ \frac{2\,(1+\bar{e})\,\Delta h}{h_e - \Delta h} \right\} + \bar{e} = e_o \simeq e_{11}$$

Given the void ratio, cohesion at D_1 is evaluated by using $c_u - r_e$ relation.

Using the $q_{sw} - c_u$ relation, lateral pressure at D_1 is given by

$$p_1 = \alpha\,c_{usl} \left[\frac{2r_{sl}}{D_s} \left\{ \frac{1}{\sqrt[3]{\left\{\dfrac{2(1+\bar{e})\,\Delta x}{\bar{H}} + \bar{e}\right\}}} \right\} \right.$$
$$\left. + \frac{C_1 D_s}{2} \left\{ \sqrt[3]{\left\{\frac{2(1+\bar{e})\,\Delta x}{\bar{H}}\right\} + \bar{e}} \right\} + C_2 \right]^{n} \qquad ..(12.0)$$

$\underline{z < h_e}$

The void ratio, cohesion and lateral pressure under no wall movement condition at J are e^*, c_u^* and p_1^* respectively and are given by :

$$e^* = e_o - c_c \log_{10} \frac{\gamma z}{\gamma h_o} \qquad ..(13.0)$$

$$c_u^* = c_{usl} \left[\frac{2r_{sl}}{D_s} \left\{ \frac{1}{\sqrt[3]{e^*}} \right\} + \frac{C_1 D_s}{2} \left\{ \sqrt[3]{e^*} \right\} + C_2 \right]^{n} \qquad ..(13.1)$$

$$p_1^* = \alpha\,c_{usl} \left[\frac{2r_{sl}}{D_s} \left\{ \frac{1}{\sqrt[3]{e^*}} \right\} + \frac{C_1 D_s}{2} \left\{ \sqrt[3]{e^*} \right\} + C_2 \right]^{n} \qquad ..(14.0)$$

Under rotation Θ, the lateral pressure equation at J_1 is evaluated by the approach mentioned above :

$$\therefore p_1 = \alpha\,c_{usl} \left[\frac{2r_{sl}}{D_s} \left\{ \frac{1}{\sqrt[3]{\left\{\dfrac{2(1+e^*)\,\Delta x}{\bar{H}_*}\right\} + e^*}} \right\} \right.$$

$$+ \frac{C_1 D_s}{2} \left\{ \sqrt[3]{\frac{2(1+e^*) \, \Delta x}{\bar{H}_*}} + e^* \right\} + C_2 \right]^n \qquad ..(15.0)$$

wherein $\bar{H}_*$ corresponds to e^*.

4.14 Effect of weight on lateral pressure equations

The soil element in expansive soil experiences reaction force equal to swelling pressure value from the wall. In other words the soil element is subjected to passive force due to reaction.

$$\sigma_p = q_{sw} \qquad ..(16.0)$$

$$\sigma_p = \gamma z \tan^2 (45+\phi/2) + 2c \tan (45+\phi/2) \qquad ..(16.1)$$

assuming $\phi = 0$

$$\gamma z = \sigma_p - 2c_u = \alpha \, c_u - 2c_u = c_u \, (\alpha-2) \qquad ..(17.0)$$

Component due to weight :

$$= \gamma z - c_u \, (\alpha-2) \qquad ..(18.0)$$

This will be additive to lateral pressure equations provided

$\gamma z - c_u \left\{ (\alpha-2) \right\}$ is a positive quantity.

4.15 Lateral pressure expressions for expansive soil with CNS thickness less than $\bar{h}$ in the backing

Fig.8b shows a rigid and smooth wall rotating at point F retaining saturated expansive soil with $\bar{h}_*$ thickness of CNS backing. At point A_1 void ratio is $e_o \simeq e_{11}$.

<u>$z > h_e$</u>

Consider any point D at depth Z. Under no wall movement condition, void ratio at D_1 is

$$e^* = \left\{ \frac{2(1+\bar{e}) \, \Delta \bar{h}_*}{\bar{H}} \right\} + \bar{e} \qquad ..(19.0)$$

Lateral pressure p_1, under no wall movement condition at D_1, is given by ;

$$p_1 = \alpha \, (\bar{c}_u - c_u^*) \qquad ..(20.0)$$

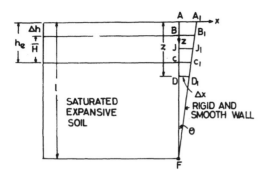

VOID RATIO AT A IS $e_0 \simeq e_{ll}$
VOID RATIO BELOW c IS $\bar{e}$
VARIATION IN e_0 TO $\bar{e}$ IS LOGARITHAMIC
IN NATURE

(a)

$\bar{h}_* =$ THICKNESS OF CNS LESS THAN $\bar{h}$,
SIMILAR TO THICKNESS OF CNS IN
VERTICAL DIRECTION.
AT POINT A VOID RATIO IS $e_0 \simeq e_{ll}$

(b)

Fig.8.(a) Rigid rotating wall retaining saturated expansive soil and
(b) Rigid rotating wall retaining saturated expansive soil with
$\bar{h}_*$ thickness of CNS

$$p_1 = \alpha \, c_{usl} \left[\left\{ \frac{2r_{sl}}{D_s} \left\{ \frac{1}{\sqrt[3]{\bar{e}}} \right\} + \frac{C_1 D_s}{2} \left\{ \sqrt[3]{\bar{e}} \right\} + C_2 \right\}^n \right.$$
$$\left. - \left\{ \frac{2r_{sl}}{D_s} \left\{ \frac{1}{\sqrt[3]{e^*}} \right\} + \frac{C_1 D_s}{2} \left\{ \sqrt[3]{e^*} \right\} + C_2 \right\}^n \right] \qquad ..(21.0)$$

After displacement, $\Delta x = (1-z)$ tan Θ, void ratio is evaluated using
the procedure mentioned earlier ; under displacement of Δx, lateral
pressure at D'_1 is evaluated as

$$p_1 = \alpha \, (c_u' - c_u^*) \qquad\qquad ..(22.0)$$

where $c_u' = c_{usl} \left[\frac{2r_{sl}}{D_s} \left\{ \frac{1}{\sqrt[3]{\dfrac{2(1+\bar{e})\Delta x}{\bar{H}} + \bar{e}}} \right\} \right.$
$$\left. + \frac{C_1 D_s}{2} \left\{ \sqrt[3]{\frac{2(1+\bar{e})\Delta x}{\bar{H}} + \bar{e}} \right\} + C_2 \right]^n$$

$\underline{z < h_e}$

Lateral pressure under no wall movement condition at J_1 is given by

$$p_1 = \alpha \, (c_u - c_u^*) \qquad\qquad ..(23.0)$$

509

where

$$c_u = c_{usl} \left[\frac{2r_{sl}}{D_s} \left\{ \cfrac{1}{\sqrt[3]{e_o - c_c \log_{10}(\frac{\gamma z}{\gamma h_o})}} \right\} + \frac{C_1 D_s}{2} \left\{ \sqrt[3]{(e_o - c_c \log_{10} \frac{\gamma z}{\gamma h_o})} \right\} + C_2 \right]^n$$

After rotation Θ, about bottom the lateral pressure at J_1' is given by

$$p_1 = \alpha \, (c_{ul} - c_u^{\ *}) \qquad\qquad\qquad ..(24.0)$$

where

$$c_{ul} = c_{usl} \left[\frac{2r_{sl}}{D_s} \left\{ \cfrac{1}{\sqrt[3]{\frac{2(1+e)\Delta x}{H} + e}} \right\} + \frac{C_1 D_s}{2} \left\{ \sqrt[3]{\frac{2(1+e)\Delta x}{H}} + e \right\} + C_2 \right]^n$$

and H corresponds to e

The stress due to weight component is given by $K\gamma z$ where K varies between K_o and K_A for CNS material.

5 COMPARISON BETWEEN THEORETICAL AND EXPERIMENTAL RESULTS

The self-equilibrating depth, heave and lateral pressure of expansive soil with and without CNS were evaluated by using the expressions developed above. These values are compared with those obtained from large-scale tests on expansive soils from different locations.

The results obtained from large-scale tests show good agreement with those obtained from the theoretical equations. A typical comparison between theoretical and experimental values for expansive soil from Malaprabha Right Bank Canal Km No.76 having the properties given in Table I with and without cohesive non-swelling soil from Byhatti are given in Table II and Fig.9.

6 SUMMARY AND CONCLUSIONS

1. On the basis of analysis of work done on montmorillonite-based saturated expansive soil media, it is realised that at any given point, there exists a binary stress condition. One of the stress conditions is generated due to internal mineralogical characteristics of expanding type clay mineral while interacting with the dipolar nature of water manifested in the form of swelling pressure. The other stress condition is due to the effect of gravity on a mass of soil manifested in the form of weight.
2. The two stress conditions are interdependent and are related to each other through the internal properties of saturated expansive soil mass such as swelling pressure of an individual expanding clay particle, swelling pressure of whole soil mass, undrained cohesion, capillary radius, index properties, grain size, compressibility index, e-logp relation and void ratio difference, $e_o - e$. The fundamental inter-property relationships are designated as

Table 1. Soil parameters

Sr.No.	Parameters	MRBC 76	Byahatti CNS
1.	D_s, cm	2.438×10^{-4}	-
2.	r_{sl}, cm	8.507×10^{-5}	-
3.	r_{ll}, cm	1.505×10^{-4}	-
4.	C_1 per cm	- 8628.1	-
5.	C_2	0.73398	-
6.	C_{usl} kg/cm^2	2.45	-
7.	c_u	0.6	-
8.	n	1.46	-
9.	e_o	1.66	-
10.	c_c	0.47	0.36 ($\bar{c}_c$)
11.	$\bar{e}$	1.0	1.6 ($\bar{h}_o$)
12.	h_o, cm	5.0	-

Table 2. a) Observed and theoretical self-equilibrating depth and heave in MRBC - 76

Sr.No.	Self-Equilibrating Depth, cm		Heave, cm	
	Observed	Theoretical	Observed	Theoretical
1.	120.0	128.03	17.27	18.133

Table 2. b) Heave under various thicknesses of CNS for MRBC 76

	Thickness of CNS, cm						
	0	20	40	60	80	100	120
Observed heave, cm	17.27	6.0	-	2.0	1.0	0.1	-
Theoretical heave, cm	18.13	7.315	4.34	2.61	1.37	0.420	0

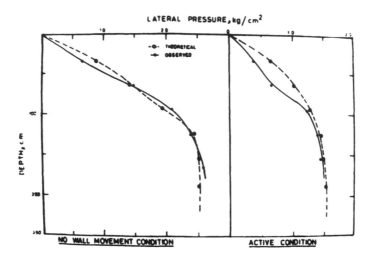

Fig.9.(a) Lateral pressure distribution of saturated expansive soil
from MRBC-76 under no wall movement condition and active
condition after a 7.5-minute rotation

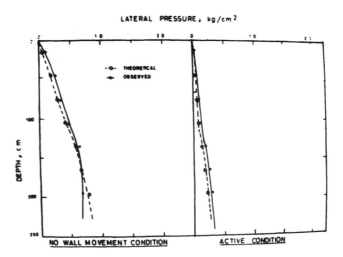

Fig.9.(b) Lateral pressure distribution of saturated expansive soil
from MRBC-76 with 100 cm CNS backing under no wall movement
condition and active condition

$$q_{swi}/\ q_{sw}-c_u-r_e-D_s/e-c_c \log_{10} p$$

$$/e_o-\bar{c}_c \log_{10}\ \bar{p}/\bar{p}_o$$

3. The fundamental relations developed are utilized in deriving expressions for heave, self-equilibrating depth, thickness of CNS, heave under various thicknesses of CNS and lateral pressure for K_o and rotation conditions, including K_A condition for saturated expansive soil with and without CNS.

4. The observed values are in close agreement with the computed theoretical values.

7 ACKNOWLEDGEMENTS

Thanks are due to the Central Board of Irrigation and Power, New Delhi for sponsoring the research on various aspects of expansive soils.

The authors are grateful to Prof. B. Nag, Director, IIT Bombay for his keen interest and constant encouragement.

Help rendered by Sarvashri D.G. Divshikar, C.P. Chavan, V.S. Khatri and staff members of Geotechnical Engineering Section is highly appreciated.

Thanks are also due to Miss Asha Shetty and Miss Ajita T. Parab for typing this paper.

REFERENCES

Finn, D.W.& Strom, B. 1958. Nature and magnitude of swell pressure, Proc. of Highway Research Board, 37:443-498.

Grim, R.E. 1953. Clay mineralogy, McGraw-Hill.

Joshi, R.P.& Katti, R.K. 1980. Lateral pressure in expansive soil with surcharges, Proc. of Fourth Inter. Conf. on Expansive Soils, Denver, Colorado, 1:223-241.

Katti, R.K., Lal, R.K., Fotedar, S.K.& Kulkarni, S.K. 1969. Depth effects in expansive clays, Proc. of Second Inter.Conf. on Expansive Clay Soils, College Station, Texas, USA, :362-373.

Katti, R.K., Desai, I.D.& Kulkarni, S.K. 1973.Field observations on variations of engineering characteristics with depth in expansive soils, Proc. of Third Inter. Conf. on Expansive Soils, Haifa, Israel, 1:115-120.

Katti, R.K.& Kate, J.M. 1975. Role of micro-particles in interaction between CNS layer and underlying expansive soil media, Proc. of Fifth Asian Regional Conf. SM and FE, India, 1:15-18.

Katti, R.K. 1979. Search for solutions to problems in black cotton soils, First Annual Lecture, Indian Geotechnical Society, Jour. 9:1-80.

Katti, R.K., Moza, K.K.& Katti, D.R. 1984. Unconventional behaviour of expansive soils, Proc. of 6th Budapest Conf. on SM and FE, Hungary, :137-146.

Katti, R.K., Moza, K.K.& Katti, Dinesh R. 1985. Further studies on shear strength of expansive soils covered with cohesive non-swelling soil layer-mineralogy to mathematical models, Final Report, submitted to Central Board of Irrigation and Power, New Delhi, :1-208.

Katti, R.K., Moza, K.K.& Katti, Dinesh R. 1985. Studies on active earth pressure with CNS backing in expansive black cotton soil (Including mathematical models based on displacements in relation to self-equilibrating boundary concept), Interim Report, submitted to Central Board of Irrigation and Power, New Delhi, :1-342.

Kulkarni, S.K. & Katti, R.K. 1973. A micro-particle and micro-anchor approach to mechanics of swelling soil media, Proc. of Third Int. Conf. on Expansive Soils, Haifa, Israel, 1:43-52.

Taylor, D.W. 1964. Fundamentals of soil mechanics, Asia Publishing House, Bombay :362-531.

Environmental Geotechnics and Problematic Soils and Rocks, Balasubramaniam et al. (eds)
© 1987 Balkema, Rotterdam. ISBN 90 6191 785 9

Comparative study of problematic soils in two areas around Cairo

Mohamed A.El-Sohby
Faculty of Engineering, Al-Azhar University, Madinet Nasr, Cairo, Egypt

S.Ossama Mazen
General Organization for Housing, Building and Planning Research, Cairo, Egypt

Azza M.Elleboudy
Faculty of Engineering, Shoubra, Banha University, Egypt

ABSTRACT: In a comparison between two areas of problematic soil around Cairo, the origin, nature and occutrrence of deposits were investigated and described. A study of the collapse and swelling characteristics was made based on undisturbed samples retained from various depths and locations in the two areas under consideration. Samples were tested for their physical, mineralogical chemical and geotechnical properties. The behaviour of these soils was then investigated and discussed in the light of test results and the geological, geomorphological and environmental features.

1 INTRODUCTION

Arid and semi-arid soils cover large parts of Egypt's land surface. In recent development plans, the arid lands around Cairo were opened up to the urban expansion of the city, which already occupied all the available areas in the narrow alluvial plain. However, the development of the new areas was controlled by ground conditions.

The present work is a geotechnical study of two representative arid zones included in the development plan; they are: Nasr City, which is located Northeast of Cairo, and the 6th of October City, which is located Southwest of Cairo (Fig. 1).

Although the soil formations of the two areas were deposited in the same geological period, they exhibit different characteristics. In Nasr City, cracks in many buildings due to the swelling of expansive soils have been observed. The 6th of October City represents another phase of the problem involved in partially saturated soils, where settlement is due to the collapse of soil when wetting occurs with higher frequency.

The level of damage caused each year by differential movement of such soils is well documented. However, the similarity in the general features of cracks in buildings as a result of settlement due to collapse of the soil underneath or heave due to expansive clayey soil makes differentiation difficult and confusion very probable. Therefore, careful study is required for proper indentification of the problem.

Geotechnical engineers practicing in arid and semi arid zones are interested in a better udnerstanding of these phenomena. This promoted the investigators to carry out the present comparative study.

2. PHYSICAL SETTING

Naser City is located between latitudes 30° 2' 00" and 30° 5' 30" N on the margin of the Mukattam Hills that lie south of it. It lies between longitudes 31° 17' 00" and 31° 22' 30" E with El-Gabal El-Ahmar forming the southern part of it. The area covers about 58 square kilometers with an altitude rangeing between 45 and 135 meters above sea level. (Fig. 2)

The 6th of October City is located at the intersection, of latitude 30° N with longitude 30° 45' E at a distance of about 32 kilometers southest of Cairo. The site is located on upland where the elevation varies from 180 to 190meters above sea level. (Fig. 3).

3. GEOLOGICAL, GEOMORPHOLOGICAL AND ENVIRONMENTAL FEATURES

The local geology of the two areas under investigation may be described within the general gology of Egypt on which the work of Hume (1925); Cuvillier (1930), Sandford and Arkell (1939); Shukry (1953); Said (1962) and Abu Shouk (1978) is the most authoritative. In Nasr City, Lower

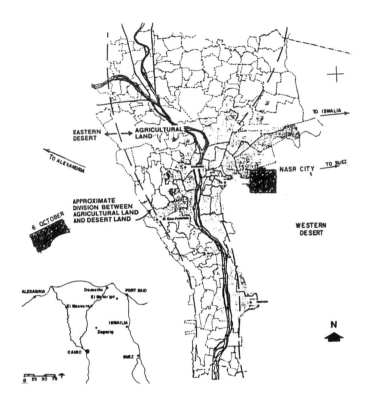

Figure 1. Map of Cairo and location of the two investigated sites.

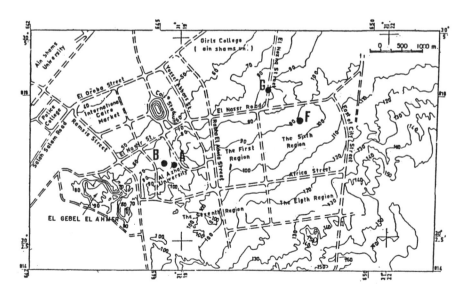

Figure 2. Nasr City area and location of boreholes.

516

Mukattam limestone of Middle Eocene age was deposited under marine conditions. Towards the end of the Upper Eocene, the basin became shallower and more detrital sandy sediments were derived. This was followed by a period of sea regression leading to erosional unconformity where the basin was dissected by a large number of drainage systems. These formations were subjected to tectonic movement which became intensive toward the end of the Oligocene time.

During early Miocene times, the area was submerged and deposition of lower Miocene shallow water sediments was received. This was followed by a period of substantial erosion and non-marine Miocene formation. After a brief episode of non-marine sedimentation, clastics entered the basin in large volumes due to the rising Alpine Orogen in the South. These sedimentary clastics are of Pliocene age. This was followed by alluvial coarse sand and gravel of Pleistocene/Recent age.

As a result, a succession of clayey and sandy beds belonging to the Oligocene Pliocene and Pleistocene/Recent ages were deposited. Four types of clayey horizons can be recognized. The first belongs to the Pliocene time and the other three are of Oligocne age.

The clayey beds are either shale, claystone, siltstone or conglomerate. Because of the mode of formation, these clayey beds are masked by sand and gravel with variable thickness and when the site is excavated they appear to occur in well defined beds sandwiched between sands and gravels.

The sandy beds appear in a burnt red and brown colour, slightly cemented but friable enough to be excavated by light earth moving equipment. The clayey formations are brown, greyish brown, redish grey and highly fissured. They vary from a firm blocky soil to a brittle clay-shale with well-defined beddings.

The depth to the first occurrence of clayey beds was found to be a maximum of 13 meters in the north eastern part of the city. In other locations it varied between 1 and 4 meters. The maximum thickness of these clayey beds varied between 4 and 12 meters. This thickness incrases in general, towards the northwest and southwest directions.

In the 6th of October city, two basic facies are recognized: 1) a fluviatile to fluvio-marine facies and 2) estuarine facies. The fluviatile and fluvio-marine facies is closely associated with a river sytem that seems to have run consequent on the uplifted Eocene bed of North Africa. This river system had drained the North African continent and built an alluvial plain that covered the area. The sediments are thick, and almost entirely clastic;

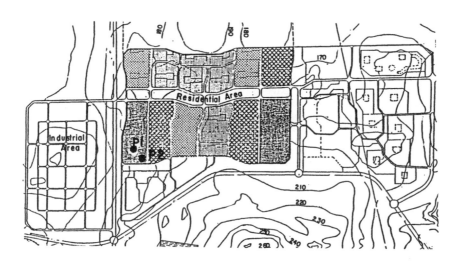

Fig. 3. The 6th of October City area and location of boreholes.

sandy to the south and shaly to the north. Where the base is exposed or logged by drilling, the lower Miocene sediments conformably overlie the Oligocene red beds which conists of sands, gravels, sandstone and clay. These Oligocene sediments have the same lithological character and probably also the same distribution as the overlying lower Miocene deposits. Estuarine facies is known to the north of the area.

The rain water in the Quaternary period, which was directed to the Nile Valley in the Pleistocene time, produced in some locations deposits at the surface represented by the clayey sands and some clay pockets at greater depths.

In contrast to the relatively simple geological features of Nasr City, the subsurface geology of the 6th of October City presents an exceedingly complicated picture. The succession of strata shows complex variations from layer to layer and place to place. Unconformities between many stages and of all magnitudes are found. This could be due to the turbulence in the sedimentation process according to the varying movement of the water courses rather than the shallow marine environment conditions encountered in Nasr City subsoil evelopment.

The turublence in the formation of the 6th of October City subsoil caused argillaceous material to intercalate between sands. Thus the sand paricles were covered with thin layers of clay which when dry, constituted the cementatious materials binding the coarse grained particles together.

4. SOIL EXPLORATION

The absence of water table in both sites and the cementation of soil provided suitable conditions for excavating open pits. Locations for these pits were selected where damage had been cuased to pavements and structure as a result of the presence of potentially expansive or collapsible soils. Figures 2 & 3 show the locations of these open pits with respect to each site.

Blocks of soil were hand cut from the side of these pits, wrapped to maintain natural moisture content and identified prior to transportation to the research testing laboratory.

Profile descriptions are given and soil samples were chosen to represent major variations in each area. (Nasr City and the 6th October City)

5. NASR CITY

5.1 Material investigated

Figure 4 give typical pit records of the four selected locations. The logging may be simplified by noting the disposition of only three basic soil types in the pit sides; they are: 1) clay-free, consisting of sand and gravel; 2) mixture of clay, silt and sand; and 3) claystone, clayshale and siltstone.

Four clayey soil samples were selected for the present study. They were chosen to differ widely in their clay content (5-59%).

5.2 Mineralogical composition

X-Ray diffraction analysis was carried out for the identification of clay and non-clay minerals in tested samples. The samples were examined in four forms: 1) normal samples; 2) oriented clay samples; 3) oriented clay samples glycolated with glycerol, and 4) oriented clay samples heated for two hours at 550°C.

Both natural and oriented clayey samples were obtained from powder passing sieve number 200 (0.074 mm). The oriented clay fraction was then obtained by separation after deposition in distilled water. From the separated clay fraction 3 slides were prepared. The first was examined untreated, the second treated with glycerol and the third heated at temperature of 550°C for two hours. Hence, four X-Ray spectrograms were obtained for each sample.

The identification of the non-clay minerals was carried out using ASTM cards (1960) for identification of minerals by X-Ray technique. The identification of clay minerals was done according to Millot (1976) through the use of characteristic peaks for each clay mineral interpreted by the X-Ray diffraction patterns of the tested soil.

Although various shades of colour could be distinguished in the samples, the mineralogy did not appear to be as varied. The clay mineralogy resulting from this analysis illustrated little variation either laterally or vertically. The predominent clay minral in all tested clayey samples was found to be either mixed layers of montmorillonite-vermiculate or montmorillonite-illite. The exchangeable ions present are either sodium, calcium or potassium. The non-clay minerals present are either silicate minerals, oxide mineral or carbonate and evaporate minerals. It is noted that all samples results indicate

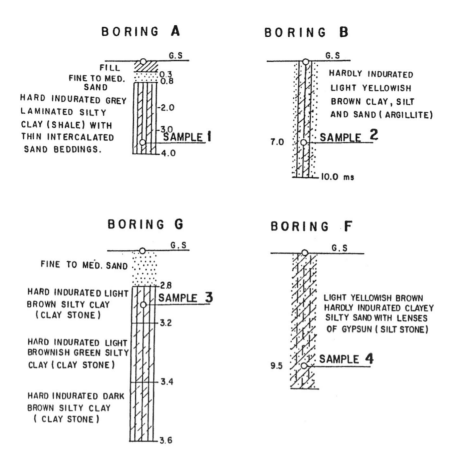

Figure 4. Soil profiles of selected borholes at Nasr City

N.B.
- The indicated ground level is the zero level of the boreholes. Reduced ground levels are indicated by the contour lines in Figure 2.

- Ground water level was not encountered in boreholes up to a depth of 30 meters.

the presence of mostly expansive clay mienrals. (see Table 1)

5.3 pH - value

From each sample, a representative part was obtained for the determintion of the pH-value. This part was divided into two divisions: the first was taken as obtained and the second as powder passing No. 200 sieve (74 micron). A suspension was then prepared by adding 25 ml of distilled water to 10 grams of each part of the sample and the pH-value of the slurry was then measured using a standard pH-meter. The measurement of the pH value was then repeated

until a fixed value could be attained. The pH-value was then determined for each sample as the average of the two readings.

Obtained results of pH tend to the alkaline side. They were found to vary between 7.40 and 8.15. This confirms that the deposition of clayey beds of Nasr City soil was under marine environmental conditions.

5.4 Geotechnical properties

A laboratory testing program of limited scope was undertaken to determine the swelling charcteristics and other soil properites.

5.4.1 Physical and index properties

Liquid limit, plastic limit, natural moisture content, specific gravity, dry unit weight and particle size distribution were determined in accorance with BS 1377:1975. Sampleswere tested in two foerms: natural samples; and oriented clay fraction obtained by separation after deposition in distilled water. A list of test results is presented in Table 2 and particle size distribution is given in Fig. 5.

The properties may be summariezed as follows:

Properties of whole samples

Liquid limit "W_L" %	Non-85.5
Plastic limit "Wp" %	Non-42.5
Plasticity index I_p	Non-54
Shrinkage limit	Non-25
Specific gravity	2.67-2.72
Dry unit weight Mg/m^3	1.48-1.98
Natural moisture content %	6.1-18.4
Clay content %	5% - 5%
Activity (Skempton, 1953)	0 - 0.93

Properties of clay fraction (part of samples less than 2 micron)

Liquid limit "W_L" %	110 - 219
Plastic limit "W_p" %	35 - 66
Platicity index I_p	73 - 153
Shrinkage limit %	14.7 - 18.4
Activity of clay fraction (El-Sohby, 1981)	0.73 - 1.53

5.4.2. Swelling characteristics

The swelling pressure was measured according to the pre-swelled method (El-Sohby and Mazen, 1980). The test measures the pressure necessary to consolidate a pre-swelled sample until it goes back to the initial voids ratio. The measurements were made on 2.5-inch (63 mm) diameter specimen sawed into discs with a thickness of 20 mm - The disc was fitted into a brass ring, and the conventional oedometer was assembled. A small axial stress, about 7 KN/m², was applied to the specimen, and the cell was flooded with distilled water to cover the top of the porous plate. The swelling heave was measured by the dial gauges. After the test specimen was allowed to swell completely, the original height of the sample was re-established by increasing the axial stress. The value of the stress, Ps, at which no further swelling strain was observed, was designated the swelling pressure.

The swelling test took about 14 days for each sample. After this time, no further swelling strain was observed and equilibrium was assumed to have been established between the pore water in the sample and the water in cell of the oedometer. The relationship between the incremental value of stress and the observed percentage swelling are represented in Fig. 6. The value of the swelling pressure is the stress at which no further swelling strain was observed. These values were found, for the tested soil, ranged between 1.6 MN/m² and 4 MN/m² which indicate extremely high swelling pressures. It may be noted that sample 3 with a relatively low dry unit weight of 1.48 Mg/m³ has a swelling pressure of 2 MN/m². Futhermore, sample 4 with a clay content as low as 5% has a swelling pressure of 1.6 MN/m². This may be explained by the mineralogical composition of the soil rather than its physical state.

6. THE 6TH OF OCTOBER CITY

6.1 Material investigated

Figure 7 gives typical pit records of the two selected loations. The loging may be summarized as mainly composed of sands with small amounts of clay, silt, carbonates, iron oxides and sometimes salt and gypsum. The cementing materials are susceptible to softening or partial removal by the arrival of water.

6.2 Mineralogical composition

Soils were tested for their mineralogical composition using x-ray diffraction analysis as indicated in 5.2. It was found that the clay fraction is composed of 52% Illite, 39% Kaolinite and only 9% Montmorillonite.

6.3 Chemical analysis

The results of the chemical analysis of the tested samples are given in Table 3.
Another detailed chemical analysis of a sample taken from test pit No. 2 was made and gave the results shown in Table 4.

Table 1: Summary of mineralogical analysis

Mineral	Soil Sample Number			
	1	2	3	4
Montmorillomite-vermiculite	Calcium (42%)	Sodium (20%)	Calcium (46%)	Potassium (5 %)
Silicate	24%	25%,	25%	45%
Carbonate and Evaporite	4%	26%	13%	18%
Oxides	13%	30%	16%	30%

Table 2: Basic physical and index properties of tested soils

Physical and Index Property	Soil Sample Number			
	1	2	3	4
Liquid Limit "W_L" %	81.5	51.5	85.5	Sandy
Plastic Limit "W_p" %	27.8	35.5	42.5	Sandy
Plasticity Index I_p	53.7	16.0	43.0	Sandy
Shrinkage Limit	13.8	24.5	14.0	Sandy
Specific Gravity	2.72	2.69	2.70	2.67
Dry Unit Weight $_d$ "Mg/m^3"	1.96	1.98	1.48	1.82
Natural Moisture Content %	6.1	13.8	18.4	12.7
Clay Content	59	20	46	5
Activity of Soil	0.9	0.8	0.93	Sandy
Liquid Limit of Particle 2 micron %	110	219	28	127
Plastic Limit of Particles 2 micron %	35	66	50	54
Plasticity Index of Particles 2 micron	75	153	78	73
Shrinkage Limit of particles 2 micron	14.7	17	16	18.4
Activity of Clay Particles	0.75	1.53	0.78	0.73

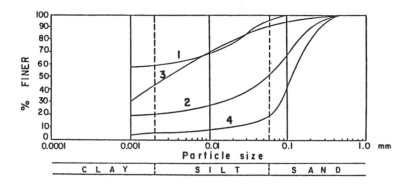

Figure 5. Grading of tested samples taken from Nasr City

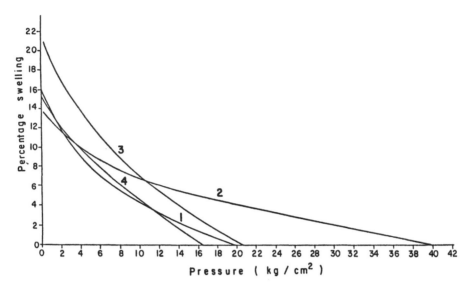

Figure 6. Swelling pressure test results of undisturbed samples from Nasr City

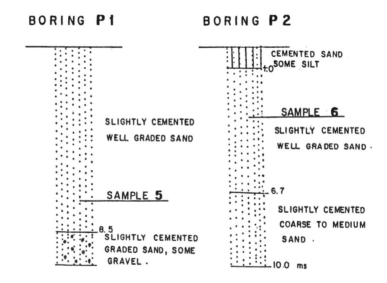

Figure 7. Soil profile of selected boreholes at the 6th of October City.

6.4 Geotechnical properties

This study concentrated on two representative samples taken from the previously
mentioned boreholes, chosen to characterize
the different types of collapsible soils
encountered at the site, regardless of
the presence of other types of clayey
soils and rock formations which are less

abundant in the investigated depth and
have no significant problems as foundation
materials.

6.4.1 Physical properties

Classification tests were carried out
on the representative samples, including
mechanical analysis, natural moisture

Table 3.   Results of the chemical analysis of tested soil.

Soil Sample Number	Depth (m)	pH	Alkalinity (p.p.m.)	Chloride (p.p.m.)	SO$_3$ (p.p.m.)	Total dissolved salts (p.p.m.)	Conductivity (mohs)
5	7	7.1	120	1700	475	3670	5790
6	3	7.4	122	1450	875	3950	6360

Table 4.   Total chemical analysis of a representative sample.

Soil Sample Number	Na$_2$O %	K$_2$O %	CaO %	MgO %	Fe$_2$O$_3$ %	Free iron oxide %	Organic matter %
6	1.29	0.17	2.17	2.96	2.22	0.29	0.25

* the amount of calcium carbonates is about 0.12%

content, Atterberg limits, and bulk density. The natural moisture content was around 2% and the bulk densities were 1.68 and 1.73 Mg/m³. The dry mechanica indicated that the soil is mainly sand with a limited amount of fines. It was noticed that the sand particles were coated with fines and cementing materials, giving a higher percentage of sand than actual in spite of the care taken in breaking the soil into its individual particles with the fingers and the rubber-tipped pestle. Hence, wet sieve analysis was used to precisely determine the grain size distribution of soil (El-Sohby and Elleboudy, 1986). Figure 8 presents the actual gradation of the tested samples. The plasticity test results shown that the tested materials are non-plastic. Table 5 summarizes the basic physical properties of the tested soils.

6.4.2 Collapsibility characteristics

In order to determine the sensitivity of this deposit to settlement due to wetting, the collapse potential was determined by using the oedometer test. The specimens used in the test were carefully hand carved from the block samples to fit neatly in the consolidometer ring. The test was conducted at the natural moisture content in a fixed-ring consolidometer using the standard procedure. At the end of a loading increment of 150 KN/m², the specimen was flooded with water and left for 24 hours, and then the oedometer test carried out to a maximum loading of 400 KN/m² under a soaked condition. Rapid reduction in volume was observed when water was

applied to the samples which revealed the tendency of this soil to collapse upon wetting. The collapsibility test results are shown in Fig. 9. The collapse settlement of sample No. 5 was 6% and of sample No. 6 7%. These are considered quite high values and might cause trouble to foundations according to the Jennings and Knight (1975) criterion for collapsible soils.

DISCUSSION

An inspection of the properties of the tested soils given in the tables and figures for the two sites provide an indication of the variations of nearly all the properties. The only property which shows little variation is the pH value. These properties appear to be a primary response to geological, geomorphological and environmental factors. Thus, the volume change may be described as the outcome of the interaction between soil and all these factors.

At the site of the 6th of October City, the grain size distribution of tested soil is mainly sand. As mentioned previously, these soils are water laid deposits. They were originally deposited by flash floods or mud flows and consist primarily of loose water deposited sediments which form alluvial fans, flows and flow-slides. This caused turbulence in the sedimentation process, thus increasing the range of coarse grained material and at the same time coating the sand particles with clay. Thus the collapsible grain structure was developed. The main constituent of the clay binding material, was found to be Kaolinite and illite which

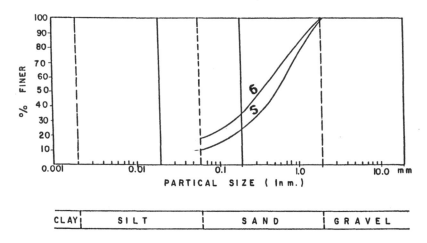

Figure 8. Grading of tested samples taken from the 6th of October City.

Table 5. Physical and index properties of the tested samples

Soil Sample Number	Depth (m)	Physical and Index Properties						
		W %	d Mg/m³	Clay content	% Fines	% Sand		
						Fine	Med	Coarse
5	7	2.01	1.70	8	9	17	32	42
6	3	2.00	1.65	14	17	18	28	37

exhibit no expansive property. Thus, on the arrival of water it becomes susceptible to softening or partial removal.

At the site of Nasr City, where the surficial process of weathering is active the topography is not flat and the environmental conditions were shallow marine, the process of the sedimentation had become different. The particles of sand, silt and clay obtained as a result of the weathering of rock are deposited in the sea bed in layerss. These layers were subjected to high overburden pressures which tend to eject water from sediment, resulting in the particles coming very close to each other. The main effect of the weathering in this case appears to be the changing of the mineralogy of the weathered soil and the dessication through which the moisture retention capacity of the material is reduced. The variations in the moisture content within this site could be, in addition to environmental factors, due to differences in drainage conditions from one location to another.

By comparing the results of oedometer tests on samples from both sites, a striking fact emerges. Sample 3 from Nasr City, with a dry unit weight of 1.48 Mg/m³, is highly expansive (swelling pressure = 2 MN/m²) whereas samples from the 6th of October City, with dry unit weight of 1.65 and 1.70 Mg/m³, are highly collapsible.

Also sample 4 from Nasr City, with a clay content of 5%, is highly expansive (swelling pressure = 1.6 MN/m²) when compared to the collapsible samples of the 6th of October City which have a clay content raning between 8 and 14%.

The above comparison indicates that it is difficult to divide partially saturated soils into swelling and collpasible in terms of physical properties. Also further investigation is needed to develop better understanding of the behaviour of such soil.

CONCLUSIONS

1--The properties of tested soil from the two sites appear to be primarily a

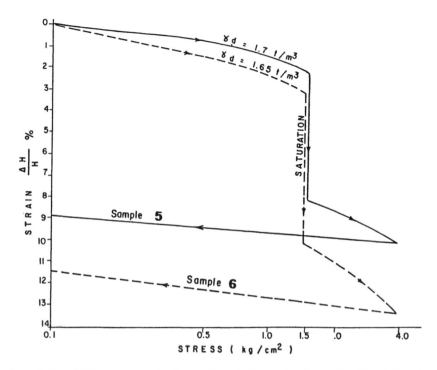

Figure 9. Collapsibility test results for undisturbed samples from the 6th of October City

response to geological, geomorphological and environmental factors.

2- Nasr City subsoil had been formed under a shallow marine deposit environment which predominated during the process of formation. As a result, a succession of clayey and sandy beds belonging to the Oligocene, Pliocene and Pleistocene/Recent ages were deposited. The beds consisted mainly of shale, claystone, silt-stone and conglomerate exhibiting swelling characteristics.

3- The 6th of October City subsoil had originally been deposited by flash floods or mud flows and consisted primarily of loose water deposited sediments. This provided the major components for collapsible soil and bulky shape material consisting mainly of sand and exhibiting collapsible characteristics It was formed mainly in the Oligocene time.

4- An inspection of the properties of tested soils from the two sites provides an indication of variations in nearly all properties. The only property which shows little variation is the pH value.

5- By comparing the results of oedometer test on samples from both sites, it is found that a sample with a relatively low dry unit weight (1.48 Mg/m³) taken from Nasr City is highly expansive whereas the two samples with higher dry unit weight (1.65 and 1.70 Mg/m³) taken from the other site were highly collapsible. Also, a sample with low clay content (5%) taken from Nasr City was found to be highly swelling whereas the samples from the other site, being highly collapsible, contains higher percentage of clay.

6- The above comparison indicates that the behaviour of expansive and collapsible soil is still not completely clear despite the efforts that have been expended for years in the study of expansive and collapsible soils.

7- This suggests a need for greater research effort in the comparative study aspects of the problem.

SELECTED REFERENCES

Abu Shouk, M.I. 1978. Geological and mineralogical factors affecting engineering properties of expansive soils at Nasr City. M.Sc. thesis. Al-Azhar University, Cairo.

Barden, L., Madedor, A.D. and Sides, G.R. 1969. Volume change characteristics of unsaturated clay. ASCE. Journal of the Soil Mechanics and Foundations Division. 95:33-51.

British Standard Specifications 1975. Method of test of soils for civil engineering purposes. No. 1377. London.

Chen, F.H. 1975. Foundation on expansive soils, Amsterdam : Elsevier.

Dudley, J.F. 1970. Review of collapsing soils. ASCE. Journal of the soil Mechanics and Foundations Division. 96:925-947.

El-Ramly A.H. 1965. Swelling characteristics of some Egyptian soils. Journal of the Egyptian Society of Engineers. pp 15-24.

El-Ramly, A.H. and El-Demery M.G. 1973. Field and laboratory studies on certain swelling soils in Egypt. Inter-African Symposium on Building materials. Lagos.

El-Sohby, M.A. 1981. Activity of soils. Proceedings of the 10th International Conference of Soil Mechanics and Foundation Engineering. Stockholm, Vol. 1 pp 587-591.

El-Sohby, M.A. and Mazen, S.O. 1980. On measuring swelling pressure by two methods. 7th Regional Conference for Africa on Soil Mechanics and Foundation Engineering. Vol. 2. Accra. pp 775-783.

El-Sohby, M.A. and Mazen, S.O. 1983. Mineralogy and swelling of expansive soils. Geotechnical Engineering. 14:79-87.

El-Sohby, M.A. and Rabba, S.A. 1981. Some factors affecting swelling of clayey soils. Geotechnical Engineering. 12:19-39.

Gillot, J.E. 1968. Clay in engineering geology. Amsterdam : Elsevier.

Gibbs, H.J. and Bara, J.P. 1967. Stability problems of collapsing soils. ASCE. Journal of the Soil Mechanics and Foundations Division. 93:577-594.

Grim, R.E. 1962. Applied clay mineralogy. New York: McGraw-Hill.

Grim, R.E. 1968. Clay mineralogy. New York: MacGraw-Hill.

Holtz, W.G. and Gibbs, H.J. 1956. Engineering properties of expansive clays. Transactions ASCE - 121:641-677.

Hume, W.F. 1925. Geology of Egypt. Cairo: Survey Department Jennings, J.E. and Knight, K.A. 1975. A guide to construction on or with materials exhibiting additional settlement due to collapse of grain structure. Proceedings, 6th Regional Conference for Africa on Soil Mechanics and Foundation Engineering. Durban, pp 99-105

Jones, D.L. and Van Alphen, G.H. 1980. Collapsing sands - a case history. Proceedings 7th Regional Conference for Africa on Soil Mechanics and Foundation Engineering. Vol. 2. Accra.

Leget, R.F. 1962. Geology and Engineering. New York : McGraw-Hill.

Matyas, E.L. and Radhakrishna, H.S. 1968. Volume change characteristics of partially saturated soils. Geotechnique. 18 : 432-448.

Millot, G. 1970. Geology of clay. London: Chapman and Hall.

Said, R. 1962. The geology of Egypt. Amsterdam: Elsevier.

Sadek, J. 1926. The geography and geology of the destrict between Gebel Ataqa and El-Galala El-Bahariya. Geological Survey of Egypt. Cairo.

Shukri, N.M. 1953. The geology of the desert Est of Cairo. Bulletin of the Desert Insitute of Egypt. pp. 89-105.

Skempton, A.W. 1953. The colloidal activity of clays. Proceedings of the 3rd International Conference on Soils Mechanics and Foundation Engineering. Zurich, Vol. 1. pp 57-61.

Youssef, M.S., Sabry, A.A. and Tewfik,M.M. 1957. Substantial consolidation and swelling of clay to cause two interesting cases of serious damages to hospital building in Egypt. Proceedings of the 4th International Confernce on Soil Mechanics and Foundation Engineering. London. Vol. 1. p 129.

Environmental Geotechnics and Problematic Soils and Rocks, Balasubramaniam et al. (eds)
© *1987 Balkema, Rotterdam. ISBN 90 6191 785 9*

Effects of particle breakage on mechanical characteristics of decomposed granite soils

K.Makiuchi, T.Miyamori & S.Iwai
Nihon University, Funabashi, Japan

ABSTRACT: Effects of a particle breakage of decomposed granite soil are studied with special reference to compaction and permeability characteristics for engineering use as soil materials. These effects are investigated through laboratory experiments, Proctor's compaction and falling-head permeability tests. It was observed that the maximum density of the soil increased considerably with increase in compaction effort. As the compaction energy is increased, the content of fine particles is increased and water-permeability is lowered considerably. The reduction of permeability of the specimen can explained in terms of compaction repetition.

1 INTRODUCTION

Decomposed granite soil (*Masa-do*) is a soil made from granite through long weathering and which has particle breakage properties. The breakage of soil particles not only relates to physical and mechanical characteristics, but also greately affects compaction properties, the permeability of compacted soil, etc. as earth materials for engineering use.

Generally, if soil particles are broken into finer particles, the permeability of the soil is reduced, that is, the water cutoff effect is heightened. In this research compaction and permeability tests were performed on the premise that decomposed granite soil would be used as embanking material for earth dams or levees or as core material for rock-fill dams. In this study, the breaking characteristics of particles, the reduction of permeability caused by the breakage of particles, its effects on compaction properties, and the difference of effects between repeating compaction and no-repeating compaction are reported.

2 PHYSICAL PROPERTIES OF SOILS

The decomposed granite soils for the experiment were sampled in a disturbed state from two places in Iwase-machi, Ibaraki prefecture, Japan (Specimens O and O').

The physical properties of the specimens are summarized in Table 1.

Since the breakage level of the soil particles depends on the sampling and handling methods, a specimen of about 500 g was put in a vinyl bag and free-dropped 30 times from a height of 1.5 m on a concrete floor to set it to the initial standard condition. To investigate the effects of difference of gradation, sample O' was classified into three types of uniform grading: A, B and C, as shown in Table 2.

3 TEST PROCEDURE

3.1 Compaction test

The specimen was compacted with an autoatic device according to No. 1 method, Japanese Industrial Standard: JIS A 1210 (Inside diameter of mold: 100 mm, height; 12.75 mm mass of rammer: 2.5 kg, dropping height: 30 cm, 3-layers compaction). Specimen O was compacted 25, 50, 75, 100, 150, 500 and 1000 blows/layar. Specimen O' was tested by no-repeating method: R = 0 time (where R is the number of repetitions). It was then tested by repeating method up to three cycles: R = 1 to 3 times.

3.2 Permeability test

This test was performed according to the

Table 1 Physical properties of soils tested

Sample				O	O'
Specific Gravity				2.65	2.64
Maximum grain size		(mm)		9.52	4.7
Texture	Gravel	(%)		23.2	32.3
	Sand	(%)		74.4	64.4
	Silt	(%)		2.4	2.3
Unified soil classification system				S-M	S-M
Color (dry state)				light pink	light brown

Table 2 Initial grading of soils used

Sieve	Specimen			
	O'(natural)	A	B	C
4.78mm	100			
2.00mm	67.7	90		
840μm	35.4	0	90	
420μm	17.6	0	0	90
250μm	11.2	6.0	6.0	6.0
105μm	4.9	4.0	4.0	4.0
74μm	2.3	0	0	0
		Mixing rate		

falling-head permeability test specified by JIS A 1218. When compacting specimen O, its water content was set near the optimum moisture content (±1.0%). On the other hand, specimen O' was set to the permeability testing device after compaction.

After the specimen had been saturated, it was vacuumed for 10 to 20 minutes depending on the degree of permeability. the gradation analysis was carried out before compacting the specimen and after over-drying it following the permeability test.

4 COMPACTION CHARACTERISTICS

The relationship betweew the optimum moisture content, w_{opt}, and the maximum dry density, ρ_{dmax}, of specimen O is shown in Fig. 1. As can be seen, the optimum moisture content changes little but the dry density increases remarkably in the sense that given compaction energy is low (25 to 75 blows/layer). The water leaks from the mold base plate in excess of approximately 12% of moisture content when the specimen is compacted. This implies that the specimen can hold 12 to 13% water. In the compaction range of 100 to 1,000 blows/layer, the dry density increases along the the zero air void curve.

Fig. 2 shows the compaction curves obtained from the no-repeating test: R = 0 and repeating test: r = 1 on specimen O'. It is evident from A to C that the dry density is increased, the optimum moisture content is lowered and the curve gets near that of natural specimen as the maximum size of grains is increased. The maximum dry density of natural specimen tested by repeating method: R = 1 is lower then that by the no-repeating method : R = 0. On the other hand, in the case of the specimen which has uniform grading initially, the increasing ratio of the maximum dry density is heightened in the order C, B and A.

5 PARTICLE BREAKAGE

5.1 Effects of compaction energy

The difference between the percentage of accumulated mass of soil particles left on the sieve after compacting the soil and that before compacting is obtained from the results of the gradation analysis and plotted in relation to grain diameter in Fig. 3. This figure shows clearly that the grain diameter gets not clearly smaller as the number of blows of compaction is increased. Generally, the soil particles of

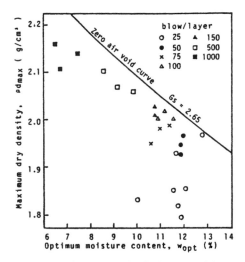

Fig. 1 Relationship between optimum
moisture content and maximum
dry density of Sample O

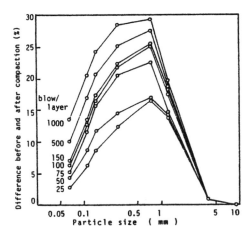

Fig. 3 Change of grain size frequency
curvr by compaction (Sample O)

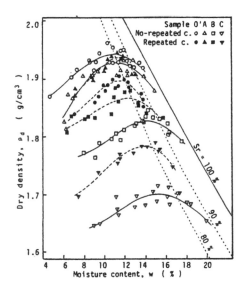

Fig. 2 Compaction curves

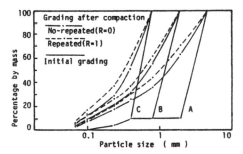

Fig. 4 Change of grading by compaction
(Samples A - C)

0.5 to 1.0 mm diameter are produced in
large amounts, and ones of lower diameter
are produced more as the number of blows of
compaction is increased.

5.2 Effects of gradation

Fig. 4 shows the change of gradation caused
by the compaction test of specimens A, B and
C, which have different maximum graing size
initially. As can be seen from the displacement of the gradation curve, the larger the
soil particles are, the more easily they are
broken into fine particles. The change of
grain diameter caused by the repeated test:
$R = 0 \rightarrow 1$, is much less than that caused by
the first compaction: $R = 0$. The breakage
of poorly graded particles is suddenly discontinued and becomes stable once they are
well graded.

Fig. 5 shows the difference of grain size
frequency curves for specimen O' obtained
from the repeated compaction test (compared
with the natural specimen before the compaction). This convex curve has a peak point
near 0.5 - 1.0 mm of the grain diameter
(equivalent to 50 ± 15% of accumlated curve).
The height of this curve shows the relationship of remaining percentage of sieve caused
by the breaking of the particles. During the
first and second cycles: $R = 1,2$ of the re-

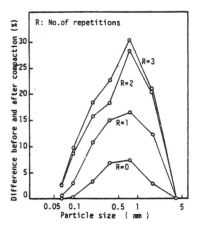

Fig. 5 Change of grain size
frequency curves by compaction
(Samples A - C)

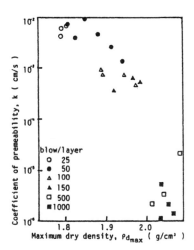

Fig. 6 Relationship between
maximum dry density and
coefficient of permeability

peated compaction test, the maximum height
of the curve is increased about 10%, and
that means a large quantity of particles
have been broken. When the third cycle is
started: R = 3, the breakage of particles
is almost finished. The point at the left
end of the left end of the curves indicates
the increase of fine particles, below 74 μm,
which is just below 3%. Thus, the breakage
of particles of the natural specimen is not
directly related to the increase in the fine
particles.

6 PERMEABILITY

6.1 Effects of compaction energy

Fig. 6 shows the relationship between the
maximum dry density and the coefficient of
permeability. As shown in this figure, the
coefficient of permeability is considerably
lowered as the maximum dry density is increased. It can be seen from Fig. 3 that
the permeability is not affected only by
the increase of density, but also by the
fineness of particles caused by the breakage.

6.2 Effects of repetition of compaction

The decrease in the coefficient of permeability caused by repeated compaction is
shown is shown in Fig. 7. Although the
coefficient of permeability is affected by
the moisture content at the time of compaction, the coefficient of the specimen
for the repeated compaction test: R = 0 → 1
tends to decrease. In general, the coefficient of permeability is in proportion to
the void ratio, $e^3/(e+1)$. Fig. 8 shows the
relationship between, k, and $e^3/(e+1)$,
under the optimum compaction condition. In
figure (a), the coefficient, k, of the
natural specimen decreases sharply as the
compaction is repeated: R = 1 → 2, and that
seems to be caused by the increase of fine
particles (See Fig. 5). In figure (b), the
coefficient of permeability decreases more
in the case of the repeated compaction test
than the no-repeated compaction test, and
this decrease is higher when the soil particles are finer. This is related to the
increase in fine particles caused by the
compaction which was described before.

6.3 Effects of fine particles

As shown above, the coefficient of permeability, k, decrease as the breakage of particles is promoted by the increase of compaction energy is plotted logarithmically
on the horizontal axis and the quantity of
particles which pass a 74 μm mesh is plotted
on the ordinate using its average and
standard deviation in Fig. 9 to show the
average of the coefficient of permeability
and its maximum and minimum ranges. From
this figure, it is found that the quantity
of fine particles decreases and the permeability is lowered as the compaction energy
is increased.
This research was conducted to see if the
decomposed granite soil can be used as dam
material, and it became evident that the

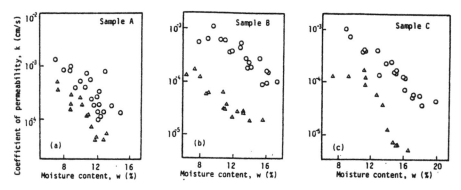

Fig. 7 Change of the coefficient of permeability by repeated compaction
(o no-repeated:R=0, ▲ Repeated:R=1)

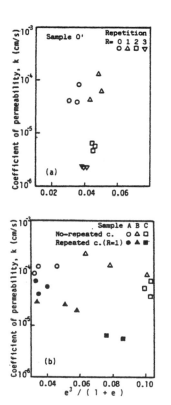

Fig. 8 Relation between the
coefficient of permeability
and $e^3/(1+e)$

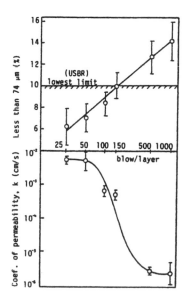

Fig. 9 Relationship among fine grain
content, coefficient of permeability
and compaction energy

less than 74 μm is more than 10%. According
the code, the decomposed granite soil can
be used after at least 500 blows/layer compacting, and its coefficient of permeability,
k, attains value of the order of 1.0×10^{-6}
cm.

soil tested, under current natural state,
was inadequate to for use in dams. The code
by the United States, Bureau of Reclamation
(USBR) provides that a soil can be employed
for a dam core if its content of particles

7 CONCLUSIONS

The particle breakage of the decomposed
granite soil has effects on its compaction
characteristics and permeability. These

effects were inverstigated by ladoratory experiment, and the results obtained from the study are summarized here.

(1) The maximum dry density of the decomposed granite soil is increased considerably as the compaction energy is increased.

(2) The above increase in maximum dry density is caused by densification accompanied by the breakage of particles.

(3) Fine particles are produced in proportion to the compaction energy, and the diameter range of most of the broken particles is 0.5 to 1.0 mm.

(4) Both compacting characteristics and permeability of the soil depend on its initial grading. The breakage of particles is promoted and the permeability is decreased by the repetition of compaction.

(5) As the compaction energy is increased, the content of fine particles is increased and the permeability is lowered significantly. According to the code by USBR, the soil tested in this study can be used as material for earth dams and rock-fill dams, after compacting under a compaction energy equivalent to 500 blows/layer.

ACKNOWLEDGEMENTS

Much of the experimental work was carried out by the students, Messrs. T. Naoe, X. Yoshihara, K. Fujinawa and T. Seki in The Dept. of Transportation Eng., Nihon University. The author's sincere gratitude is expressed to them.

REFERENCES

Ed. by Japanese soc. of soil mech. and found. eng. 1979. Engineering properties and its application of weathered granite and decomposed granite soil (in Japanese).

Matsuo, S. and T. Fukumoto 1976. Factors affecting the crushability of decomposed granite soil (in Japanese). Jour. of soil mech. and found. eng. Vol.16, No.4, pp. 85-92.

Ed. by Japanese soc. of soil mech. and found. eng. 1979. Soil testing methods - 2nd ed. - (in Japanese).

Taylor, D.W. 1948. Fundamentals of soil mech. John Wiley.

Ed. by Japanese soc. of soil mech. and found. eng. 1982. Handbook of soil engineering - 2nd ed. - (in Japanese).

Environmental Geotechnics and Problematic Soils and Rocks, Balasubramaniam et al. (eds)
© *1987 Balkema, Rotterdam. ISBN 90 6191 785 9*

Geotechnical properties of the very sensitive Ariake clay in Saga plain

R. Nakamura
Kyushu National Agricultural Experiment Station, Saga, Japan
K. Onitsuka, G. Aramaki & N. Miura
Department of Civil Engineering, Saga University, Japan

ABSTRACT: This paper describes the physical and mechanical properties of the Ariake clay in Saga Plain and the effect of salt content on both these properties. The natural water content of the clay is considerably higher than the liquid limit; the sensitivity ratio is so high that it is comparable with the quick clay of northern Europe. Good correlations exist between the salt content and the plasticity index, remoulded strength and sensitivity ratio.

1. INTRODUCTION

Saga Plain lies north of the Ariake Sea in central Kyushu (Fig.1) in the western area of the Chikugo River. It is a lowland of elevation of less than +5m, with an area of about 40,000 ha (400km^2). The Ariake Sea has a tidal range as wide as 6m, with a mean high-water level of +2.89m, which means that half of the Saga Plain is lower than the mean high-water level of the Ariake Sea. Thus, when typhoon hits during high tide, the dikes break and causes severe damage to Saga city.

Three rivers, the Chikugo, Kase and Rokkaku Rivers flow through the Saga Plain and into the Ariake Sea (Fig.2). The Chikugo River is the largest one in Kyushu. The Kase River flows from the granite area with a large amount of drifting sand, and the bed of the river, in some places, is higher than the

Fig.1 Location of Saga Plain

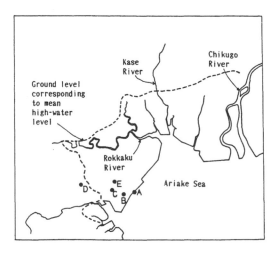

Fig.2 Environmental circumstances of the Saga Plain

neighboring plain. The periphery of the Kase River is geologically complicated, with traces of a sandy layer due to the inflow of soils caused by floods. There are a lot of mountains sandstone and andesite in the upriver area. The Rokkaku River has a gentle slope and meanders with muddy flow.

Saga Plain was made by the alluvial process of these rivers, the so-called Jyomon marine transgression of about 10,000 years ago. The tidal flat of the Ariake Sea even now develops 5 - 6 cm per year in thickness. Reclamation in Saga Plain has rapidly developed in the past several hundreds years though the gradual construction of dikes step, as shown in Fig.3. The Saga Plain has a very small gradient of 1/1,500 - 1/3,000. Extremely large amounts of mud have sedimented on the seaside part of the dike, and the seaside is higher than the inside of the dike. Thus this area has a serious problem of poor drainage (Nakamura,1984). The area around the Ariake Sea is covered with a blue clay or silt layer, called the Ariake clay layer. This layer is 15 - 20 m deep from the soil surface, with a maximum depth of 30 m. Below this layer is a layer of gravel containing diluvium sand and the pumice.

Figure 4 is a map showing lines of same thickness in the Ariake clay layer. This layer loosely sedimentes and is very soft, with N value of zero, as shown in Fig.5. The natural water content of the clay is

in general higher than its liquid limit hence the liquidity index is heigher than unity. Ariake clay has such serious problem-soil that the settlement due to consolidation caused by relatively light surface loads is considerable and slip failure in dikes occurs very easily (Onitsuka,1983). Furthermore, the lowland of this area continues to subside due to the pumping up of ground water. Ariake clay is one of the most problematic soils in Japan.

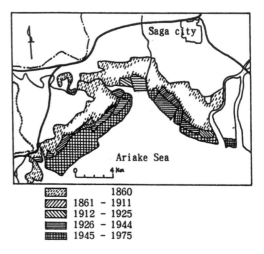

▨	1860
▧	1861 - 1911
▨	1912 - 1925
▤	1926 - 1944
▦	1945 - 1975

Fig.3 Development of reclamation in Saga Plain

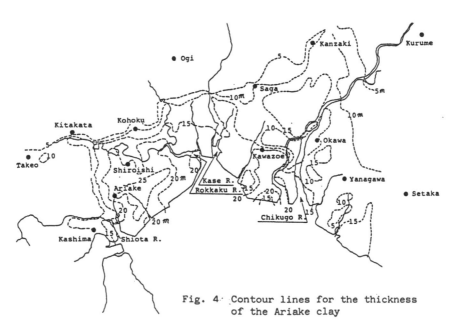

Fig. 4 Contour lines for the thickness of the Ariake clay

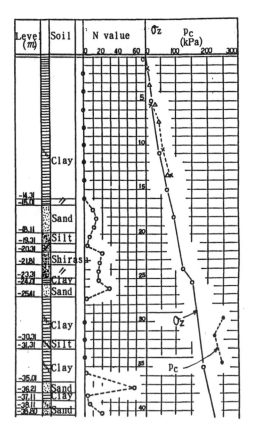

σ_z : Effective overberden pressur
P_c : Pre-cosolidation stress

Fig.5 Typical soil profile of Ariake clay

2. PHYSICAL AND MECHANICAL PROPERTIES OF ARIAKE CLAY

2.1 Physical properties

Table 1 shows the physical properties of Ariake clay. The clay minerals contained in are montmorillonite, illite, hydrohalloysite, and metahalloysite (Onitsuka et al.,1976). Ariake clay contains a lot of diatom remains.

The confidence interval of 95% for the distribution of the specific gravity of particles is 2.56 - 2.70 for clay particles, and 2.61 - 2.76 for silt particles. The clay ($<5\mu$m) content of the soil on the north shore, between the estuaries of the Kase River and the Chikugo River, is approximately the same as that of the soil on the east shore (south of Chikugo River), averaging 37% on ;the average for the soil on the west shore (west of Kase River, including Isahaya Bay) is high, of about 52%. Such a geological difference in grain size may result from the counterclockwise gulf flow of the Ariake Sea (Yamaguchi et al.,1964).

It is reported that the mechanical properties of the Ariake clay layer can be divided into two layers, the upper and lower layers, in terms of difference in environment at the time of sedimentation (Ariake Sea Research Group,1965). Fujikawa and Takayama (1980) classified the Ariake clay layer into the above layer and the below layer at a depth of 10m, and showed that there exist distinct differences between their basic properties. Some typical results for consistency and activity are shown in Table 1. One of the

Table 1   Physical properties of Ariake clay
(Fujikawa and Takayama, 1980)

	Ariake clay			Usual marine clay in Japan
Specific gravity Gs	Clay : 2.56 ~ 2.70 Silty soil : 2.61 ~ 2.76			2.65 ~ 2.75
Clay content (%)	West shore : 25.4 ~ 78.4 East and north shore : 11.4 ~ 63.4			
	Clay of upper layer	Clay of lower layer	Silty soil	
Liquid limit w_L (%) Plastic limit w_p (%)	60.1 ~ 138.7 30.5 ~ 59.1	53.5 ~ 111.7 28.0 ~ 54.2	27.9 ~ 97.3 24.0 ~ 50.0	45 ~ 100 20 ~ 40
Activity A	0.77 ~ 2.21	0.57 ~ 1.79	0.35 ~ 2.13	> 1.25
Natural water content w_n (%)	. 50 ~ 200*			.
Natural void ratio e_n	1.0 ~ 5.0*			
Liquidty index I_L	> 1			< 0.8

* by the authors

characteristics of Ariake clay is, as stated previously, that the natural water content is higher than the liquid limit and hence the liquidity index is larger than unity, as shown in Fig.6. Most of the Ariake clays are classified into CH, and have an activity A of over 0.75. In particular, the activity of the clay in the upper layer above 10m often exceeds the 1.25, of an active clay.

2.2 Mechanical properties

Table 2 shows the mechanical properties of Ariake clay.

1) Consolidation properties

The relationship between the compression index C_c and the natural void ratio e_n or natural water content w_n for Ariake clay

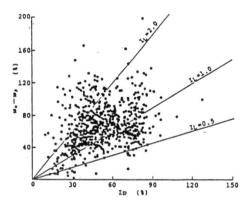

Fig.6 Liquidity index of Ariake clay

can be expressed by the following equation (Onitsuka et al.,1985);
$C_c = 0.49 (e_n - 0.41)$, $C_c = 0.013(w_n - 10)$.
For a consolidation pressure lower than 200 kPa, the coefficient of volume compressibility m_v is in the order of 10^{-3} (1/kPa), and for consolidation pressure higher than 200 kPa the coefficient is in the order of 10^{-4}(1/kPa).

The m_v corresponding to the consolidation yield stress p_c is mostly in the order of 10^{-3} (1/kPa). The coefficient of consolidation c_v is between $3.5 * 10^1$ and $1.5 * 10^3$cm2/d. Values in the order of 10^1 cm2/d are mostly observed in the west shore of the Ariake Sea including the Isahaya Bay.

2) Strength properties

The unconfined compressive strength, q_u, is $3 \sim 30$ kPa for the upper layer of about 10m depth and $30 \sim 100$ kPa for the lower layer. The axial strain at failure, ε_f, is 2 – 4 % for the undisturbed soil.

The relationship between the modulus of deformation E_{50} and the undrained shear strength $c_u (=q_u /2)$ is plotted in Fig.7. The values of both q_u and c_u increase linearly with increase in depth $z(m)$. Assuming that the q_u – z relationship can be expressed by a linear equation, q_u(kPa) = a+b*z, the constants a and b are determined as:a=4 $\sim$ 16 and b=14 $\sim$ 36.

When the specimens of the Ariake clay are remoulded, most of them become liquid and unconfined compression tests can no longer be carried out. It is then necessary to evaluate the remoulded strength by vane shear test. Thus the sensitivity ratio may be determined by

Table 2   Mechanical properties of Ariake clay

Compression index C_c	$C_c = 0.49(e_n - 0.41)$ $C_c = 0.013(w_n - 10)$		
Coefficient of volume compressibility m_v (1/kPa)	Consolidation pressure		
	p<200kPa	p>200kPa	p=Pc
	10^{-3}	10^{-4}	10^{-3}
Coefficient of consolidation c_v (cm²/d)	$3.5 \times 10^1 \sim 1.5 \times 10^3$		
Unconfined compressive strength q_u (kPa)	Upper layer	Lower layer	
	3 – 30	30 – 100	
Strain at failure in unconfined compression test ε_f (%)	2 – 4		
Constants a,b (q_u=a+bz)	a : 40 – 160 b : 14 – 36		
Rate of strength increase Cu/p	$\frac{1}{3}$		
Sensitivity ratio S_t	>8, >16 (most of Ariake clay)		

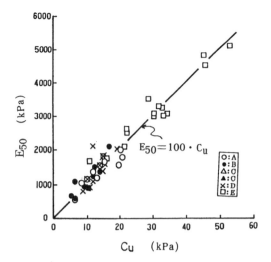

Fig.7 Relationship between undrained shear strength c_u and modulus of deformation E_{50}

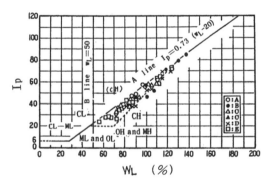

Fig.8 Plasticity chart for the tested samples

$S_t = q_u / 2c_r$, where c_r is the vane shear strength. The values of S_t for Ariake clay are shown in Table 2. The sensitivity ratio is mostly above 16, and sometimes as high as several tens or hundreds, as shown below. Ariake clay is indeed a very sensitive clay and may well be comparable with the quick clay in northern Europe. The rate of strength increase in consolidation, c_u / p, was about 1/3, as shown in Table 2, and this value is not so different from that of normal alluvial clays in Japan.

3. EFFECT OF ELAPSED TIME AFTER RECLAMATION ON MECHANICAL PROPERTIES

3.1 Sampling points

The clay near the coast of the Ariake Sea has a relatively short history of reclamation and most samples behave as normally consolidated clay. On the other hand, the clay of the inland area has a longer history of reclamation and there are many sites where the clay deposits in an overconsolidated state. In fact, the clays are sometimes overconsolidated by the effect of changes in ground water level, or the clays near the surface dried due to cultivation. The leaching effect seems to be changed with distance from the shore and the time since the reclamation. The time elapsed after reclamation and the salt content of the clay are considered to greatly influence the mechanical properties of Ariake clay. For example, stabilizing treatment using quicklime is known to be affected greatly by salt content (Miura et al.,1986).

At the five different times after reclamation in the Shiroishi area, clay samples were taken by thin-wall sampler, and the physical and mechanical properties of each sample was investigated, especially for time after reclamation and salt content.

The descriptions for the five points are as follows (the locations of each point are shown in Fig.2.).

A: within the Ariake Sea on the edge of the reclamation area
B: 30 years after reclamation
C: 250 years after reclamation
D: 1,500 years after reclamation, at the foot of a mountain
E: under a road which was constructed near the point C.

The thicknesses of the clay layers for the five points are 17, 22, 20, 10 and 21m for A, B, C, D and E, respectively. Figure 8 is the plasticity chart and shows the samples classified into CH like most Ariake clays. Figs. 9-13 show the soil profiles and the results of the physical and shear tests for these samples. All the soils consist mainly of fine particles, clay and silt, accounting for 90 ~ 98 % of the total weight. The plastic limit w_p is 35 ~ 40 %, irrespective of the location or depth of the sampling point. The liquid limit w_L is about 60 ~ 140 % at all points, and their values tend to decrease with increase in depth.

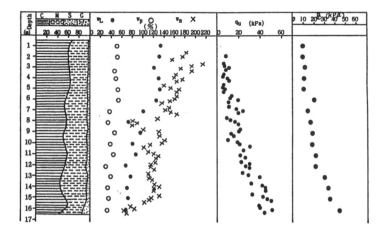

Fig.9 Soil profile of point A (within the Ariake Sea)

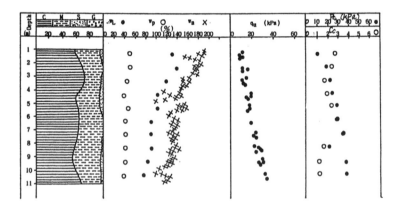

Fig.10 Soil profile of point B (30 years after reclamation)

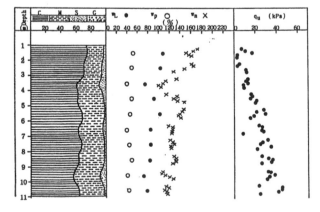

Fig.11 Soil profile of point C (250 years after reclamation)

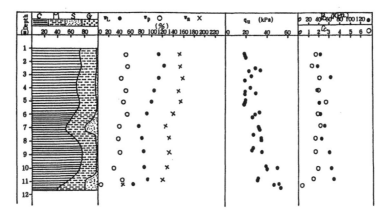

Fig.12   Soil   profile   of point   D   (1500 years after reclamation)

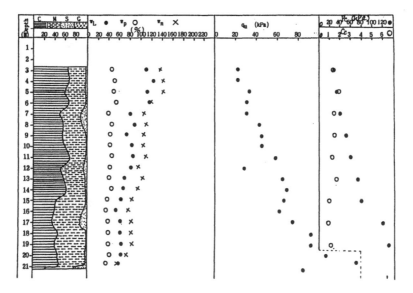

Fig.13   Soil profile of point E (under a road constructed near point C)

3.2   Effect of salt content   on   physical properties

Figure   14   shows   the   salt   content   at various   depths   at   the   five   sampling points.   The salt content was measured by the   following   method   (Manual   of Experiments in Agricultural   Chemistry, 1957): 50g   water   is   added   to   a   clay sample of 10g in dry state and the mixture is agitated for 1 hour.

Four drops of 5% potassium chromate   are added   to   10g   of   the   filtrate   of   this mixture   and   agitated.   This   solution   is titrated   by   0.02N   silvernitrate   solution and   the   amount   required   to   make   the

solution   red   brown,   M,   is   determined. Thus,   the   salt   content   is   calculated   by NaCl   (%)   = 0.001169 × M.   The   amount Cl(%)   is   obtained   by   multiplying   this value by 0.6067.

As shown in Figure 14,   the salt content is generally high near the surface, and it lowers   with   depth   z.   Considering   the distance   from   the   seashore and also   the time   elapsed   after   reclamation,   it   is natural   that   the salt content of point   A is   the highest among the five   samples. The greater the distance from the seashore and   the   longer   the   time   elapsed   since reclamation,   the   lower   the   salt   content. The salt content at point D,   the furthest

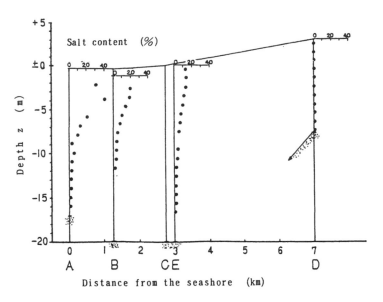

Fig.14   Distribution of salt content for the five points

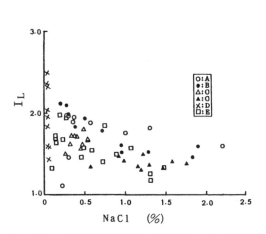

Fig.15   Relationship between liquidity index I_L and salt content NaCl

Fig.16   Relationship between plasticity index I_p and salt content NaCl

away from the seashore, is almost zero.
   The relationship between liquidity index and salt content is indicated in Fig.15, in which no correlation could be found between the two.   Figure 16 shows the relationship between plasticity index and salt content (since the plastic limit is almost constant, the change of plasticity index is regarded as the change in liquid limit).   The correlation between the two is good, i.e., the plasticity index decreases with decrease in salt content. Figure 17 indicates the relationship between activity and salt content, which seems to be relatively close.

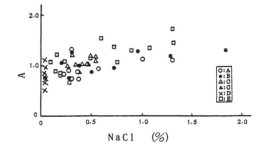

Fig.17   Relationship between activity A and salt content NaCl

540

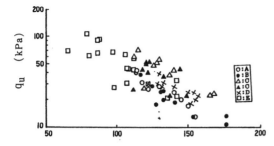

Fig.18 Relationship between unconfined compressive strength q_u and natural water content w_n

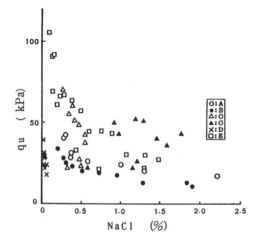

Fig.19 Relationship between unconfined compressive strength q_u and salt content NaCl

3.3 Effect of salt content on undrained shear strength

As shown in Figs. 9–13, the unconfined compressive strength q_u (=2c_u), increases roughly in proportion to the depth z and so does the consolidation yield stress p_c. As suggested by these data, except for sample E, Ariake clay is generally seen as a normally consolidated clay. As shown in Fig.18, there is a close relationship between the unconfined compressive strength and the natural water content. However, the unconfined compressive strength is not well correlated with salt content, as indicated in Fig.19. These results suggest that the undrained shear strength of undisturbed soil is strongly dependent on water content, and hardly at all dependent on salt content.
Figure 20 gives the plots of water content and remoulded strength determined

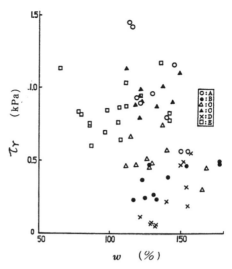

Fig.20 Plots of remoulded strength and salt content NaCl

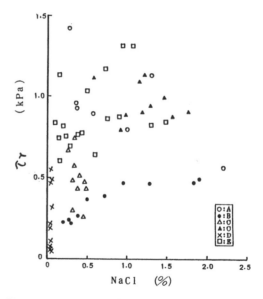

Fig.21 Relationship between remoulded strength and salt content NaCl

by vane shear test. The points are widely dispersed, and no clear relationship can be detected between the two. On the other hand, the vane shear strength of the remoulded soil plotted against the salt content shows, as indicated in Fig.21 for each clay, that the strength increases with the increase in salt content. That is, undrained shear strength of remould clay is dependent on the salt content but hardly at all on the water content,

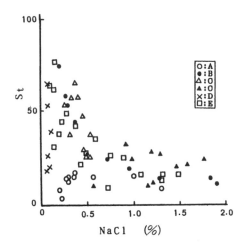

Fig.22 Relationship between the sensitivity ratio S and salt content NaCl

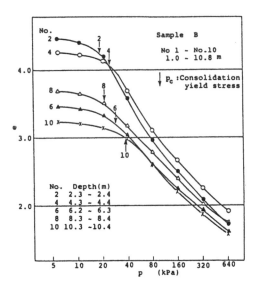

Fig.23 e – log p curves for the tested Ariake clays

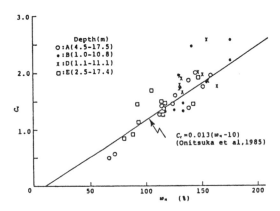

Fig.24 C_c – w_n curves for the tested Ariake clays

unlike the case with the undisturbed sample.

The relationship between the sensitivity ratio and salt content is presented in Fig. 22, showing that the sensitivity ratio decreases with increasing salt content. The points with high sensitivity ratios of 200 ~ 300, are the ones for the clay sampled near the foot of the mountains, point D. The salt content of clays sampled at this point is near zero, probably due to leaching. Careful work is required in earth moving in this area because the strength of the clay can easily decrease. Sensitivity ratios as high as 300 ~ 500 were found in the area 2 ~ 3 km away from point D. On the other hand, the sensitivity ratio at point A, located within the sea, is the lowest of all five points because the remoulded strength is the highest due to the high salt content (Nakamura,1984).

The sensitivity ratio is known to increase with increase in liquidity index (Skempton,1953), but no clear relationship between these two parameters can be found for the above five point samples.

3.4 Consolidation properties

1) e – logp characteristics and compression index

Figure 23 shows an example of a e – logp curve obtained for Ariake clay at point B at a depth of 1.0 – 10.8 m. It can be seen that the natural void ratio decreases and consolidation yield strength increases with the increase in depth. The e – logp curve appears to contract to a bundle with

the increase in consolidation pressure. A similar tendency was seen in the samples at other points.

The compression index C_c for the Ariake clay at the four points, A, B, D and E were plotted against natural water content w_n in Figure 24. The straight line in the figure was determined statistically from many data on Ariake clay, as mentioned in section 2.2. The data obtained for the four point samples in the Shiroishi area also fitted this line and the C_c – w_n can also be expressed by $C_c = 0.013 (w_n - 10)$.

The four samples from points A, B, D and E are different not only in the depth at which the samples were obtained but also

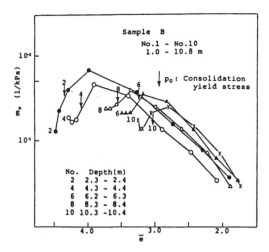

Fig.25 Change of the coefficient of volume compressibility m with increasing mean void ratio e

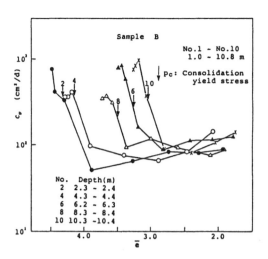

Fig.26 Change in the coefficient of consolidation c_v with increasing mean void ratio $\bar{e}$

in their different w_n and e_n values. The time elapsed after reclamation and the amount of salt content are also different. However, neither the e – logp characteristics nor the compression index seem to be affected by the differences in the environmental and historical conditions.

2) Consolidation parameters

From the standard consolidation test the coefficient of volume compressibility, m_v (1/kPa), the coefficient of consolidation c_v (cm^2/d), and the coefficient of permeability k(cm/sec) were determined.

The relationship between the coefficient of volume compressibility and the mean void ratio $\bar{e}$ is shown in Fig.25. The value of m_v increases with decrease in $\bar{e}$ when the consolidation pressure is smaller than the consolidation yield stress (i.e.,overconsolidation region). In the normal consolidation region, on the other hand, m_v decreases with decrease in e. The $m_v - \bar{e}$ curves are parallel to each other irrespective of the difference in natural void ratio. As shown in this figure, the coefficient of volume compressibility m_v ranges between $1.5 \cdot 10^{-4}$ and $5.0 \cdot 10^{-3}$ (1/kPa).

The coefficient of consolidation c_v decreases with the decrease in the mean void ratio in an overconsolidation region, 1/10 – 1/50 of the c_v for the low pressure region (Fig.26). In the normal consolidation range, the value of c_v increases slightly with the decrease in mean void ratio. For the whole region

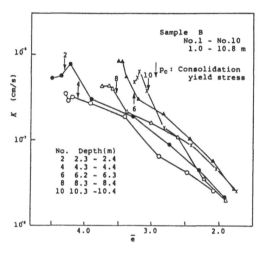

Fig.27 Change of the coefficient of permeability k with increasing mean void ratio $\bar{e}$

tested, the coefficients of consolidation for the samples of four points ranged between 30 and 3,000 (cm^2/d), but as far as normal consolidation pressure is region concerned, the range of values is restricted to 30 $\sim$ 200 (cm^2 /d).

The coefficient of permeability k, decreases linearly with the decrease in the mean void ratio $\bar{e}$ (see Fig.27). The lines are almost in parallel irrespective of the value of the natural void ratio. The coefficient of permeability for the clay at the four points ranges from 1×10^{-8}

to 2×10^{-6}(cm/sec). When the curves of k-$\bar{e}$ for the four points are plotted on the same figure, it is found that the coefficient for points A, C and D are nearly the same, but that for point E it is slightly higher.

From the above discussion on the consolidation parameters, it may be said that the time elapsed after reclamation and salt content do not affect the consolidation constants.

4. CONCLUSIONS

1) The natural water content of Ariake clay is, in general, considerably higher than the liquid limit. The sensitivity ratio of this clay is so high that it is comparable with the quick clay of northern Europe.

2) The salt content of the clay is generally high near the surface, and decreases with the increase in depth. The further inland from the seashore, and the longer the time elapsed since reclamation, the smaller the salt content.

3) There exists a good correlation between salt content and plasticity index. The sensitivity ratio has a close relationship with the salt content.

4) The undrained shear strength of the undisturbed clay depends strongly on the water content and that of the remoulded clay depends on the salt content.

5) Neither the time elapsed after reclamation nor the salt content affect consolidation characteristics such as the e - logp property, compression index, coefficient of consolidation and coefficient of permeability.

REFERENCES

Ariake Bay Research Group 1965. Quaternary system of the Ariake and the Shiranui bay areas, with special reference to the Ariake soft clay. The Association for the Geological Collaboration in Japan. pp.1-86 (in Japanese)

Fujikawa,t. and Takayama,M. 1980. Physical properties of Ariake marine clay - Geotechnical properties of Ariake marine clay (1). Transactions of the Japanese Society of irrigation, Drainage and Reclamation Engineering. pp.51-61 (in Japanese).

Manual of Experiments in Agricultural Chemistry. Vol.1,Sec.4,Department of Agricultural Chemistry. Kyoto University. Sangyo Tosho Pub. Co.,Ltd. 1957. (in Japanese).

Miura,N.,Koga,Y. and Nishida,K. 1986. Application of a deep mixing method with quicklime for Ariake-clay ground. Tsuchi-to-Kiso, JSSMFE, Vol.34, No.4, pp.5-11 (in Japanese).

Nakamura,R. 1984. Mechanical properties of the Ariake clay for the construction and the investigation. Report of the Society for lowland Disaster Prevention Technology. pp.23-73 (in Japanese).

Onitsuka,K.,Hayashi,S.,Hirata,T. and Murata,S. 1976. Studies on anisotropy of Ariake alluvial clay. Journal of the Japanese Soil Mechanics and Foundation Engineering. Vol.16,No.3. pp.111-121 (in Japanese)

Onitsuka,K. 1983. Ariake clay - Problem soils in Kyushu and Okinawa -. Kyushu Branch of the Japanese Society of Soil Mechanics and Foundation Engineering. pp.23-39 (in Japanese).

Onitsuka,K. and Yoshitake,S. 1985. Correlation between compression index and index properties of Ariake marine clay. Tsuchi-to-kiso, JSSMFE, Vol.33, No.4. pp.57-59 (in Japanese).

Skempton,A.W. 1953. The Colloidal 'Activity' of Clays. Proceedings of the 3rd Int.Conf.Soil.Mech.,Vol.1. pp.57-61.

Yamaguchi,E.,Nanba,N.,Oka,A.,Nagaishi,Y. and Muraoka,Y. 1964. Geotechnical properties of alluvial clays in the tidal shore of Ariake bay. Bulletin of the Agricultural Engineering Research Station. No.2. pp.509-566 (in Japanese).

Environmental Geotechnics and Problematic Soils and Rocks, Balasubramaniam et al. (eds)
© *1987 Balkema, Rotterdam. ISBN 90 6191 785 9*

Influence of initial water content and soaking on compressibility and shear strength of decomposed granite soil

K.Onitsuka & S.Yoshitake
Department of Civil Engineering, Saga University, Japan

ABSTRACT : Decomposed granite soil, a representative residual soil in Japan, is widely distributed in the western part of the country. Natural and man-made slopes of decomposed granite soil, which are stable under normal dry conditions, fail easily due to rainfall. On the basis of many data from consolidation and shear tests on various decomposed granite soils, the influence of initial water content and soaking on compressibility and shear characteristics were clarified.

1 INTRODUCTION

Decomposed granite soil, a representative residual soil in Japan, is widely distributed in the western part of the country. Large deposits of decomposed granite soil are also found in the northern part of Kyushu (the western-most island). The distribution of granite in Japan is shown in Fig.1(JSSMFE,1982) and occupies an area of 13% of the country's area.

Because of compaction efficiency and ease of acquisition, this soil is often used as an embankment material for roads, housing and earth dam construction. However, it is classified as a "problem soil" in Japan for the following reasons :
  1. its physical and mechanical properties depend on the degree of weathering
  2. the soil particles break easily
  3. it shows settlement due to soaking
  4. it is vulnerable to water-induced changes due to non-cohesive sandy soil
It is well-known that natural and man-made slopes, stable under normal conditions, fail easily due to rainfall. Soil structure strength decreases because of decrease in effective stress by soaking. Thus, soil collapse results from shear deformation. On the basis of our many data from many tests, the influence of initial water content and soaking on compressibility and shear strength of decomposed granite soil were elucidated.

2 SAMPLES

The samples were obtained from five sites in Saga prefecture and from two sites in Fukuoka prefecture in Kyushu, Japan. They were tentatively labeled A, B, C, D, E, K and L . Their respective soil properties are shown in Table 1. Fig.2 shows grain size accumulation curves of the samples. Consolidation and shear tests (direct shear test, triaxial compression test) were carried out under "Unsoaked" conditions(US) and "Soaked" conditions(S). The details of specimen preparation and the test procedure have already been presented in other papers (Uchida et al., 1968, Onitsuka 1974, Onitsuka and Hara 1975, Onitsuka and Nanri 1982, Onitsuka 1983, Onitsuka et al., 1985).

3 INFLUENCE OF SOAKING ON COMPRESSIBILITY

3.1 Mechanism of settlement due to soaking

Fig.3 shows particles in state of equilibrium under an uniaxial constant load. The particles have menisci at each contact point. Consider the settlement of partially saturated decomposed granite soil due to soaking. As is well-known, Bishop's effective stress equation for partially saturated soils is,

$$\sigma' = \sigma - \overline{u} = \sigma - u + \chi (u_a - u_w) \quad \cdots\cdots (1)$$

where

σ' :effective normal stress
σ :total normal stress
$\overline{u}$:equivalent pore pressure
u_a :pore air pressure
u_w :pore water pressure
χ :parameter, which is unity for saturated soils and zero for dry soils.

One of the authors designated the increase in effective stress $\chi (u_a - u_w)=p$, as intergranular stress due to suction and the value of p , that is , $\chi u_w =p^*$, when pore air pressure is zero ($u_a=0$) , as intergranular stress due to capillary water. The relation among σ' , σ , u_a and u_w can be expressed by Bishop's equation, stated above. Usually, pore air pressure is equal to the atmospheric pressure ($u_a=0$) and hence,

$$\sigma' = \sigma + \chi (- u_w) = \sigma + p^* \quad \cdots\cdots\cdots (2).$$

Consequently, shearing resistance at this state is given by equation (3)

$$\tau = c' + (\sigma + p^*) \tan\phi' \quad \cdots\cdots\cdots (3).$$

Assuming the strength parameter c' and ϕ' are constant regardless of water content, shearing resistance under soaked conditions is given by equation (4) since the intergranular stress due to capillary water disappears:

$$\tau = c' + \sigma\tan\phi' \quad \cdots\cdots\cdots\cdots (4).$$

Shearing resistance after soaking is smaller by a value of $p^*\tan\phi'$ at least than that before soaking. Slides of particles to condense state, namely settlement, may occur due to a decrease in shearing resistance with soaking.

3.2   Settlement due to soaking

Fig.4 shows the relation between settlement due to soaking and initial water content. After consolidation of statically compacted decomposed granite soil($\phi6.0\times2.0$cm) under constant normal stress, water is supplied to the consolidometer, and the sample is soaked. The settlement should show a peak value at a water content of 5% where intergranular stress p^* is maximum. But the dry samples showed maximum settlement at a 0% water content due to a decrease in shearing resistance resulting from the lubrication effect of water. Settlement after soaking of partially decomposed granite soil, which is sandy soil with a non-plastic or nearly non-plastic consistency, should occur with the decrease in shearing resistance due to the disappearance of intergranular stress from capillary water and due to lubrication.

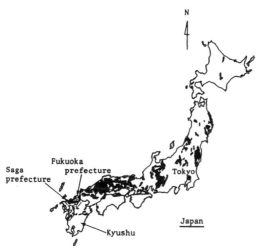

Fig.1   Distribution of granite

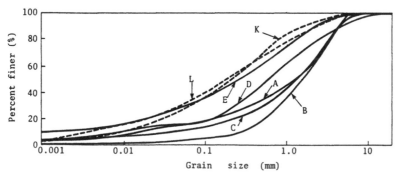

Fig.2   Grain size accumulation curves

Table 1　　　　Soil properties

Sample		A	B	C	D	E	K	L
Sampling　　sites		Saga city Kawakubo	Karatsu city Yabuta	Saga prefecture Kanzaki(ozaki)	Saga prefecture Kawakami	Saga city Kawakubo(kita)	Fukuoka city Kanayama	Fukuoka city Kanakuma
Specific gravity　G_s		2.64	2.61	2.64	2.60	2.67	2.63	2.65
Apparent specific gravity G_a		2.22	2.20	—	2.34	—	—	—
Distribution (%)	Gravel	37	40	36	24	9	8	8
	Sand	45	56	51	63	56	63	58
	Silt	12	2	7	9	20	21	26
	Clay	6	2	6	4	15	8	8
Consistency (%)	W_L	49	45	48	NP	46	36	39
	W_p	29	24	31	NP	32	31	29
	I_p	20	21	17	NP	14	5	10
Classification		SM	SM	SM	SM	SM	SM	SM
Natural water content　$W_n(\%)$		21.5	20.7	10.6	—	—	—	—
Natural void ratio　　e_n		0.86	0.70	0.75	—	—	—	—

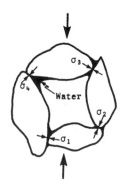

Fig.3　Menisci at each contact
　　　　point of particle

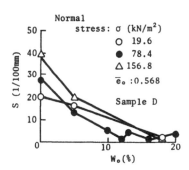

Fig.4　Settlement due to soaking

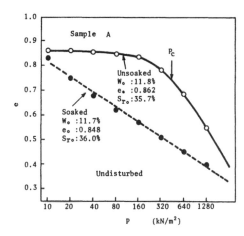

Fig.5　e-log p curves of undisturbed decomposed
　　　　granite soil (Unsoaked , Soaked)

The e-log p curve obtained from the consolidation test ($\phi6.0\times2.0$cm) under unsoaked and soaked conditions is now considered. In the case of the unsoaking test, specimens were consolidated for 30 minutes at each loading step. Other consolidation test procedures were based on the standard consolidation test. In the case of the soaking test, water was supplied to a consolidometer and the specimens were consolidated after being saturated. Loading time and other steps in the soaking test were the same as in the unsoaking test. Fig.5 shows e-log p curves of undisturbed decomposed granite soil. After soaking, the consolidation yield stress p_c disappeared and e-log p curves became a straight line such as the normally consolidated clay. The e-log p curve under unsoaked conditions was very similar to the curve for undisturbed clay. The e-log p curve under soaked conditions was very similar to that for disturbed clay. We designated the straight line after soaking, the soaking line. However, when the specimen's void ratio was small, for example with a natural void ratio e_o, or compacted void ratio e_n was below the soaking line in the e-log p curve, no large settlement due to soaking occurred. The relation of the e-log p curve for soaked specimens of comparative high density did not show a straight line such as that shown in Fig.6. Other soil showed

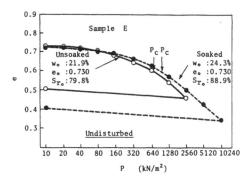

Fig.6   e-log p curves of undisturbed decomposed granite soil (Comparative high density, Unsoaked , Soaked)

this phenomenon as well as decomposed granite soil. Settlement due to soaking, i.e., collapse, has been observed in other kinds of soil(Onitsuka and Hara 1975).

4   SHEAR CHARACTERISTICS

4.1   Shear strength and intergranular stress of partially saturated decomposed granite soil

We have previously stated that the effective stress of partially saturated soil in relation to shear strength could be expressed by Bishop's equation. The increase in effective stress $\chi(u_a - u_w) = p$ was termed intergranular stress due to suction ; the value of p, i.e., $-\chi u_w = p^*$ was termed intergranular stress due to capillary water for zero pore air pressure ($u_a = 0$) :
  Here, the influence of intergranular stress on shear strength is considered. Generally, the lower the degree of saturation (the smaller the radius of curvature of the meniscus between particles), the greater the suction ($u_a - u_w$) becomes. Under saturated conditions, since the pore air pressure u_a is equal to the pore water pressure u_w, the value of suction is zero. The value of χ, as stated earlier, is a parameter dependent on the degree of saturation, with a value of 1.0 for 100% saturation and 0 for dry soil. It is considered that the relationship between the suction ($u_a - u_w$), the value of χ and the degree of saturation is given as in Fig.7(a). Therefore, the intergranular stress p ($p = \chi(u_a - u_w)$) due to suction, that is, the product of multiplication suction ($u_a - u_w$) and χ is 0 for saturated soil and 0 for dry soil, as shown

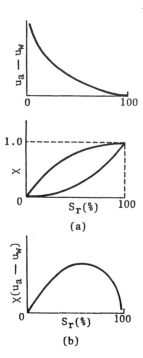

(a)

(b)

Fig.7   Relation between intergranular stress due to suction and degree of saturation

in Fig.7(b). The intergranular stress exhibits a peak value at some degree of saturation. It is estimated that intergranular stress $-\chi u_a = p^*$ due to capillary water in the case of $u_a = 0$ shows a similar tendency in intergranular stress due to suction for the degree of saturation.
Consideration was given to non-cohesive material whose mechanical properties are not so influenced by soil structure or electro-chemical force despite difference in initial water content. Disregarding the influence of suction between particles and considering only pore air pressure, effective stress is given as

$$\sigma'' = \sigma - u_a \quad \cdots\cdots\cdots\cdots\cdots(5).$$

(σ'' is different from σ' in Eq.(1))

Assuming the strength parameters of soil, c', ϕ', to be constant, shearing resistance in termes of effective stress is given by

$$\tau = c' + (\sigma - u_a)\tan\phi' = c' + \sigma''\tan\phi' \cdots(6).$$

If the intergranular stress due to suction is considered, $(\sigma - u_a + \chi(u_a - u_w) = \sigma'' + p)$ acts on soil skeltons. Shearing resistance can be obtained from the following equation in the same way as shown above.

$$\tau = c' + (\sigma'' + p)\tan\phi'$$
$$= c' + p\tan\phi' + \sigma''\tan\phi'$$
$$= (c' + \Delta c) + \sigma''\tan\phi'$$

where, $\Delta c = p\tan\phi'$.

Thus, it is understood that cohesion apparently increases by intergranular stress due to suction. The same consideration applies to intergranular stress due to capillary water in the case of $u_a = 0$.

The results indicate that in the case of non-cohesive material whose angle of shearing resistance ϕ' is unchangeable regardless of differences in water content, intergranular stress p or p* due to suction or capillary water mainly affected the apparent cohesion Δc. If effects of intergranular stress are great, the apparent cohesion Δc of non-cohesive material may indicate a peak value at some degree of saturation against water content, as already shown in Fig.7(b).

In soaking, the intergranular stress due to suction or capillary water disappears as does the apparent cohesion Δc. The disappearance of intergranular stress and apparent cohesion may be related to the shear strength of decomposed granite soil, which is non-plastic or nearly non-plastic.

4.2 Influence of soaking

4.2.a Relationship between shear stress and displacement

Decomposed granite soil exhibits less strength for soaking. Fig.8(a) and (b) shows the relationship between shear stress, vertical displacement and horizontal displacement of partially saturated decomposed granite soils under unsoaked and soaked conditions. When decomposed granite soils are soaked, shear strength becomes less than for unsoaked soil and no peak point is evident in the stress-displacement curve. Moreover, vertical displacement shows a tendency to shrink because of soaking. This is attributed to the ease of reorientation because of the disappearance of suction, increased particle breakage activity (Matsuo and Fukumoto 1976, Miura 1979) and the lubrication effects of water due to soaking.

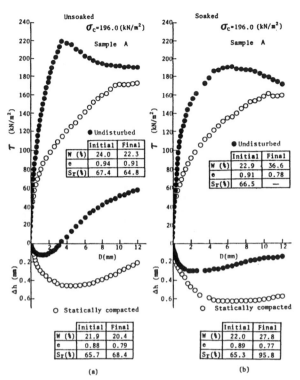

Fig.8 Relationship between shear stress τ, vertical displacement Δh and horizontal displacement (Constant normal stress, Unsoaked, Soaked)

4.2.b Pore pressure

In the case of partially saturated decomposed granite soil, the pore air pressure in equation(1) was measured. The pore air pressure of specimens in the triaxial compression test (Onitsuka 1973) increases rapidly at a degree of saturation of more than 70%. This tendency was found not only for decomposed granite soil but also for other fine-grained soils. The pore air pressure of saturated soil is given by,

$$\Delta u_a = B_a (\Delta\sigma_3 + A_a (\Delta\sigma_1 - \Delta\sigma_3))\ldots\ldots\ldots(8)$$

where A_a, B_a are pore air pressure coefficient.

Fig.9 shows the relationship between pore air pressure coefficients A_a, B_a and initial degree of saturation from measurments on decomposed granite soil from Kanayama in Fukuoka prefecture. For both decomposed granite soil and other fine-grained soils compacted at or near optimum water content, B_a was about 0.3 at maximum and increased rapidly over

optimum water content. When the compacted
density was high, these soils expanded
during shear. A_a became negative. When
the density was low, A_a was positive
and a peak value at a certain degree of
saturation was noted. This tendency was
also observed for other fine-grained
soils.

4.2.c   Shear strength and dilatancy

The constant normal stress direct shear
test was carried out under unsoaked and
soaked conditions after consolidation of
decomposed granite soil compacted at
different initial water contents. Fig.10
shows the relationship between shear
stress, vertical displacement and
horizontal displacement. Fig.11 shows the
relationship between shear strength and
initial water content. Shear
strength was estimated so as to
find a peak value at a water
content of 5% where intergranular
stress may be maximum. Since the
intergranular stress due to
capillary water disappeared with
the soaking, shear strength showed
the same value regardless of the
initial water content. However,
compared with unsoaking, shear
stress showed a considerable
decrease. Fig.12 shows a decrease
in shear strength due to soaking of
undistured decomposed granite soil
with a natural water content and
statically compacted specimens

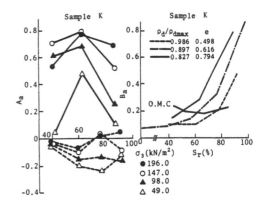

Fig.9   Relationship between pore air pressure
coefficient and the initial degree of
saturation

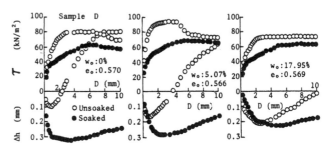

Fig.10   Relationship between shear stress τ , vertical
displacement Δh and horizontal diplacement D
(Compacted decomposed granite soil , Unsoaked,
Soaked , σ_c=78.4kN/m^2))

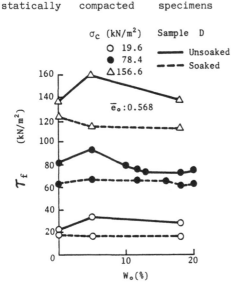

Fig.11   Relationship between shear strength
and initial water content (Constant
normal stress test)

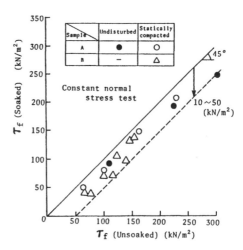

Fig.12   Influence of soaking on shear
strength (Constant normal stress)

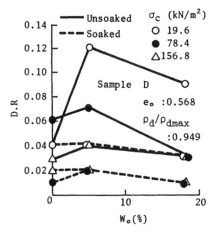

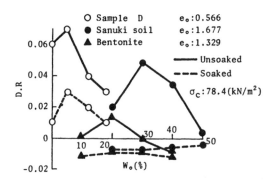

Fig.13 Relationship between dilatancy
rate and initial water content

Fig.14 Relationship between dilatancy rate
and initial water content
(Other compact fine-grained soil ,
Unsoaked , Soaked)

having the same water content and density as the undisturbed specimens. The shear strength decreased by $10 - 50(kN/m^2)$ due to soaking. Thus, disturbance of soil during sampling and compaction, regardless of whether it is undisturbed soil or compacted soil, does not have much effect on the decrease in shear strength due to soaking.

Next, consider dilatancy. Regarding volume change during shear in the case of unsoaking, a compacted specimen of dry side of optimum water content and of high density showed distinct expansion. However, a specimen of wet side of optimum water content did not show the same tendency. Since the effects of intergranular stress due to capillary water are considered the same as the role of interlocking, dry specimens of optimum water content undergo expansion. The dilatancy rate ($D.R=d(\Delta h / \Delta D)$) . i.e., gradient of vertical displacement Δh and horizontal displacement ΔD at failure showed the peak value at the dry side of optimum water content(w_{opt} =13.5%), as shown in Fig.13. In the soaking of specimens of decomposed granite soil compacted so as to have the same water content and density, there was no such expansion even though the density was high. The dilatancy rate of soaked specimens was about $1/2 - 1/3$ that for unsoaked ones. A similar tendency was noted for other

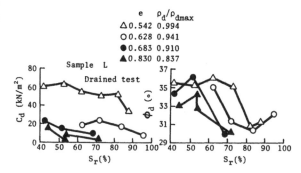

Fig.15 Relationship between apparent
cohesion, angle of shearing
resistance and initial degree
of saturation (Drained test)

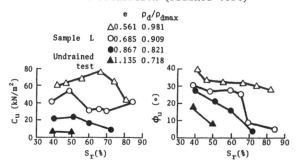

Fig.16 Relationship between apparent
cohesion, angle of shearing
resistance and initial degree
of saturation (Undrained test)

compacted fine-grained soils, as shown in Fig.14, although there were some differrnces.

4.2.d Strength parameter

The sample used was decomposed granite

soil from Kanayama in Fukuoka prefecture. The specimens were compacted in a mould so as to have a specified void ratio and degree of saturation. For these specimens, a triaxial compression was carried out under consolidated-undrained and consolidated-drained conditions. The results in Fig.15 and 16 show the relationship between cohesion c, angle of shearing resistance ϕ and degree of saturation. Cohesion c tended to decrease with saturation under both undrained and drained conditions. Cohesion becomes nearly zero when the compacted density is low. In the drain test, the angle of shearing resistance was virtually unaffected by a decrease in degree of saturation ; there was only a decrease from 36° to 30° . However, in the undrained test, the degree of saturation influenced the angle of shearing resistance, and in particular, when the compacted density was low, the angle of shearing resistance became less than 10°, approaching 0°.

Fig.17 shows the effect of initial water content on the most coarse-grained decomposed granite soil in the drain direct shear test. In the unsoaking test, cohesion c_d showed a peak value at the dry side of optimum water content as is also evident in Fig.15 and 16 for decomposed granite soils from Kanayama in Fukuoka prefecture and other fine-grained soils. Cohesion c_d following soaking was nearly the same value regardless of differences in initial water content, being $1/2 - 1/3$ that of cohesion c_d of unsoaked soils. In Fig.17, differences in cohesion of unsoaked and soaked specimens are the apparent cohesion (Δc) by intergranular stress, and the maximum Δc was obtained at dry side of optimum water content. This indicated that there is a relationship between apparent cohesion and degree of saturation, as mentioned in 4.1, and that the conclusions drawn from Fig.7(b) are correct. The angle of shearing resistance decreased by only 3°— 4° with initial water content, regardless of unsoaking or soaking. Thus, the water content does not have so much influence on the angle of shearing resistance, as also evident from Fig.15. Fig.18 shows a decrease in cohesion of various soils due to soaking. The meaning of the symbols in Fig.18 are as follows :

CS :consolidated constant normal stress direct shear test

CV :consolidated constant volume direct shear test

UD :undisturbed specimen

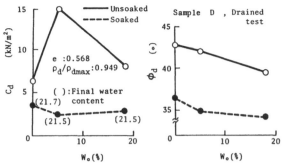

Fig.17 Relationship between apparent cohesion, angle of shearing resistance and initial water content (Constant normal stress test)

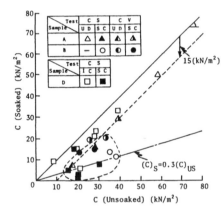

Fig.18 Decrease in apparent cohesion due to soaking

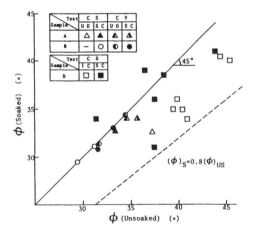

Fig.19 Decrease in angle of shearing resistance due to soaking

552

SC :static compaction
IC :impact compaction.

The undisturbed decomposed granite soil samples showed a decrease of about $15 (kN/m^2)$ at most. Specimens prepared by static compaction underwent a marked decrease due to soaking and their cohesion after soaking was about $15 (kN/m^2)$. That is, the cohesion of soaked soil was about 30% of that of unsoaked soil. The effect of soaking on the angle of shearing resistance is given in Fig.19. As mentioned before, the angle of shearing resistance was virtually unaffected by soaking and was on the 45° line for each soil. This angle of shearing resistance for compacted decomposed granite soils with different initial water content and comparatively high density may decrease by 20% due to soaking. The decrease in shear strength due to soaking depends mainly on the decrease in apparent cohesion regardless of the angle of shearing resistance.

5   CONCLUSIONS

Consolidation and shear tests were conducted on various decomposed granite soils. The effect of initial water content and soaking on compressibility and shear characteristics were clarified.

1) Settlement of partially saturated decomposed granite soil after soaking is caused by the decrease in shearing resistance due to the disapperance of intergranular stress from capillary water and due to lubrication.
2) In the case of different initial water contents, the shear strength shows a peak value at the dry side of optimum water content where intergranular stress due to capillary water may be maximum. Soaking the specimens, shear strength becomes less than that without soaking and shows the same value regardless of the initial water content.
3) Volume undergoes distinct expansion at the dry side of optimum water content where intergranular stress due to capillary water may be maximum. This may be due to the fact that the effect of intergranular stress from capillary water is similar to that of interlocking. Also, the dilatancy rate shows a peak value at the dry side of optimum water content. In soaking, there was not very much expansion, and the dilatancy rate was small. This value becomes 1/2 – 1/3 that of unsoaked specimens.
4) Cohesion shows a considerable decrease

due to soaking in the case of the drained test. This value is about 30% that of unsoaked specimens. The angle of shearing resistance is virtually unaffected by soaking. The decrease in shear strength due to soaking depends mainly on the decrease in cohesion.
5) The above conclusions, (1) – (4), can be explained by Bishop's equation.

6   REFERENCES

JSSMFE 1982, Handbook of Soil Mechanics and Foundation Engineering, pp.1459.
Matsuo, S. and Fukumoto, T. 1976, Factors affecting the crushability of decomposed granite soil. JSSMFE, Vol.16, No.4, pp.84-92(in Japanese).
Miura, N. 1979, A mechanism of deteroration of decomposed granite soil due to submergence. Journal of National Disaster Science, Vol.1, No.1, pp.67-76.
Onitsuka, K. 1973, Experimental study on pore pressure and mechanical characteristics of compacted partially saturated soils. Dr. Eng. Thesis presented to Kyushu University, Japan(in Japanese).
Onitsuka, K. 1974, Effect of water on the Shear Characteristics of cohesionless Materials. Report of the Faculty of Science Engineering Saga University, Japan, No.2, pp.7-15(in Japanese).
Onitsuka, K. and Hara, Y. 1975, Effect of soaking on the mechanical characteristics of compacted soils. Proc. 5th Asian Reg. Conf. SMFE, Vol.1, pp.117-124.
Onitsuka, K. and Nanri, M. 1982, Shear strength of undisturbed and compacted decomposed granite soils. Report of the 17th Japan Annual meeting on Soil Mechanics and Foundation Engineering, JSSMFE :461-464(in Japanese).
Onitsuka, K. 1983, Mechanical properties of decomposed granite soil. Proc.7th Asian Reg. Conf. SMFE, Vol.1, pp.352-357.
Onitsuka, K., Yoshitake, S. and Nanri, M. 1985, Mechanical properties and strength anisotropy of decomposed granite soil. Soils and Foundations, Vol.25, No.2, pp.14-35.
Uchida, I., Matsumoto, R. and Onitsuka, K. 1968, Shear characteristics of partially saturated soils. Soils and Foundations, Vol.8, No.3, pp.33-45.

Environmental Geotechnics and Problematic Soils and Rocks, Balasubramaniam et al. (eds)
© 1987 Balkema, Rotterdam. ISBN 90 6191 785 9

Instability detection in rock

Blahoslav Kostak
Czechoslovak Academy of Sciences, Institute of Geology and Geotechnics, Praha

ABSTRACT: Instability detection in rock may solve environmental, technical and scientific problems. There are simple means of crack monitoring like mechanical and moire gauges. Three-dimensional measurement is desirable to recognize the effects of shearing, and climatic reversible effects must be discerned in it. Crack monitoring proved to be useful in finding creep movements as well as in checking different theories about fracturing and stability prognostication on natural slopes, roadcuts and structures.

Crack monitoring

There are many factors that change the natural balance in our environment, which is often marked by deformation or even destruction of natural as well as engineering objects. Besides, there are often indications of a natural process, the activity of which is not clear, and a decision is needed with regard to the stability of the objects involved. The instability can often be detected as displacement. Generally, geodesy would be applied with all its instrumentation and requirements, and these involve limitations with respect to the number of measurements which can be made during a year. A cheaper and often more appropriate method can be used to check cracks, fissures and fracture zones, which usually react in a very sensitive way to any instability. A number of cracks on natural rock slopes, in caves and galleries, on tectonic zones, in foundations of historical buildings, etc. was successfully investigated, and the movement was not only detected but also classified in terms of the process which was occurring in the natural environment.

To make a succesful check of this kind calls for a well designed project, suitable instrumentation, and a reliable interpretation of the results. The examples that follow may serve to guide research in this field.

Observation project

The problem to be solved regards the measurements of very slow displacements on cracks. It should be arranged in three dimensions, and sensitive enough to prove long-term permanent displacements as well as the character of the movement.

The most important condition for the instrumentation is in the high stability needed to detect slow displacements in the aggressive conditions of the natural environment. Many natural deformation processes have the character of creep, in which displacement proceeds at rates as low as 10^{-1} to 10^{-2} mm per year. The processes are to be observed by displacement measurements under temperature variations, which in Europe easily reach 10°C per day or 30°C per year, even in shaded places like open cracks on mountainous rock slopes. Under such conditions, one often observes seasonal rock dilatations in cracks of the order of 1 to 2 mm, which cannot be directly connected with deep natural processes in rock. One comes to the conclusion, therefore, that when the observation takes place close to the surface of the terrain, the observation must be designed, and the instrumentation made capable discerning permanent deformation 10 to 100 times lower than the reversible deformations due to temperature. To produce reliable results under such circumstances, a period of about 3 years of observations is needed. Only then trends in wobbling diagrams can be distinctly recognized.

The selection of the particular cracks to be observed has to be concordant with the engineering-geological survey of the area. The survey will generally come to a conclusion with regard to the deformational mechanism expected as well as to the activity and hierarchy of the cracks. We should prefer cracks of the primary system to be monitored, i.e. cracks directly affected by the process investigated, to avoid the interference of secondary factors. Of course, this is not always possible, and often we cannot avoid the selection of secondary cracks, simply because they are accessible, well exposed, or not extremely wide. In any case, the monitored crack must be representative regarding the process under investigation, Sometimes, a series of cracks can be monitored, either to check the function of a structure in the field or to make the check in a cross section. Cracks well sheltered in natural caves or fissures are preferable, being less affected by temperature variations and ice.

Measurements in galleries offer special problems, as caving and other relaxation effects in rock may interfere. Therefore, natural openings are always preferable if a natural process is investigated.

The field project must obviously be individual. However, there should be clear idea in advance not only about the general aims of the measurement but also about the use of data in the framework of investigation. Some field data may be of little use when unproperly obtained. This is why simplicity may be recommended, just to avoid later discussions about the factors involved.

Crack instrumentation

Crack monitoring can be instrumented with commercially or individually produced equipment of all the different designs found in rock mechanics (Kovari 1977, Franklin, Denton 1973) with many innovations. Many of the available instruments have a sophisticated design with the implicated elements of electrical gauges, ensuring high accuracy when used carefully. Considering, however, the remote location of most natural objects without any power supply, and only limited protection in severe outdoor conditions only simple and extremely stable methods are usually applicable (Kostak, Rybar 1978).

The following systems can therefore be recommended:

a) Simple mechanical contact extensometers, or ball deformeters with dial gauges, to monitor two point crack widening, because of their simplicity. Only signs (balls) are left outdoors while the instruments function.

b) Vibrating-wire electrical gauges (Dreyer, 1977), because of their accuracy and long-term stability, when remote sensing is necessary and the instrument is to be built into the structure.

c) Mechanical optical gauges using a moire effect (Kostak 1969), because of their long-term stability and simplicity in three-dimensional monitoring even under severe outdoor conditions. The instruments are built in, and have been found very stable in outdoor conditions.

Results which will be reported later were obtained mainly with the moire gauges. These were mounted on cracks from 10 mm to wide open (1 m approximately), with the use of a steel tube bridge between the opposite rock walls. The steel tubes are rigidly connected with the walls so that the displacements can be observed as an eccentricity of circular grids in the gauge, where all the three space components plus rotation can be indicated. A moire gauge is shown in Fig. 1, equipped with two moire indicators in two planes-vertical and horizontal. The registration is either automatic battery-powered or, more frequently, periodically manual. The gauge works as a spatial displacement indicator which displays moire fringe patterns on spiral grids in two orthogonally arranged units (Fig. 2). A picture is taken and the temperature registered for later evaluation in a laboratory. The gauge is fixed at the spot in a box and no special equipment except a flash is carried to make the reading, and no special maintenance is necessary. The gauges proved capable of surviving such extremal situations as floods in shafts, freezing in ice, and earthquakes, with no or little damage. The gauges designed and produced by the Institute of Geology and Geotechnics of the Czech. Academy of Science in Prague are able to prove trends in all the three space displacement components by 0.1 mm/year and rotation by 10^{-4}. ¶/4 rad, significantly.

Rock movements

Rock movements are often evident in nature due to human activity, for natural reasons or because of genetic processes in the past. Recent rock movements are usually connected with some gravitational action. Typical situations with important environmental applications are found on slopes. Here, block fields are often described, and the stability, origin, and present reactions should be investigated.

A frequent situation in nature consist of a series of individual blocks separated by cracks parallel with the tectonically predisposed valley possibly eroded by a river (Fig. 3). The blocks are situated on a plateau margin, some of them sliding down the slope. There are scientific and practical reasons to look for the stability of the blocks as well as for the process involved.

An engineering geological survey may recognize further details which suggest a theoretical explanation of the process. Very often the blocks are supported by ductile beds (Pasek, Kostak 1977) creeping under the weight of rock. Other causes trigger the process: caving in carst or mining areas, deep fracturing, sliding along inclined beds, suffossion, etc.

Examples of finished monitoring projects may be the best way showing particularities and helping us to understand how to interpret the results.

Blocks on ductile beds

The Bulgarian Black Sea coast is carved by many giant coastal deformations that affect towns, highways and recreation areas, and cause damage, and bring about serious environmental problems. Some coastal sections show fast movements in terms of meters per year; other sections show movements 10 times less rapid, and some much slower displacements. The section at Taukl'iman (Kostak, Avramova-Taceva 1981) north from the cape Kaliakra, where Sarmatian limestones overlay Oligocene clays, does not show any apparent activity; however, its slope is several kilometers in length, developed in huge steps of limestones (Fig. 4), some above, others under the sea level. Neither the process, nor its activity is clear; but according to the slope morphology, it is evidently sliding towards the sea. Such sliding could be the result of earthquakes, which are frequent in the area, or due to a process of more favourable conditions for sliding in the past with high sea level variations, and the present role of sea abrasion could not be avoided either.

The monitoring started in 1973 after negative results of geodetic measurements with one moire gauge in a marginal crack of the plateau.

Fig. 1   A moire gauge installed for crack monitoring. The displacement
         in two orthogonally arranged plane indicators is registered by
         moire patterns.

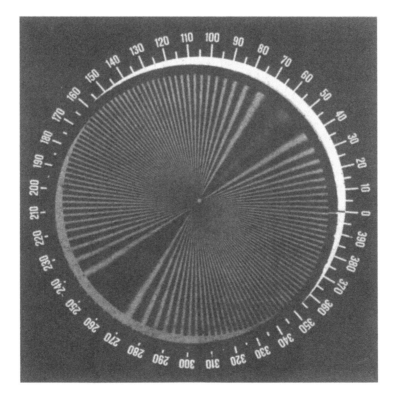

Fig. 2   A moire fringe pattern indicating displacement as the vector of
         eccentricity of spiral grids in superposition. The vector is
         given by the number of fringes and its orientation red on the
         angular scale from its axis of symmetry.

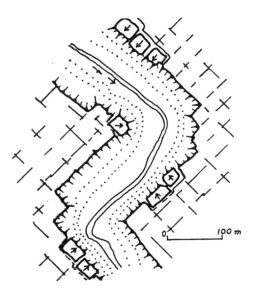

Fig. 3 A typical situation of a block-field unstable on slopes of a
river cut into a valley tectonically predisposed.

Fig. 4 Huge steps of limestone blocks along the coastal slopes of the
Black Sea at Taukliman, Bulgaria. The origin and present
activity of the deformation was investigated by the means of
the crack monitoring (Kostak, Avramova-Taceva 1981).

Fig. 5   Monitoring in a marginal crack of the Taukliman deformation.
The moire gauge between two limestone walls about 3 m under the
plateau. The thickness of limestones about 40 m, cracks open
void to 10-16 m.

Later, two other gauges were
mounted into the next fissures
which separated the highest steps
from the plateau (Fig. 5). In
three years of monitoring, the
results displayed movements which
were not in any doubt, and
continued monitoring gave a chance
to gain more detailed evaluation of
the process. It has been shown
that there is a continous
separation of blocks, at a rate of
about 0.25 to 0.70 mm/year, as well
as sinking to the extent of 0.25 to
0.60 mm/year due to creep in bed
layers under the blocks. The
continuity of the process was not
seriously changed by earthquakes
(e.g. Vrancea 1977) that induced
block vibrations of the order of a
centimeter. In the winter of
1980, a sudden drop of about 16 mm
occurred in a period when there
were no earthquakes. The
monitoring made it possible to
conclude that the primary process
is not that of earthquake, since
these may only accelerate the
primary, deep seated process of
creep in clayey sediments under the
plateau. There are signs that
creep at present comes to later
(tertiary) stages of the fracturing
of bonds. It also indicates that
the process will next enter sliding
stages and that the transport of
the remoulding clayey beds is
inhibited by the sea abrasion.
Even other more detailed
conclusions about the origin and
history of the deformation could be
made. A schematic cross section of
the marginal area of the Taukliman
deformation is given in Fig. 6,
including a typical diagram of
displacements.

Steep valley slopes in brittle rock

Cretaceous sandstones form a
plateau in northwest Bohemia cut by
the river Labe into a canyon with a

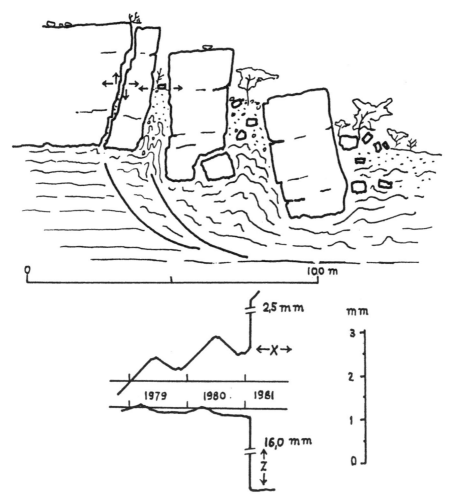

Fig. 6   Block-type deformation caused by creep of clayey beds under the weight of blocks of limestones. Taukliman - Black Sea coast, Bulgaria.   The displacement diagram shows steady trends with sudden drops.

present denivelation of about 200 m. Walls and towers of sandstone blocks are often unstable and rockfalls endanger highway and railway transport along the river. There are numerous individual objects on the slopes that need observation with regard to their stability.   On the other hand, there is a need for an evaluation of the process taking place generally in the massif and resulting in the instability of the brittle rock.

A survey project in recent years provided data about the risk in individual sections of the canyon, and even supplied a successful prognosis of a massive rockfall (Zvelebil 1984).

An engineering geological survey, including crack monitoring, with the use of mechanical contact extensometers as well as the moire gauges was carried out in the formation of open cracks behind the front walls and deep in the toe of the exposed massif.   It was discovered that only very unstable sections show trends in producing opening cracks high at the heads of the towers, while climatic

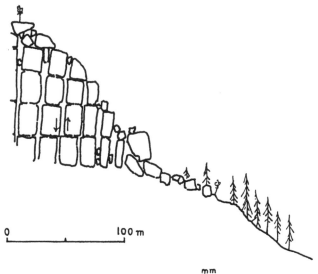

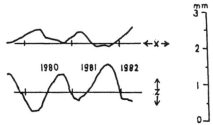

Fig. 7 Deformations of brittle sandstones in Labe canyon near Hrensko, Czechoslovakia. Very slow bending of towers is indicated in cracks as shear. Instability starts with fracturing in overstressed tower bases.

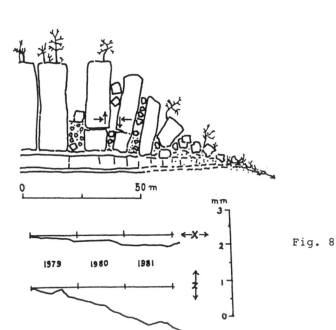

Fig. 8 Marginal section "Pekielko" of a tabular mauntain Szczeliniec W. in Klodzko region, Poland. The process is probably activated by suffosion in marlstones under sandstone blocks.

reversible movements at the top of 100 m high walls reach several cm per year without decrease of their stability. A long-term deformational process in the massif is connected with a very small downslope rotation indicated as shearing of the order of 10^{-2} mm per year deep in a crack. Instability is indicated by the opening of new cracks at a depth, and results from the fracturing of the sandstones which are highly stressed in the toe parts. Walls of old quarries are found to be in most danger due to the acceleration of the process, which in natural conditions proceeds very slowly. A typical cross section in the canyon is shown in Fig. 7, including a diagram registered by long-term crack monitoring.

Marginal slopes of a tabular mountain

Tabular mountains on the Polish-Czech border in the Klodzko region reach up to 1000 m above sea level. Their tops are formed by about 37 m thick beds of deeply cut brittle sandstones sitting on alternating layers of clayey and sand intercalations of marls or marlstones about 80 m thick (Pasek, Pulinowa 1976). Old block movements are evident all round the mountain and the cracks are widely opened to labyrinths at a place called "Pekielko". A tourist dormitory and a cable lift are situated in an area where the stability is doubtful.

Crack monitoring and geodetic measurements were carried out in the area (Cacon, Kostak 1976). While geodetic results were found to be mostly within error limits, the moire gauge found stability at the cable lift and steady movements in the "Pekielko" region. Here, vertical movements like the sinking of blocks have been found, while the spacing of cracks seems almost stable.

The process can be explained by suffosion which results in the caving of small openings in the deep layers of marls. The stability in the crack spacing, on the other hand, is probably due to wedged boulders frequently found in cracks. Such a filling keeps the

cracks compressed, leaving little chance for free dilatations.

A schematic cross section of the "Pekielko" area is given in Fig. 8. It is important to note that the process is localized in the "Pekielko" area only, as just here water comes out from a spring, found lower in the slope. The movement diagrams show clearly the trends, which show little seasonal variations when compared with those illustrated in Fig. 7. This is because the observation point was localized in a well-shaded place, and a spot where the temperature dilatations will not interfere to any great extent. Good localization of the monitoring instruments will simplify the interpretation.

CONCLUSIONS

The stability of natural objects like block fields and caves, can be now checked with relatively simple technical means, even in remote areas, by crack monitoring. In given examples a moire gauge mounted in cracks was successfully used together with other methods of investigation. A well-planned research program can solve important problems of environmental engineering. Deep processes like creep in rock, with systematically increasing and persistent effects, must be separated from the effects of climatic origin and notably those of reversible character. Therefore, observations must proceed over wider periods of time, covering at least one whole climatic period. Three years of monthly observations are usually the optimum needed to obtain significant results. Methods unable to discern the climatic reactions, and also instruments not designed to resist continually the severe conditions in the natural environment, should therefore not be used in view of the long-term character of the measurements.

Experience shows that estimates made by geologists as to the movements of objects in nature usually enhance the reality when checked by correct measurements. Even well expressed cracks may prove to be in a stable position. Quantitative checks with the monitoring of cracks is usually the only way to make valuable decisions.

Given examples of crack monitoring show the kind of diagrams that are usually obtained, and these should be carefully interpreted. They also demonstrate some of the mechanical processes found in nature that result in the formation of caves.

REFERENCES

CACON, S. and KOSTAK, B. (1976). Displacement registration of sandstone blocks in the Stolowe Gory Mts., Poland. - **Bulletin IAEG, 13,** pp. 117-122.

DREYER, H. (1977). Long-term measurements in rock mechanics by means of Maihak vibrating-wire instrumentation. - In:Kovari, K.: **Field measurements in rock Mechanics,** A.A. Balkema, Rotterdam, pp. 109-122.

FRANKLIN, J.A. and DENTON, P.E. (1973). Monitoring of rock slopes. - **Quarterby Journal Engineering Geology,** 6, London, pp. 259-286.

KOSTAK, B. (1969). A new device for in-situ movement detection and measurements. - **Exp. Mechanics, 9, 8,** Easton, Pa., pp. 374-379.

KOSTAK, B. and Avramova - Taceva, E. (1981). Propagation of coastal slope deformation at Taukliman, Bulgaria. - **Bulletin IAEG, 23,** pp. 67-73, Krefeld.

KOSTAK, B. and Bybar, J. (1978). Measurements of the activity of very slow slope movements. - In: Natau, O., Fecker, E., Reik, G.: **Grundlagen und Anwendung der Felsmechanik Kolloquium Karlsruhe 1978,** Trans. Tech. Publications, Clausthal, pp. 191-205.

KOVARI, K. (1977). Field measurements in rock mechanics - A.A. Balkema, Rotterdam.

PASEK, J. and Kostak, B. (1977). Svahove pohyby blokoveho typu. (Block-type slope movements.) - Rozpr. Cs. Akad. Ved, R. Mat. Prir. **Ved, 87, 3,** p. 58, Academia Praha (In Czech, Engl. summary).

PASEK, J. and Pulinowa, M.Z. (1976). Block movements of Cretaceous sandstones in the Stolowe Gory Mts., Poland. - **Bull. IAEG, 13,** pp. 79-82.

ZVELEBIL, J. (1984). Skalni zriceni u Hrenska a jeho prognoza. (Hrensko rockfall and its prognosis). - **Geol. Pruzk., 26, 10,** pp. 294-296, Praha (In Czech).